Dough Rheology
and
Baked Product Texture

DOUGH RHEOLOGY
AND
BAKED PRODUCT
TEXTURE

Edited by

Hamed Faridi, Ph.D.
Nabisco Brands, Incorporated
East Hanover, New Jersey

Jon M. Faubion, Ph.D.
Kansas State University
Manhattan, Kansas

An **avi** Book
Published by Van Nostrand Reinhold
New York

An AVI Book
(AVI is an imprint of Van Nostrand Reinhold)

Copyright © 1990 by Van Nostrand Reinhold
Library of Congress Catalog Card Number 89-5594
ISBN 0-442-31796-4

Printed in the United States of America

Van Nostrand Reinhold
115 Fifth Avenue
New York, New York 10003

Van Nostrand Reinhold International Company Limited
11 New Fetter Lane
London EC4P 4EE, England

Van Nostrand Reinhold
480 La Trobe Street
Melbourne, Victoria 3000, Australia

Nelson Canada
1120 Birchmount Road
Scarborough, Ontario M1K 5G4, Canada

16 15 14 13 12 11 10 9 8 7 6 5 4 3 2 1

Library of Congress Cataloging-in-Publication Data
Dough rheology and baked product texture/edited by Hamed A. Faridi
and Jon M. Faubion.
   p. cm.
  Bibliography: p.
  ISBN 0-442-31796-4
  1. Dough — Mechanical properties. 2. Baked products — Texture.
I. Faridi, Hamed. II. Faubion, Jon M.
TX560.D68D68 1989
664'.752 — dc19
                                                   89-5594
                                                   CIP

Contents

Foreword

Cereal chemists are interested in rheology because the dough undergoes some type of deformation in every phase of the conversion of flour into baked products. During mixing, dough is subjected to extreme deformations, many that exceed the rupture limit; during fermentation, the deformations are much smaller and therefore exhibit a different set of rheological properties; during sheeting and molding, deformations are at an intermediate level; and, finally, during proofing and baking, the dough is subjected to a range of deformations at varying temperatures. Accordingly, the application of rheological concepts to explain the behavior of dough seems a natural requirement of research on the interrelationships among flour constituents, added ingredients, process parameters, and the required characteristics of the final baked product. At any moment in the baking process, the rheological behavior, that is, the nature of the deformation, exhibited by a specific dough derives from the applied stress and how long the stress is maintained. The resulting deformation may be simple, such as pure viscous flow or elastic deformation, and therefore easy to define precisely. Moreover, under some conditions of stress and time (i.e., shear rate), doughs behave as ideal materials and their behavior follows theory derived from fundamental concepts. Under usual conditions encountered in baking, however, the rheological behavior is far from ideal; shear rates vary widely and sample size and dimensions are ill-defined. Because of the highly variable and complex conditions of deformation and sample dimensions, a wide gap persists between technologically practiced and theoretical rheology as applied to bread doughs.

Significant progress has been made in the development of theory and appropriate instrumentation for research on the fundamental aspects of dough rheology; this progress is the subject of the first six chapters of Dough Rheology and Baked Product Texture. New fundamental equations have been developed, and instruments have been perfected to measure the necessary parameters. Computers have become an integral part of the instruments and are used for data processing; the ability to process the results rapidly facilitates on-line measurements. The remaining chapters deal with aspects of rheology applied to the processing of cereals into specific consumer products.

Developments in applied dough rheology have followed innovations in instrumentation. Because the common approach to instrument development has been to imitate the breadmaking process, the instruments are highly empirical. Many of the empirical instruments have been very useful in practical quality control, but they have had limited value in generating fundamental data. On the other hand, instruments that lead to the determination of fundamental parameters have not found general use in the mill or bakery laboratories. Accordingly, research is needed to fill the gap between theory and practice.

In reviewing the history of applied dough rheology, one cannot help but be impressed by the breadth of the contributions of one person, Carl Wilhelm Brabender (1897–1970). He has left a legacy of a series of high-quality instruments designed to measure physical properties of doughs at all stages of the baking process. His contributions to cereal chemistry go far beyond the development of instruments. His three-phase concept of breadmaking, formulated in the 1950s and published in *Cereal Foods World* (1965), has withstood the test of time. For each of the three phases—mixing, fermentation and machining, and baking—he developed an instrument to measure the rheological properties of the dough: the farinograph for the mixing phase, the extensograph for the fermentation and machining phase, and the amylograph for the baking phase. Brabender postulated that if two or more flours gave the same results by the three instruments, they would give the same result by the baking test. Accordingly, it became possible to define flours in terms of quantitive parameters determined with each instrument. This approach has become common practice in many countries.

Furthermore, Brabender showed that deviations from specifications can readily be corrected by changing the wheat mix to adjust the farinogram, by adding oxidizing improvers to adjust the extensigram, and by adding diastatic malt or changing the wheat mix to adjust the amylogram. Subsequent modifications in the farinograph with the development of the high-shear model and the Do-Corder made it a useful instrument in the modern mechanical development of breadmaking processes. The extensograph and the amylograph have remained virtually unchanged for over 50 years.

So far, the complexity of the dough system and the variability of the baking process have prevented the rheologists from developing a rigorous equation of state based on composition, deformation, time, and temperature. This is the challenge for future dough rheologists.

W. BUSHUK
University of Manitoba

Preface

This book contains 16 chapters of concise and up-to-date review of various aspects of dough rheology and baked product texture. It is the most comprehensive review ever published on the subject and we hope that it will be used worldwide as a reference by researchers, baking technologists, food scientists, engineers, quality control staff of bakeries, students, and others working in the baked product manufacturing industries.

Since the 1950s, the production plants of the large bakers have become bigger, more automated, and, most importantly, less tolerant of variations in the functionality of ingredients and of changes in processing conditions. Variabilities that could have been absorbed easily 30 years ago today cause serious manufacturing problems that affect both product quality and operating profits. This volume is intended to fill a technical gap that exists in our industry. We hope that it will help the baking industry to appreciate the science of dough rheology and to apply the know-how needed to minimize day-to-day manufacturing problems.

Our sincere thanks to the distinguished experts from industry and academia who contributed manuscripts to this volume. Our thanks are also extended to the editorial staff of Van Nostrand Reinhold, Inc., and Beehive Production Services for their excellent suggestions and editing of the text.

Contributors

Malcolm C. Bourne, Institute of Food Science and Technology, Cornell University, Geneva, New York 14456

Jean-Louis Doublier, Laboratoire de Physicochemie des Macromolecules, Institut National de las Recherche Agronomique, B.P. 27, Nantes, Cedex 03, France

Bruce A. Drew, 4425 Abbott Avenue South, Minneapolis, Minnesota 55410

Ann-Charlotte Eliasson, Department of Food Technology, University of Lund, Box 124, S-221 00 Lund, Sweden

Hamed Faridi, RMS Technology Center, Nabisco Brands Incorporated, 200 Deforest Avenue, P.O. Box 1943, East Hanover, New Jersey 07936

Jon M. Faubion, Department of Grain Science and Industry, Shellenberger Hall, Kansas State University, Manhattan, Kansas 66506

David H. Hahn, Hershey Foods Corporation, Hershey, Pennsylvania 17033

Laura M. Hansen, Department of Foods and Nutrition, Kansas State University, Manhattan, Kansas 66506

R. Carl Hoseney, Department of Grain Science and Industry, Shellenberger Hall, Kansas State University, Manhattan, Kansas 66506

Rudolph Leschke, 5550 Via Vallarta, Yorba Linda, California 92686

Harry Levine, Fundamental Science Group, Nabisco Brands Incorporated, P.O. Box 1943, East Hanover, New Jersey 07936

Leon Levine, Levine & Associates Incorporated, 2665 Jewel Lane, Plymouth, Minnesota 55447

Jimbay Loh, General Foods Corporation, 555 South Broadway, Tarrytown, New York 10591

Wesley Mannell, General Foods Corporation, 555 South Broadway, Tarrytown, New York 10591

Juan A. Menjivar, RMS Technology Center, Nabisco Brands Incorporated, 200 Deforest Avenue, P.O. Box 1943, East Hanover, New Jersey 07936

Robert Y. Ofoli, Department of Agricultural Engineering, 102 Farrall Hall, Michigan State University, East Lansing, Michigan 48824

Carole S. Setser, Department of Foods and Nutrition, Kansas State University, Manhattan, Kansas 66506

Louise Slade, Fundamental Science Group, Nabisco Brands Incorporated, P.O. Box 1943, East Hanover, New Jersey 07936

Ronald Spies, Continental Baking Company, 1 Checkerboard Square, St. Louis, Missouri 63164

Bernhard van Lengerich, Fairlawn Development Center, Nabisco Brands Incorporated, Fairlawn, New Jersey 07410

Dough Rheology
and
Baked Product Texture

Chapter 1

Fundamental Aspects
of Dough Rheology

Juan A. Menjivar

The rheology of wheat flour dough has been a topic of much interest to cereal chemists for several decades (Schofield and Scott Blair, 1932). The flow and deformation behavior of doughs are recognized to be central to the successful manufacturing of bakery products. A natural consequence of this long-lasting interest is that the technical literature related to dough rheology is extensive and, for newcomers to the field, a challenge to cover.

Even though the rheology of dough has been much studied, the challenge of understanding the physical properties that control flow and deformation still remain (Bloksma, 1988). This challenge is particularly true for those whose main interest is in the physicochemical basis of dough rheology, as well as for those who would like to use this basic understanding to tailor-make innovative products and processes.

In the bakery products industry, a better quantitative understanding is needed of the factors that control the rheology of wheat flour dough. One of the reasons for this need is illustrated in Figure 1–1, which indicates how much is expected from dough rheology. Mechanical and/or rheological measurements are used at numerous points in the development of new products and processes, during the optimization stage, or during manufacture to ensure consistent production. Many expectations exist for dough rheology, but the understanding and ability to reliably measure *material properties* of dough lag far behind those expectations. These lags are understandable, as wheat flour dough is a most complex composite biological material (Bloksma and Bushuk, 1988). Because of this complexity, the scientific study of dough rheology involves many different disciplines (Fig. 1–2). Mastering all of these disciplines is a big task. In addition, controlled rheological measurements on wheat flour doughs are difficult and time consuming. Consequently, making reproducible measurements on dough is often a humbling experience. Despite the challenges, however, the progress in understanding flour and its biochemical components is obvious, as it is in the fields of colloid chemistry,

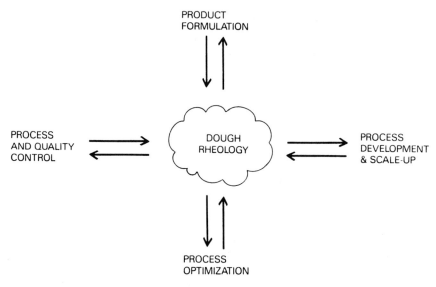

Figure 1–1. Role of dough rheology in cereal-based industry.

rheometry, polymer rheology, and new instrumentation. Therefore, the challenge ahead is to use advances in all these supporting sciences to increase our understanding of dough rheology.

To date, empirical mechanical testing (also referred to as physical testing) has provided tremendous help in evaluating dough additives/conditioning agents such as oxidizing and reducing agents, enzymes, and emulsifiers (Anderson, 1956; Smith and Andrews, 1952), as well as in evaluating flour quality and functionality (Brabender, 1973). The advantages of empirical physical testing have been and remain numerous:

- They are relatively easy and fast to perform. This makes them practical.

- They provide a great deal of heuristic knowledge, which has been accumulating over the past 50 years.

- The instruments needed to perform the tests are inexpensive.

- Their use does not require much training in the physical sciences, since the methods are already established.

The primary limitations of physical testing are that the results can not be described in terms of fundamental rheological properties, and the results represent only a point in a spectrum of the important rheological properties of dough. These limitations become more relevant as mechanization and automation of baking processes start to become prevalent. Qualitative ideas about the relationships between dough rheology and dough structure, and of the link

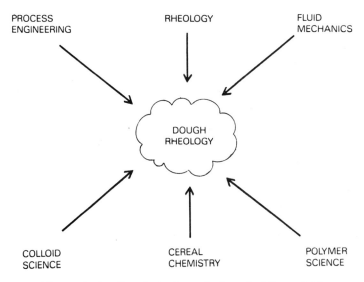

Figure 1–2.   Dough rheology: An interdisciplinary field.

between the rheological properties of dough and its behavior during mechanical handling and baking are not sufficient in such high-speed, high-capacity situations. It is at this point that a clear and quantitative understanding of the rheological properties of dough is no longer a luxury but a necessity.

The intent of this chapter is to provide a conceptual framework of the basic principles of rheology as they apply to practical problems common in the baking industry. The reader is referred to more comprehensive reviews on dough rheology (Bloksma and Bushuk, 1988) and authoritative treatments on rheology and rheometry (Ferry, 1980; Walters, 1975, 1980) for more detailed discussions and information. Specifically, the aims of this chapter are the following:

1. To introduce a practical view of analyzing problems commonly encountered in baking processes, using the leverage provided by a small set of simple physical concepts.
2. To illustrate that the fundamentals of rheology are an additional tool in the understanding and solution of complex dough processing problems. When properly applied, these fundamental concepts can lead to successful insights and results.
3. To emphasize the concept that the rheology of complex materials, such as wheat flour dough, can not and should not be characterized by single-point measurements. Instead, a variety of conditions should be explored, and a match between the kinematic conditions of interest and experimental rheological measurements should be attempted to obtain more satisfactory results.

STRESS, STRAIN, AND STRAIN RATE

Rheology, the science of deformation and flow of matter, has as its specific objective the investigation of the properties of materials that govern their flow and deformation under external forces. In order to accomplish this objective, rheologists study the load-deformation behavior of materials under controlled experimental conditions. A number of concepts are necessary in order to begin the process of rheological characterization. The most basic are the concepts of stress, strain, and strain rate. The rheological response of any material is physically expressed by stresses, which in turn are mathematically expressed as functions of ither strain and strain rate, or strain and time.

In physical terms, stress is a measure of force concentration on a material. Under external forces, a body can respond with as many as nine stresses, which are represented by the stress tensor (Fig. 1–3). In practice, only six components of the stress tensor are considered to be independent and

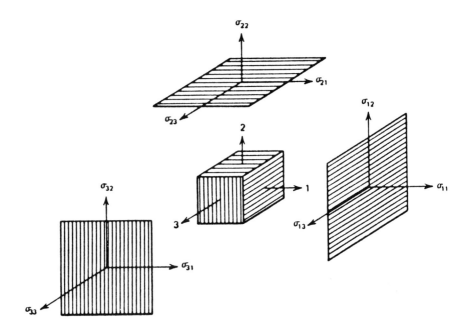

— $\sigma_{11}, \sigma_{22}, \sigma_{33} \rightarrow$ NORMAL STRESSES
— $\sigma_{ij}, i \neq j \rightarrow$ SHEAR STRESSES
— AS MANY AS 6 INDEPENDENT COMPONENTS ASSOCIATED WITH EVERY POINT OF MATERIAL

Figure 1–3. Components of stress tensor.

associated with every point of a material. There are two types of stresses: those that act in a direction parallel to the material surfaces they deform, called *shear stresses*, and those that act in a direction perpendicular to the surfaces of material they deform, called *normal stresses*. The advantage of defining stresses, as opposed to forces, is that stresses characterize the ability of material surfaces to respond to external forces, independently of sample size or shape. This property is essential for determining the rheological properties of a material.

Virtually any material is deformed under external forces. The rheological quantity associated with deformation is *strain*, a measure of the relative displacement between the particles of a material. Strain is also a tensorial quantity and, being a relative measure of deformation, is always defined as the ratio of two quantities with units of length. Therefore, it is dimensionless. When external forces produce *flow* of a material, another quantity of rheological interest is the rate at which the deformation is occurring. This quantity is characterized by another tensorial quantity, the *strain rate*, with units of inverse time. In mathematical terms, the strain rate is simply the time derivative of the applied strain.

The formal definitions of strain and strain rate can be found in any textbook on rheology; the reader is referred to Ferry (1980). In later sections specific cases are cited that may help the reader better understand the physical meaning of these important terms. Still, some comments as to the value of information on strain and strain rates are relevant at this point.

It is well documented that the applied *deformation history* has a dramatic effect on the rheological behavior of wheat flour doughs (Frazier et al., 1985). However, this experimental fact is seldom taken into consideration when a process is being scaled up from the bench to the pilot plant or plant scale. Almost invariably, neglecting this important consideration results in the following problem: doughs prepared from each of three different processing scales, the laboratory, pilot, and plant, "feel different" and most likely have different physical properties. The final product will not be equivalent in most cases.

In this scale-up exercise, one uncontrolled variable is that the deformation history applied to the dough is different for each of the three different scales, even when the equipment looks similar. Consequently, the rheological properties of the resulting doughs are different. The concepts of strain and strain rate are the most basic rheological tools for characterizing this deformation history in ways that could be repeated or approximated from the bench to the pilot plant, and eventually to the plant scale; yet these concepts have not been consistently applied for wheat flour doughs.

Dough mixing is a good illustration of the possible differences in deformation histories at different processing scales. Histories applied by laboratory mixers are different from those created by pilot plant mixers. Likewise, pilot plant mixers are different from plant scale mixers. Generally, the speed of the mixer impeller is known, but impeller and bowl configurations are most likely to be very different for different mixers. These two factors, impeller and bowl configurations, will make two given mixers apply a different "deformation history" to dough, even though they operate at the same impeller speed. In rheological terms, the mixers apply different strains and strain rates to the dough. In principle, however, if the same strains and strain rates were applied to the dough, regardless of the size of the mixer, the resultant dough should have the equivalent physical properties. Unfortunately, the mechanism of calculating strains and rates for mixers is not readily available, and such calculations are far from common.

A sound basis to characterizing strain rates, strain rate distribution, and total strain applied by different types of mixers is clearly needed. Power and energy requirements for optimum bread dough mixing have been reported by Frazier and co-workers (1975). These results are quite valuable in guiding the development of bread dough mixing processes; however, they do not address the issue of equivalent dough mixing in two different types of mixers. It is still unclear whether their findings are general enough to be used in scaling up dough mixing processes.

MATERIAL PROPERTIES AND THE RHEOLOGICAL EQUATION OF STATE

One of the advantages of defining deformation and flow processes in terms of stresses and strains is that it is then possible to determine fundamental rheological properties experimentally. Figure 1–4 illustrates three basic and general definitions that are used to determine viscosity, modulus, and compliance. Several types of viscosities, moduli, and compliances can be referred to, depending on the kinematics of the experiment under consideration (Bird et al., 1987). For instance, the most common type of viscosity reported in the literature is *shear viscosity*. Shear viscosity provides a measure of the difficulty encountered in making a material flow under shear conditions. For macromolecular fluids, the shear viscosity is highly dependent on molecular weight, molecular weight distribution, and polymer concentration (Graessley,

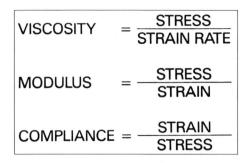

Figure 1–4. Rheological material properties.

1984; Ferry, 1980). For colloidal systems it is highly dependent on particle concentration and shape, and on the type and strength of interparticle forces (Russel, 1987). Shear viscosity is also indispensable in specifying pipeline systems and mixing and sheeting/calendaring equipment. Knowledge of shear viscosity is also necessary to optimize oven rise and baking conditions, although for these two latter applications, *extensional viscosity* would theoretically be more suitable.

Extensional viscosity provides a measure of the difficulty encountered in making materials flow under stretching conditions. As has been suggested, not all flow processes involve shear deformations alone. One example of this is dough sheeting, in which the dough is stretched significantly in addition to the shear flow between the rolls. A more dramatic example is gas cell expansion during dough fermentation and oven rise. This is an exclusively biaxial (radial and tangential) stretching flow. Successful empirical instruments such as the Brabender Extensigraph and Chopin Alveograph operate in the extensional mode. Figure 1–5 illustrates the difference between shear and extensional deformation.

Characterizing forces and deformations in terms of stresses and strains allows the determination of physical properties inherent to the material. In rheology these fundamental physical properties are termed *material properties* or *material functions*. As for any other physical property, *rheological material properties* are independent of factors such as testing geometry and sample size. A large portion of the literature is reported in rheological terms that are difficult to relate to physical properties. This difficulty has been and continues to be a major problem in the way wheat flour doughs are characterized. It consequently is difficult to compare results obtained in different laboratories. Also, it is difficult to relate these results to other areas of science that could be of great assistance to dough rheology, such as the physical chemistry of colloids and polymer physics.

Another advantage of expressing rheological information in terms of well-defined stresses and strains is that the resulting information lends itself to expression within the realm of a *rheological equation of state*. Rheological equations of state are intended to compile information about a material such that its behavior could be predicted under different deformation and flow conditions. Rheological equations of state are expressed as mathematical functions that relate stress to strain and time (or strain rate). The best known and oldest rheological equations of state are those for a purely elastic solid (Hooke's Law) and a purely viscous fluid (Newton's Law). These equations have formed the basis of the way solids and liquids have been characterized for several centuries.

Figure 1–6 presents the one-dimensional expression for Hooke's and Newton's laws. For a Newtonian liquid, all the energy input necessary to make it flow at a given rate is dissipated as heat, while for a Hookean body, the energy necessary to deform it is stored as potential energy, which is fully recoverable. Wheat flour doughs exhibit both types of behavior, and along

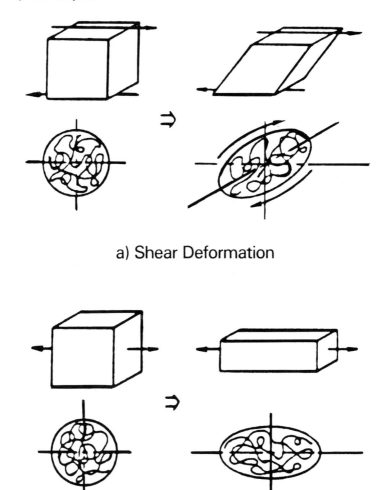

a) Shear Deformation

b) Extensional Deformation

Figure 1–5. Diagramatic representation of shear and extensional deformation of an isolated macromolecule.

with many other types of macromolecular and colloidal materials, are referred to as *viscoelastic*. Viscoelastic materials have the ability to spend the applied energy in both processes; that is, heat generation (viscous process) and energy storage (elastic process). The unique viscoelastic properties of wheat flour dough are often related to its unique performance in bakery applications. However, as will become obvious in later chapters, their characterization still remains a challenge.

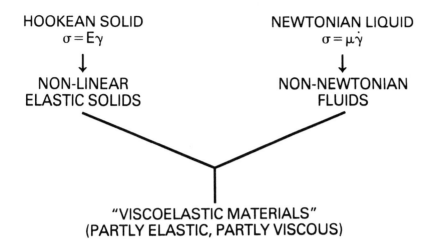

HOOKEAN SOLID
$\sigma = E\gamma$

⬇

NON-LINEAR
ELASTIC SOLIDS

NEWTONIAN LIQUID
$\sigma = \mu\dot{\gamma}$

⬇

NON-NEWTONIAN
FLUIDS

"VISCOELASTIC MATERIALS"
(PARTLY ELASTIC, PARTLY VISCOUS)

Where: σ = Stress

E = Elastic Modulus

μ = Newtonian Viscosity

γ = Strain

$\dot{\gamma}$ = Strain Rate

Figure 1–6. Ideal elastic and viscous behavior.

DEVIATIONS FROM IDEAL BEHAVIOR

Deviations from Hookean or Newtonian behavior are the rule in most food materials of interest, including dough. Figure 1–7 shows two of the most common deviations that food fluids show: decrease in viscosity with shear rate (shear thinning, pseudoplastic behavior) and the presence of an apparent yield stress: that is, a minimum stress required to make a material flow. These deviations are among those that are characteristic of wheat flour doughs. Figure 1–8 shows stress-strain curves for wheat flour doughs. These diagrams illustrate striking deviations from Hooke's Law and many similarities to the elastomeric behavior of natural rubber (Aklonis et al., 1972) (Fig. 1–9). Thus, wheat flour doughs exhibit apparent yield stress, shear thinning behavior, and unique strain hardening effects. In addition to these deviations, the rheological properties of dough are markedly dependent on deformation history. Due to all of these complications, a rheological equation of state for dough is far from being developed. Mechanical models that describe specific deformations have been proposed (Schofield and Scott Blair, 1933), and a conceptually useful expression has been developed by Tshoegl and co-workers (1970)

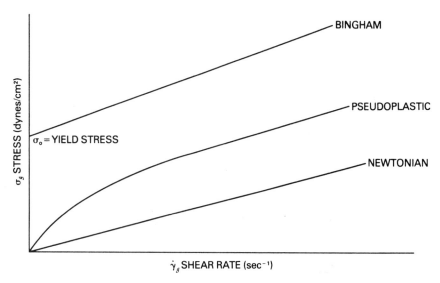

Figure 1–7. Specific deviations from Newtonian behavior.

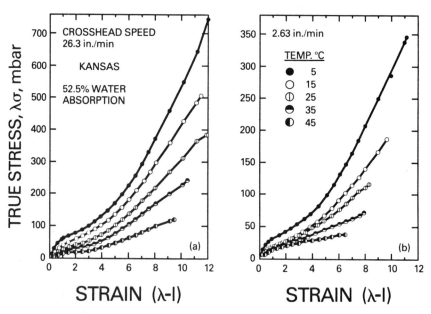

Figure 1–8. True stress-strain curves for wheat flour dough. (*From Tschoegl et al., 1970*)

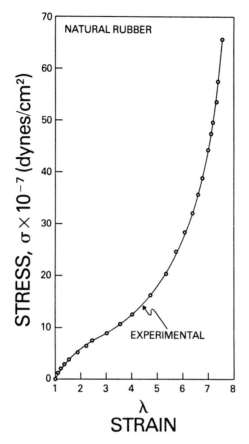

Figure 1–9. Large deformation behavior of natural rubber. (*From Aklonis et al., 1972*)

that describes the behavior of dough in simple tensile deformation. Still, extensive research is needed in this area.

RHEOMETRY

Rheometry is the science that deals with making controlled rheological measurements. Rheologists have spent considerable time and effort in trying to develop flow and deformation configurations that make stresses, strains, and strain rates easy to calculate (rheometric flows/deformations). Rheometric deformations are then used to determine the rheological properties of materials whose rheological equations of state are unknown. Figure 1–10 shows

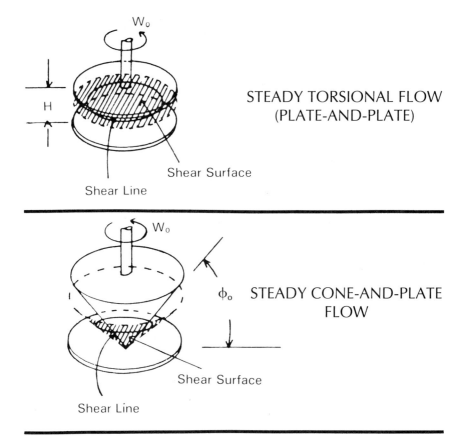

STEADY TORSIONAL FLOW
(PLATE-AND-PLATE)

Shear Surface

Shear Line

ϕ_o STEADY CONE-AND-PLATE
FLOW

Shear Surface

Shear Line

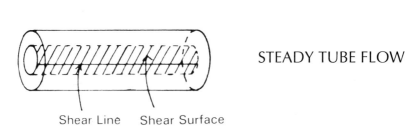

STEADY TUBE FLOW

Shear Line   Shear Surface

Figure 1–10. Representative viscometric flows. (*After Bird et al., 1987*)

schematics of a number of shear viscometric flows often used for synthetic polymers. Table 1–1 presents a sample of the material functions that can be experimentally determined by using these viscometric geometries (Bird et al., 1987; Walters, 1975). It is important to realize that using rheometric deformations becomes even more important with wheat flour doughs, since their material properties are strong functions of strain and strain rate. For

Table 1-1. Examples of Material Functions

Simple Shear Flow	Material Function
Steady shear	Viscosity function
	Primary and secondary normal stress coefficients
Small-amplitude oscillatory	Storage modulus and loss modulus
Stress relaxation	Relaxation modulus

these configurations, the applied strain and strain rate are known and can be controlled; thus controlled studies of the fundamental rheological properties of dough are possible.

In practice, the baking industry uses a variety of methods to characterize the rheology of doughs. Table 1–2 shows one classification of these methods and points out some of the advantages and disadvantages of each type. In reality, empirical and imitative methods have been preferred over fundamental rheometric methods. As has been mentioned, the main drawback of empirical and imitative methods is that it is difficult to know the strains and rates applied to the dough during the experiment. Also, it generally is not possible to calculate the applied stresses. Therefore, the application of the equations given in Figure 1–4 for the determination of material properties is not possible. Table 1–3 lists some of the most used methods of each type.

APPLICATIONS OF FUNDAMENTAL RHEOLOGY CONCEPTS

Some examples are presented in this section to provide the reader with an idea of how and where these simple concepts can be applied. More work

Table 1-2. Different Types of Rheological Measurements

Fundamental
Well-defined physical properties
Useful for structure–property relationships
Useful for process and engineering calculations

Empirical
Parameters proven useful but not well defined
Inexpensive, fast, rugged

Imitative
Parameters meaured under conditions that simulate application
Usually associated with empirical parameters
Usefulness based on experience

Table 1-3. Examples of Rheological Methods

Type	Method	Typical Rheological Property/Attribute
Fundamental	Tube viscometer	Viscosity function
	Cone-and-plate	Elastic Modulus
	Parallel plate	Creep compliance
Empirical	Bostwick Consistometer	Consistency
	Brookfield Viscometer	Consistency
	Back extrusion	Consistency, viscosity
	Extensigraph	Resistance, extensibility
Imitative	Food texture profile	Textural attributes
	Farinograph	Mixing time, mixing tolerance
	Chopin Alveograph	Resistance, extensibility

clearly is needed to show the practical value of these concepts. Still, it is hoped that this section will lead to an appreciation of the potential benefits of their application.

Strain Rates during Fermentation and Baking Processes

A useful exercise for trying to understand the rheological events involved in any process is to try to get better acquainted with the kinematic conditions. In other words, what types of deformation, and what magnitudes of strains and strain rates, are involved in the process. This approach has not necessarily been common in what is generally considered to be dough rheology, yet some estimates exist in the literature. Bloksma (1988) reported some estimates of the shear and extensional rates involved in typical breadmaking operations and physical tests. The following is a list of some of his estimates:

1. Bread dough fermentation: extensional rates from 10^{-4} s^{-1} to 10^{-3} s^{-1}.
2. Oven rise: extensional rates on the order of 10^{-3} s^{-1}.
3. Brabender Extensigraph: extensional rates on the order of 10^{-1} s^{-1} to 1 s^{-1}.
4. Chopin Alveograph: extensional rates between 10^{-1} s^{-1} and 1 s^{-1}.
5. Mixing (bulk fermentation): shear rates of approximately 10 s^{-1}.
6. Brabender Farinograph: shear rates on the order of 10 s^{-1}, for a dough consistency of 500 Brabender Units.
7. Mixing (mechanical dough development): shear rates in the range of 10 s^{-1} to 100 s^{-1}.

These estimates immediately point out a mismatch between strain rates for practical baking situations and physical testing conditions. For instance, extensional processes such as bread dough fermentation and oven rise occur at rates on the order of 10^{-4}–10^{-3} s^{-1}, while instruments used to test doughs under extensional conditions (Brabender Extensigraph and Chopin Alveograph), operate at rates between 10^{-1} s^{-1} and 1 s^{-1}. These rates are at least 100-fold higher than the actual application. It is well known that dough is a shear-thinning material, so it is only reasonable to question whether extrapolating conclusions from measurements at kinematic conditions that are so different to those of the practical application will produce the desired information. At least some attempt should be made to obtain rheological measurements at the appropriate rates.

Even more dangerous than the above-mentioned situation is the practice of trying to predict the baking behavior of a particular dough based on mechanical information obtained with a recording mixograph. Not only do recording mixographs operate at higher rates (approx. 10 s^{-1}), but the total strains involved are several orders of magnitude higher than those that take place during fermentation and oven rise. Total strains during fermentation and oven rise are in the neighborhood of 10–20, while in dough mixing strains could be as high as 3,000–5,000. This is not to say that useful rheological information cannot be obtained from these devices, but clearly there are important reasons to question the validity of the practice.

Dynamics of Gas Cell Expansion

Another situation, relevant to the baking industry, is the deformation of dough under the pressure of expanding gas bubbles, as occurs during the development of gas cell structures in breadmaking. Although not used within the context of dough rheology, the kinematics of a single spherical bubble, immersed in a fluid and expanding under the pressure of a gas, has been analyzed (Bird et al., 1987). Figure 1–11 diagrams the growth of a spherical gas bubble of radius $R(t)$ under an increasing gas pressure P_G.

The governing equations for the biaxial strain and strain rates involved are also illustrated in Figure 1–11 (see the boxed expressions). A practical use of these expressions is in the determination of the strain and strain rate conditions likely to happen during dough fermentation or baking. These estimates provide an idea of the rheological conditions at which dough should be evaluated in the laboratory to try to predict its performance in either of these two processes. For instance, if as an average the radii of gas bubbles increase tenfold during oven rise, the average total strain on the dough would be 10. Also, if the assumption is made that it takes 15–20 minutes to accomplish this increase, the average strain rate involved is, according to the equation $\dot{\gamma} \approx 1/\Delta t$,

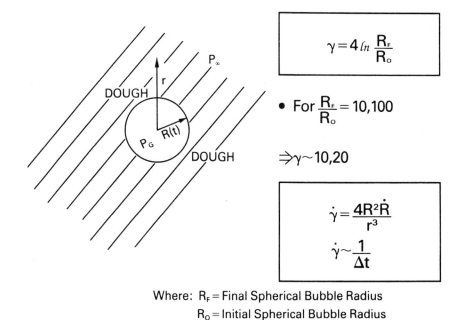

$$\gamma = 4 \ln \frac{R_F}{R_o}$$

• For $\frac{R_F}{R_o} = 10,100$

$$\Rightarrow \gamma \sim 10,20$$

$$\dot{\gamma} = \frac{4R^2\dot{R}}{r^3}$$

$$\dot{\gamma} \sim \frac{1}{\Delta t}$$

Where: R_F = Final Spherical Bubble Radius
R_o = Initial Spherical Bubble Radius
γ = Total Strain
$\dot{\gamma}$ = Strain Rate
Δt = Time Interval

Figure 1–11. Dynamics of bubble growth: kinematics.

roughly one thousandth per second. Notice that this is on the same order of magnitude as Bloksma's (1988) estimates. Therefore, in order to simulate the mechanical behavior of dough during this oven rise cycle, measurements should be taken in the range of strains and rates mentioned above.

Another aspect of this analysis on a growing gas bubble within dough is the quantitation of the stresses involved. Figure 1–12 shows the equation that governs the balance of forces on the dough surrounding an expanding spherical gas bubble. One piece of information that is missing in order to solve this differential equation is the rheological equation for dough; that is, the expression that relates how the radial extensional stresses change as a function of the radial position in the dough. Such an equation would provide the needed information on how dough responds as the bubble grows and would allow the determination of a direct relationship between the kinetics of gas cell development and rheological properties. This result could then form a physical basis for determining the relationship between particular rheological properties of dough and dough performance during gas cell formation.

$$(P_G - P_\infty) = \underbrace{\frac{2T}{R}}_{} + \underbrace{3\int_{R}^{\infty} \frac{\sigma_{rr}}{r} dr}_{}$$

GAS PRESSURE	SURFACE TENSION	DOUGH RHEOLOGY

Where: P_G = Gas Pressure Inside Bubble

P_∞ = Equilibrium Pressure Far from Bubble

T = Dough-Gas Interfacial Tension

R = Bubble Radius

σ_{rr} = Radial Extensional Stress on Dough

r = Radial Position

Figure 1–12. Dynamics of bubble growth: Balance of forces.

Viscometric Measurements on Dough: Stress-Strain Behavior

When little is known about the rheological response of a material, one of the most effective ways to start learning about its material properties is to conduct rheometric measurements. This procedure has often been done for wheat flour doughs by using a technique that has been successful in the characterization of synthetic polymers: dynamic oscillatory shear measurements (DOSM) (Ferry, 1980).

There are several reasons for using DOSM when trying to understand the rheology of dough. The reader is referred to Faubion et al. (1985) for a recent review. However, one of the drawbacks that is often cited is that the kinematic conditions during DOSM are mechanistically different from any type of deformation found during baking. In essence, DOSM are designed to be small strain experiments, generally involving strains on the order of 0.1% to 5%, while the strains experienced by dough in the breadmaking process can range from 100% during sheeting, to 1,000% during fermentation and oven rise, and up to 500,000% during mixing. Perhaps the two most successful instruments in terms of attempting to match kinematic conditions during breadmaking are the extensigraph and the alveograph. Their mode of deformation is similar to the extension that takes place during fermentation and oven rise. Their drawbacks are that the rates of deformation they apply on dough are higher than those found in practice, and that elaborate analyses and calculations are required to obtain a limited idea of the material properties of dough (Muller et al., 1961; Bloksma, 1958).

In this section an alternative approach is presented (Menjivar, 1986a; Menjivar and Kivett, 1986). This approach consists of the rheometric measurement of the kinetics of stress development in a dough sample contained between two plates. The configuration used is shown in Figure 1–13, where the expressions for the velocity distribution, shear rate, and shear stress at the edge of the plate are shown as well. In such a configuration, a dough sample is prepared in the shape of a disk 2.5 cm in diameter and 0.25 cm in height. The dough sample is loaded between the two plates as shown in Figure 1–13, and enough time is allowed for normal stresses to relax (generally 15 min for bread dough). The lower plate is rotated at a constant

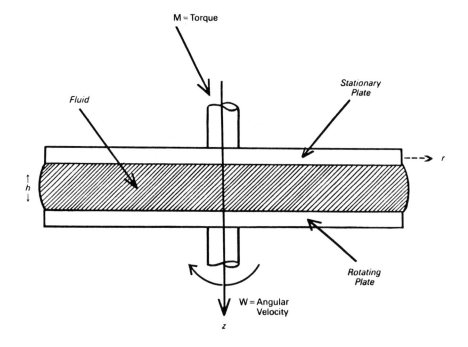

$$\text{Velocity Field} = V_\theta = \frac{rzW}{h}$$

$$\text{Shear Rate at } R = \dot{\gamma}_R = \frac{RW}{h}$$

$$\text{Shear Stress at } R = \tau = \frac{2M}{\pi R^3}$$

Figure 1–13. Shear deformation between parallel plates.

angular velocity (W), which in turn gives rise to a constant shear rate at the plate edge. Torque is measured at the upper plate, and the shear stress can be calculated at the plate edge following the equation $\tau = 2M/\pi R^3$ (see Fig. 1–13). Since this is a constant shear rate experiment, the total shear strain applied on the dough sample can be obtained by multiplying the shear rate by the elapsed time. When this is done, a stress-strain diagram such as the one illustrated in Figure 1–14 is obtained. The data shown in Figure 1–14 were obtained at a shear rate of 5. 0 × 10^{-2} s^{-1}, representing the shear counterpart to the extensigraph diagram shown in Figure 1–15, which is obtained under a tensile deformation of the dough.

The data illustrated in Figure 1–14 have two advantages. First, the stress-strain diagram can be obtained easily and expressed in fundamental (as opposed to empirically defined) units. Second, the deformation history of dough during the experiment is well defined. Therefore, fundamental rheological properties can be obtained directly from this experiment. These fundamental rheological properties should, in turn, be correlatable to extensigraph or alveograph data. For instance, from the stress-strain diagram shown in Figure 1–14 the following rheological properties can be identified: apparent yield stress, elastic modulus (or strain-hardening modulus), failure stress, and failure strain. These terms have empirical counterparts in the extensigram shown in Figure 1–15, most notably the maximum resistance (R_m), which is similar to the failure stress, and the extensibility (E), which is similar to the failure strain.

To further illustrate the use of this type of fundamental information, the effect of gluten concentration on the prefailure portion of the stress-strain diagram shown in Figure 1–14 is shown in Figure 1–15. Here the percent gluten figures represent the weight percentage of gluten that was dry-blended with an 8% protein soft flour. Then flour-water doughs were prepared at a constant flour to water ratio, and mixed to peak in a Brabender Farinograph at 63 rpm.

It is interesting to note in Figure 1–16 that increasing the gluten content in flour has a much more pronounced effect on the elastic modulus than on the apparent yield stress. When the elastic modulus is plotted on a double logarithmic scale against the percent of gluten added to the flour (Fig. 1–17), a clear-cut change in the slope is observed at gluten addition levels between 7% and 10%. Above 10% gluten addition, the elastic modulus increases noticeably, more steeply than at lower levels.

This behavior is similar to that exhibited by the viscosity of polymer solutions as a function of polymer concentration (Fig. 1–18) (Graessley, 1984; Menjivar, 1986b). For polymer solutions, the steeper slope beyond a critical polymer concentration is associated with the onset of physical entanglements between polymer molecules. In bread doughs, the gas-holding ability of dough is generally associated with the formation of a continuous gluten network. This protein network provides a balance of elastomeric properties that enable

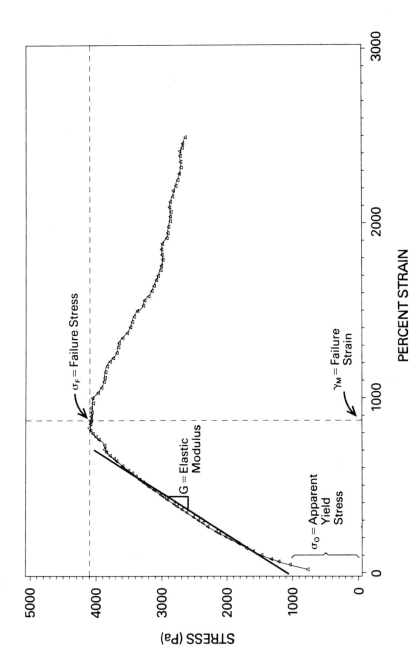

Figure 1–14. Representative shear stress-strain diagram obtained with plate-and-plate configuration in a Rheometrics Fluids Rheometer. *(After Menjivar, 1986a)*

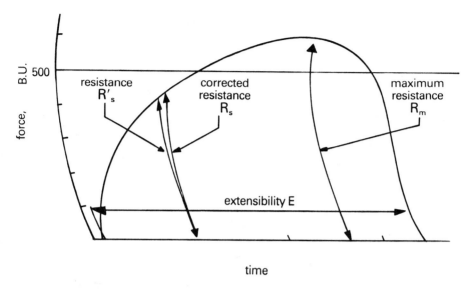

Figure 1–15. Representative load-deformation diagram obtained with Brabender Extensigraph.

dough to stretch without early breakage during the formation of gas cells. It is possible that this change in the slope of the modulus versus gluten concentration illustrated in Figure 1–17 does, in fact, represent a qualitative change in the nature of the gluten network such that the strain-hardening elastometric properties of dough are substantially improved beyond 7% gluten addition. More research in this area would prove valuable in the understanding of the fundamental rheological properties of dough required for good performance during baking.

Another relevant aspect of the viscoelastic/elastomeric behavior of wheat flour dough is the effect of strain rate on its rheological properties. This relevance derives from the fact that different processing operations take place at different rates (see the section on the strain-rates during fermentation and baking processes). Consequently, there is a need to understand the rheological behavior of dough within the broad range of strain rates (10^{-4}–10^2 s^{-1}) encountered in actual practice.

The effect of shear rate on the stress-strain response of bread dough is illustrated in Figure 1–19, where stress-strain curves are shown at rates of 0.01 s^{-1}, 0.05 s^{-1}, and 0.10 s^{-1}. This narrow range of strain rates already illustrates the significant effect of rate on the rheological behavior of dough. Our knowledge of this effect is relatively limited and it warrants more effort, given the broad range of deformation rates involved during dough processing.

Moreover, Figure 1–20 compares the stress-strain diagrams obtained with the plate and plate configuration to those obtained with an extensigraph

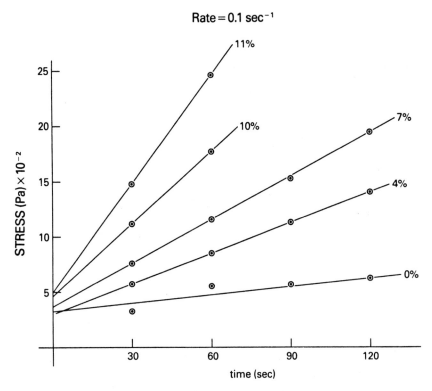

Figure 1–16. Effect of gluten addition on stress-strain curves of flour/water doughs. (*After Menjivar and Kivett, 1986*)

and corrected by using Muller's approach (Muller et al., 1961). Both sets of data were obtained by using the same flour, flour-to-water ratio, and mixing procedure. The results in Figure 1–20 are very valuable. It has been mentioned throughout this chapter that extensional deformations are key to the processes of dough fermentation and oven rise. However, the method presented in this section applies a shear deformation to the dough. Hence, the question arises as to how shear rheological properties will relate to the tensile properties of interest. Figure 1–20 represents the first evidence that, although in principle not the preferred mode of testing, shear rheological properties of dough seem to be closely related to its extensional rheological properties.

CONCLUDING REMARKS

The emphasis of this chapter has been threefold: (1) to illustrate the need to use basic physical concepts in obtaining sound insights into the rheological

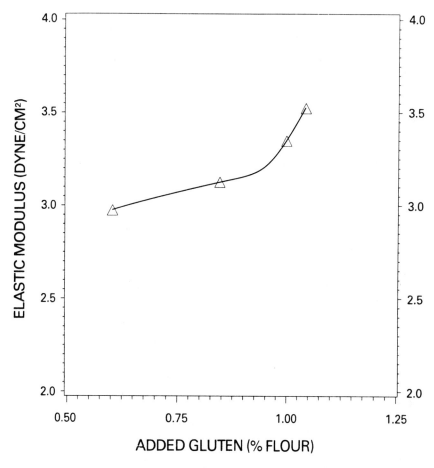

Figure 1–17. Double logarithmetic plot of the effect of gluten addition on the elastic modulus of wheat flour/water doughs. (*After Menjivar and Kivett, 1986*)

behavior of dough, (2) to illustrate the relationship between rheology and dough structure, and (3) to examine the behavior of dough during processing. These objectives have been briefly examined under the light of three concepts: stress, strain, and strain rate. It is apparent that the use of these concepts has generally been disregarded in the technical literature, even though they represent fundamental concepts important in advancing dough rheology as a physical science.

These notions of stress, strain, and strain rate have been discussed, as the needed basis to characterize the rheological behavior of dough in terms of material properties. Tangible advantages of using these basic rheological concepts have been presented in an effort to provide an idea of their use in practical situations. Overall, the understanding and advancement in the

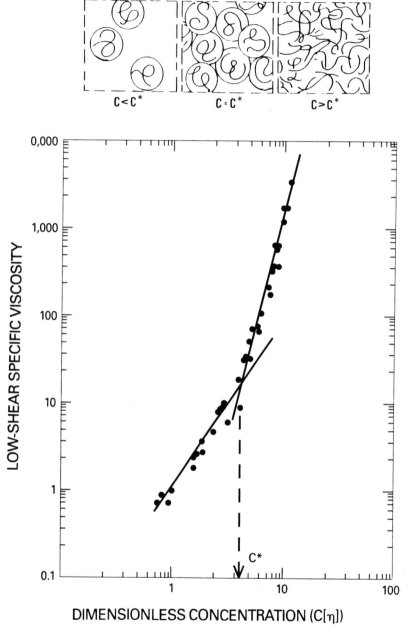

Figure 1–18. Effect of polymer concentration on the low-shear viscosity of a linear polyssaccharide (Hydroxy-propyl-guar) solution. (*After Menjivar, 1986b*)

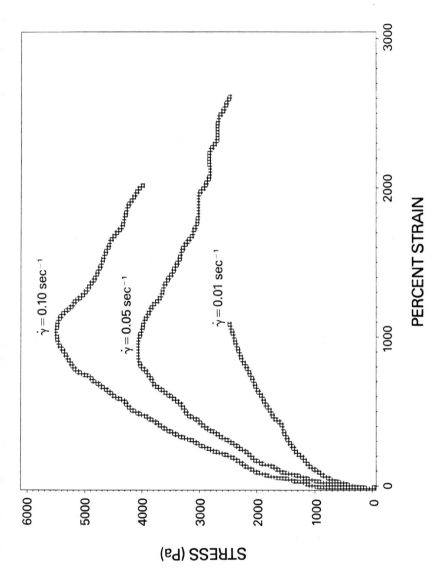

Figure 1–19. Effect of shear rate on stress-strain behavior on flour/water doughs. (*After Menjivar and Kivett, 1986*)

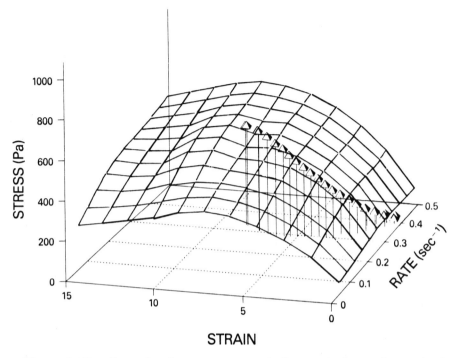

Figure 1–20. Comparison between stress-strain diagrams in shear and extensional deformations. (*After Menjivar, 1986a*)

knowledge of wheat flour dough rheology has also been limited by our ability to make meaningful fundamental rheological measurements. Experimental measurements of dough are difficult, and creative, sound, relevant techniques are needed to characterize the rheology of dough.

In the past few decades, cereal research has focused successfully on the chemical and biochemical aspects of bakery products manufacture. With the need to automate and update production processes, the rheologist's chief challenge is to apply fundamental scientific principles to better describe, predict, and control rheological changes during processing. The systematic application of sound physical concepts should provide a strong aid in this endeavor.

REFERENCES

Aklonis, J. J., MacKnight, W. J., and Shen, M. 1972. *Introduction to Polymer Viscoelasticity*, Chap. 6. New York: Wiley-Interscience.

Anderson, J. A. 1956. Comparative study of the improving action of bromate and iodate by baking data, rheological measurements, and chemical analyses. *Cereal Chem.* **33**:221–239.

Bird, R. B., Armstrong, R. C., and Hassager, O. 1987. *Dynamics of Polymeric Liquids*, Vol. 1: *Fluid Mechanics*, 2d ed. New York: Wiley.

Bloksma, A. H. 1988. Rheology of the breadmaking process. Paper presented at 8th International Cereal and Bread Congress, May 30–June 3, Lausanne, Switzerland.

Bloksma, A. H. 1986. Rheological aspects of structural changes during baking. In *Chemistry and Physics of Baking*, ed. J. M. V. Blanshard, P. J. Frazier, and T. Galliard, pp. 170–178. London: Royal Society of Chemistry.

Bloksma, A. H. 1958. A calculation of the shape of the alveograms of materials showing structural viscosity. *Cereal Chem.* 35:323–330.

Bloksma, A. H., and Bushuk, W. 1988. Rheology and chemistry of dough. In *Wheat: Chemistry and Technology*, Vol. 2, Chap. 4, ed. Y. Pomeranz. St. Paul, MN: Am. Assoc. Cereal Chemists.

Brabender, M. 1973. Resistography. A dynamic quick method to classify wheat and flour quality. *Cereal Sci. Today* 18:206–210.

Bushuk, W. 1985. Flour proteins: structure and functionality in dough and bread. *Cereal Foods World* 30:447–448, 450–451.

Faubion, J. M., Dreese, P. C., and Diehl, K. C. 1985. Dynamic rheological testing of wheat flour doughs. In *Rheology of Wheat Products*, ed. H. Faridi, pp. 91–115. St. Paul, MN: Am. Assoc. Cereal Chemists.

Ferry, J. D. 1980. *Viscoelastic Properties of Polymers*. New York: Wiley.

Frazier, P. J., Fitchett, C. S., and Russell Eggitt, P. W. 1985. Laboratory measurement of dough development. In *Rheology of Wheat Products*, ed. H. Faridi, pp. 151–175. St. Paul, MN: Am. Assoc. Cereal Chemists.

Frazier, P. J., Daniels, N. W. R., and Russell Eggitt, P. W. 1975. Rheology and the continuous breadmaking process. *Cereal Chem.* 52:106r–130r.

Graessley, W. W. 1984. Viscoelasticity and Flow in Polymer Melts and Concentrated Solutions. In *Physical Properties of Polymers*, ed. J. E. Mark, A. Eisenberg, W. W. Graessley, L. Mandelkern, J. L. Koenig, pp. 97–153. Washington, D.C.: American Chemical Society.

Halton, P. 1949. Significance of load-extension tests in assessing the baking quality of wheat flour doughs. *Cereal Chem.* 26:24–45.

Menjivar, J. A. 1986a. A new model to characterize the rheological properties of dough: Practical implications. Paper presented at Annual AIChE Meeting, November 2–7, Miami Beach, Florida.

Menjivar, J. A. 1986b. Use of gelation theory to characterize metal cross-linked polymer gels. *Adv. Chem Series* 213:209–226.

Menjivar. J. A. and Kivett, C. 1986. Rheological properties of dough in shear deformation and their relation to gas holding capacity. Paper presented at the 71st AACC Annual Meeting, October 5–9, Toronto, Canada.

Muller, H. G., Williams, M. V., Russell Eggitt, P. W., and Coppock, J. B. M. 1961. Fundamental studies on dough with the Brabender extensigraph. I. Determination of stress-strain curves. *J. Sci. Food Agric.* 12:513–523.

Rasper, V. F. 1975. Dough rheology at large deformation in simple tensile mode. *Cereal Chem.* 52(3,II):24r–41r.

Rasper, V., Rasper, J., and De Man J. 1974. Stress-strain relationships of chemically improved unfermented doughs. I. The evaluation of data obtained at large deformations in simple tensile mode. *J. Text. Studies* 4:438–466.

Russel, W. B. 1987. *The Dynamics of Colloidal Systems.* London: University of Wisconsin Press.

Schofield, R. K., and Scott Blair, G. W. 1933. The relationship between viscosity, elasticity, and plastic strength of a soft material as illustrated by some mechanical properties of flour dough. III. *Proc. Roy. Soc. (London)* **A141**:72.

Schofield, R. K., and Scott Blair, G. W. 1932. The relationship between viscosity, elasticity, and plastic strength of a soft material as illustrated by some mechanical properties of flour doughs. I. *Proc. Roy. Soc. (London)* **A138**:707–718.

Smith, D. E., and Andrews, J. S. 1952. Effect of oxidizing agents upon dough extensograms. *Cereal Chem.* **29**:1–17.

Tschoegl, N. W., Rinde, J. A., and Smith, T. L. 1970. Rheological properties of wheat flour doughs. II. Dependence of large deformation and rupture properties in simple tension on time, temperature, and water absorption. *Rheol. Acta* **9**:223–238.

Walters, K. 1975. *Rheometry.* London: Chapman and Hall.

Walters, K. 1980. *Rheometry: Industrial Applications.* New York: Research Studies Press.

Chapter 2

The Viscoelastic Properties of Wheat Flour Doughs

J. M. Faubion and R. Carl Hoseney

It should be evident that cereal scientists need to be able to measure and understand the fundamental mechanical properties of wheat flour doughs. Restated more precisely, the goal is to understand the relationships between the forces acting on dough, its subsequent deformation, and time. This goal has been the impetus for a great deal of research over the past 60 years. Several recent reviews (Hlynka, 1970; Hibberd and Parker, 1975b; Baird, 1983; Faubion et al. 1985; Faubion and Faridi, 1986) present the rationale for applying fundamental rheological tests to investigate the mechanical properties of dough. Bushuk (1985) sums up this rationale concisely:

> In breadmaking, the dough undergoes some type of deformation in every phase of the process. During mixing, dough undergoes extreme deformations beyond the rupture limits; during fermentation the deformations are much smaller; during sheeting and shaping, deformations are of an intermediate level; and finally during proofing and baking, dough is subjected to more deformations. Accordingly, the application of rheological concepts to the behavior of doughs seems a natural requirement of research on the interrelationships among flour composition, added ingredients, process parameters and the characteristics of the loaf of bread.

The process of generating the data necessary to characterize the rheology of dough is far from complete, because of the difficulty of determining the material properties of systems as complex as wheat flour doughs. If determining the material properties were simple, most (if not all) of the required information would now be in hand. Of the large body of research that exists on the rheological properties of wheat flour doughs, the great portion is empirical rather than fundamental in nature. It is important to bear in mind, however,

Contribution No. 89-328-B from the Kansas Agricultural Experiment Station.

that this information is not worthless simply because it is empirical. On the contrary, the information has provided scientists with great insight into the behavior of dough. Most of the phenomena (relaxation and dough development to name only two) currently being investigated with fundamental rheological techniques have been described empirically, and the information deserves review and consideration.

The preponderance of empirical rather than fundamental data reflects the fact that doughs are difficult to analyze. Dough is viscoelastic (Schofield and Scott Blair, 1932), combining the properties of a Hookean solid with those of a non-Newtonian viscous fluid. In wheat flour doughs, stress is not a function of applied strain or strain rate alone but of a combination of both. In addition, the viscoelastic behavior of dough is nonlinear. The ratios of stress/strain (apparent modulus) and stress/strain rate (apparent viscosity) are not constants, but are functions of stress. Therefore, the characterization of nonlinear viscoelastic systems such as dough requires that their viscous and elastic components be determined as functions of testing rate *and* strain level (Bagley and Christianson, 1986). These behaviors and the exigencies they impose on testing are not recent discoveries. In fact, they were identified in the first series of studies designed to test fundamental properties (Schofield and Scott Blair, 1932, 1933a, 1933b). It is these behaviors that make determining the relationships between stress, strain, and time extremely challenging.

Knowledge of the fundamental mechanical properties of a specific dough (or even of all doughs) has limited value alone. Such knowledge is more valuable if it provides insight into the structure or chemistry of the dough, to explain previously observed phenomena or to allow modeling of responses to variations in processing or formulation. That saltation can occur only after the body of fundamental knowledge is combined with models and insights from empirical studies. This chapter provides a brief overview of both fundamental studies, those attempting to characterize the viscoelastic properties and those describing well-known but empirically measured dough properties.

MIXING

As has been explained, empirical investigations of the physical properties and behavior of wheat flour dough should not be dismissed out of hand. They have provided the current understanding of dough. In many cases, they have provided the first descriptions of phenomena currently under study with fundamental rheological tests. Consequently, they should be discussed here.

In order to create a dough, flour and water must be combined at a rather specific ratio. Using too much water results in a slurry lacking any of the properties of a dough. On the other hand, if too little water is added, the result is a slightly cohesive powder. When intermediate amounts of water are added, those suitable to form a viscoelastic dough, the result is a paste or sticky mass that behaves much like a glue. Because dough formation is not spontaneous, this mass must be mixed to develop a cohesive, viscoelastic dough.

In the laboratory, dough formation can be followed by using the mixograph or farinograph, recording mixers developed specifically for the purpose (Hoseney, 1985). Mixograms of four different flours are shown in Figure 2-1. Although the mixograph and farinograph are empirical rheological instruments, they do provide much useful information. The height (y axis) of the mixogram curve is a measure of the dough's resistance to the extension caused by the passage of the mixing pins. Curve length is related to the time the dough has been mixed. (Each vertical line represents 1 minute.) The width of the curve is related to the cohesiveness and elasticity of the dough. As the mixograms in Figure 2-1 show, flours can have widely differing mixing properties.

In general, as the flour–water paste is mixed, the system becomes more resistant to extension; that is, the mixogram (or farinogram) curve height increases. This process continues only up to a specific point, however. The dough then becomes less resistant to extension, giving a mixing peak. Research has shown that to produce the best loaf of bread from a flour, the dough must be mixed to the peak of the mixing curve. The time required to develop a dough with a maximum resistance to extension is, then, the

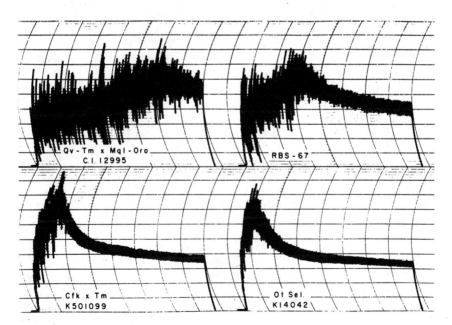

Figure 2–1. Mixogram curves of hard wheat flours demonstrating differing mixing times: long (C.I. 12995), medium (RBS-67), short (K501099), and very short (K14042). (*From Hoseney and Finney, 1974, fig. 1, with permission.*)

optimum mixing time. From the curves in Figure 2-1, it is evident that flours can and do differ widely in their optimum mixing times.

Why should this increase in resistance to extension take place early in the mixing process? Flour is composed of discrete particles. Although they are small (average 150 μm for hard wheat flour), flour particles are quite large as compared to their constituent starch granules and protein molecules (Figure 2-2). When water is added to dry flour, the surfaces of the particles hydrate rapidly. Bernardin and Kasarda (1973) demonstrated that when flour particles and water come into contact, gluten protein fibrils form spontaneously and extend from their surfaces into the surrounding water. Gluten, when dry, is a glassy polymer, but as it takes up water, it undergoes a glass transition (Hoseney et al., 1986). This transition renders it mobile and, therefore,

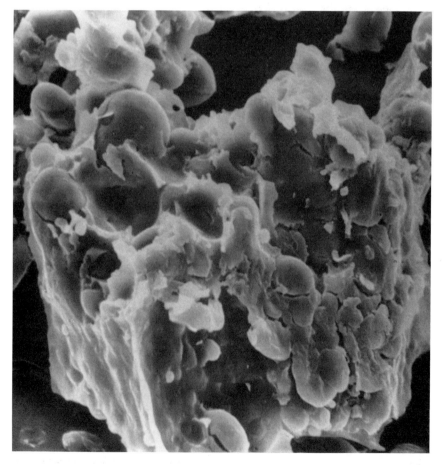

Figure 2–2. Scanning electron micrograph of a single particle of hard wheat flour. (*From Hoseney and Seib, 1973, fig. 7, with permission.*)

able to interact with other gluten polymers to form a dough. In a system with limited water, such as a bread dough, water must penetrate the flour particle before the protein can undergo the transition, interact, and form a dough. In systems where water content is limited even more severely (e.g., saltine cracker doughs), dough formation is also severely limited (Rogers and Hoseney, 1987).

As mixing begins, water rapidly wets the outer surfaces of the flour particles. Because the surface area of the particles is limited, however, water is present in excess. The large excess of water means that most of the water is free and gives the flour–water mixture mobility. The system's resistance to extension is small and the mixogram curve is low. As mixing continues, the hydrated protein fibrils on the particle surfaces are wiped away by contact with the mixer blades, the sides of the bowl, or other flour particles. The resulting new particle surface is then hydrated rapidly. This is a continuous process in which flour particles are rapidly worn away, creating a continuous system of hydrated protein fibrils with starch granules dispersed throughout. As more and more of the protein becomes hydrated, the amount of free water in the system decreases. The resistance to extension of the system increases, therefore, causing the observed increase in mixogram curve height. Once all of the protein in the flour is hydrated, continued mixing will not cause the mixing curve to increase in height.

This explanation implies that dough mixing is essentially a hydration process. Dough development is then a random entanglement of hydrated protein fibrils.

The incorporation of air is a second important consequence of dough mixing, and is often overlooked in rheological studies of mixed doughs. The incorporation of air is an important process, because this air provides nuclei for the cells forming the crumb grain in the final loaf of bread. It therefore has a significant effect on the mechanical properties of both the dough and finished product. Air incorporation can be demonstrated by following the changes in the density of a dough during mixing (Fig. 2-3) or by scanning electron microscopy of cryofractured dough (Fig. 2-4).

OVERMIXING

At optimum mixing, the dough system is hydrated and developed to the point of maximum resistance to extension. This does not, however, produce the mixing peak seen in Fig. 1-1. For that to happen, the physical properties of the dough must be changed by additional mixing that decreases the resistance to extension.

What is observed when a dough is mixed past its point of minimum mobility (overmixed) depends to a great extent on the flour being tested. Some flours resist the changes associated with overmixing, and their mixograms show

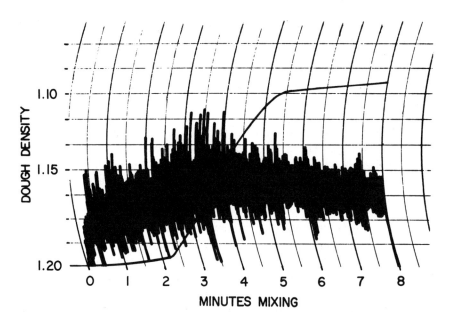

Figure 2–3. Mixogram of hard wheat flour with solid line included showing the change in dough density as a function of mixing. (*From Junge et al., 1981, fig. 1, with permission.*)

little apparent change after optimum mixing. Others, those at the bottom of Figure 2-1, for example, undergo rapid and severe changes. The amount and rate of these changes indicate the doughs' mixing tolerance. Thus, a flour that resists change with continued mixing past optimum is considered to have good mixing tolerance.

The large changes in mechanical (and functional) properties induced by overmixing have stimulated considerable research interest. The first attempt to explain these changes was that of Hlynka (1970), which explained the phenomena as the result of shear thinning of the gluten protein matrix. Continued mixing (shearing) of large molecules, such as the proteins making up gluten, can lead to molecular alignment in the direction of flow and reduced resistance to mixing (shear). The entanglement network (MacRitchie, 1980) model of dough structure and breakdown is somewhat similar, proposing that time and stress cause slippage at the contact points of the entangled gluten network.

———————————→

Figure 2–4. Scanning electron micrographs of cryofractured, freeze-dried, flour–water doughs. *Top*, dough containing 0.5% sodium stearoyl lactylate; *lower right*, dough containing 3% shortening; *lower left*, control flour–water dough. (*From Junge et al., 1981, fig. 5, with permission.*)

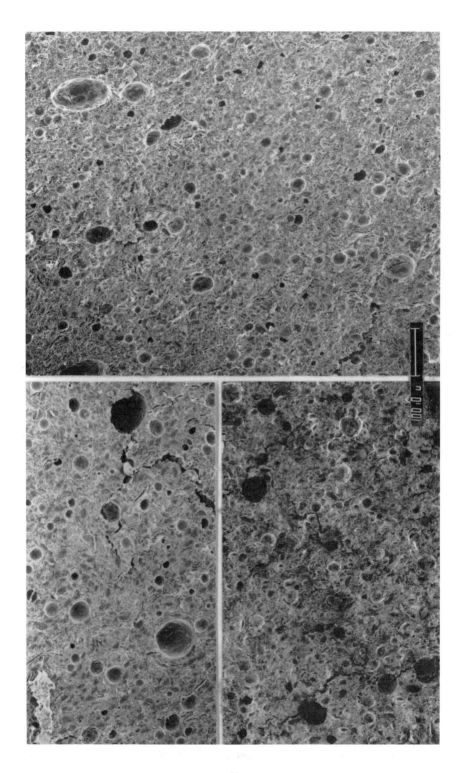

Here, excessive mixing would decrease the degree of entanglement and cause a dough to break down.

Although clearly plausible, shear thinning apparently is not the correct model. Doughs do not break down if they are overmixed in a non-oxygen atmosphere such as N_2 (Baker and Mize, 1937) or if the water-soluble fraction of the flour is first removed (Schroeder and Hoseney, 1978). These two factors would not be expected to have any effect on shear thinning of the protein matrix.

The fact that doughs mixed in a nitrogen atmosphere do not break down suggests that overmixing involves oxidation. The fact that removing water solubles from flour prevents overmixing indicates that a component of the water-soluble fraction is involved in the oxidation. Conn and Kichline (1971) provided clues to the nature of the active compound in the water solubles by demonstrating that fumaric acid and related compounds not only reduced mixing time but also increased the rate of dough breakdown during overmixing. N-ethylmaleimide (NEMI), a thiol blocking agent, is also known to affect overmixing, presumably by interfering with thiol–disulfide interchange. Schroeder and Hoseney (1978) reported that α, β unsaturated compounds such as fumaric, maleic, and ferulic acids affected overmixing in a manner similar to NEMI. Unlike NEMI, these compounds do not interact with cysteine during mixing (Weak et al., 1977), suggesting that their action cannot be explained by an effect on thiol–disulfide interchange.

The action of activated (α, β unsaturated) double-bond compounds can, however, be reversed or blocked by the enzyme lipoxygenase, which is present in enzyme-active soy flour, or by free radical scavenging antioxidants (Fig. 2-5). MacRitchie (1975) suggested that high-speed, high-shear mixing could break disulfide bonds between gluten polypeptides and create thiyl free radicals. Dronzek and Bushuk (1968) have presented evidence indicating the formation of free radicals during mixing. Sidhu and colleagues (1980) used ^{14}C-fumaric acid to show that, during mixing, small amounts of the activated double-bond compounds became covalently linked to gluten. The data obtained by these investigators led to the hypothesis that the effect of activated double-bond compounds is caused by their reaction with thiyl free radicals created on gluten protein during mixing.

If disulfides are broken during mixing, then the gluten proteins will be partially depolymerized. This depolymerization should result in changes in two basic physical properties, solubility and viscosity, increasing the former and decreasing the latter. Increases in protein solubility as a result of dough mixing have been shown by a number of investigators (Tsen, 1967; Danno and Hoseney, 1982; Graveland et al., 1984). Danno and Hoseney (1982) also demonstrated that the viscosity of protein extracted from overmixed doughs or doughs mixed with NEMI, fumaric acid, or ferulic acid was lower than the viscosity of protein from optimally mixed doughs. If extracted gluten is treated with β-mercaptoethanol (BME), breaking all the disulfide bonds, its relative

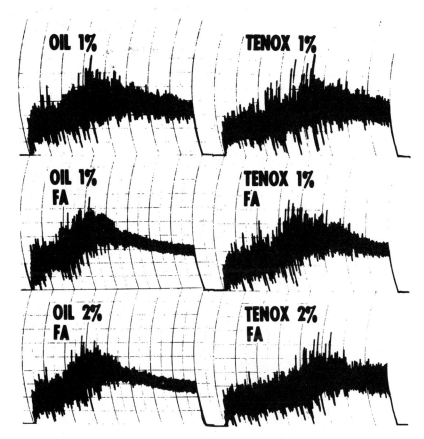

Figure 2–5. Mixograms demonstrating the effect of the activated double-bond compound fumaric acid (FA) on overmixing and the effect of the free-radical scavenging antioxidant Tenox on its effect. (*From Schroeder and Hoseney, 1978, fig. 7, with permission.*)

viscosity is reduced. When protein from optimally mixed doughs, overmixed doughs, and doughs containing activated double-bond compounds (ADB) is treated with BME, the resulting relative viscosities are equivalent. Taken together, these findings are consistent with the theory that disulfide bonds are broken during mixing and that the above changes in viscosity are the result of those bonds being broken.

Ferulic acid, the predominant activated double-bond compound known to be in flour, is present at levels much lower than those required to accelerate dough breakdown. Reconstitution studies performed by Jackson and Hoseney (1986) showed that the rate of dough breakdown during overmixing was controlled by a factor(s) in the gluten and starch fraction and not in the water-soluble fraction. The water-soluble fraction contains the ferulic acid.

Fast-acting oxidizing flour improvers such as KIO_3 or Azodicarbonamide (ADA) induce a rapid dough breakdown similar to that found with activated double-bond compounds (Fig. 2-6). An attractive possibility is that the oxidants are oxidizing an as yet unidentified molecule to produce activated double-bond compounds.

FERMENTATION

Yeast-containing doughs obviously are not transformed (baked) into their final products immediately after mixing. Rather, they are subjected to panary fermentation. The changes in the mechanical properties of a dough that take place during fermentation are large and have significant effects on the handling properties of the dough and the quality of the final product. Assessing these changes in fundamental terms has proved exceedingly difficult, however. When it is freshly mixed, a dough is sticky, relatively inelastic, and resistant to extension. If the dough is allowed to "rest," this resistance decays rapidly

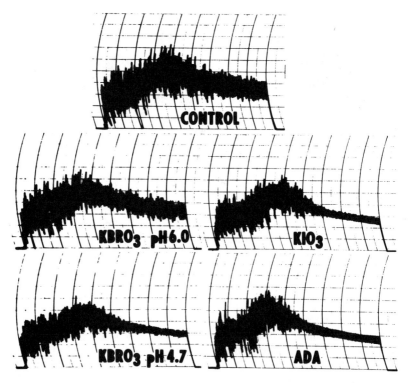

Figure 2–6. Mixograms demonstrating the effects of certain oxidants on overmixing. (*From Weak et al., 1977, fig. 1, with permission.*)

to yield a very extensible system. If that dough is allowed to undergo fermentation, even more changes occur. The dough loses its apparent stickiness, and becomes drier, less extensible, and much more elastic. These changes are readily apparent and can be measured subjectively by manipulating the dough. Quantitating these changes in fundamental terms has yet to be accomplished, however.

Fermenting dough presents a number of analytical challenges, the foremost of which is the fact that as it ferments, neither its shape nor its density remains constant. Fundamental rheological properties become nearly impossible to measure if the geometry of the test piece changes in an uncontrolled manner during the test. In addition, dough relaxes as a function of time after being worked or manipulated (e.g., after being formed into a specific geometry for testing). However, if a fermenting dough is given time to relax after being manipulated, its geometry and density will change, making any resulting data difficult to interpret.

A significant advance in the empirical measurement of the mechanical properties of fermenting dough was the development of the "spread test" (Hoseney et al., 1979). As is outlined in Figure 2-7, a fermenting dough is molded into a cylinder of known dimensions. The dough is then placed on a smooth plate in a fermentation cabinet (30°C, 90% relative humidity). After 60 minutes, the width (W) and height (H) of the cylinder are measured and the W/H ratio is calculated as the spread ratio. Because the spread ratio is independent of the volume of the dough piece (Hoseney et al., 1979), it can indicate (albeit in nonfundamental terms) the rheology of a dough that is changing in size during fermentation.

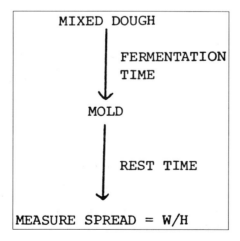

Figure 2–7. Experimental scheme for the spread test. (*From Hoseney et al., 1979, fig. 1, with permission.*)

This test has shown that oxidants decrease (Fig. 2-8) and reducing agents increase spread ratio (Table 2-1) (Lillard et al., 1982). Yeast was shown to have a strong oxidizing effect on the flow properties of the dough (Fig. 2-9). In addition, this effect was shown not to have been caused by the metabolic products of yeast fermentation. When properly oxidized and fermented, bread doughs do not spread (their elastic forces exceed the force of gravity). It is clear that during the proof period, dough expands rather than flows to fill the baking pan. Loaves that are incompletely oxidized or fermented (referred to by bakers as being "green") have sharp corners and edges because the necessary rheological changes have not occurred and the dough has flowed to fill the pan.

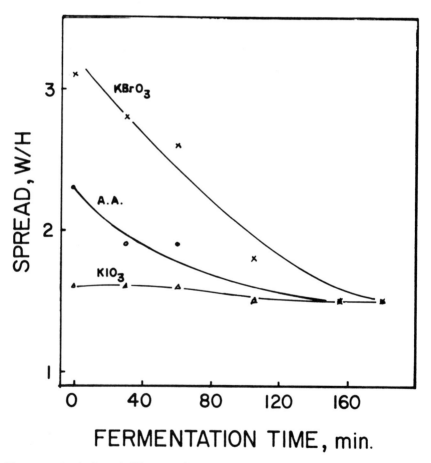

Figure 2–8. Action of different oxidants as measured by the spread test. (*From Hoseney et al., 1979, fig. 3, with permission.*)

Table 2-1. Effect of Oxidants and Reducing Agents on Spread Ratio

	Spread Ratio (W/H)	
Compound	No Fermentation	Full Fermentation
Flour–water control	2.8	3.8
Cysteine (15 ppm)	3.1	4.0
Azodicarbonamide (20 ppm)	2.0	1.9

Note: Abstracted from Hoseney et al. (1979).

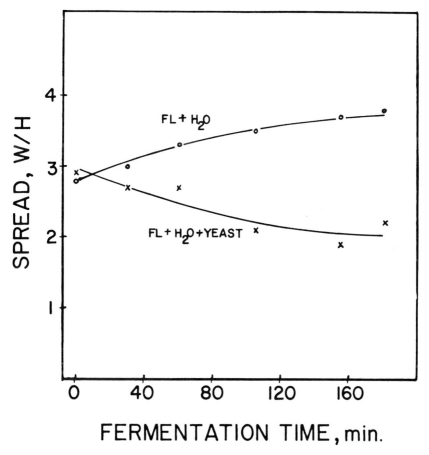

Figure 2–9. Effect of yeast on the spread ratio of a fermenting dough. (*From Hoseney et al., 1979, fig. 5, with permission.*)

HEATING

The temperature-triggered rheological changes that take place in a dough as it is heated are very important (Hoseney, 1986). This is made clear by the fact that proofing doughs made from good- and poor-quality flours to the same height does not result in the same loaf volume. Stated slightly differently, flours that are very different in quality may show those differences only after the doughs are heated in the oven. Cake batters and cookie doughs produced from good- and poor-quality flours behave similarly; that is, batters of the same density and dough pieces from the same cutter will not necessarily result in cakes of equivalent volume or cookies of equal size.

Recent work with bread doughs (Moore and Hoseney, 1986) demonstrates that heat-triggered changes in dough rheology are important in determining the volume of the final baked loaf (Fig. 2-10). Of particular importance is a change that occurs at $\sim 55°C$, which appears to be related to the effects of lipids and formula shortening.

A more obvious change in mechanical properties caused by heating is the conversion of dough to bread. The mechanical properties of dough and bread are quite different and, if the properties of the former persisted throughout heating, the baked dough would collapse when cooled. The chemical changes responsible for this "setting" are not yet known. Schofield and co-workers (1983) showed that gluten proteins become polymerized during heating. However, this would not be expected to change their rheological properties. The rheological properties of gluten do not change during heating (Dreese et al., 1988c). If gluten does not change, then it is reasonable to consider that starch may. However, in bread dough systems, starch gelatinizes at a much lower temperature than the setting of the dough.

FUNDAMENTAL
STUDIES

In attempting to make fundamental measurements of the rheological properties of nonlinear materials such as dough, it is critical to use sample shapes and loading patterns that cause the strains in the sample to be as constant and homogeneous as possible (Hibberd and Parker, 1975b). The results produced by many popular dough testing instruments are unique to each instrument and cannot be expressed in fundamental physical terms, because, in part, the geometries of the samples under analysis are uncontrolled and unpredictable. Consequently, the strains imposed on the sample are complex and nonuniform.

The loading or deformation patterns appropriate for use in measuring the viscoelastic properties of doughs can be categorized as transient or dynamic

(Hibberd and Parker, 1975b). In all cases, the reason for knowing these patterns is to be able to relate stress with strain and time.

Among the transient loading patterns, the most common are as follows. *Creep* occurs when a stress is applied suddenly and is maintained at the applied level. The strain resulting from that stress is measured as a function of time. *Stress relaxation* occurs when a sample is rapidly deformed to a predetermined strain level. The stress resulting from the maintenance of this strain is measured as a function of time. Deformation at a constant strain rate measures the stress required to maintain a constant strain rate in the sample. Stress is measured as a function of time. Deformation with a constant stress loading rate requires that stress on a sample be increased at a constant rate. Strain is measured as a function of time.

As has been pointed out by certain authors (Hibberd and Parker, 1975b; Ferry, 1970), the data resulting from the third and fourth test methods are very difficult to analyze; therefore, most transient loading studies have utilized creep or stress relaxation techniques.

Testing procedures utilizing dynamic loading patterns, generally adaptations of methods widely used in the study of polymers (Bagley and Christianson, 1986; Smith et al., 1970), are now widely applied to foods (Rao, 1984). The most common loading pattern is a sinusoidally varying strain of fixed frequency with measurement of the resulting stress.

In this chapter studies are grouped according to the type of loading pattern or test method employed rather than the instrument used.

CREEP AND CREEP RECOVERY

The earliest studies performed to systematically investigate the viscoelastic properties of doughs were those of Scott Blair and colleagues (Schofield and Scott Blair, 1932, 1933a, 1933b, and 1937). These excellent papers, still well worth reading for the logic of their approach and the clarity of their explanations, reported studies conducted using a mercury bath extensometer that subjected cylinders of dough to controlled stretching (stress). After varying periods of stretching, the stress was released, the dough was allowed to recover for various times, and recovery was measured.

With this technique, it proved to be possible to separate the resultant strain into two parts: unrecoverable (viscous or plastic flow) and recoverable (elastic deformation). Because the rate of deformation and stress were known, it was possible to calculate a coefficient of viscosity of $0.55-15 \times 10^6$ poise. The applied force and elastic deformation allowed calculation of a modulus of elasticity for the dough of $1.6-5.5 \times 10^4$ cgs units.

In making these calculations, the assumption was made that dough was behaving as a Newtonian liquid and a Hookean solid, which, of course, was an oversimplification. Apart from being the first work to rigorously investigate the fundamental rheological properties of doughs, Scott Blair's work is significant in that it clearly demonstrated that dough is viscoelastic and, specifically, that its viscoelasticity is nonlinear.

Glucklich and Shelef (1962) used results from creep and creep relaxation tests (also employing a mercury bath extensometer) to conclude that, for the doughs tested, there was no region of linear behavior. Later studies by Bloksma (1962) employed creep and creep relaxation techniques but utilized truncated cone and plate geometry. As was the case with Scott Blair's work, it was possible to calculate an apparent modulus of elasticity of approximately 10^4 dynes/cm^2. Bloksma also calculated the shear compliance, $J(+)$ (shear strain/shear stress), of his doughs and found that it varied from 10^5/g·cm·s^2 to 10^{-3}/g·cm·s^2. In these studies, shear compliance was nonlinear in that there were large (greater than tenfold) changes in compliance as stress increased from 20 N/m^2 to 500 N/m^2.

Smith and Tschoegl (1970) also used creep and creep relaxation measurements but with novel instrumentation. Dough rings of defined geometry were suspended in a fluid of equal density and subjected to extensional deformation. Unlike most other studies, these investigators found a region of steady state flow; that is, one where the rate of increase in strain with time was constant. In addition, creep compliance was found to be independent of stress (linear in its response) up to stresses of 328 N/m^2.

Hibberd and Parker (1975a) also found a region over which the behavior of the dough was linear. Creep and creep relaxation tests using parallel plate geometry indicated that linearity existed but at lower stresses (1.5–40 N/m^2) than those reported by Smith and Tschoegl (1970). At stresses above that level, doughs behaved as nonlinear viscoelastic materials.

Finally, Muller and co-workers (Muller et al., 1961, 1962) sought to derive stress-strain curves from the load extension curves produced by the extensigraph. By calibrating their instrument in cgs units and applying Scott Blair's creep and creep recovery technique, the researchers sought to separate viscous flow from elastic deformation. By using this method, it was possible to calculate a coefficient of viscosity of 0.3–3.6 $\times$ 10^4 poise and an apparent modulus of elasticity of 0.5–6.7 $\times$ 10^4 cgs units.

The latter work has been strongly criticized, on two primary points. As was the case with Scott Blair's approach, in calculating the coefficient and modulus, dough was assumed to be both a Newtonian liquid and a Hookean solid even though the earlier work clearly showed it to be nonlinear. In addition, because the extensigraph was used to impose the stress, the stress on and strain in the test sample were not homogeneous and could only be approximated numerically. This latter criticism has proved extremely difficult to overcome in transient tests (Smith and Tschoegl, 1970; Muller et al., 1962; Hibberd and Parker, 1975a).

STRESS RELAXATION

A second major type of transient loading, stress relaxation, has been the basis for a number of investigations. Hlynka and co-workers (Hlynka and Anderson, 1952; Cunningham et al., 1953; Cunningham and Hlynka, 1954), in a series of classic investigations, utilized a specially designed split-pin "relaxometer" operating in tension. Stress relaxation of dough was found not to be highly dependent on initial strain, and the calculated values of N (2.5×10^6 poise) and G (0.64×10^4 cgs) generally agreed with earlier determinations of Schofield and Scott Blair (1933a). Tension did not decay as a simple exponential function, and the results indicated that dough possessed a spectrum of relaxation times. Work carried out slightly later, using a mercury bath extensometer (Glucklich and Shelef, 1962; Shelef and Bousso, 1964) supported these conclusions. Stress decayed in a nonlinear manner and indicated the existence of a spectrum of relaxation times. Shelef and Bousso's (1964) data indicated that, under the conditions used, stress relaxation was fairly independent of initial stress.

Although Hibberd and Parker (1975b) suggested that, with the relaxometer employed by Hlynka, strain is fairly uniform over most of the sample, it is not perfectly uniform by any means. Rasper (1975), after comparing extension testing methods utilizing the Brabender Extensigraph and the Instron Universal Testing Machine, pointed out an inherent problem with extensional testing of doughs with either instrument; that is, the necessity of being able to calculate the "mass of the sample actually involved in the development of the resistance to extension." Because the area, in cross section, of the sample does not change proportionally with its increase in length, this problem is formidable.

Testing procedures (including stress relaxation) involving uniaxial compression should alleviate the need to calculate the effective mass involved in resisting extension. While this technique has only recently been used to any great extent in testing wheat flour doughs, Chatraei and colleagues (1981) used compression testing to analyze viscoelastic polymer fluids and interpreted the results in terms of biaxial extensional viscosities. In a recent study Bagley and Christianson (1986) pointed out that the analysis of extensional or elongational flow (if it can be measured precisely) is relevant to the study of doughs, because this type of flow occurs during fermentation-induced bubble growth.

Although uniaxial compression is attractive as a method to measure the force-deformation behavior of foods (Olkku and Sherman, 1979), particularly doughs, the method tended to give higher moduli than those measured by torsional or simple shear deformations (Bagley et al., 1985a). This phenomenon, determined to be due to the existence of frictional effects between the compressing surfaces and the material undergoing compression, could be minimized by utilizing Teflon®-coated platens and lubricating the platen/sample interface with paraffin oil (Bagley et al., 1985a, 1985b).

Bagley and Christianson (1986) used this technique (lubricated uniaxial compression) in a series of experiments designed to measure the apparent biaxial elongational viscosity of commercial, chemically leavened wheat flour doughs. Comparison of the data derived for a series of doughs, all tested at the same crosshead speed and deformation, revealed that the standard deviation for either peak stress or stress after 600 s was 10%. This figure was no better than that obtained by Muller and co-workers (1961), who estimated the total error in stress readings using the extensigraph (9.5%). Because the compressional technique does not require that variation in effective mass during compression be calculated, as does the extensigraph method, the latter should, in theory, be more accurate.

Log–log plots of apparent biaxial elongational viscosity versus radial extension rate showed that the limiting values reached were strongly affected by the speed of the compressing platen. Thus, as Bagley and Christianson (1986) concluded, elongational viscosity data for doughs "are not determined by radial extension rate alone."

When Bagley and Christianson (1986) represented their data as a stress growth function (Dealy, 1984), stress growth coefficients appeared to be dependent on both time and crosshead speed. As a consequence, doughs could not be characterized by a single curve relating stress growth coefficient versus time. This is an important conclusion both in its own right and as a clear illustration of the general unsuitability of single point measurements for characterizing the rheological properties of wheat flour doughs.

Peleg (1979, 1980) has developed a method for treating stress relaxation curves for solid foods such as doughs. Relaxation data are normalized, and a decay parameter $Y(t)$ is calculated as shown in Equation 1.

$$Y(t) = [F(o) - F(t)]/F(o) \qquad (2\text{--}1)$$

$F(o)$ is the force at the start of relaxation, and $F(t)$ the force at a time t after relaxation begins. $Y(t)$ is plotted against time, and two constants are calculated that relate to the rate and extent of stress relaxation. The a constant describes the level to which stress decays during the specified time, and the b constant the rate at which stress decays (Cullen-Refai et al., 1988). Thus, as the a constant approaches zero, the material behaves increasingly as an elastic solid. For viscoelastic solids, lower values of b indicate slower stress relaxation.

Cullen-Refai and co-workers (1988) applied this method to data derived from the lubricated uniaxial compression of fermenting wheat flour doughs and generated the curves illustrated in Figure 2-10. Comparison of a and b constants for the various treatments showed that the overmixed and undermixed doughs varied significantly in their a constants. Oxidation caused doughs to behave significantly more like a solid (larger a constants) than unoxidized

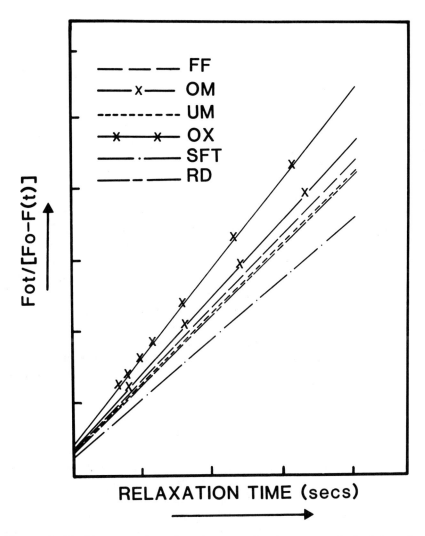

Figure 2–10. Linear transformation of stress relaxation curves obtained from fermenting doughs. The doughs tested included fermentation (FF), overmixed (OM), undermixed (UM), oxidized (OX), reduced (RD), and short fermentation time (SFT). (*From Cullen-Refai et al., 1988, fig. 2, with permission.*)

controls. Incomplete fermentation caused doughs to behave more like a liquid (lower *a* constants) than their respective controls. These last two observations are encouraging for several reasons. They are in agreement with previous studies employing the spread test (Hoseney et al., 1979), which concluded that fermentation causes doughs to become more elastic and that oxidation reduces the ability of a dough to flow under the force of gravity.

DYNAMIC LOADING

As has been mentioned, tests to measure dough viscoelasticity in which the loading pattern is oscillatory, or dynamic rather than transient, have become popular during the past 25 years. The theory underlying these tests, their adaptation from methods in polymer rheology (Ferry, 1970), and the instrumentation used have been the subject of recent reviews (Faubion et al., 1985; Rao, 1984). In most of the studies, the instrumentation utilized parallel plate sample geometry and sinusoidally oscillating simple shear stress, although earlier studies (Hibberd and Wallace, 1966; Hibberd, 1970a, 1970b) used coaxial cylinder geometry.

The tests are capable of imposing low stresses of low magnitudes on the samples as well as operating at low strains and strain rates. Data from such tests are used to calculate the dynamic storage (G') and loss moduli (G''), respectively, the energy elastically stored and dissipated during each deformation cycle. The loss tangent (G''/G') is taken as a measurement of the ratio of viscous and elastic response of the material being tested. The absolute validity of these calculations requires that the samples be linearly viscoelastic (stated more correctly, that the testing be carried out in a region of linear sample behavior). Because the ability of wheat flour doughs to behave linearly (see below) has not yet been established, this condition may only be approached rather than being met. In spite of this caveat, dynamic measurements of the viscoelastic properties of doughs have been used to study most, if not all, of the factors known to affect dough behavior.

NONLINEAR BEHAVIOR

Most food materials that display linear viscoelasticity behave in this manner only when the strain is quite low. The level at which nonlinear behavior occurs is variable, further complicating analysis. Transient loading studies of doughs (see above) have for the most part been carried out at relatively large deformations. The results have been somewhat inconclusive but generally show that the viscoelastic behavior of wheat flour dough is nonlinear. However, if strain levels exist at which a dough would be linearly viscoelastic, analysis of its properties would be simplified greatly. Because dynamic methods can operate at extremely low strains, it is not surprising that most studies using dynamic loading have included an examination of the ability of the dough to behave as a linear viscoelastic material.

Early studies by Hibberd and Wallace (1966) determined that for dough produced from a flour of a single wheat variety, viscoelastic behavior was linear at very low strain amplitudes (less than 0.0022%). At higher strain amplitudes, both storage and loss moduli were highly strain dependent. Hibberd (1970a,

1970b) used this "linear range" in further studies (see below). However, subsequent work by the same research group (Hibberd and Parker, 1975a) omitted any reference to a region of linear behavior and treated the doughs under analysis as being nonlinear materials at any strain.

Such treatment is consistent with later reports (Smith et al., 1970; Navickis et al., 1982; Szczesniak et al., 1983), which concluded that both G' and G'' decrease with increasing strain and that this nonlinearity exists over all strains.

In addition to this general nonlinearity, the amplitude of the deforming strain also affects the relative magnitudes of the storage and loss moduli. Both dynamic tests (Szczesniak et al., 1983) and transient, tensile analyses (Funt Bar-David and Lerchenthal, 1975) have shown that at low (less than 0.1%) strains, G' exceeds G''. At higher strains, the relative magnitude is reversed. Thus, increasing strain causes the behavior of a dough to change from that of a viscoelastic solid to that of an elastoviscous liquid.

Frequency of oscillation also affects the dynamic storage and loss moduli of wheat flour doughs. In this case, the frequency dependence is positive, so G' and G'' increase with frequency (Hibberd and Wallace, 1966; Smith et al., 1970; Cumming and Tung, 1977). Dreese and colleagues (Dreese et al., 1988a, 1988b) confirmed the general observation of a positive dependence of G' on frequency. Tests of a small number of samples suggested (Fig. 2-11) that frequency dependence was the same for both hard and soft wheat flour doughs.

WATER CONTENT

The amount of water present in a wheat flour dough is known, empirically, to significantly affect both the rheological properties of the dough and the quality of the finished baked product (Abdelrahman and Spies, 1986). There seems to be no question that, as the water content of a dough increases, both G' and G'' decrease (Hibberd, 1970b; Hibberd and Parker, 1975b; Navickis et al., 1982; Dreese et al., 1988b). Results of recent studies of hard wheat flour doughs are presented in Figure 2-12. Whether or not an interaction exists between the effects of testing frequency and water content is unresolved. Navickis and colleagues (1982) found such an interaction, while Hibberd and Parker (1975b) concluded that no interaction exists and the effects measured in the ranges of frequencies and moisture contents tested are separable.

Abdelrahman and Spies (1986) further investigated the inverse relationship between G' and G'' and water content. Optimally mixed doughs, tested immediately and then again 30 minutes after mixing, showed that increased absorption caused both the viscous and elastic responses of the dough to change but at different rates. Similar behavior (i.e., a change in the loss tangent) is shown in Figure 2-12. The value of the change in G' after the

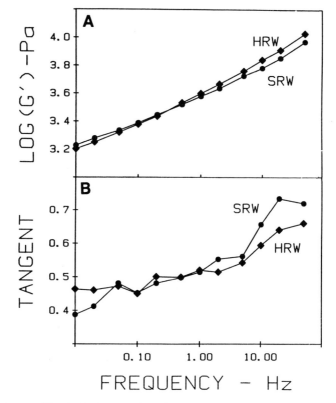

Figure 2–11. Rheological properties of doughs from hard red and soft red winter wheat flours. (*From Dreese et al., 1988b, fig. 3, with permission.*)

30-minute rest, as a function of dough water content, reaches a minimum at a value close to the optimal absorption level empirically determined by the mixograph. This result implies that doughs at optimum absorption show the least change in their dynamic storage moduli over time after mixing.

MIXING TIME

Like absorption, the mixing times required by doughs have clear optima. The optimum has been related to the gross textural properties of a dough and its finished product quality. Bohlin and Carlson (1980) concluded that both G' and G'' were dependent on mixing time. As might be expected, this dependence differed among flours produced from different wheat varieties. It was not clear whether this difference resulted from inherent differences in the

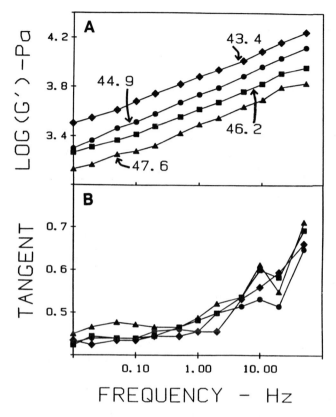

Figure 2–12. Effects of moisture content in flour–water doughs. Numbers indicate percent moisture. (*From Dreese et al., 1988b, fig. 8, with permission.*)

optimum mixing time or stability to overmixing of the flours tested. Dreese and co-workers (1988b) concluded that mixing doughs past optimum caused them to behave similarly to optimally mixed doughs with increased water contents (Fig. 2-13). Thus, increases in mixing time resulted in reduced G' at all frequencies tested. Pointing out the similarity in appearance and "feel" and dynamic rheological behavior between overmixed doughs and doughs containing excess water, the authors speculated that mixing past optimum may reduce the water binding capacity of the gluten in the dough.

Abdelrahman and Spies (1986) compared the dynamic storage moduli of doughs mixed for different times before and after a 30-minute rest. Plots of changes in G' versus mixing time demonstrated no change in storage modulus because of resting at a mixing time approximately 1 minute greater than the mixograph optimum. No explanation was advanced to account for the fact that this change occurred at a point past the empirically determined optimum.

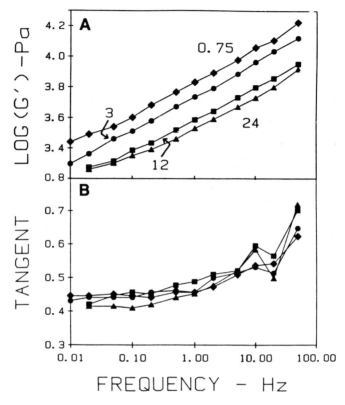

Figure 2–13. Effect of mixing time on flour–water doughs at 44.9% moisture. Numbers indicate mixing time in minutes. Optimum mixing time for this flour was 3.0 minutes. (*From Dreese et al., 1988b, fig. 9, with permission.*)

FLOUR QUALITY

A significant challenge inherent in understanding how (or if) fundamental mechanical properties of wheat flour doughs relate to the quality of the final product is the existence of a wide range in flour protein quality. While empirically determined mixing characteristics are generally good indicators of major differences in baking quality, they are not generally able to indicate subtle differences (Abdelrahman and Spies, 1986). Abdelrahman and Spies subjected two flours with similar chemical compositions and mixing properties, but different baking qualities, to dynamic rheological tests; the tests showed that the higher quality flour had lower G', G'', and loss tangent values than did the poor-quality flour. Because all three values were lower, differences in water absorption between the two flours could not account for the observed differences. The additional water required to reduce G' would have resulted

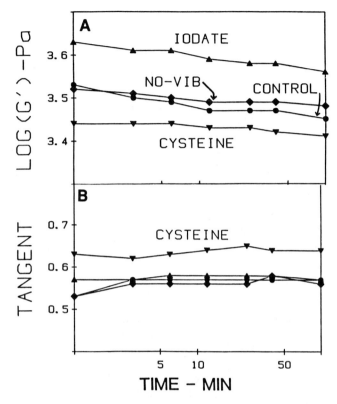

Figure 2–14. The effect of time in the rheometer at 0.5 Hz and 1% strain on flour–water doughs. Cysteine and iodate doughs contained 30 ppm (flour wt. basis) of L-cysteine HCl and KIO_3, respectively. (*From Dreese et al., 1988b, fig. 2, with permission.*)

in increased loss tangents. The authors concluded that both the absolute levels of the viscous and elastic components of a dough and the ratio of those components are important in controlling the baking quality of a flour.

OXIDIZING AND REDUCING AGENTS

Adding the thiol compound cysteine to a flour–water dough reduces G' and increases the tangent slightly relative to control doughs (Fig. 2-14). The fast-acting oxidant KIO_3 increases the G' value for a flour–water dough slightly without significantly changing the loss tangent. While these changes are in the direction that might be anticipated from the known effects of these agents on empirically measured dough properties (Bloksma, 1964; Bloksma and Nieman,

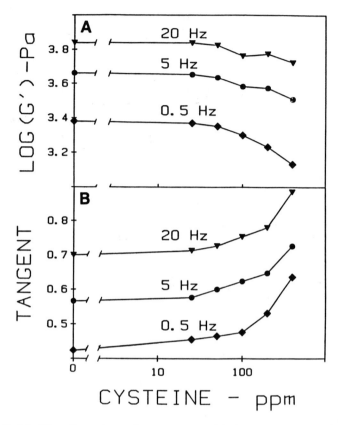

Figure 2–15. The effect of cysteine on commercial gluten–water doughs at 58.9% moisture. (*From Dreese et al., 1988b, fig. 10, with permission.*)

1975), their magnitudes are small. Based on this result, Dreese and co-workers (1988b) concluded that dynamic tests were no better than transient empirical tests in demonstrating the effects of oxidants and reductants on doughs.

In a simplified test system of commercial gluten–water doughs, cysteine decreased G' and increased the tangent (Fig. 2-15), changes consistent with a reduction in polymer cross-links in the system. The results, therefore, provide indirect support for the commonly held view that cysteine breaks or inhibits the formation of disulfide cross-links between gluten polypeptides.

EFFECTS OF FERMENTATION

Yeast-leavened products must, by definition, undergo fermentation. This process results in changes in the rheological properties of the dough as measured

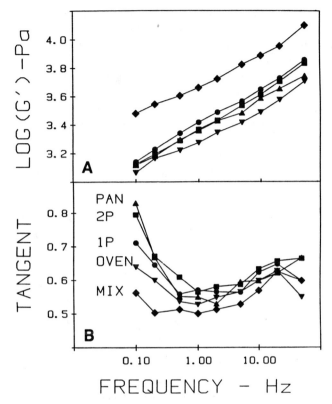

Figure 2–16. The effect of fermentation time on G′ and tangent of bread doughs. Fermentation times for the treatments are: 1P = first punch (113 minutes); 2P = second punch (163 minutes); pan = 180 minutes; oven = 243 minutes. (*From Dreese et al., 1988b, fig. 4, with permission.*)

by several transient testing methods (Hoseney et al., 1979; Cullen-Refai et al., 1988). Testing fully formulated doughs during fermentation, Dreese and co-workers (1988b) showed that large reductions in G′ and the loss tangent occurred between mixing and the first punch (113 minutes of fermentation) (Fig. 2-16). Changes in G′ and the loss tangent with increased fermentation were small. At testing frequencies of 0.1–1.0 Hz, the loss tangent was inversely related to frequency. Up to the oven stage of the process, this frequency dependence was positively correlated with fermentation time.

EFFECTS OF HEATING

The transformation of a fully proofed dough to a baked loaf involves profound changes in the rheological properties of the wheat flour dough system. The fact that two apparently equivalent doughs, proofed to the same height, can produce loaves of significantly different volumes suggests that the temperature-dependent changes that occur during baking may define or determine flour quality. Despite the fact that there is, a priori, no reason to presume that only those changes occurring prior to baking are relevant, little data exist on the fundamental rheological properties of doughs during heating (Bloksma and Nieman, 1975; LeGrys et al., 1980; Dreese et al., 1988a, 1988c). This lack of data may be because of the inherent difficulty in heating a dough mass uniformly while at the same time testing it.

Conventional oven baking heats by conduction. Such heating produces internal temperature gradients in the dough so that any temperature-triggered changes in dough rheology occur at many different times as each point heats

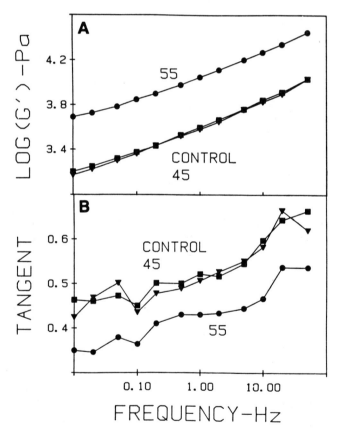

Figure 2–17. Frequency scans of flour–water doughs. All doughs were tested at 25°C. Control doughs had no previous heating. 45°C and 55°C doughs were heated to those temperatures respectively prior to cooling to 30°C and testing. (*From Dreese et al., 1988a, fig. 1, with permission.*)

to the required temperature. Dreese and colleagues (1988a) overcame this limitation by adapting the electrical resistance oven technique (Junge and Hoseney, 1981) to the dynamic rheometer. By using the instrument's top and bottom plates as electrodes and the dough as the resistor, it proved possible to heat all points of the sample uniformly while measuring G' and G''. Using this technique on flour–water doughs before and after heating to 85°C, Dreese and colleagues (1988a) found the increase in G' and decrease in tangent (Fig. 2-17) previously speculated to be due to starch gelatinization (Bloksma and Nieman, 1975) or protein cross-linking (LeGrys, 1980). Heating to temperatures below 45°C resulted in no irreversible changes, while heating at temperatures above 45°C caused the above-described irreversible changes in G' and loss tangent. Measurements during heating (Fig. 2-18) demonstrated that the rapid, irreversible changes in moduli occurred between 55°C and 75°C.

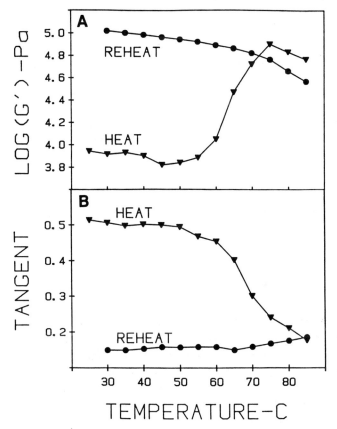

Figure 2–18. The effect of heating on G' (@5 Hz) and tangent of flour–water doughs. Heat = first heating. Reheat = heating after dough had been heated to 90°C and cooled. (*From Dreese et al., 1988a, fig. 2, with permission.*)

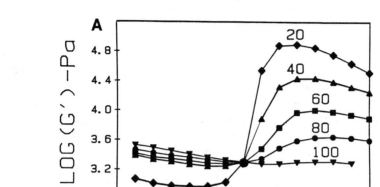

Figure 2–19. Effect of first heating on G' (@2 Hz) and tangent of doughs made from blends of commercial starch and commercial gluten. Figures on the plots correspond to the percent gluten in the final blend. (*From Dreese et al., 1988a, fig. 3, with permission.*)

The facts that these changes were proportional in magnitude to the starch content of the gluten–starch doughs (Fig. 2-19), were essentially absent in gluten–water doughs (Fig. 2-20), and could be mimicked by the addition of pregelatinized starch to gluten–water doughs (Fig. 2-21) suggested that the changes resulted from starch gelatinization during heating. Because the temperature-dependent changes could not be mimicked by reducing moisture content, the authors speculated that starch gelatinization was affecting dough rheology in a way other than by simply absorbing water. One possible effect of gelatinization may be to increase the opportunity for hydrogen bonding between starch and gluten molecules.

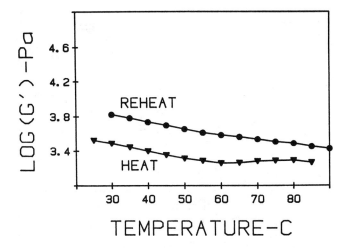

Figure 2–20. The effect of first heating and reheating on G′ (@2 Hz) of gluten–water dough. (*From Dreese et al., 1988a, fig. 8, with permission.*)

FLOUR COMPONENTS

There is no question that gluten protein is the predominant factor defining or controlling the viscoelastic properties of wheat flour doughs. Because of this, research on the effect(s) of flour components has focused on two principal questions: (1) how does the behavior of doughs produced from intact flour differ from those of gluten plus starch, and (2) what is the effect of protein (gluten) content on viscoelastic properties?

These questions have proved difficult to study independently. Work designed to determine the effect(s) of gluten content on dough behavior have employed, nearly without exception, mixtures of previously isolated gluten plus refined wheat starch combined to reach predetermined final protein contents. Because of this, the relationship studied has been less the effect of flour protein content on dough behavior than the effect of gluten:starch ratio on the behavior of gluten–starch doughs. In such studies, reconstituted, although incomplete, systems have been compared with intact flour of equivalent protein content. The behavior of the protein itself (i.e., gluten–water doughs), which is quite different from that of either gluten–starch or intact flour doughs, is somewhat outside the scope of this chapter. This subject has been treated in a recent review (Dreese et al., 1988c).

Several studies have demonstrated that gluten–starch doughs exhibit more linear behavior than do doughs created from intact flour (Szczesniak et al., 1983; Hibberd, 1970b) although less linearity than doughs composed of 100%

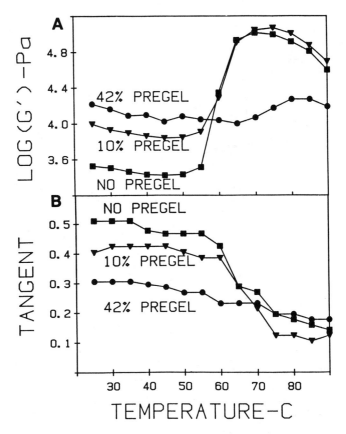

Figure 2–21. G′ and tangent of doughs made from blends of commercial gluten (15%), commercial and pregelatinized wheat starch (shown in the figure), and commercial unmodified wheat starch (remainder). Moisture content for all doughs was 44.9%. (*From Dreese et al. 1988a, fig. 6, with permission.*)

gluten plus water (Smith et al., 1970; Dreese et al., 1988c). The linearity of modulus versus frequency plots decreases as the starch content in the gluten–starch blend increases. Hibberd (1970b) suggests that this variation in frequency dependence indicates that starch in these doughs is not simply an inert filler, but rather an active participant in determining the viscoelastic properties. Matsumoto (1979) reached a similar conclusion based on stress relaxation measurements of what he termed synthetic doughs. Again, starch granules were responsible for large increases in nonlinear behavior.

Beyond its effects on linear behavior, the effects of protein or gluten:starch ratio on the physical properties of a dough appear somewhat contradictory.

Earlier research (Hibberd, 1970b; Smith et al., 1970; Navickis et al., 1982) showed positive relationships between the gluten content of gluten–starch systems and both dynamic moduli (G' and G''). Hibberd's work demonstrated that the frequency dependence of both G' and G'' increased as the protein:starch ratio of the "doughs" increased.

It is important to note that all of the above studies were carried out using systems in which the dough moisture content was constant, regardless of protein:starch ratio. Because gluten absorbs more water than starch, an artifact could be created causing doughs to behave as if they were drier (increase in G') as protein content increased.

Dreese and co-workers (1988c) tested this possibility by preparing gluten–starch doughs in which the water absorption was 103% of the gluten weight and 59% of the starch weight in the blend. Under these conditions, log G' versus frequency plots increased in slope as the gluten contents of the system increased. At gluten contents equal to or less than 40%, G' varied inversely with gluten content, contrary to earlier reports. Increasing gluten content in the blend above 40% had no further effect on G'. Apparently, at 60% starch or less, starch no longer contributes significantly to the dynamic viscoelastic properties of a dough and the system behaves akin to 100% gluten. Alternatively, and more likely, the changes caused by alterations in protein content above 40% cannot be measured by the dynamic stress–strain technique employed in this work.

Studies comparing the dynamic properties of an intact flour (12.4% protein, db) with those of a starch plus 15% gluten blend (13.3% protein, db) found that plots of G' versus frequency were significantly different (Dreese et al., 1988c). The slope of the former was steeper and resembled that produced by a 100% gluten–water dough. This result suggested that flour components in addition to gluten and starch were affecting the rheological properties of the dough.

By fractionating this flour into its gluten, starch, and water solubles, it was possible to assess the effect of each component as well as the fractionation/reconstitution procedure on G', G'', and the loss tangent. Reconstituted gluten/starch/water solubles differed little in behavior from that of the parent flour. Tangents (at all but the lowest testing frequencies) were equal to those of intact flour doughs. At all testing frequencies, the dynamic storage modulus of the reconstituted flour was roughly 0.1 log unit lower than that of previously unfractionated flour.

Doughs reconstituted without the water-soluble fraction had significantly lower tangent values and were more elastic (had higher G' values). In this instance, the dynamic test reflected the empirical observation that the presence of water solubles in a dough causes it to be "slacker" and more viscous. Abdelrahman and Spies (1986) addressed the same question in a slightly different manner. Supplementing intact flour with isolated water solubles resulted in reductions in both G' and G'' relative to unsupplemented flour. In this case, addition of water solubles caused the loss tangent to decrease, contrary to the findings of Dreese and colleagues (1988c).

Supplementing native flour with either isolated gluten or starch resulted in increases (relative to native flour) in the moduli and reductions in the tangents of the supplemented doughs (Abdelrahman and Spies, 1986). The nature of the responses differed, depending on the component added. Consistent with its effects in isolated gluten–starch systems, starch supplementation resulted in nonlinear responses in both moduli and the tangent. Changes in G' and G'' due to gluten supplements were linear.

Gluten–starch doughs, created from laboratory isolated/lyophylized components, possessed lower tangents and higher G' values than did those produced from commercially isolated gluten–starch doughs (Dreese et al., 1988c). This result is surprising, because polymer theory (Ferry, 1970) suggests that these results indicate that the lyophylized system, which was subjected to very mild heating, contained more effective cross-links than its commercially isolated counterpart. Commercially isolated gluten is heated during isolation, and heating is believed to result in increased cross-linking between gluten polypeptides (Bale and Muller, 1970; Schofield et al., 1983). It may be that additional chemical or physical differences exist between commercial and lab-isolated gluten that override the effects of heating.

REFERENCES

Abdelrahman, A. A., and Spies, R. D. 1986. Dynamic rheological studies of dough systems. In *Fundamentals of Dough Rheology*, ed. H. Faridi and J. M. Faubion, pp. 87–103. St. Paul, MN: AACC.

Bagley, E. B., and Christianson, D. D. 1986. Response of commercial chemically leavened doughs to uniaxial compression. In *Fundamentals of Dough Rheology*, ed. H. Faridi and J. M. Faubion, pp. 27–36. St. Paul, MN: AACC.

Bagley, E. B., Christianson, D. D., and Wolfe, W. J. 1985a. Frictional effects in compressional deformation of gelatin and starch gels and comparison of material response in simple shear, torsion and lubricated uniaxial compression. *J. Rheol.* **29**:103–108.

Bagley, E. B., Wolfe, W. J., and Christianson, D. D. 1985b. Effect of sample dimensions, lubrication and deformation rate on uniaxial compression of gelatin gels. *Rheol. Acta* **24**:265–271.

Baird, D. G. 1983. Food dough rheology. In *Physical Properties of Foods*, ed. M. Peleg and E. B. Bagley, pp. 343–350. Westport, CT: AVI.

Baker, J. C., and Mize, M. D. 1937. Mixing dough in vacuum and in the presence of various gases. *Cereal Chem.* **14**:721–734.

Bale, R., and Muller, H. G. 1970. Application of statistical theory of rubber elasticity to the effect of heat on wheat gluten. *J. Food Tech.* **5**:295–300.

Bernardin, J. E., and Kasarda, D. D. 1973. Hydrated protein fibrils from wheat endosperm. *Cereal Chem.* **50**:529–536.

Bloksma, A. H. 1962. Slow creep and recovery of wheat flour doughs. *Rheol. Acta* **2**:217.

Bloksma, A. H. 1964. Oxidation by potassium iodate of thiol groups in unleavened wheat flour doughs. *J. Sci. Food Agric.* **15**:83–88.

Bloksma, A. H., and Nieman, W. 1975. The effects of temperature on some properties of wheat flour doughs. *J. Text. Stud.* **6**:343–361.

Bohlin, L., and Carlson, T. L. G. 1980. Dynamic viscoelastic properties of wheat flour dough: dependence on mixing time. *Cereal Chem.* **57**:175–181.

Bushuk, W. 1985. Rheology: theory and application to wheat flour doughs. In *Rheology of Wheat Products*, ed. H. Faridi, pp. 1–26. St. Paul, MN: AACC.

Chatraei, Sh., Macosko, C. W., and Winter, H. H. 1981. Lubricated squeezing flow: a new biaxial extensional rheometer. *J. Rheol.* **25**:433–443.

Conn, J. F. and Kichline, T. P. 1971. Reduction of mixing requirements for yeast leavened bread doughs. U.S. Patents 3,556,804 and 3,556,805.

Cullen-Refai, A., Faubion, J. M., and Hoseney, R. C. 1988. Lubricated uniaxial compression of fermenting doughs. *Cereal Chem.* **65**:401–403.

Cumming, D. B., and Tung, M. A. 1977. Modification of the ultrastructure and rheology of rehydrated commercial wheat gluten. *J. Can. Inst. Food Sci.* **10**:109–113.

Cunningham, J. R., and Hlynka, I. 1954. Relaxation time spectrum of dough and the influence of temperature, rest and water content. *J. Appl. Phys.* **25**:1075–1081.

Cunningham, J. R., Hlynka, I., and Anderson, J. A. 1953. An improved relaxometer for viscoelastic substances applied to the study of wheat dough. *Can. J. Technol.* **31**:98–108.

Danno, G., and Hoseney, R. C. 1982. Effects of dough mixing and rheologically active compounds on relative viscosity of wheat proteins. *Cereal Chem.* **59**:196–198.

Dealy, J. M. 1984. Official nomenclature for material functions describing the response of a viscoelastic fluid to various shearing and extensional deformations. *J. Rheol.* **28**:181–195.

Dreese, P. C., Faubion, J. M., and Hoseney, R. C. 1988a. Dynamic rheological properties of flour, gluten and gluten-starch doughs. I. Temperature-dependent changes during heating. *Cereal Chem.* **65**:348–353.

Dreese, P. C., Faubion, J. M., and Hoseney, R. C. 1988b. Dynamic rheological properties of flour, gluten and gluten-starch doughs. II. Effect of various processing and ingredient changes. *Cereal Chem.* **65**:354–359.

Dreese, P. C., Faubion, J. M., and Hoseney, R. C. 1988c. The effect of different heating and washing procedures on the dynamic rheological properties of wheat gluten. *Cereal Foods World* **33**:225–228.

Dronzek, B., and Bushuk, W. 1968. A note on the formation of free radicals in dough during mixing. *Cereal Chem.* **42**:286.

Faubion, J. M., Dreese, P. C., and Diehl, K. C. 1985. Dynamic rheological testing of wheat flour doughs. In *Rheology of Wheat Products*, ed. H. Faridi, pp. 91–116. St. Paul, MN: AACC.

Faubion, J. M., and Faridi, H. 1986. Dough rheology: Its benefits to cereal chemists. In *Rheology of Wheat Products*, ed. H. Faridi and J. M. Faubion, pp. 1–9. St. Paul, MN: AACC.

Ferry, J. D. 1970. *Viscoelastic Properties of Polymers.* New York: Wiley.

Funt Bar-David, C. B., and Lerchenthal, C. H. 1975. Rheological and thermodynamic properties of gluten gel. *Cereal Chem.* **52**:154r–169r.

Glucklich, J., and Shelef, L. 1962. An investigation into the rheological properties of flour dough. Studies in shear and compression. *Cereal Chem.* **39**:242–255.

Graveland, A., Bosveld, P., Lichtendonk, W. J., and Moonen, J. H. E. 1984. Structure of glutenins and their breakdown during mixing by a complex oxidation-reduction system. In *Proceedings of the 2d Workshop on Gluten Proteins*, ed. A. Graveland and J. H. E. Moonen, pp. 59–68. The Netherlands: TNO Wageningen.

Hibberd, G. E. 1970a. Dynamic viscoelastic behavior of wheat flour doughs. II. Effects of water content in the linear region. *Rheol. Acta* **9**:497–500.

Hibberd, G. E. 1970b. Dynamic viscoelastic behavior of wheat flour doughs. III. The influence of starch granules. *Rheol. Acta* **9**:501–505.

Hibberd, G. E., and Parker, N. S. 1975a. Dynamic viscoelastic behavior of wheat flour doughs. IV. Non-linear behavior. *Rheol. Acta* **14**:151–157.

Hibberd, G. E., and Parker, N. S. 1975b. Measurement of the fundamental rheological properties of wheat flour doughs. *Cereal Chem.* **52**:1r–23r.

Hibberd, G. E., and Parker, N. S. 1975c. Parallel plate rheometer for measuring the viscoelastic properties of wheat flour doughs. *Cereal Chem.* **55**:102.

Hibberd, G. E., and Wallace, W. J. 1966. Dynamic viscoelastic behavior of wheat flour doughs. I. Linear Aspects. *Rheol. Acta* **5**:193–198.

Hlynka, I. 1970. Rheological properties of dough and their significance in the breadmaking process. *Baker's Dig.* **44**:40–57.

Hlynka, I., and Anderson, J. A. 1952. Relaxation of tension in stretched dough. *Can. J. Technol.* **30**:198–210.

Hoseney, R. C. 1985. The mixing phenomenon. *Cereal Food World* **30**:453–457.

Hoseney, R. C. 1986. Component interaction during heating and storage of baked products. In *Chemistry and Physics of Baking*, ed. J. M. V. Blanschard, P. J. Frazier and T. Galliard, pp. 216–226. London: Royal Chemistry Society.

Hoseney, R. C., and Finney, P. L. 1974. Mixing—a contrary view. *Baker's Dig.* **48**:22–24, 26, 28, 66.

Hoseney, R. C., and Seib, P. A. 1973. Structural differences in hard and soft wheats. *Baker's Dig.* **47**:26–28, 56.

Hoseney, R. C., Hsu, K. H., and Junge, R. C. 1979. A simple spread test to measure the rheological properties of fermenting dough. *Cereal Chem.* **56**:141–143.

Hoseney, R. C., Zeleznak, K., and Lai, C. S. 1986. Wheat gluten: A glassy polymer. *Cereal Chem.* **63**:285–286.

Jackson, G. M., and Hoseney, R. C. 1986. Effects of endogenous phenolic acids on the mixing properties of wheat flour doughs. *Cereal Chem.* **64**:79–85.

Junge, R. C., and Hoseney, R. C. 1981. A mechanism by which shortening and certain surfactants improve loaf volume in bread. *Cereal Chem.* **58**:408–412.

Junge, R. C., Hoseney, R. C., and Varriano-Marston, E. 1981. Effect of surfactants on air incorporation in dough and the crumb grain of bread. *Cereal Chem.* **58**:338–342.

LeGrys, G. A., Booth, M. R., and Al-Baghdadi, A. 1980. The physical properties of wheat proteins. In *Cereals, a Renewable Resource*, ed. L. Munck and Y. Pomeranz, pp. 243–264. St. Paul, MN: AACC.

Lillard, D. W., Seib, P. A., and Hoseney, R. C. 1982. Isomeric ascorbic acids and derivatives of L-ascorbic acid: their effects on the flow of dough. *Cereal Chem.* **59**:291–296.

MacRitchie, F. 1980. Physicochemical aspects of some problems in wheat research. In *Advances in Cereal Science and Technology*, Vol. 3, ed. Y. Pomeranz, pp. 271–326. St. Paul, MN: AACC.

MacRitchie, R. 1975. Mechanical degradation of gluten proteins during high speed mixing of doughs. *Polymer Sci.* **49**:85–90.

Matsumoto, S. 1979. Rheological properties of synthetic flour doughs. In *Food Texture and Rheology*, ed. P. Sherman, pp. 291–302. London: Academic Press.

Moore, W. R., and Hoseney, R. C. 1986. The effects of flour lipids on the expansion rate and volume of bread baked in a resistance oven. *Cereal Chem.* **63**:172–174.

Muller, H. G., Williams, M. V., Russell Eggitt, P. W., and Coppock, J. B. M. 1961. Fundamental studies on dough with the Brabender extensigraph. I. Determination of stress-strain curves. *J. Sci. Food Agric.* **12**:513–523.

Muller, H. G., Williams, M. V., Russell Eggitt, P. W., and Coppock, J. B. M. 1962. Fundamental studies on dough with the Brabender extensigraph. II. Determination of the apparent elastic modulus and coefficient of viscosity of wheat flour dough. *J. Sci. Food Agric.* **13**:572–577.

Navickis, L. L., Anderson, R. A., Bagley, E. B., and Jasburg, B. K. 1982. Viscoelastic properties of wheat flour doughs: Variation of dynamic moduli with water and protein content. *J. Text. Stud.* **13**:249–259.

Olkku, M., and Sherman, P. 1979. Compression testing of cylindrical samples with an Instron universal testing machine. In *Food Texture and Rheology*, ed. P. Sherman, pp. 157–175. New York: Academic Press.

Peleg, M. 1979. Characterization of the stress relaxation curves of solid food. *J. Food Sci.* **44**:277–281.

Peleg, M. 1980. Linearization of relaxation and creep curves of biological materials. *J. Rheol.* **25**:451–463.

Rao, V. N. M. 1984. Dynamic force-deformation properties of foods. *Food Tech.* **21**:103–109.

Rasper, V. 1975. Dough rheology at large deformations in simple tension. *Cereal Chem.* **52**:24r–41r.

Rogers, D. E., and Hoseney, R. C. 1987. Tests to determine the optimum water absorption for saltine cracker doughs. *Cereal Chem.* **64**:370–372.

Schofield, R. K., and Scott Blair, G. W. 1932. The relationship between viscosity, elasticity and plastic strength of soft materials as illustrated by some mechanical properties of flour doughs. In *Proc. Roy. Soc. (London)* **A138**:707–718.

Schofield, R. K., and Scott Blair, G. W. 1933a. The relationship between viscosity, elasticity and plastic strength of soft materials as illustrated by some mechanical properties of flour doughs. II. *Proc. Roy. Soc. (London)* **A139**:557–566.

Schofield, R. K., and Scott Blair, G. W. 1933b. The relationship between viscosity, elasticity and plastic strength of soft materials as illustrated by some mechanical properties of flour doughs. III. *Proc. Roy. Soc. (London)* **A141**:72–85.

Schofield, R. K., and Scott Blair, G. W. 1937. The relationship between viscosity, elasticity and plastic strength of soft materials as illustrated by some mechanical properties of flour doughs. IV. The separate contributions of gluten and starch. *Proc. Roy. Soc. (London)* **A160**:87.

Schofield, J. D., Bottomley, R. C. Timms, M. F., and Booth, M. R. 1983. The effect of heat on wheat gluten and the involvement of sulfhydryl–disulfide interchange reactions. *J. Cereal Sci.* **1**:241–253.

Schroeder, L. F., and Hoseney, R. C. 1978. Mixograph studies. II. Effect of activated double bond compounds on dough mixing properties. *Cereal Chem.* **55**:348–360.

Shelef, L., and Bousso, D. 1964. A new instrument for measuring relaxation in flour dough. *Rheol. Acta* **3**:168–172.

Sidhu, J. S., Nordin, P., and Hoseney, R. C. 1980. Mixograph studies. III. Reaction of fumaric acid with gluten proteins during mixing. *Cereal Chem.* **57**:159–163.

Smith, T. L., and Tschoegl, N. W. 1970. Rheological properties of wheat flour doughs. IV. Creep and creep recovery in simple tension. *Rheol. Acta* **9**:339.

Smith, J. R., Smith, T. L., and Tschoegl, N. W. 1970. Rheological properties of wheat flour doughs. III. Dynamic shear modulus and its dependence on amplitude, frequency and dough composition. *Rheol. Acta* **9**:239–252.

Szczesniak, A. A., Loh, J., and Manell, W. R. 1983. Effect of moisture transfer on the dynamic viscoelastic properties of flour–water systems. *J. Rheol.* **27**:537–556.

Tsen, C. C. 1967. Changes in flour proteins during mixing. *Cereal Chem.* **44**:308–317.

Weak, E. D., Hoseney, R. C., Seib, P. A., and Baig, M. 1977. Mixograph studies. I. Effect of certain compounds on mixing properties. *Cereal Chem.* **54**:794–802.

Chapter 3

Rheological Properties of Cereal Proteins

Ann-Charlotte Eliasson

The protein content of intact cereals ranges from about 5% in millet to over 15% in oat (Wieser et al., 1980). The protein content of wheat flour can vary from 7% to 17% (Redman, 1971). Besides contributing to the nutritional value of the cereal, proteins also give functional properties to foods. Among the cereal proteins, effects on functional properties are most evident in the case of wheat proteins, due to the unique viscoelasticity of gluten. Similar properties are shown to only a very small extent by proteins in rye and barley, and not at all by proteins in the other cereals. It is not surprising, therefore, that so much work has been done to find out what makes the wheat proteins unique.

Several chemical and physicochemical methods, from determinations of amino acid sequences to measurements of rheological properties, have been used to obtain a better understanding of the wheat proteins. For most methods, however, a protein fraction is required that is essentially pure before measurements can begin. A wide range of fractionation procedures for wheat proteins has therefore been developed during the years.

This chapter begins with a description of some chemical and physicochemical properties of cereal proteins. However, the main subject is the *rheological* properties of cereal proteins. These properties have not been investigated for all cereals. In fact, they have been investigated almost exclusively for wheat proteins. As a consequence, several important cereals are omitted (rice, millet), whereas others are included because they have been studied as additives to gluten (e.g., oat).

The rheological properties of cereal proteins in solution are first discussed, followed by a treatment of A-gliadin, the cereal protein most thoroughly investigated by rheological methods (both in solution and in more concentrated systems). Thereafter the rheological properties of protein dispersions (again gluten) are discussed.

CLASSIFICATION, COMPOSITION, AND PHYSICOCHEMICAL PROPERTIES OF CEREAL PROTEINS

Classification According to the Osborne Scheme

Cereal proteins are usually divided into four groups according to the solubility fractionation scheme of Osborne (Ewart, 1968a), in which albumins are those proteins soluble in distilled water, globulins are soluble in dilute salt solutions; prolamines, in aqueous ethanol; and glutelins, in dilute acids and alkalis. This fractionation scheme may be adapted to any cereal. The distribution of proteins among the solubility classes differs among the cereals. One example is given in Table 3-1. The solubility classes are not homogeneous. This became evident when different electrophoretic methods were used to characterize the proteins. Prolamines from wheat contain 30 or more individual proteins, and glutelins from wheat contain at least 40–50 unique polypeptides (Bietz, 1979).

Albumins and Globulins

While the albumins are water soluble and the globulins are salt soluble, both groups of proteins are similar in that they are characterized by relatively high amounts of charged amino acids and the presence of thiol-containing amino acids. The molecular weight of albumins is low, about 13,000, whereas molecular weights within the range of 25,000–100,000 have been reported for globulins (Ewart, 1972b; Pyler, 1983). Albumins, as well as globulins, are metabolic proteins and their electrophoretic patterns are constant among varieties. These proteins are not thought to play an essential role in bread-

Table 3-1. Distribution of Protein between Different Solubility Classes in the Osborne Fractionation

| Cereal | Percent of Total Protein Extracted as | | | |
	Albumins	Globulins	Prolamines	Glutelins
Wheat	14.7	7.0	32.6	45.7
Rye	44.4	10.2	20.9	24.5
Barley	12.1	8.4	25.0	54.5
Oat	20.2	11.9	14.0	53.9
Rice	10.8	9.7	2.2	77.3
Millet	18.2	6.1	33.9	41.8
Maize	4.0	2.8	47.9	45.3

Source: Data from Wieser et al. (1980), with permission.

making, except, of course, those globulins that happen to be enzymes such as the α- or β- amylases (Redman, 1971; Ewart, 1972b; Pyler, 1983).

Prolamines

The prolamines, together with the glutelins, are the storage proteins in cereals; that is, their function is to provide the seedling with nitrogen, carbon, and sulfur. In contrast to the metabolic proteins, a change in amino acid composition is not critical to the function of these proteins. The result has been the evolution of a great number of prolamines.

The amino acid composition of prolamines is characterized by a low level of charged amino acids. Typically 5–10% of the amino acids are charged (Krull and Wall, 1969). As is shown in Table 3-2, the proportions of glutamine and proline are high. The disulfide bridges present are intramolecular. Amino acid sequences have been determined for peptides from prolamines (Bietz, 1979; Wieser et al., 1987), and several similarities have been found in the prolamines from wheat, rye, and barley.

Glutelins

The glutelins are those proteins remaining after the albumins, globulins, and prolamines have been extracted. The glutelins comprise from 25% of the total protein in rye to 77% of the total protein in rice (Table 3-1). The remaining glutelin fraction can be further fractionated with acetic acid to give an acid-insoluble residue (Ewart, 1968a).

The partial amino acid compositions of glutelins are given in Table 3-3. The glutelins are similar to the prolamines in that the levels of charged amino acids are low, and the levels of glutamine and proline are high. The disulfide bridges present are thought to be intramolecular as well as intermolecular.

The glutelins are the most difficult proteins in cereals to investigate due to their high molecular weights and their low solubility. To extract the glutelins

Table 3-2. Partial Amino Acid Composition of Cereal Prolamines

Amino Acid (mol-%)	Wheat	Rye	Barley	Oat	Maize
Glx	37.7	36.0	35.9	34.6	19.7
Pro	16.9	18.7	23.4	10.4	10.3
Gly	3.0	4.6	2.3	2.7	2.6
Cys	2.2	2.2	1.9	3.4	1.0

Source: Data from Wieser et al. (1980), with permission.

Table 3-3. Partial Amino Acid Composition of Cereal Glutelins

Amino Acid (mol-%)	Wheat	Rye	Barley	Oat	Maize
Glx	30.7	20.1	24.7	19.4	16.3
Pro	12.2	9.6	14.5	5.6	11.7
Gly	8.1	9.4	6.5	8.1	7.0
Cys	1.4	0.8	0.5	1.2	1.8

Source: Data from Wieser et al. (1980), with permission.

completely, or to study them by electrophoresis, they must be reduced. However, much attention has been focused on the glutelins, and especially to the glutelins in wheat, since they are claimed to be responsible for the unique viscoelastic properties of wheat flour dough (Ewart, 1972b). Several models have been presented to explain the structure and behavior of the wheat glutelins. These models will be discussed in connection with rheological measurements on wheat glutelins.

Description of Some Wheat Protein Fractions

Wheat Prolamines—Gliadins

The gliadins are soluble in 70% aqueous ethanol, and they have been divided into subgroups (α-, β-, γ-, and ω-gliadins) according to their electrophoretic mobility at low pH values. The ω-gliadins show the slowest electrophoretic mobility at low pH values. The disulfide bridges present are all intramolecular, so whether electrophoresis is performed with or without a reducing agent does not change the number of proteins detected. The ω-gliadins are unique in that they do not contain sulfur-containing amino acids (Shewry and Miflin, 1984).

Gel filtration chromatography of gliadins shows three distinct fractions (Bietz, 1979). The α-, β-, and γ- gliadins have molecular weights in the range 30,000–40,000 whereas ω-gliadins have higher molecular weights (60,000–80,000). There is also a high molecular weight (HMW) fraction of gliadin with a molecular weight in the range 100,000–200,000. This HMW gliadin is an oligomer composed of smaller subunits, which are probably similar to low molecular weight (LMW) glutenin subunits (Bietz and Wall, 1980).

The electrophoretic pattern shown by gliadins after sodium dodecyl sulfate polyacrylamide gel electrophoresis (SDS–PAGE) is a characteristic of the variety, and is not affected by growing conditions (Redman, 1971). Attempts have been made to find a correlation between the SDS–PAGE pattern for gliadins and the quality of flour. Such a correlation has not been found, however, except in the case of durum wheat and pasta quality (Damidaux et al., 1980).

The partial amino acid compositions of the gliadins are given in Table 3-2. The gliadins are characterized by high amounts of proline and glutamine, and low amounts of charged amino acids. The isoelectric points of most gliadins are in the range pH 5–9. (Shewry and Miflin, 1984). The hydrophobicity of gliadins has been calculated from the amino acid composition as 1,109 cal (Pomeranz et al., 1970). When gliadins are separated by reversed phase high-performance liquid chromatography (RP–HPLC), they are fractionated into three groups of differing hydrophobicity (Seilmeier et al., 1987). The most hydrophilic group corresponds to the ω-gliadins, the intermediate group to the α-gliadins, and the most hydrophobic group to the γ-gliadins.

Partial amino acid sequences have been determined for several gliadins (Bietz, 1979; Wieser et al., 1987; Shewry and Miflin, 1984), and many similar sequences have been found. The N-terminal sequences of the S-rich prolamines are rich in proline, whereas the C-terminal sequences are rich in cysteine and methionine (Shewry et al., 1986).

The conformations of several cereal proteins have been determined by circular dichroism (CD) spectroscopy. Some of these results are summarized in Table 3-4. The α-, β-, γ-gliadins have similar contents of α helix (33–37%). The structure of ω-gliadin is rich in the β turn, whereas the α helix or β sheet has not been detected in these proteins. Heating of γ-gliadin to 80°C causes a partial loss of the α helix structure. The changes are completely reversible on cooling (Tatham and Shewry, 1985). When ω-gliadin is heated, a conformational change was observed at about 60–70°C, which was interpreted as an increase in class B β turns. Tatham and Shewry suggested that the main stabilizing forces in the α-, β-, and γ-gliadins are covalent disulfide bonds and noncovalent hydrogen bonds, whereas ω-gliadins are stabilized by hydrophobic interactions.

Table 3-4. Conformations of Wheat Proteins Determined by Circular Dichroism Spectroscopy

Protein	α Helix (%)	β Structure (%)	β Turns and Unordered (%)
A-gliadin[a]	24	33	44
α-gliadin[b]	36–37	11–12	52–53
β-gliadin[b]	36–37		
γ-gliadin[b]	33–34		
LMW subunits of glutenin[c]	32–36	19–24	41–49

[a] pH 3 in water (Purcell et al., 1988)
[b] In 70% (v/v) aqueous ethanol (Tatham and Shewry, 1985).
[c] In 50% (v/v) aqueous propan-1-ol with 50 mM dithiothreitol, 20° C (Tatham et al., 1987).

Changes in the gliadin fraction during breadmaking have been studied by HPLC and SDS–PAGE (Menkovska et al., 1987). No changes in the gliadin fraction were detected during dough mixing or fermentation. The heating during baking led to interactions among the gliadins, especially in the crust.

Wheat Glutelins—Glutenins

The molecular weight of native glutenin is estimated to be several million (Bietz, 1979). The glutenins are, therefore, excluded to a great extent from all gels in gel chromatography unless they are first reduced. The molecular weights of the reduced glutenin monomers have been calculated through SDS–PAGE to be in the range 10,000–130,000 (Bushuk, 1984). The glutenin subunits give rise to typical patterns in SDS–PAGE (Payne et al., 1980), and a correlation between the electrophoretic pattern and quality has been sought. The presence of certain combinations of subunits is correlated with an increased dough resistance, increased extensibility, and decreased dough breakdown (Payne et al., 1987; Lagudah et al., 1988). Other subunits are correlated with poor mixing properties.

The glutenin subunits have been numbered according to their mobility in SDS–PAGE (Payne et al., 1980). They are also classified according to their apparent molecular weights (by SDS–PAGE) as HMW subunits, middle molecular weight (MMW) subunits, and LMW subunits (Seilmeier et al., 1987).

The hydrophobicity of the total glutenin fraction has been calculated to be 1,016 cal from its amino acid composition. This is lower than the value for gliadin (1,109 cal) (Pomeranz et al., 1970).

When the reduced glutenins were studied by RP–HPLC, about 20 major peaks were detected (Seilmeier et al., 1987). These peaks were divided into three groups of differing hydrophobicity, with the most hydrophilic group corresponding to MMW glutenins, the intermediate group corresponding to HMW glutenins, and the most hydrophobic group corresponding to LMW glutenins. Partial amino acid compositions for these groups are given in Table 3-5. The HMW glutenin contains a particularly high level of glycine.

Table 3-5. Partial Amino Acid Composition of Reduced Glutenins

Amino Acid (mol-%)	Glutenin Subunits		
	HMW	MMW	LMW
Glx	29.2–34.9	44.0–45.1	33.1–37.3
Pro	10.9–14.6	15.4–15.7	12.6–13.9
Gly	13.5–18.4	6.4–7.4	4.7–7.1

Source: Data from Seilmeier et al. (1987), with permission.

The conformation of LMW subunits of glutenin has been studied (Tatham and Shewry, 1985), and found to be similar to the conformations of α-, β-, and γ-gliadins (Table 3-4). When LMW subunits of glutenins were heated, a considerable part of the α helix was lost, probably due to the absence of stabilizing disulfide bonds in the reduced LMW subunits (Tatham et al., 1987).

Wheat Gluten

The most important property of gliadin and glutenin is their ability to form gluten, the dough-forming protein fraction in wheat. Gluten is prepared by forming a dough that is washed in excess water to remove starch and water solubles. The remaining gluten contains roughly 67% water and 33% dry matter. The dry matter is composed not only of proteins, but also of starch, pentosans, and lipids (if nondefatted flour is used). For the purposes of this discussion, gluten is considered to be a concentrated dispersion of gliadins and glutenins. Gluten prepared in this way has been used both for rheological measurements and for other physicochemical studies. Here the effects of heating on gluten properties will be described, and also the surface properties of gluten proteins.

As has been discussed, heating gliadins results in some loss of structure. This loss has been observed in CD spectra (Tatham and Shewry, 1985). When gluten is analyzed by differential scanning calorimetry (DSC), however, only very small endotherms are observed (Eliasson and Hegg, 1980). The lack of a denaturation endotherm when gluten is heated might be explained by endothermic and exothermic events canceling each other, or by the absence of significant cooperativity in the protein structure. However, changing the polarity of the solvent used, which was thought to change the balance between exothermic and endothermic processes, did not result in a denaturation peak (Hoseney et al., 1987). Although a denaturation peak has not been detected by DSC, a transition temperature corresponding to a glass transition in amorphous polymers (T_g) has been observed (Hoseney et al., 1986). The subject of the T_g of water-plasticized gluten in discussed in detail in Chapter 5.

When heated gluten samples were used in baking experiments, a reduction in loaf volume was obtained for gluten heated to 60°C or above (Schofield et al., 1983). It was suggested that the glutenins were unfolded on heating up to 75°C. This unfolding facilitated sulfhydryl/disulfide interchange and increased polymerization of glutenin. At temperatures above 75°C gliadin proteins were affected.

The behavior of wheat gluten at the air/water interface was first studied by Tschoegl and Alexander (1960a), using the surface balance technique. These investigators found that gluten films were unusually stable, and that the films showed a considerable compressibility. Protein polymerization has been shown to occur during repeated compressions and expansions of a wheat protein film at the air/water interface (Lundh et al., 1988). When the protein spread at the air/water interface contained only gliadins, polymerization was not observed. The surface properties of gluten proteins have also been studied in relation to their foam stability (Mita et al., 1977). A gluten foam was much more stable than foams from other proteins (e.g., egg albumin, gelatin, and peptone). Moreover, the stability of the gluten foam was increased if defatted gluten was used.

RHEOLOGICAL PROPERTIES OF INDIVIDUAL PROTEINS OR PROTEIN FRACTIONS IN SOLUTION

Definition of Viscosity

The study of the rheological properties of a protein solution involves the measurement of viscosity, either the flow behavior of the solution or its hydrodynamic properties. Flow behavior is studied over a range of shear rates ($\dot{\gamma}$), and the measured shear stress (σ) is plotted against $\dot{\gamma}$ to construct a flow curve. If the protein solution behaves like a Newtonian fluid, the following relation holds:

$$\sigma = \eta \cdot \dot{\gamma} \qquad (3\text{--}1)$$

where η is the viscosity. However, deviations from Newtonian behavior are often found. These deviations are to be expected at high protein concentrations due to interactions between protein molecules. In these instances, viscosity will no longer be independent of $\dot{\gamma}$. In this case, an apparent viscosity (η_{app}) can be calculated at each shear rate. To describe the flow curve mathematically, the so-called power law is used:

$$\sigma = k \cdot \dot{\gamma}^{\eta} \qquad (3\text{--}2)$$

Here, k is the consistency coefficient and n is the flow behavior index; $n = 1$ for a Newtonian liquid. Shear thinning behavior is encountered when $n < 1$, and shear thickening when $n > 1$.

The viscosity of a solution of macromolecules is affected by several factors, including the shape and size of solute molecules, the molecular weight distribution of the macromolecules, the presence of electric charges in the system, and the concentration of the systems (Frisch and Simha, 1956). Viscosity measurements offer a possibility to obtain information concerning the conformation of macromolecules in solution. The parameter used to characterize the hydrodynamic behavior of macromolecules is the intrinsic viscosity $[\eta]$.

The effects of a solute on the viscosity of a solution was first studied by Einstein, who derived the relation

$$\eta = \eta_s(1 + \alpha \cdot \phi) \qquad (3\text{-}3)$$

where η is the viscosity of the solution, η_s is the viscosity of the solvent, ϕ is the volume fraction occupied by the solute, and α is a constant that depends on the nature of the solute. The relation holds only in the very dilute solutions, and in this concentration range, the relative viscosity shows a linear dependence on concentration ($\eta_r = \eta/\eta_s$). A specific viscosity, η_{sp}, is defined as:

$$\eta_{sp} = \frac{\eta - \eta_s}{\eta_s} \qquad (3\text{-}4)$$

The intrinsic viscosity $[\eta]$, finally, is obtained by measuring viscosity at different concentrations, calculating η_{sp}, and plotting this against concentration. The intrinsic viscosity is obtained after extrapolation to zero concentration, and is expressed as m^3/kg (or ml/g). The intrinsic viscosity $[\eta]$ is thus defined as

$$[\eta] = \lim_{c \to 0} \left(\frac{\eta_{sp}}{c} \right) \qquad (3\text{-}5)$$

where c is the concentration.

Molecular weights of macromolecules in solution can be determined from the following relation:

$$[\eta] = K \cdot M^a \qquad (3\text{-}6)$$

where M is the molecular weight and K and a are constants that depend on the nature of the solvent, the polymer, and temperature, but not on concentration or molecular weight (Frisch and Simha, 1956).

In many applications, measurements are often simplified to involve only η_r; that is, the viscosity of the solution after a certain treatment (e.g., heating) is compared to the viscosity of the solution before treatment. Any changes in η_r might then be interpreted as being due to changes in conformation or molecular weight (as has been discussed).

Hydrodynamic Behavior of Cereal Proteins

The conformation of a protein is stabilized by covalent as well as noncovalent bonds. The primary covalent bond other than the peptide bond is the disulfide bond, whereas noncovalent bonds include salt linkages, hydrogen bonds, and hydrophobic interactions. Knowledge concerning the importance of these stabilizing forces can be obtained from measurements of $[\eta]$ in solvents that either disrupt or strengthen a certain type of bond, as will be exemplified below.

Gluten, Glutenin, and Gliadin

The hydrodynamic behaviors of gluten, gliadin, and glutenin have been compared, and the influences of pH and ionic strength on $[\eta]$ have been investigated (Taylor and Cluskey, 1962; Wu and Dimler, 1963; Wu and Dimler, 1964; Wu et al., 1967). Some results are given in Table 3-6. These results all show that gluten, glutenin, and gliadin behave as asymmetric molecules and not as compactly folded globular molecules. The $[\eta]$ values for globular proteins range from 3 to 4 ml/g (Field et al., 1987), and the values for glutenin as well as for gliadin exceed this range. Gliadin might be regarded as more folded than glutenin due to its lower $[\eta]$ values; that is, 16–24 ml/g compared with 43–124 ml/g for the glutenins (Table 3-6). Gliadin was shown to have $[\eta]$ values in the range 15–21 ml/g measured in aluminum lactate at 25°C (Beckwith et al., 1966). These values correspond to a molecular weight (MW) of 46,000. The gliadin preparation could be separated by gel chromatography into a fraction I with $[\eta] = 30$ ml/g and a fraction III with $[\eta] = 9.3$ ml/g. The corresponding molecular weights (determined from sedimentation analysis) were 104,000 and 30,000 respectively (Beckwith et al., 1966).

The influence of ionic strength on $[\eta]$ is much greater for the glutenins than for the gliadins, indicating that the glutenin molecules are more extended at low ionic strength. This observation has been explained by the somewhat higher amount of ionizable groups in glutenin relative to gliadin (Wu et al., 1967).

Support for this view is found in studies on durum glutenin. The intrinsic viscosity of a specific HMW glutenin subunit from durum wheat was measured in different solvents. In 0.05 M acetic acid/0.01 M glycine at 30°C, an $[\eta]$ value of 48.4 ml/g was obtained (Field et al., 1987). In 50% (v/v) aqueous propan-1-ol, $[\eta]$ was 44.7 ml/g. The calculated molecular dimensions of the glutenin subunit in these two solvents were 504 Å and 492 Å (length), and 17.5 Å and 17.9 Å (diameter), respectively. Based on viscosity measurements, a specific hydrodynamic volume can be calculated for glutenin subunits (Kacskowski et al., 1968). This volume was shown to increase in 7 M urea, and to decrease in D_2O, compared with its value obtained in water. The effect of D_2O was explained by the substitution of weak hydrogen bonds by the stronger deuterium bonds, resulting in a more compact tertiary structure. In urea, the molecules unfold due to the breaking of hydrogen bonds.

Table 3-6. Intrinsic Viscosity [η] Obtained for Gluten, Gliadin, and Glutenin in Buffers of Varying pH and Ionic Strength

Buffer System		Gluten [η] (ml/g)	Gliadin [η] (ml/g)	Glutenin [η] (ml/g)
Sodium lactate buffer, pH 3.1, 25°C				
	0.003 M	57	21	114
	0.03 M	25	16	43
Al-lactate buffer	0.0167 M		16	88
3 M urea–0.11 M KCl, 25°C				
	pH 3.3–3.5	27.6		
	pH 4.1–4.3		18	58
	pH 10.1–10.3	31	19	82
3 M urea + KCl, pH 4.8–5.5				
	I = 0.0035[a], 20°C		24	124
	I = 0.5		17	51

Source: Data compiled from Taylor and Cluskey (1962); Wu and Dimler (1963); Wu and Dimler (1964); Wu et al., (1967).
[a]The ionic strength was 0.0025 for the gliadin preparation.

Molecular weights for glutenins have been calculated from Equation 3-6. To apply this equation, the [η] value corresponding to the random coil conformation for the protein must be used. When η_{sp} was measured as a function of time for glutenin dispersed in different solvents (Hamauzu and Yonezawa, 1971), η_{sp} changed with time in 0.01 M acetic acid (HAc), in 6 M dimethylformamide/0.01 M HAc, and in 6 M urea/0.01 M HAc. The η_{sp} value did not change with time in 6 M guanidine-HCl/0.01 M HAc or in 3 M guanidine-HCl/0.01 M HAc. In 6 M guanidine-HCl/0.01 M HAc, which was regarded as the best dispersing agent for glutenin, [η] was measured as 98 ml/g. With this technique the [η] value for a HMW glutenin subunit obtained in 6 M guanidine-HCl ([η] = 48.9 ml/g) was used to calculate its molecular weight (Field et al., 1987). The value obtained, 65,000, was much lower than the value calculated from electrophoretic mobility measurements. When [η] was then determined in a stronger denaturing agent, guanidine thiocyanate, the resulting value, 57.7 ml/g, yielded a calculated molecular weight of 84,000.

The importance of disulfide cross-links in maintaining the conformation of certain gluten proteins has been demonstrated by measuring [η] before and after the addition of reducing agents such as sodium bisulfite or thioglycolic acid (Udy, 1953). When either reducing agent was added to a gluten dispersion, its viscosity immediately decreased and finally reached a constant value. The [η] value was 30.9 ml/g in sodium salicylate and decreased to 24.8 ml/g in the reduced samples (Udy, 1953). For glutenin and gliadin dispersions (in 0.0617 M aluminum lactate, pH 3.0), [η] values were 68 and

Table 3-7. Intrinsic Viscosity [η] after Reduction and Alkylation, and After Reoxidation of Gliadin and Glutenin

Preparation	Glutenin[a] $[\eta]$ ml/g	Gliadin[a] $[\eta]$ ml/g
Native	77.0	22.8
Reduced		
Alkylated	23.0	27.0
Reoxidized at 0.1%	20.0	23.0
Reoxidized at 5.0%	45.0	37.9

[a] $[\eta]$ was measured at pH 3.1 in aluminium lactate–lactic acid buffer for glutenin, and at pH 3.1 in aluminum lactate–3 M urea buffer for gliadin.
Source: Data compiled from Beckwith and Wall (1966); Beckwith et al., (1965).

15 ml/g, respectively. When the proteins were reduced and alkylated, the [η] values changed to 22 ml/g in the case of glutenin and to 18 ml/g in the case of gliadin (Nielsen et al., 1962). These results show that when the formation of intermolecular cross-links is inhibited, as in the case of alkylated glutenin subunits, the [η] values of glutenin and gliadin are similar. The [η] value of gliadin did not change significantly after reduction and alkylation, indicating that intermolecular cross-links play no role for the conformation of gliadin.

The changes in [η] with time after the addition of a reducing agent were measured for gluten dispersed in SDS-HAc-urea (Ewart, 1980). The initial value of 109 ml/g decreased to 47 ml/g after 3 days. After 11 days, [η] had decreased to 39 ml/g. For another gluten dispersion the largest relative decrease in [η] (from about 210 to 94 ml/g) was obtained when about 17% of the disulfide bridges were broken (Ewart, 1988). Complete reduction resulted in an [η] value of 48 ml/g for this gluten dispersion. The changes in conformation due to the presence of the disulfide bridges have been studied by measuring [η] after both reduction and akylation, and after reoxidation at different protein concentrations (Table 3-7). The results indicate that the glutenin molecules contain intermolecular disulfide bonds, whereas gliadin contains only intramolecular disulfide bonds. However, provided that reoxidation of the reduced samples occurs at a concentration high enough, intramolecular as well as intermolecular disulfide bonds are formed (Beckwith et al., 1965; Beckwith and Wall, 1966).

Barley Prolamines

The conformation of the barley prolamine hordein C was studied by Field and colleagues (1986), and intrinsic viscosity was measured in different solvents and at different temperatures (Table 3-8). The C-hordein is very similar to the

Table 3-8. Intrinsic Viscosities and Molecular Dimensions of Barley C-Hordein

Solvent System	Temperature (°C)	[η] (ml/g)	Molecular Dimensions	
			Length (Å)	Diameter (Å)
0.1 M acetic acid	15.3	19.31	301	18.5
	30.0	18.84	282	19.1
	50.0	14.68	265	19.7
70% (v/v) ethanol	20.0	28.93	352	17.3
	30.0	30.62	363	17.0
	50.0	28.05	350	17.2
70% (v/v) ethanol/10 mM NaCl	30	26.79	346	17.4
5.9 M guanidine hydrochloride	25	40.81	—	—

Source: Adapted from Field et al. (1986).

HMW glutenin subunit discussed previously. This protein is more compact in acetic acid than in ethanol, presumingly due to electrostatic effects. The heating evidently led to a more compact conformation, most evident in acetic acid, where the increased stabilization due to hydrophobic interactions is expected to be more pronounced (Field et al., 1986). For a barley "gluten" preparation, [η] was 87 ml/g (Ewart, 1980).

The molecular weight was calculated for the barley hordein C, and a value of 54,300 was obtained for the protein in 6 M guanidine-HCl (Field et al., 1986). This was in good agreement with the value of 52,570 obtained by sedimentation equilibrium ultracentrifugation in 6 M urea, indicating that the C-hordein was completely unfolded in 6 M guanidine-HCl.

Viscosity and Flow Curves of Gluten Proteins

Flow curves (σ versus $\dot{\gamma}$) have been constructed for gliadin and glutenin solutions in 70% ethanol (Wall and Beckwith, 1969). Both 4% and 10% (w/w) gliadin solutions showed Newtonian behavior; that is, σ was linearly related to $\dot{\gamma}$. The 4% glutenin solution showed a much higher σ than the corresponding gliadin solution at the same $\dot{\gamma}$. The glutenin solution was non-Newtonian and showed shear thinning behavior. Flow properties of gluten dispersions in the concentration range 2–16% (w/w) were investigated by Mita and Matsumoto (1981) (Fig. 3-1). When the flow curves were fitted to the power law (Eq. 3-2), n was found to decrease from 1.00 at a concentration of 4% to 0.81 at a concentration of 16%. The apparent viscosity (η_{app}) calculated for $\dot{\gamma} = 110$ s^{-1} was found to be 6 mPas at 4%, 30 mPas at 8%, and about 150 mPas at 12%.

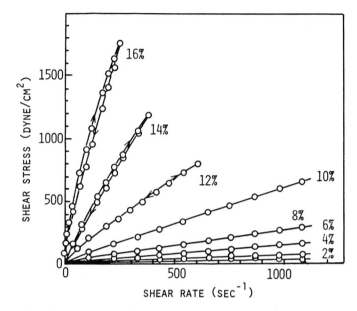

Figure 3-1. Flow curves obtained for gluten dispersions of various concentrations (percent based on dry weight). (*From Mita and Matsumoto, 1981, with permission*)

In 3 M urea, 1% gluten dispersions decrease in viscosity as pH increases from 4 to 7. From pH 7 to 9, the viscosity remains constant (Mita et al., 1977).

Reduction and Oxidation of Gluten Proteins

The change in viscosity of gluten, glutenin, and gliadin dispersions after the addition of the reducing agent 2-mercaptoethanol has been investigated (Ewart, 1979). Viscosity decreases rapidly in gluten and glutenin dispersions. The viscosity of the gluten dispersion decreased from about 115 mPas to less than 80 mPas in 3 hr. In contrast, the viscosity of a gliadin dispersion of the same concentration changed little (from 14.3 to 13.1 mPas) during the same period of time (Ewart, 1979). These results show the importance of the disulfide bridges for the viscosity of glutenin dispersions.

Reduction of disulfide bridges in gliadin and glutenin proteins also results in changes in the appearance of their flow curves. Reduced alkylated gliadin solutions (10%) showed non-Newtonian behavior that was absent in the sample prior to reduction (Wall and Beckwith, 1969). Higher σ values are obtained in reduced alkylated than in native (unreduced) gliadin solution. Reduced alkylated glutenin solutions (4%) were also non-Newtonian. Lower σ values were observed for this type of solution compared to the unreduced glutenin solution.

Interactions Between Gluten Proteins and Other Components

Gliadin and glutenin structures depend on both covalent and noncovalent bonds. In the previous section several examples were given of the changes in viscosity that occur when covalent disulfide linkages are broken. Glutenin conformation, and consequently the viscosity of it in solution, might also be changed by altering the noncovalent linkages; for example, the solubilization of glutenin by salts of a fatty acid (Kobrehel and Bushuk, 1977). Hydrophobic interactions between protein molecules are thought to be replaced by hydrophobic interactions between protein molecules and the fatty acid salt. This theory can be investigated rheologically. When samples composed of glutenin (5 mg/ml) and sodium stearate (4 mg/ml) were heated to 30, 40, and 50°C, a decrease in viscosity was observed with increasing temperature (Wasik et al., 1979). The viscosity decreased with time at 50°C, which was interpreted to result from dissociation of glutenin into smaller aggregates.

Chemical Modification of Gluten Proteins

Chemical modification of gluten proteins has been used to show the importance of specific amino acids in determining the properties of gluten proteins (Aranyi and Hawrylewicz, 1972). The effects of some chemical modifications of gluten on $[\eta]$ are shown in Table 3-9. Deamidation caused a decrease in $[\eta]$, as did esterification. Acetylation seemed not to greatly affect the $[\eta]$ value.

The flow properties of gluten methyl ester dispersions, with 65% of the amide groups removed, have been studied (Mita and Matsumoto, 1981). The gluten methyl ester dispersions showed Newtonian behavior at concentrations below 8% (w/w), whereas the gluten dispersions deviated from Newtonian behavior at even lower concentrations. The η_{app} value was lower for the gluten methyl ester dispersion than for the gluten dispersion at the same concentration and the same $\dot{\gamma}$.

An increase in viscosity has been observed after succinylation of gluten. Here, the change was explained by the existence of a more dissociated protein in the case of the succinylated sample compared to the control (Grant, 1973).

When gluten is hydrolyzed its solution viscosity is reduced. Hydrolysis of a 2% gluten dispersion in 0.1 N HCl or 4 N acetic acid resulted in a rapid initial

Table 3-9. The Intrinsic Viscosity of Gluten Proteins After Chemical Modification

Sample	Type of Chemical Modification	Conditions	$[\eta]$ (ml/g)
Gluten[a]	Deamidation	—	51.9
		6 hr 30°C, 0.6 N HCl 17% conversion	28.1
		48 hr 30°C, 1.2 N HCl 86% conversion	20.0
Glutenin[a]	Deamidation	—	87.5
		24 hr 30°C, 0.6 N HCl 40% conversion	12.9
Gliadin[a]	Deamidation	—	21.7
		24 hr 30°C, 0.6 N HCl 49% conversion	14.5
Gluten[b]	Deamidation	10% conversion	42.5–50.2
		100% conversion	42.9–53.7
Gluten[b]	Esterification	—	42.0–53.0
		methanol, 2 mmol/g	25.8–32.1
Gluten[b]	Acetylation	0% conversion	39.7–51.0
		40% conversion	36.8–48.2
		100% conversion	36.5–46.1

[a] $[\eta]$ was determined in 0.017 M aluminum lactate, pH 3.1 (Beckwith et al., 1963)
[b] $[\eta]$ was determined in 0.05 N acetic acid (Lasztity, 1980).

decrease in viscosity (Aranyi and Hawrylewicz, 1972). The decrease in η was even more pronounced if disulfide bonds in the protein were broken prior to hydrolysis.

Viscosity of Gluten Proteins in Relation to the Baking Process

The rheological experiments described so far give information concerning protein conformation and the contribution to that conformation of different types of bonds and interactions. Rheological investigations can also give information about changes in proteins during bread production. The influence of mixing and of increased temperature during baking on rheological properties of gluten proteins will be described.

Mixing

It is well known that protein solubility increases during mixing. This increase in solubility has been interpreted as the result of a depolymerization process (Tanaka and Bushuk, 1973; MacRitchie, 1975). The changes in η_r of SDS extracts from doughs mixed to different levels have been investigated (Danno and Hoseney, 1982). A linear relationship between η_r and the protein concentration in the dough extract was found. The η_r value was higher for dough extracts than for similar extracts from flour, but the addition of 2-mercaptoethanol to the extracting solution gave η_r values that were the same for extracts from dough and flour. Protein extracts from overmixed doughs, which have radically different handling properties, showed significantly lower η_r values than extracts from doughs mixed to optimum. These differences in η_r were abolished by the addition of 2-mercaptoethanol (Danno and Hoseney, 1982). The results support the view that increases in molecular weight of gluten proteins occur during mixing, whereas overmixing causes depolymerization.

Heating

At ambient temperature, the relationship between log η_r and gluten concentration in acetic acid is linear (Udy, 1953). This was also shown to be true for dispersions that had been heated to boiling, although the slopes of these lines were reduced. Interestingly, after boiling, differences in log η_r versus concentration between varieties were evident (Udy, 1953). In other studies, SDS extracts of doughs heated to 100°C showed decreases in η_r with time (Danno and Hoseney, 1982). For a 12% gluten dispersion, η_{app} decreased with temperature over the range of 20 to 50°C (Mita and Matsumoto, 1981). The η_{app} value for a gluten methyl ester dispersion was found to decrease over the same temperature interval, although the reduction was smaller.

The same analysis can be applied to the investigation of temperature effects during the drying of gluten (Dalek et al., 1970). Table 3-10 shows the [η] of dried gluten dispersed in sodium salicylate, with or without urea. As might be expected, spray drying and drum drying caused considerable changes in [η] compared to freeze drying. The lower values are consistent with the increased cross-linking due to heating (Schofield et al., 1983). The differences in [η] obtained with urea showed that spray drying and drum drying change the gluten considerably.

Table 3-10. Intrinsic Viscosity of Gluten Dispersions in Sodium Salicylate—Influence of Gluten Drying Method

Drying Method	[η] Without Urea (ml/g)	[η] In Urea (ml/g)
Freeze-drying	30.5–39.0	41.5–54.5
Vacuum-drying	32.5–41.0	35.0–66.0
Spray-drying	28.2	35.0
Drum-drying	28.0	34.5

Source: Adapted from Dalek et al. (1970).

Correlation Between Rheological Properties and Flour Quality

The possibility of a correlation between the viscosity of a gluten extract and its parent wheat variety was first investigated by Udy (1953). This investigator plotted log η_r against protein concentration and obtained straight lines with slopes that were independent of wheat variety if the gluten was dispersed in sodium salicylate. Slope was dependent on wheat variety when the gluten was dispersed in acetic acid, however. Gluten extracts from hard wheats were found to give $[\eta]$ values 13% higher than those obtained from soft wheat extracts (Cluskey et al., 1961). For other wheat varieties a correlation between hydrodynamic behavior of glutenins and gluten quality exists only for samples of extremely different quality (Bartoszewicz et al., 1972). When extracts of acetic acid–urea–cetyltrimethylammonium bromide (AUC) from different wheat varieties are compared, large differences in relative viscosities between varieties can be found. The stronger wheat gave the highest value and the weak wheat the lowest value (Butaki and Dronzek, 1979b). Furthermore, when the AUC extracts were reduced, there were differences in η_r values that depended on the wheat variety.

Ewart (1980) determined $[\eta]$ values for gluten and glutenin for 36 wheat samples; a high correlation was found between $[\eta]$ and the loaf volume per gram protein. There was no difference in $[\eta]$ for the same variety grown at different locations. It was concluded that genetic factors were more important than environmental factors in determining intrinsic viscosity. In a similar study (Dexter and Matsuo, 1980), thirty durum lines were investigated for possible correlations between the $[\eta]$ values for their glutens and pasta quality. A correlation was found between gluten strength and $[\eta]$ (correlation coefficient = 0.53) for the AUC extract of gluten, whereas the correlation between pasta cooking quality and $[\eta]$ was poor (correlation coefficient = 0.14). The differences between durum samples grown at different locations were also investigated, and they were shown to be similar to those for bread wheats (Ewart, 1980); that is, no significant differences were found. The $[\eta]$ values for gluten extracts from durum wheats were in the range 61–131 ml/g, and those for bread wheats were in the range 75–121 ml/g (Dexter and Matsuo, 1980; Ewart, 1980).

Glutenin Models

The structure of gliadins is not a subject of controversy, and it is generally agreed that they are more or less globular proteins stabilized by intramolecular disulfide bonds.

There is a certain amount of controversy regarding the structure of glutenins, however. Several hypotheses have been proposed for the structure of glutenins, and due to experimental difficulties connected with studies of glutenins, it is difficult to prove conclusively which model is right or wrong. In this section some models of glutenin structure will be discussed in relation to rheological properties of glutenin.

The most important question concerning the structure of glutenin is the role of the disulfide bonds. After the peptide bond, the disulfide bond may be regarded as the most important bond in glutenin. The glutenin molecule can be viewed as a three-dimensional permanent network (Muller, 1969). A second possibility is that the disulfide bonds cross-link glutenin polypeptides into linear glutenin polymers, which are entangled and form transient networks. A third possible role for the disulfide bridges might be to give the glutenin molecules a conformation suitable for intermolecular interactions through hydrogen bonding, ionic linkages, and hydrophobic interactions. In this last model, intermolecular covalent bonds therefore are not necessary.

In glutenin models based on a permanent three-dimensional network, the theory of statistical rubber elasticity has been used to explain the rheological behavior of glutenin and gluten (Muller, 1969). The existence of a permanently cross-linked network explains the elasticity of gluten or glutenin but not viscous flow (Ewart, 1968b). The concept of thiol–disulfide interactions was introduced to explain this viscous flow.

In his first hypothesis on glutenin structure, Ewart (1968b) assumed that the glutenin molecule was composed of about 50 peptide chains as subunits. Of these subunits about one-third contained three disulfide bridges, and the remaining two-thirds contained two disulfide bridges. The most favored configuration of this polymer was postulated to be a somewhat contracted conformation, which could explain the elasticity of glutenins. The rapid decrease in the viscosity of glutenin dispersions after the addition of reducing agents was taken as evidence that the glutenin polymers are linear and not branched or cross-linked. In later modifications of the initial hypothesis, Ewart (1972a) suggested that the glutenin polymers are laid down in the seed; that is, they are not artifacts of wetting and mechanical treatment of the flour.

The term concatenation was introduced to describe the long linear glutenin molecules, which consist of polypeptide chains attached to one another by disulfide bonds (Ewart, 1972a; Greenwood and Ewart, 1975; Ewart, 1977) (Fig. 3-2). The possibility that these concatenated proteins are not involved in extensive entanglement coupling, but rather are arranged in roughly parallel structures has been recognized (Ewart 1977). The number of disulfide bonds between neighboring subunits has been analyzed, and the most recent suggestion is that at least two-thirds, and possibly all, of the junctions are likely to consist of a single disulfide bond (Ewart, 1988). The concatenation hypothesis is based on the polymerization of glutenin subunits into large molecules. This

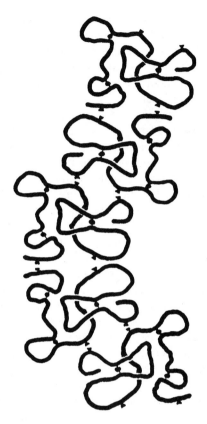

Figure 3-2. Part of a linear glutenin molecule showing polyeptide chains linked to neighbors by disulfide bonds. (*From Ewart, 1972a, with permission*)

polymerization has been proposed in several works (Khan and Bushuk, 1979; Graveland et at., 1985; Belitz et al., 1986). Graveland and co-workers (1985) used SDS fractionation and electrophoresis to elucidate which subunits are linked into different aggregates.

Glutenin molecules observed with the scanning electron microscope (SEM) appear to be fibrous structures (Orth et al., 1973). Globular A-gliadin molecules are known to interact to form fibrous structures without involving intermolecular disulfide bonds (Bernardin, 1975). Bernardin (1978b) later suggested that all protein–protein interactions during dough mixing occur through the formation of noncovalent bonds. In this model, the role of disulfide bonds would be to maintain the glutenin molecules in such a conformation that these noncovalent bonds are possible (Fig. 3-3).

Ewart (1968b) proposed that the elasticity of glutenin molecules is related to their native conformation, which was supposed to be somewhat contracted.

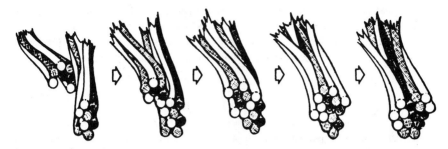

Figure 3-3. Fibril–fibril interactions during dough mixing. The stronger the interaction, the darker the coloring of the fibril. The fibrils in the center of the figure, where the average interaction between all fibrils is at a maximum, would correspond to an optimally mixed dough. (*From Bernardin, 1978b, with permission*)

If the molecule is extended it then prefers to return to the energetically more favorable contracted conformation. Its viscous flow depends on the presence of noncovalent bonds, which allow molecular slippage to occur during deformation. Recently, it was suggested that the elasticity of HMW glutenin subunits is similar to the elasticity of elastin, the mammalian connective tissue (Tatham et al., 1985). The repetitive central domains present in both HMW subunits and in elastin are repetitive β turns. These investigators suggested that β turns form the β spiral, a helical structure consisting of repetitive β turns, with 13.5 amino acid residues per turn and a translation of 9.45 Å per turn. Several molecules containing β spirals may then aggregate to form elastic fibrils, stabilized by hydrogen bonding and hydrophobic interactions.

RHEOLOGICAL BEHAVIOR OF A-GLIADIN

In some wheat varieties an α-gliadin has been found that possesses special aggregation properties. This α-gliadin, which aggregates at pH 5.0 and 0.005 M KCl, has been termed A-gliadin (Bernardin, 1975). The aggregation is reversible, and at pH 3.1 and low ionic strength A-gliadin is monomeric with a molecular weight of about 31,000 (Cole et al., 1983). Increasing pH or ionic strength causes the monomeric A-gliadin to aggregate into larger aggregates due to hydrogen bonding. The aggregates dissociate when the temperature increases, and at 50°C A-gliadin exists only as monomers (even in 0.001 M HAc, pH 5.0). The A-gliadin, therefore, offers a possibility to study the rheological properties of wheat proteins without the complication of intermolecular disulfide bonds. A-gliadin is also unique in that it is the purest wheat protein fraction studied by rheological methods. The rheological properties of other wheat protein fractions have not been so thoroughly investigated. The rheological properties of A-gliadin have been studied in solutions as well as in dispersions; A-gliadin, therefore, is a suitable connection between the preceding section concerning the viscosity of protein solutions and the following discussion concerning protein dispersion; that is, mainly gluten.

A-Gliadin in Solution

When the intrinsic viscosity values of A-gliadin in different aggregation states were determined (Cole et al., 1984), $[\eta]$ values obtained at pH 4.0 in 0.001 M HAc and 0.01 M NaCl (4.0 ml/g) were close to values obtained for globular proteins (3–4 ml/g). Since A-gliadin is present partly as aggregates and partly as monomers at these conditions (Cole et al., 1983), this result strongly suggests that A-gliadin is a globular protein. Its reduction and carboxymethylation had only minor effects on $[\eta]$. In 0.1 M HAc at pH 2.9, $[\eta]$ increased from 12.0 ml/g in the unreduced sample to 13.7 ml/g in the reduced sample.

A-Gliadin Dispersions

The intrinsic viscosity of A-gliadin has, of course, been determined in solutions at very low concentrations (10.0 mg/ml and below). When more concentrated solutions (i.e., above 5% w/w) were studied, such dispersions formed a nematic mesophase (Bernardin, 1975; Bernardin, 1978a); that is, the concentrated dispersion was birefringent in polarized light. Stress (σ) was found to increase with shearing time until a plateau value was reached, after which σ decreased with further shearing. A higher plateau value shown for a shorter time was obtained when $\dot{\gamma}$ was increased. At $\dot{\gamma} = 51$ s^{-1} the plateau value was stable during 7–8 minutes of shearing, whereas at $\dot{\gamma} = 170$ s^{-1} the stability was less than 1 minute (Bernardin, 1975). These results were due to a gelation induced by shearing. With prolonged shearing (or at very high $\dot{\gamma}$), however, the gel formed was broken down. This breakdown was observed as protein precipitation and free solvent expressed from the gel. Such changes in σ with increasing and decreasing shear rates characterize the system as rheopectic (Bernardin, 1978a).

When the temperature of the A-gliadin dispersion was increased from 20 to 35°C and, further, to 40°C, gelation occurred after a progressively shorter period of shearing. At 20°C and $\dot{\gamma} = 84.8$ s^{-1}, gelation did not occur at all (Bernardin, 1975; Bernardin, 1978a).

Linear viscoelastic behavior, in which calculated moduli are independent of strain, was observed when small-amplitude oscillatory shear was applied to solutions of A-gliadin fibrils (Bernardin, 1978a). This is contrary to the situation in dough, where nonlinear behavior is observed even at the smallest strain possible to measure (Smith et al., 1970). At higher strain levels viscous flow was observed. When the phase transition from solution to gel (described above) occurred, the shear storage modulus (G') for a 6.2% A-gliadin dispersion increased from 0.99 Pa to 21.2 Pa and continued to increase to 1.63×10^3 Pa after the phase transition. The dynamic shear loss tangent (tan δ) was constant at 0.017 in the frequency range 0.01–10 Hz, which indicates a nearly complete elastic response (Bernardin, 1978a).

RHEOLOGICAL PROPERTIES OF PROTEIN DISPERSIONS

Concentrated protein dispersions show viscoelastic behavior. Several methods have been designed to differentiate between elastic recovery and viscous flow. These methods are described in other chapters in this volume.

Gluten

Principal Behavior of Gluten

Before any measurements can be performed on gluten, a sample must be prepared. Some investigators prefer to prepare gluten from wheat dough immediately prior to rheological measurements, whereas others prefer to hydrate dry isolated gluten prior to the rheological measurements. A dough kneader (e.g., the farinograph) is used to mix gluten and water into a homogeneous mass (Doguchi and Hlynka, 1967). Gluten can be prepared from a wheat flour dough, either by hand-washing or by automatic gluten washing devices.

The influence of the gluten preparation procedure on the rheological properties of gluten has been investigated by using a penetrometer (Prugar, 1969). Increased washing time, from 8 minutes to 16 minutes, increased the penetrometer value; that is, the gluten became softer. The influence of preparation procedures (i.e., washing time) was less evident with gluten from a strong flour. Differences between hand-washed and commercially prepared glutens were observed by oscillatory measurements (Dreese et al., 1988). These differences could be related in part to the drying procedure, but also to the flour:water ratio during washing. The hand-washed gluten gave a higher G' and a lower ton δ value than the commercially produced gluten. However, G' of the commercial gluten could be increased by further washing at increased water:gluten ratios. This result was explained by the removal of marginally soluble components that otherwise caused the gluten to soften. It is thus very important to standardize the gluten preparation method for rheological measurements as well as for baking experiments (MacRitchie, 1985).

The farinograph has been used to study the rheological properties of rehydrated gluten (Doguchi and Hlynka, 1967). The consistency (i.e., the maximum value in the farinogram) changed linearly with water content, but with a change in rate at a specific water content. The water content played a greater role in defining the rheological properties at the highest water contents.

The similarities and differences between gluten and wheat flour doughs have been investigated by using "synthetic" doughs made from gluten, starch, and water and tested with oscillatory measurements (Smith et al., 1970; Hibberd, 1970) or stress relaxation (Matsumoto, 1979). The moduli calculated were strongly dependent on strain for doughs but not for gluten. These relationships have also been observed in creep tests provided that the strain levels have been low (Funt Bar-David and Lerchenthal, 1975), and in stress-relaxation tests when a wheat flour dough is compared to its gluten (Bohlin and Carlson, 1981). When the strain level is high enough, gluten also shows nonlinear behavior (Funt Bar-David and Lerchenthal, 1975; Bohlin and Carlson, 1981).

Introducing starch into gluten results in nonlinear behavior. Another consequence is that G or G' and G'' become higher (Smith et al., 1970; Hibberd, 1970; Matsumoto, 1979; Szczesniak et al., 1983). The starch granules thus act as fillers that reinforce the gluten. Values for G increased with increasing volume fraction of starch granules from about 2×10^3 Pa for gluten to about 1×10^5 Pa for a gluten–starch mixture in which the volume fraction of starch granules was 0.5 (Funt Bar-David and Lerchenthal, 1975).

The G and G'' of both gluten and flour dough increase with frequency (ω) (Smith et al., 1970; Cumming and Tung, 1975; Szczesniak et al., 1983; Mita and Matsumoto, 1984). The ratio between G' and G'' (tan δ) and its dependence on ω seem to be governed by the water content of the system. Both G' and G'' decrease with increasing water content (Smith et al., 1970; Cumming and Tung, 1975; Mita and Matsumoto, 1984). Values obtained for gluten in oscillatory measurements are presented in Table 3-11. The high value of δ observed in some investigations (e.g., $\delta = 71°$ in Table 3-11) indicates a predominance of the viscous behavior in gluten (Mita and Matsumoto, 1984).

Stress relaxation measurements have shown that 3% of the initial σ remained after 10^4 s (Bohlin and Carlson, 1981). Other researchers have found a complete decline to zero in σ (Funt Bar-David and Lerchenthal, 1975). A complete relaxation of the stress is typical of liquids, and it might thus be concluded that gluten behaves as a viscoelastic liquid. Two relaxation processes have been observed for both dough and gluten. The first process occurs at short times, 0.1–10 s, and the second process occurs at longer times, 10–10^4 s (Bohlin and Carlson, 1981). The implications of the two relaxation processes will be discussed in relation to gluten models.

Table 3-11. Oscillatory Measurements on Gluten

Experimental Conditions	Sample	Results
f^a = 1 Hz^b	Gluten	G' = 4 × 10^3 Pa
	55.2 w-% water	G'' = 2 × 10^3 Pa
	77.0 w-% protein	δ = 27°
f = 0.05 Hz^c	Gluten, 20.3% gluten	G' = 1 Pa
		G'' = 3 Pa
		δ = 71°
	36.9% gluten	G' = 500 Pa
		G'' = 630 Pa
		δ = 52°
f = 1 Hz^d	Gluten, handwashed	G' = 6.3 × 10^3 Pa
		δ = 20°
	commercial gluten	G' = 2.5 × 10^3 Pa
		δ = 23°

^{a} f = frequency
^{b}Data from Smith et al., (1970).
^{c}Data from Mita and Matsumoto (1984).
^{d}Data from Dreese et al., (1988).

When gluten was subjected to creep and recovery tests, the imposed deformation was completely recoverable up to a critical tensile stress (Funt Bar-David and Lerchenthal, 1975). The recoverable deformation could be divided into immediate and delayed components. When the tensile stress exceeded the critical value, there was an unrecoverable viscous flow in addition to recoverable deformation. The yield stress was defined as the lowest σ that caused such irrecoverable deformation in the sample (Lerchenthal and Funt, 1970; Funt Bar-David and Lerchenthal, 1975).

Elastic recovery (ER, the change in sample thickness after compression is removed) values were determined for dough, gluten and for various glutenin preparations (Prasada Rao and Nigam, 1987). Gluten gave higher ER values than dough, and isolation of glutenin with different solvents gave fractions with even higher ER values, indicating that glutenin is the elastic component in gluten.

The rheological properties of gluten have been investigated by capillary viscosimetry, and a shear thinning behavior has been observed (Kieffer et al., 1982).

At low strain levels gluten shows linear viscoelastic behavior (Smith et al., 1970; Hibberd, 1970). When the strain is increased, nonlinear behavior is obtained (Funt Bar-David and Lerchenthal, 1975), and if the strain is increased further, rupture will finally occur (Rinde et al., 1970). The rupture properties of gluten have been studied, and failure envelopes, that is, log–log plots of rupture stress versus strain at rupture, have been constructed (Rinde et al., 1970). Gluten was found to be stronger and less extensible than doughs.

The Influence of Additives on the
Rheological Properties of Gluten

The significance of disulfide bonds for the rheological properties of gluten is usually studied by the addition of an oxidant or a reducing agent (Doguchi and Hlynka, 1967). An alternative method is to exclude oxygen (and therefore oxidation) during the preparation of gluten. When gluten is prepared from doughs under N_2, load deformation curves for gluten from strong wheat varieties show lower resistance and higher extensibility compared with those for samples prepared in air (Belitz et al., 1986). Gluten prepared under N_2 but in the presence of the oxidant potassium bromate showed the same rheological behavior as gluten prepared in air. The influence of the atmosphere during preparation was much less pronounced for wheat varieties of poor baking performance, in which only slight increases in extensibility were observed. It was concluded that for gluten formation proteins suitable for oxidative polymerization must be available (Belitz et al., 1986).

To interpret the rheological behavior of gluten in the presence of oxidants or reducing agents, the statistical theory of rubber elasticity has been applied (Muller, 1969; Bernardin, 1978a). The molecular weights between cross-links are calculated from the following relation:

$$M_c = \rho RT/G \qquad (3\text{--}7)$$

The average molecular weight between cross-links, M_c is related to the polymer solution density ρ, the gas constant R, the absolute temperature T, and the shear modulus G. Because two cross-linked chain segments contribute to each cross-link, the number of cross-links per unit mass might be calculated as $1/(2\,M_c)$ (Muller, 1969). When these relationships were applied to gluten, M_c for a strong gluten was found to be 8.9×10^6, whereas for a soft gluten, M_c was 10.9×10^6 (Muller, 1969). The corresponding values of $1/(2M_c)$ were 5.6×10^5 and 4.6×10^5, respectively (Muller, 1969). The addition of a fast-acting oxidant (100 ppm KIO_3) to a strong gluten caused M_c to decrease from 9.3×10^6 to 7.4×10^6, which corresponded to a change in $1/(2M_c)$ from 5.4×10^5 to 6.8×10^5. The results were interpreted as showing that a strong gluten is more cross-linked than a weak gluten, and that the effect of the oxidant KIO_3 was to increase the cross-linking even further. It should be borne in mind, however, that the cross-links calculated in this way need not be covalent cross-links (Bernardin, 1978b). In the case of A-gliadin Bernardin assumed that no covalent cross-links were present, as the estimated energy of interaction between A-gliadin fibrils was far too low for a covalent bond. The calculation of M_c gave approximately one cross-link for two to three A-gliadin subunits.

In contrast to the results described above, the addition of iodate does not significantly change the maximum consistency or development time of gluten in the faringograph (Doguchi and Hlynka, 1967). The addition of the thiol blocker N-ethylmaleimide (NEMI), on the other hand, reduces the mixing time and increases the rate of breakdown in the faringograph (Doguchi and Hlynka, 1967). The addition of glutathione, a LMW thiol containing peptide that might interfere with gluten disulfide formation, caused a gluten–water dough to soften. The effect was especially evident at increased temperatures (Funt Bar-David and Lerchenthal, 1975). When the thiol containing amino acid cysteine was added to gluten, the maximum force in load–deformation curves was decreased (Kieffer et al., 1983), as were both relaxation time ($t_{1/2}$) and G in shear stress relaxation experiments (Mita and Bohlin, 1983). The oxidant $KBrO_3$ was found to have the opposite effect: G as well as $t_{1/2}$ increased (Table 3-12). The rheological properties of gluten are thus related to the formation of protein–protein disulfide bridges.

An alternative method to study cross-links in gluten is to induce permanent cross-links between the protein molecules with glutaraldehyde, which cross-links proteins between ϵ-amino groups of lysine and terminal α-amino groups. The addition of glutaraldehyde to gluten resulted in an increase in G as well as in $t_{1/2}$ (Mita and Bohlin, 1983) (see Table 3-12). Cross-linking by glutaraldehyde or by oxidation, therefore, affects the rheological properties of gluten in the same way.

The pH and ionic strength affect the rheological properties of gluten. Gluten consistency in the faringograph increased with pH over the pH range 4.5–6.0 (Doguchi and Hlynka, 1967). The effects of different salts were investigated in extension (Kieffer et al., 1983). The effect of NaCl was found to be concentration dependent such that when NaCl was present at less than 4%,

Table 3-12. The Influence of Certain Additives on the Shear Stress Relaxation Parameters of Gluten

Sample	$t_{1/2}(s)$	$G(Pa \times 10^3)$
Gluten without additives	0.66 ± 0.10	1.10 ± 0.05
	0.53 ± 0.05	1.18 ± 0.05
Glutaraldehyde		
1,000 ppm	3.57 ± 0.05	4.43 ± 0.29
2,000 ppm	8.10 ± 2.52	8.41 ± 0.20
$KBrO_3$		
30 ppm	0.67 ± 0.01	1.52 ± 0.03
300 ppm	0.88 ± 0.06	1.41 ± 0.10
1,500 ppm	1.15 ± 0.05	1.63 ± 0.11
Cysteine		
50 ppm	0.46 ± 0.02	1.19 ± 0.05
200 ppm	0.34 ± 0.01	1.10 ± 0.09

Source: Data from Mita and Bohlin (1983), with permission.

a weakening effect was obtained. The addition of NaCl at greater than 5% resulted in strengthening. Kieffer and co-workers investigated several Cl⁻ salts at a level of 3.7%, and gluten strength was decreased in the following order: $K^+ < NH_4^+ = Na^+ < Ba^{2+} < Li^+ < Ca^{2+} < Mg^{2+}$. The compounds NaF, Na_2SO_4, and NaH_2PO_4 increased gluten strength. The effects of different anions on dough properties have been discussed in relation to the influence of these anions on water structure (Kinsella and Hale, 1984). The ions F^- and Cl^- should thus be expected to enhance the hydrophobic interactions between gluten proteins due to their structuring effect on water. This effect causes the strengthening effect of NaF and of Cl^- salts at high concentrations. The weakening effect of NaCl at low levels is explained by decreased electrostatic repulsion (Kieffer et al., 1983).

To evaluate the contribution of hydrogen bonding to the rheological properties of gluten, the rupture properties of gluten prepared in H_2O and in D_2O, respectively, have been studied (Inda and Rha, 1981). No effect of D_2O (which strengthens hydrogen bonds) on rupture properties was observed. The addition or urea (which breaks hydrogen bonds) caused rupture at much lower strains. The effects on rheological behavior of breaking hydrogen bonds were much greater than those of strengthening hydrogen bonds. A decrease in the extensibility of urea-treated gluten was observed by Doguchi and Hlynka (1967), who also found that gluten consistency measured in the faringraph decreased with increasing urea concentration. An even greater effect was obtained when guanidine-HCl replaced urea.

To evaluate the importance of hydrophobic interactions to the rheological properties of gluten, compounds that disrupt hydrophobic interactions between the proteins have been added to gluten (Doguchi and Hlynka, 1967). When gluten was treated with the organic solvent acetone or dioxane, its consistency (measured in the farinograph) decreased. The extensibility (as measured by stretching tests) of acetone-treated gluten also decreased. When hydrocarbons were added to gluten, the stress relaxation time $t_{1/2}$ decreased (Lasztity, 1971). This effect increased with increasing hydrocarbon chain length. Fatty acids decrease the $t_{1/2}$ of gluten; in this case the shortest acids were found to exert the largest effect. When the addition of a hydrocarbon (which disrupts hydrophobic interactions) was combined with the addition of urea (which disrupts hydrogen bonds), the formation of gluten eventually was totally inhibited (Lasztity, 1971).

Chemical and Enzymatic Modification of Gluten

The effects of adding reducing agents have already been described. With only the thiol groups blocked, no changes in stress relaxation parameters have been observed (Lasztity, 1969), indicating that free thiol groups have no significant effect on the rheological properties of gluten.

The amide groups in gluten proteins can be replaced by ester groups, which decrease the possibility for hydrogen bonding. Gluten dispersions at concentrations below 17% (w/w) showed higher G' and G'' values in oscillatory measurements than dispersions of gluten methyl ester (Mita and Matsumoto, 1984). At concentrations above 17%, however, the gluten methyl ester gave the highest G' and G'' values. Moreover, the δ values were very low for the gluten methyl ester compared with those for the unmodified gluten, indicating a more elastic response. Gluten methyl ester dispersions formed loosely packed gel structures at high concentrations, whereas gluten dispersions behaved as viscoelastic liquids. Esterification of gluten was found to decrease $t_{1/2}$ in stress relaxation, with larger decreases at higher degrees of esterification (Lasztity, 1969). The properties of deamidated gluten were investigated with a penetrometer (Lasztity, 1969), and increased penetration depths were observed with increasing degrees of deamidation. Acetylation of gluten resulted in increased penetration depths; that is, the gluten became softer with increasing degrees of acetylation. The $[\eta]$ values for extracts of these esterified, deamidated, and acetylated gluten samples are shown in Table 3-9. These results show that the rheological properties of gluten depend not only on its disulfide bridges, but also on the ability to form hydrogen bonds.

When arginine residues in gluten were modified with 1,2-cyclohexanediene, the dough-forming properties of gluten were lost, as judged from the mixogram of a reconstituted flour of starch and modified gluten (Batey, 1980). Modification of lysine residues with potassium cyanate caused only slight changes in the mixogram, that is, a reduced peak height and a somewhat increased mixing time were observed (Batey, 1980). The arginine residues thus play a more critical role for the gluten structure than do the lysine residues.

The rheological properties of gluten can also be modified by the action of proteolytic enzymes (naturally occurring or added). An increase in penetration depth was observed for gluten treated with proteases (Gabor et al., 1982a). This effect is interesting in that changes in rheological properties were observed at one-third of the enzyme concentration necessary to affect the molecular weight distribution (as observed by gel chromatography). There was no correlation between enzyme activity (as determined from their effects on casein) and the influence of the enzymes on the rheological properties of gluten (Kruger, 1971; Gabor et al., 1982b). Pronase, fungal protease, trypsin, and papain (proteases) all had an effect on the rheological properties of gluten, whereas β-amylase, lipase, lipoxidase, and catalase had no effect (Kruger, 1971). When protease was added to gluten, the force component of load–deformation curves was found to decrease. At 180 minutes after treatment, the extensibility of these samples was very much increased (Kieffer et al., 1981). If the proteases naturally present in flour were inhibited, the reverse effect was observed; that is, the force was increased and the extensibility decreased.

Rheological Properties of Gluten in Relation to
Its Behavior During the Baking Process

During the baking process, gluten proteins are exposed to intensive mechanical work as well as to heat, conditions that might be expected to change their rheological properties profoundly. Moreover, interactions may occur between gluten and other components; that is, those present in flour as well as those added during dough mixing. These other components might also change the rheological properties of gluten. In this section the influences of mechanical work and heat on the rheological properties of gluten will be discussed, as well as the interactions between gluten and other components that might be present in the dough. Correlations between the rheological properties of gluten and its performance in baking or pasta production will be discussed.

The consistency of a wheat flour dough changes with the amount of mixing. The same is true for gluten (Heaps et al., 1968). The relaxation time in tensile tests was measured for gluten prepared from doughs after different levels of work input, and relaxation time increased with the level of work input until a maximum value was obtained. Thereafter, the relaxation time decreased. It was suggested that the increase in relaxation time was due to a decrease in sulphydryl groups as the gluten network was created. The decrease in relaxation time could thus be due to an increase in the number of sulphydryl groups, because of mechanical damage to the gluten matrix (Heaps et al., 1968). In fact, the behavior of gluten proteins during dough mixing has been explained by invoking a theory of mechanical degradation of polymers in shear (MacRitchie, 1975). The shear strain rate during mixing was estimated by MacRitchie at 10^2 s^{-1}, which should be enough for mechanical breakdown to become significant. In this model a prerequisite for degradation in shear is that a large number of polymer molecules are entangled. Degradation would then occur because the polymer molecules could not disentangle sufficiently in response to the shear stress. For large molecules, such as are found in mixed gluten, the highest tension would build up at the center of the chain and the chain would thus break preferentially at the center.

An increase in tensile stress at a constant strain has been observed as gluten has been heated from 20°C to 40°C (Funt Bar-David and Lerchenthal, 1975). When gluten was heated to 70, 80, and 90°C, respectively, a decreased extensibility was observed. The extension was least for the sample that had been heated to the highest temperature (Doguchi and Hlynka, 1967). Extensibility also decreased with holding time at a given temperature. In a related study, the compressibility of gluten that had been heated for different periods of time in a boiling water bath decreased with increasing heating time (Jeanjean et al., 1980). Great differences between varieties in their response to heat were also observed. Gluten from durum wheat seemed to be more sensitive to heat than gluten from bread wheat (Doguchi and Hlynka, 1967; Jeanjean et al., 1980).

When gluten was steamed at 100°C for 15 minutes, a change in its dynamic rheological properties was observed (Dreese et al., 1988). The dynamic compression storage modulus (E') increased and δ decreased, a rheological behavior indicating a more cross-linked gluten after heating. Prolonged steaming caused the E to increase further. When gluten was heated to 80°C in the rheometer, cooled, and reheated (Dreese et al., 1988), the G' was higher and tangent was lower during the reheating. The magnitude of these changes was smaller than when the gluten was treated by steaming. It was suggested that the changes were not due to denaturation of the protein but rather to gelatinization of starch remaining in the gluten.

Bale and Muller (1970) used the statistical theory of rubber elasticity to interpret the changes that occur when gluten is heated. In this treatment of the data, a decrease in M_c was found when gluten was heated to above 50°C. The percentage elastic deformation and viscous deformation were calculated, and at 30°C these values were 88.8 and 11.2%, respectively, for a crude gluten preparation. After heating to 60°C these figures had changed to 90.6 and 9.4%, respectively, and after heating to 90°C the viscous deformation was lost completely (Bale and Muller, 1970).

The effects of sugars (maltose, sucrose, fructose, and glucose) on the rheological properties of gluten have been investigated (Kwa et al., 1976). Generally η_{app} increased when sugars were present. Phospholipids at low levels were found to have a strengthening effect on gluten in extension tests (Kieffer et al., 1983).

When ascorbic acid, a common dough improver, was added to dried gluten during rehydration in the faringraph, a decrease in consistency was observed (Zetner, 1968). The function of ascorbic acid was shown to be due neither to the reduction of disulfide bonds nor to blocking of thiol groups. In stress relaxation experiments ascorbic acid was found to increase $t_{1/2}$ as well as G (Mita and Bohlin, 1983). However, ascorbic acid did not function in the same way as another improver, $KBrO_3$. Also, when gluten was treated with a great excess of $KBrO_3$ the addition of ascorbic acid still increased $t_{1/2}$ and G.

The importance of the gliadin/glutenin ratio for the rheological properties of gluten has been investigated (Belitz et al., 1987). Increasing the proportion of gliadin present in the gluten increased the extensibility and decreased the resistance of gluten. By capillary viscosimetry, Belitz and co-workers found that viscosity decreased with increasing proportions of gliadin in the gluten.

The rhelogical properties of gluten with added prolamines from cereals other than wheat were also investigated (Belitz et al., 1986; Belitz et al., 1987). Of the added prolamines, those from sorghum, oat, barley, and rye all caused extensibility to increase and resistance to decrease. Only maize prolamine had the opposite effect; that is, it increased resistance and decreased extensibility. When these results were compared to the hydrophobicity of the added prolamines, it was noted that the maize prolamine was the most hydrophobic prolamine added.

Interactions between gluten and starch have been investigated by using small-amplitude oscillatory techniques (Lindahl and Eliasson, 1986). Gluten was mixed with starch (wheat, triticale, rye, barley, maize, or potato), and the rheological behavior after gelatinization of the starch–gluten mixture was examined. Lindahl and Eliasson found that the G' increased when gluten was added to wheat and rye starches, decreased when gluten was added to maize, and was unaffected when gluten was added to triticale, barley, or potato starch.

In order to identify those rheological parameters that correlate to the quality of the flour, it is first necessary to establish whether there are differences at all between the glutens from different wheat varieties. Large differences in the rheological properties of the glutens from bread and durum wheats have been reported in several studies (Doguchi and Hlynka, 1967; Matsuo, 1978; Kieffer et al., 1981). Durum gluten was found to be more extensible than gluten from bread wheat.

Glutens from different durum varieties also differed in their rheological properties. The elastic recovery after compression of durum gluten has been found to correlate with pasta cooking quality (Damidaux et al., 1980). Varieties with high cooking quality showed high ER values, whereas the varieties of low cooking quality showed low ER values. Moreover, the varieties of high absolute elastic recovery contained "band 45" gliadin in 59 of the 66 tested varieties. The varieties with low absolute elastic recovery contained "band 42" gliadin in 46 of 47 tested varieties (Damidaux et al., 1980).

A stretching test was used to rank durum varieties, and the calculated breaking force was in the range $3.72–12.54 \times 10^3$ Pa, with the weak varieties giving the lower values (Matsuo, 1978). Gluten from triticale was tested similarly, and the breaking forces were in the range $6.8–15.5 \times 10^3$ Pa, values that were lower than those for wheat gluten ($10.3–13.8 \times 10^3$ Pa) (Pena and Ballance, 1987). Stretching tests show great differences in extensibility for glutens from different wheat varieties (Butaki and Dronzek, 1979a; Kieffer et al., 1981; Ram and Nigam, 1981).

It therefore is possible to differentiate between wheat varieties by using rheological tests. It is much more difficult to explain how the differences in rheological properties depend on the molecular structure of gluten, however. Oscillatory measurements of wheat flour doughs show that protein content affects G' and G'' (Navickis et al., 1982). At a given moisture level higher protein levels give higher values of G' and G''. Stress relaxation measurements have been used to find a correlation between the rheological properties and the content of certain amino acids in gluten (Lasztity, 1970; Telegdy-Kovats and Lasztity, 1970). As an example, a correlation between rheological properties and cysteine content was investigated by using 202 gluten samples. The correlation obtained for these samples was poor (0.206 for $t_{1/2}$), although a better correlation (0.593) was obtained for a single wheat variety (Lasztity, 1970).

The correlation between percentage HMW protein (as determined by gel filtration) and stress relaxation measurements was investigated by using 120 gluten varieties (Lasztity, 1970). A maximum in $t_{1/2}$ as well as in G was found as a function of percentage HMW proteins.

The rheological properties of gliadin, glutenin, and acid insoluble residue were investigated in compression tests (Ram and Nigam, 1983). Gliadin and residue protein from three different wheat varieties (one weak, one medium strong, and one strong) all showed similar rheological behavior, whereas the properties of the glutenin fraction differed between varieties. When gluten was reconstituted from glutenin, gliadin, and residue protein, and stretching tests were performed (Ram and Nigam, 1981), the reconstituted gluten was found to be less extensible than the native gluten. Protein fractions were interchanged between varieties, and the extensibility of the reconstituted gluten was found to be governed by the glutenin fraction (Ram and Nigam, 1981).

Gluten Models

The glutenin molecules are often regarded as determining the viscoelastic properties of gluten. Because gluten contains other components, however, a model that attempts to describe gluten should include the localization and function of these other constituents (gliadins, lipids, and carbohydrates). So far, no complete model has been proposed. This is not surprising. Difficulties are encountered in describing glutenin alone, which is only one of the gluten constituents. The models that exist in the literature describe aspects of gluten behavior. These models have been derived from both rheological experiments and other techniques; for example, microscopic techniques. The models seem to be constructed along three lines of reasoning, each representing a different level of organization. The molecular approach discusses interactions between gluten proteins and, for example, lipids on a strictly molecular and individual basis (e.g., Pomeranz et al., 1970; Chung, 1986). In other models the polymeric behavior of the protein molecules is emphasized (e.g., Greenwood and Ewart, 1975). Gluten has also been described on a colloidal level; that is, the interactions between groups of molecules are described (Grosskreutz, 1961).

As is the case for glutenin models, gluten models consider the significance of disulfide bridges; that is, if intermolecular disulfide bridges are necessary for gluten structure and behavior. In this respect the stress relaxation results are unambiguous: gluten behaves as a viscoelastic liquid and not as a solid (Funt Bar-David and Lerchenthal, 1975; Bohlin and Carlson, 1981; Mita and Bohlin, 1983).

The polymeric behavior of glutenin was described in a model recently presented by Graveland and Henderson (1987). In this model gluten is composed of huge glutenin molecules with MWs above 20 million, and from smaller glutenins with MWs in the range from 1 to 5 million. The smallest glutenin is proposed to be a monomer with a MW of 1 million. The other glutenin molecules are, therefore, polymers thought to be composed of different numbers of the monomer. Insoluble glutenins, observed in the transmission electron microscope, were seen to be compact globular particles with a diameter of 10 nm (Graveland and Henderson, 1987). In this model the gliadins are described as a diluent that functions to reduce the stiffness of the gluten. The viscoelastic behavior of gluten would therefore depend on the gliadin:glutenin ratio. Similar models, in somewhat different forms, have been discussed earlier (Ewart, 1972b, Belitz et al., 1986).

The first event that occurs during dough development is the hydration of protein (Meredith and Wren, 1969; Graveland and Henderson, 1987). When wheat endosperm is wetted on a microscope slide, protein strands are observed to spread out from the broken endosperm cells (Bernardin and Kasarda, 1973a). The fibrils exhibit elastic recovery as well as viscous flow. The fibrils are about 0.5 μm in diameter, but composed of microfibrils with diameters of 50 to 100 Å (Bernardin and Kasarda, 1973b). Investigations have shown that defatting the flour prior to wetting does not change its fibril forming ability. Different cereals showed this ability to a greater or lesser extent; for example, triticale and rye formed fewer fibrils than wheat. No fibrils at all were formed by maize, rice, or barley (Bernardin and Kasarda, 1973a).

Grosskreutz (1961) described gluten as a continuous protein matrix interrupted by lipid bilayers. Protein platelets were described in which a hydrophobic center was surrounded by a polar surface. This colloidal model was further developed by Carlson (1981). In his model the polar lipids are oriented at the air/gluten interface with the hydrophobic parts of the lipids in the air phase, and the polar parts oriented toward polar parts of the gluten proteins. The structure of gluten was discussed in relation to results obtained by electron microscopy and X-ray diffraction techniques by Hermansson and Larsson (1986). Here gluten structure was described as consisting of globular aggregates with hydrophobic cores and hydrophilic surface zones forming a continuum. The similarity between this gluten model and the L2 phase formed in aqueous systems of simple amphiphilic substances was pointed out by Hermansson and Larsson (1986).

Gluten was found to exhibit a spacing at 55 Å in X-ray diffraction (also observed in other works [Grosskreutz, 1961]), when it contained the maximum amount of water. When gluten was dried, a spacing at 44 Å was observed. This behavior was discussed in relation to the behavior of polar lipids, and it was suggested that the spacings were present because polar lipids in gluten formed the lamellar liquid–crystalline phase (Hermansson and Larsson, 1986). The presence of this phase in wheat gluten has also been shown by phosphorus magnetic resonance spectroscopy and freeze-fracture electron microscopy (Marion et al., 1987). The structure of gluten observed in the electron microscope was independent of the lipids present; that is, gluten formation did not depend on interactions between proteins and lipids (Hermansson and Larsson, 1986).

The rheological behavior of gluten has been interpreted according to the theory of cooperative flow (Bohlin, 1980; Mita and Bohlin, 1983). The observed macroscopic flow is regarded as the consequence of cooperative rearrangements of a large number of flow units. These flow units may be any aggregate of the constituents of the sample. It is possible to obtain information concerning how many units cooperate during flow. The theory, however, is not able to elucidate the nature of the cooperative units. In some cases, the correlation between flow units (indicated by the coordination number z) and structure elements seems to be quite straightforward. In the hexagonal liquid–crystalline phase, where one lipid cylinder is surrounded by six neighbors, z was found to be six. In the lamellar liquid–crystalline phase, where each lipid bilayer is surrounded by two other bilayers, z was two (Bohlin and Fontell, 1978). When the structure is not known, it might therefore be possible to obtain some guidance from the analysis of cooperative flow. In the case of gluten, however, the relation between z and structure is not as simple as described above for polar lipids. Gluten exhibits two flow processes: a primary process with a relaxation time of about 1 s and a z number of four, and a secondary process with a relaxation time of about 500 s and a z number of two (Mita and Bohlin, 1983). The primary process was interpreted as occurring in a fibrillar structure, whereas the secondary process was proposed to occur in a lamellar superstructure. Cross-linking, chemical or noncovalent, arrests the flow in the fibrillar structure and the flow in the lamellar structure then becomes more important.

Maize Prolamines

The viscoelastic properties of concentrated ethanol dispersions of zein were studied by measuring how σ and primary normal stress difference ($N_1 = \sigma_{11} - \sigma_{22}$) varied with $\dot{\gamma}$ and protein concentration (Menjivar and Rha, 1980). Shear viscosity was found to increase with concentration, but at a concentration of 42% (w/w) the slope changed significantly. It was further found that the zein dispersions in ethanol showed shear thinning. The results were fitted to the power law (Eq. 3-2), and both n and k varied with the concentration, with n being 0.81–0.83 (Menjivar and Rha, 1980). Also, N_1 was found to vary with concentration and $\dot{\gamma}$, although the effects were less than in case of σ. The changes in rheological behavior at certain concentrations were related to an "overlap" concentration, where crowding of molecules became important.

Changes in rheological behavior were also observed at certain concentrations when the expansion of zein solutions was studied at the exit of capillaries

(Menjivar and Rha, 1981). The die swell ratios observed were characteristic for viscoelastic behavior when the zein concentration was 35.7% (w/w) or above. At lower concentrations the behavior was practically Newtonian.

SURFACE RHEOLOGY OF WHEAT PROTEINS

A wheat flour dough is built up from a continuous gluten phase in which starch granules, air cells, and other dough constituents are dispersed. Different interfaces are present in the dough, the most important being the air/water interface. The air/water interface is important because of its relation to the gas holding capacity of the dough (Larsson, 1983). Surface active molecules, polar lipids, and proteins orient themselves at the air/water interface and thus affect its properties. The rheological properties of different components at an interface can be studied by methods corresponding to those used to characterize bulk rheology (see example in Criddle, 1960).

The surface oscillation technique has been used to characterize the rheological properties of gluten films at the air/water interface and at the oil/water interface (Tschoegl and Alexander, 1960b). The surface viscosity (η_s), as well as the surface rigidity (G_s), were found to increase with time for these gluten films. Furthermore, both η_s and G_s were strongly dependent on the type of interface. Measurable values of η_s and G_s were not obtained until the film was compressed to an area of about 0.5 m^2/mg at the air/water interface. At the oil/water interface, measurable values of η_s and G_s were obtained when the film was compressed to 1.0 m^2/g (Table 3-13).

Gluten films at the oil/water interface showed a maximum η_s value as well as a maximum G_s at pH values around pH 7.5. This pH value was, consequently, interpreted as being the isoelectric point of gluten. Both η_s and G_s decreased with increasing ionic strength. The rheological properties of a gluten film on 24% urea were very similar to those of the control.

Table 3-13. Gluten Film Spreading on Different Substrates[a]

	Area (m^2/mg)	
Substrate	η_s	G_s
Buffer pH 6.8, I = 0.1 oil/water	1.3	0.9
Buffer pH 6.8, I = 0.1 air/water	0.5	0.4
10% Sodium salicylate	0.5	0.3
24% Urea	1.0	0.6

Source: Adapted from Tschoegl and Alexander (1960b).
[a]The area to which gluten films spread at different substrates had to be compressed in order to give measurable values of surface viscosity (η_s) and rigidity (G_s)

When gluten instead was spread on 10% sodium salicylate, a considerable decrease was observed in η_s as well as in G_s. The area/gram of the protein film at which measurable values of η_s and G_s were obtained is given in Table 3-13. The rheological properties of the gluten films were thus changed radically when the interactions between protein molecules were inhibited by the presence of sodium salicylate. This result was interpreted as showing that the viscoelasticity of the gluten film depended on both hydrogen bonds and ionic linkages (Tschoegl and Alexander, 1960b).

REFERENCES

Aranyi, C., and Hawrylewicz, E. J. 1972. Preparation and isolation of acid-catalyzed hydrolysates from wheat gluten. *J. Agric. Food Chem.* **20**:670–675.

Bale, R., and Muller, H. G. 1970. Application of the statistical theory of rubber elasticity to the effect of heat on wheat gluten. *J. Food Technol.* **5**:295–300.

Batey, I. L. 1980. Chemical modification as a probe of gluten structure. *Ann. Technol. Agric.* **29**:363–375.

Bartoszewicz, K., Kaczkowski, J., and Liss, W. 1972. Hydrodynamic properties of gluten and glutenin obtained from flour of different baking quality. *Bull. Acad. Pol. Sci. Ser. Sci. Biol.* **20**:827–831.

Beckwith, A. C., Nielsen, H.-C., Wall, J. S., and Huebner, F. R. 1966. Isolation and characterization of a high-molecular-weight protein from wheat gliadin. *Cereal Chem.* **43**:14–28.

Beckwith, A. C., and Wall, J. S. 1966. Reduction and reoxidation of wheat glutenin. *Biochim. Biophys. Acta* **130**:155–162.

Beckwith, A. C., Wall, J. S., and Dimler, R. J. 1963. Amide groups as interaction sites in wheat gluten proteins: effects of amide–ester conversion. *Arch. Biochem. Biophys.* **103**:319–330.

Beckwith, A. C., Wall, J. S., and Jordan, R. W. 1965. Reversible reduction and reoxidation of the disulfide bonds in wheat gliadin. *Arch. Biochem. Biophys.* **112**:16–24.

Belitz, H.-D., Kieffer, R., Seilmeier, W., and Wieser, H. 1986. Structure and function of gluten proteins. *Cereal Chem.* **63**:336–341.

Belitz, H.-D., Kim, J.-J., Kieffer, R., Seilmeier, W., Werbeck, U., and Wieser, H. 1987. Separation and characterization of reduced glutelins from different wheat varieties and importance of the gliadin/glutelin ratio for the strength of gluten. In *Proceedings of the 3rd International Workshop on Gluten Proteins*, ed. R. Lasztity and F. Békés, pp. 189–205. Singapore: World Scientific.

Bernardin, J. E. 1975. Rheology of concentrated gliadin solutions. *Cereal Chem.* **52**(3, Pt. II):136–145.

Bernardin, J. E. 1978a. Effect of shear on the nematic mesophase of the wheat storage protein A-gliadin. *J. Text. Stud.* **9**:283–297.

Bernardin, J. E. 1978b. Gluten protein interaction with small molecules and ions—the control of flour properties. *Bakers' Dig.* **52**(4):20–23.

Bernardin, J. E., and Kasarda, D. D. 1973a. Hydrated protein fibrils from wheat endosperm. *Cereal Chem.* **50**:529–536.

Bernardin, J. E., and Kasarda, D. D. 1973b. The microstructure of wheat protein fibrils. *Cereal Chem.* **50**:735–745.

Bietz, J. A. 1979. Recent advances in the isolation and characterization of cereal proteins. *Cereal Foods World* **24**:199–207.

Bietz, J. A., and Wall, J. S. 1980. Identity of high molecular weight gliadin and ethanol-soluble glutenin subunits of wheat: relation to gluten structure. *Cereal Chem.* **57**:415–421.

Bohlin, L. 1980. A theory of flow as a cooperative phenomenon. *J. Colloid Interface Sci.* **74**:423–434.

Bohlin, L., and Carlson, T. L. G. 1981. Shear stress relaxation of wheat flour dough and gluten. *Colloids Surfaces* **2**:59–69.

Bohlin, L., and Fontell, K. 1978. Flow properties of lamellar liquid crystalline lipid-water systems. *J. Colloid Interface Sci.* **67**:272–283.

Bushuk, W. 1984. Functionality of wheat proteins in dough. *Cereal Foods World* **29**:162–164.

Butaki, R. C., and Dronzek, B. 1979a. Comparison of gluten properties of flour wheat varieties. *Cereal Chem.* **56**:159–161.

Butaki, R. C., and Dronzek, B. 1979b. Effect of protein content and wheat variety on relative viscosity, solubility and electrophoretic properties of gluten proteins. *Cereal Chem.* **56**:162–165.

Carlson, T. L.-G. 1981. Law and order in wheat flour dough. Colloidal aspects of the wheat flour dough and its lipid and protein constituents in aqueous media. Thesis. Lund: Lund University.

Chung, O. K. 1986. Lipid-protein interactions in wheat flour dough, gluten, and protein fractions. *Cereal Foods World* **31**:242–256.

Cluskey, J. E., Taylor, N. W., Carley, H. and Senti, F. R. 1961. Electrophoretic composition and intrinsic viscosity of glutens from different varieties of wheat. *Cereal Chem.* **38**:325–335.

Cole, E. W., Kasarda, D. D., and Lafiandra, D. 1984. The conformational structure of A-gliadin. Intrinsic viscosities under conditions approaching the native state and under denaturing conditions. *Biochem. Biophys. Acta* **787**:244–251.

Cole, E. W., Torres, J. V., and Kasarda, D. D. 1983. Aggregation of A-gliadin: gel permeation chromatography. *Cereal Chem.* **60**:306–310.

Criddle, D. W. 1960. The viscosity and elasticity of interfaces. In *Rheology Theory and Applications*, Vol. 3, ed. F. R. Eirich, pp. 429–442. New York: Academic Press.

Cumming, D. B., and Tung, M. A. 1975. Dynamic shear behavior of commercial wheat gluten. *Can Inst. Food Sci. Technol. J.* **8**:206–210.

Dalek, V., Liss, W., and Kacskowski, J. 1970. Indexes of the degree of denaturation in wheat gluten. *Bull. Acad. Polon. Ser. Sci. Biol.* **18**:743–747.

Damidaux, R., Autran, J. C., and Feillet, P. 1980. Gliadin electrophoregrams and measurements of gluten viscoelasticity in durum wheats. *Cereal Foods World* **25**:754–756.

Danno, G., and Hoseney, R. C. 1982. Effects of dough mixing and rheologically active compounds on relative viscosity of wheat proteins. *Cereal Chem.* **59**:196–198.

Dexter, J. E., and Matsuo, R. R. 1980. Relationship between durum wheat protein properties and pasta dough rheology and spaghetti cooking quality. *J. Agric. Food Chem.* **28**:899–902.

Doguchi, M., and Hlynka, I. 1967. Rheological properties of crude gluten mixed in the farinograph. *Cereal Chem.* **44**:561–575.

Dreese, P. C., Faubion, J. M., and Hoseney, R. C. 1988. The effect of different heating and washing procedures on the dynamic rheological properties of wheat gluten. *Cereal Foods World* **33**:225–228.

Eliasson, A.-C., and Hegg, P.-O. 1980. Thermal stability of wheat gluten. *Cereal Chem.* **57**:436–437.

Ewart, J. A. D. 1968a. Fractional extraction of cereal flour proteins. *J. Sci. Food Agric.* **19**:241–245.

Ewart, J. A. D. 1968b. Hypothesis for the structure and rheology of glutenin. *J. Sci. Food Agric.* **19**:617–623.

Ewart, J. A. D. 1972a. A modified hypothesis for the structure and rheology of glutelins. *J. Sci. Food Agric.* **23**:687–699.

Ewart, J. A. D. 1972b. Recent research and dough viscoelasticity. *Bakers' Dig.* **46(4)**:22–26, 28.

Ewart, J. A. D. 1977. Re-examination of the linear glutenin hypothesis. *J. Sci. Food Agric.* **28**:191–199.

Ewart, J. A. D. 1979. Glutenin structure. *J. Sci. Food Agric.* **30**:482–492.

Ewart, J. A. D. 1980. Loaf volume and the intrinsic viscosity of glutenin. *J. Sci. Food Agric.* **31**:1323–1336.

Ewart, J. A. D. 1988. Studies on disulfide bonds in glutenin. *Cereal Chem.* **65**:95–100.

Field, J. M., Tatham, A. S., Baker, A. M., and Shewry, P. R. 1986. The structure of C-hordein. *FEBS Lett.* **200**:76–80.

Field, J. M., Tatham, A. S., and Shewry, P. R. 1987. Determination of prolamin configuration and dimensions by viscometric analysis. In *Proceedings of the 3rd International Workshop on Gluten Proteins*, ed. R. Lasztity and F. Békés, pp. 478–489. Singapore: World Scientific.

Frisch, H. L., and Simha, R. 1956. The viscosity of colloidal suspensions and macromolecular solutions. In *Rheology Theory and Applications*, Vol. 1, ed. F. R. Eirich, pp. 525–613. New York: Academic Press.

Funt Bar-David, C. B., and Lerchenthal, CH. H. 1975. Rheological and thermodynamic properties of gluten gel. *Cereal Chem.* **52**:154r–169r.

Gabor, R., Täufel, A., and Ruttloff, H. 1982a. Veränderungen der molmassen von Weizenkleberproteinen nach schwacher Proteolyse. *Z. Lebensm. Unters. Forsch.* **175**:399-402.

Gabor, R., Täufel, A., and Ruttloff, H. 1982b. Zur Wirkungsweise verschiedener mikrobieller Proteasepräparate auf Weizengluten und-mehl. *Die Nahrung* **26**:37–46.

Grant, D. R. 1973. Modification of wheat flour proteins with succinic anhydride. *Cereal Chem.* **50**:417–428.

Graveland, A., Bosveld, P., Lichtendonk, W. J., Marseille, J. P., Moonen, J. H. E., and Scheepstra, A. 1985. A model for the molecular structure of the glutenins from wheat flour. *J. Cereal Sci.* **3**:1–16.

Graveland, A., and Henderson, M. H. 1987. Structure and functionality of gluten proteins. In *Proceedings of the 3rd International Workshop on Gluten Proteins*, ed. R. Lasztity and F. Békés, pp. 238–246. Singapore: World Scientific.

Greenwood, C. T., and Ewart, J. A. D. 1975. Hypothesis for the structure of glutenin in relation to rheological properties of gluten and dough. *Cereal Chem.* **52**(3, Pt II): 146–153.

Grosskreutz, J. C. 1961. A lipoprotein model of wheat gluten structure. *Cereal Chem.* **38**:336–348.

Hamauzu, Z., and Yonezawa, D. 1971. Behaviors of glutenin in several kinds of dispersing solvents. *Bull. Univ. Osaka Pref. Ser. B.* **23**:1–7.

Heaps, P. W., Webb, T., Rusell Eggitt, P. W., and Coppock, J. B. M. 1968. Rheological testing of wheat glutens and doughs. *Chem. Ind. (London)* **32**:1095–1096.

Hermansson, A. M., and Larsson, K. 1986. The structure of gluten gels. *Food Microstruct.* **5**:233–239.

Hibberd, G. E. 1970. Dynamic viscoelastic behavior of wheat flour doughs. III. The influence of the starch granules. *Rheol. Acta* **9**:501–505.

Hoseney, R. C., Dreese, P. C., Doescher, L. C., and Faubion, J. M. 1987. Thermal properties of gluten. In *Proceedings of the 3rd International Workshop on Gluten Proteins*, ed. R. Lasztity and F. Békés, pp. 518–528. Singapore: World Scientific.

Hoseney, R. C., Zeleznak, K., and Lai, C. S. 1986. Wheat gluten: a glassy polymer. *Cereal Chem.* **63**: 285–286.

Inda, A. E., and Rha, C. 1981. Rupture properties of wheat gluten in simple tension: the role of hydrogen bonds. *J. Food Sci.* **47**: 177–180.

Jeanjean, M. F., Damidaux, R., and Feillet, P. 1980. Effect of heat treatment on protein solubility and viscoelastic properties of wheat gluten. *Cereal Chem.* **57**:325–331.

Kacskowski, J., Vakar, A. B., Demidov, V. S., and Zabrodina, T. M. 1968. Influence of deuterated water on the viscosimetric properties of wheat gluten dispersions. *Bull. Acad. Pol. Sci. Ser. Sci. Biol.* **16**:473–478.

Khan, K., and Bushuk, W. 1979. Studies of glutenin. XII. Comparison by sodium dodecyl sulfate–polyacrylamide gel electrophoresis of unreduced and reduced glutenin from various isolation and purification procedures. *Cereal Chem.* **56**:63–68.

Kieffer, R., Kim, J.-J., and Belitz, H.-D. 1981. Zugversuche mit Weizenkleber im Mikromass-stab. *Z. Lebensm. Unters. Forsch.* **172**:190–192.

Kieffer R., Kim, J.-J., and Belitz, H.-D. 1983. Einfluss niedermolekularer ionischer Verbindungen auf die Löslichkeit und die rheologischen Eigenschaften von Weizenkleber. *Z. Lebensm. Unters. Forsch.* **176**:176–182.

Kieffer, R., Kim, J., Kempf, M., Belitz, H.-D., Lehmann, J., Sprössler, B., and Best, E. 1982. Untersuchungen rheologischer Eigenschaften von Teig und Kleber aus Weizenmehl durch Capillar-viscosimetrie. *Z. Lebensm. Unters. Forsch.* **174**:216–221.

Kinsella, J. E., and Hale, M. L. 1984. Hydrophobic associations and gluten consistency: effect of specific anions. *J. Agric. Food Chem.* **32**:1054–1056.

Kobrehel, K., and Bushuk, W. 1977. Studies of glutenin. X. Effect of fatty acids and their sodium salts on solubility in water. *Cereal Chem.* **54**:833–839.

Kruger, J. E. 1971. Effects of proteolytic enzymes on gluten as measured by a stretching test. *Cereal Chem.* **48**:121–132.

Krull, L. H., and Wall, J. S. 1969. Relationship of amino acid composition and wheat protein properties. *Bakers' Dig.* **43**(4): 30–39.

Kwa, W. H. W., Tock, R. W., and Osman, E. G. 1976. The effects of selected sugars on the rheological properties of rehydrated vital gluten. *Proc. Iowa Acad. Sci.* **83**: 28–34.

Lagudah, E. S., O'Brien, L., and Halloran, G. M. 1988. Influence of gliadin composition and high molecular weight subunits of glutenin on dough properties in an F_3 population of a bread wheat cross. *J. Cereal Sci.* **7**:33–42.

Larsson, K. 1983. Physical state of lipids and their technical effects in baking. In *Lipids in Cereal Technology*, ed. P. J. Barnes, pp. 237–251. London: Academic Press.

Lasztity, R. 1969. Rheological properties of gluten. II. Viscoelastic properties of chemically modified gluten. *Acta Chim. Acad. Sci. Hung.* **62**:75–85.

Lasztity, R. 1970. Zur Frage des Zusammenhangs zwishen chemischer Struktur und rheologischen Eigenschaften von Klebereiweissen 2. Mitt. Die Rolle der quantitativen Verhältnisse von Klebereiweissfraktionen in der Ausbildung der rheologischen Eigenschaften. *Die Nahrung* **14**:569–577.

Lasztity, R. 1971. Investigation of the rhelogical properties of gluten. III. Role of hydrophobic bonds in the rheological properties of gluten. *Acta Chim. Acad. Sci. Hung.* **68**:411–419.

Lasztity, R. 1980. Correlation between chemical structure and rheological properties of gluten. *Ann. Technol. Agric.* **29**:339–361.

Lerchenthal, Ch. H., and Funt, C. B. 1970. Yield function in an unstable viscoelastic material (gluten gel). *Isr. J. Technol.* **8**:317–323.

Lindahl, L., and Eliasson, A. C. 1986. Effects of wheat proteins on the viscoelastic properties of starch gels. *J. Sci. Food Agric.* **37**:1125–1132.

Lundh, G., Eliasson, A.-C. and Larsson, K. 1988. Cross-linking of wheat storage protein monolayers by compression/expansion cycles at the air/water interface. *J. Cereal Sci.* **7**:1–9.

MacRitchie, F. 1975. Mechanical degradation of gluten proteins during high-speed mixing of doughs. *J. Polym. Sci. Polym. Symp.* **49**:85–90.

MacRitchie, F. 1985. Studies of the methodology for fractionation and reconstitution of wheat flours. *J. Cereal Sci.* **3**:221–230.

Marion, D., LeRoux, C., Akoka, S., Tellier, C., and Gallant, D. 1987. Lipid–protein interactions in wheat gluten: a phosphorus nuclear magnetic resonance spectroscopy and freeze-fracture electron microscopy study. *J. Cereal Sci.* **5**:101–115.

Matsumoto, S. 1979. Rheological properties of synthetic flour doughs. In *Food Texture and Rheology*, ed. P. Sherman, pp. 291–301. New York: Academic Press.

Matsuo, R. R. 1978. Note on a method for testing gluten strength. *Cereal Chem.* **55**:259–262.

Menjivar, J. A., and Rha, C. K. 1980. Viscoelastic effects in concentrated protein dispersions. *Rheol. Acta* **19**:212–219.

Menjivar, J. A., and Rha, C. K. 1981. Extrudate expansion of concentrated protein solutions. *J. Rheol.* **25**:237–249.

Menkovska, M., Lookhart, G. L., and Pomeranz, Y. 1987. Changes in the gliadin fraction(s) during breadmaking: isolation and characterization by high-performance liquid chromatography and polyacrylamide gel electrophoresis. *Cereal Chem.* **64**:311-314.

Meredith, O. B., and Wren, J. J. 1969. Stability of the molecular weight distribution in wheat flour proteins during dough making. *J. Sci. Food Agric.* **20**:235–237.

Mita, T., and Bohlin, L. 1983. Shear stress relaxation of chemically modified gluten. *Cereal Chem.* **60**:93–97.

Mita, T., and Matsumoto, H. 1981. Flow properties of aqueous gluten and gluten methyl ester dispersions. *Cereal Chem.* **58**: 57–61.

Mita, T., and Matsumoto, H. 1984. Dynamic viscoelastic properties of concentrated dispersions of gluten and gluten methyl ester: contributions of glutamine side chain. *Cereal Chem.* **61**: 169–173.

Mita, T., Nikai, K., Hiraoka, T., Matsuo, S., and Matsumoto, H. 1977. Physicochemical studies on wheat protein foams. *J. Colloid Interface Sci.* **59**: 172–178.

Muller, H. G. 1969. Application of the statistical theory of rubber elasticity to gluten and dough. *Cereal Chem.* **46**:443–446.

Navickis, L. L., Anderson, R. A., Bagley, E. G., and Jasberg, B. K. 1982. Viscoelastic properties of wheat flours dough: variation of dynamic moduli with water and protein content. *J. Text. Stud.* **13**:249–264.

Nielsen, H. C., Babcock, G. E., and Senti, F. R. 1962. Molecular weight studies on glutenin before and after disulfide-bond splitting. *Arch. Biochem. Biophys.* **96**:252–258.

Orth, R. A., Dronzek, B. L., and Bushuk, W. 1973. Studies of glutenin. IV Microscopic structure and its relations to breadmaking quality. *Cereal Chem.* **50**:688–696.

Payne, P. I., Holt, L. M., Harinder, K., McCartney, D. P., and Lawrence, G. J. 1987. The use of near-isogenic lines with different HMW glutenin subunits in studying breadmaking quality and glutenin structure. In *Proceedings of the 3rd International Workshop on Gluten Proteins*, ed. R. Lasztity and F. Békés, pp. 216–226. Singapore: World Scientific.

Payne, P. I., Law, C. N., and Mudd, E. E. 1980. Control by homologous group 1 chromosomes of the high-molecular-weight subunits of glutenin, a major protein of wheat endosperm. *Theor. Appl. Genet.* **58**:113–120.

Pena, R. J., and Ballance, G. M. 1987. Comparison of gluten quality in triticale: a fractionation-reconstitution study. *Cereal Chem.* **64**:128–132.

Pomeranz, Y., Finney, K. F., and Hoseney, R. C. 1970. Molecular approach to breadmaking. *Bakers' Dig.* **44**(3):22–28, 62.

Prasada, Rao, U. J. S., and Nigam, S. N. 1987. Gel filtration chromatography of glutenin in dissociating solvents: effects of removing noncovalently bonded protein components on the viscoelastic character of glutenin. *Cereal Chem.* **64**:168–172.

Prugar, J. 1969. Ermittlung der Backfähigkeit von Weizen und Weizenmehl mit dem Penetrometer. *Die Nahrung* **13**:687–696.

Purcell, J. M., Kasarda, D. D., and Wu, C.-S. C. 1988. Secondary structures of wheat α- and ω-gliadin proteins: Fourier transform infrared spectroscopy. *J. Cereal Sci.* **7**:21–32.

Pyler, E. J. 1983. Flour proteins role in baking performance. *Bakers' Dig.* **57**(3)24–25, 27–28, 33.

Ram, B. P., and Nigam, S. N. 1981. Stretchability of wheat gluten in relation to gluten composition and varietal differences. *Can. Inst. Food Sci. Technol. J.* **14**:326–328.

Ram, B. P., and Nigam, S. N. 1983. Texturometer as a tool for studying varietal differences in wheat flour doughs and gluten proteins. *J. Text. Stud.* **14**:245–249.

Redman, D. G. 1971. Wheat proteins. *Chem. Ind.* **38**:1061–1068.

Rinde, J. A., Tschoegl, N. W., and Smith, T. L. 1970. Large-deformation and rupture properties of wheat flour gluten. *Cereal Chem.* **47**:225–235.

Schofield, J. D., Bottomley, R. C., Timms, M. F., and Booth, M. R. 1983. The effect of heat on wheat gluten and the involvement of sulphydryl-disulphide interchange reactions. *J. Cereal Sci.* **1**:241–253.

Seilmeier, W., Wieser, H., and Belitz, H. D. 1987. High-performance liquid chromatography of reduced glutenin: amino acid composition of fractions and components. *Z. Lebensm. Unters. Forsch.* **185**:487–489.

Shewry, P. R., and Miflin, B. J. 1984. Seed storage proteins of economically important cereals. In *Advances in Cereal Science and Technology*, Vol. 7, ed. Y. Pomeranz, pp. 1–83. St. Paul, MN: American Association of Cereal Chemists.

Shewry, P. R., Tatham, A. S., Forde, J., Kreis, M., and Miflin, B. J. 1986. The classification and nomenclature of wheat gluten proteins: a reassessment. *J. Cereal Sci.* **4**:97–106.

Smith, J. R., Smith, T. L., and Tschoegl, N. W. 1970. Rheological properties of wheat flour doughs. III. Dynamic shear modulus and its dependence on amplitude, frequency and dough composition. *Rheol. Acta* **9**:239–252.

Szczesniak, A. S., Loh, J., and Mannell, W. R. 1983. Effect of moisture transfer on dynamic viscoelastic parameters of wheat flour/water systems. *J. Rheol.* **27**:537–556.

Tanaka, K., and Bushuk, W., 1973. Changes in flour proteins during dough-mixing. II. Gel filtration and electrophoresis results. *Cereal Chem.* **50**:597–605.

Tatham, A. S., Field, J. M., Smith, S. J., and Shewry, P. R. 1987. The conformations of wheat gluten proteins. II. Aggregated gliadins and low molecular weight subunits of glutenin. *J. Cereal Sci.* **5**:203–214.

Tatham, A. S., Miflin, B. J., and Shewry, P. R. 1985. The beta-turn conformation in wheat gluten proteins: relationship to gluten elasticity. *Cereal Chem.* **62**:405–412.

Tatham, A. S., and Shewry, P. R. 1985. The conformation of wheat gluten proteins. The secondary structures and thermal stabilities of α-, β-, γ- and ω-gliadins. *J. Cereal Sci.* **3**:103–113.

Taylor, N. W., and Cluskey, J. E. 1962. Wheat gluten and its glutenin component: viscosity, diffusion and sedimentation studies. *Arch. Biochem. Biophys.* **97**:399–405.

Telegdy-Kovats, L., and Lasztity, R. 1970. Chemical structure of gluten proteins. *Jena Rev.* **15**:116–120.

Tschoegl, N. W., and Alexander, A. E. 1960a. The surface chemistry of wheat gluten. I. Surface pressure measurements. *J. Colloid Sci.* **15**:155–167.

Tschoegl, N. W., and Alexander, A. E. 1960b. The surface chemistry of wheat gluten. II. Measurements of surface viscoelasticity. *J. Colloid Sci.* **15**:168–182.

Udy, D. C. 1953. Effect of bisulfite and thioglycolic acid on the viscosity of wheat gluten dispersions. *Cereal Chem.* **30**:288–301.

Wall, J. S., and Beckwith, A. C. 1969. Relationship between structure and rheological properties of gluten proteins. *Cereal Sci. Today* **14**:16–18, 20–21.

Wasik, R. J., Daoust, H., and Martin, C. 1979. Studies of glutenin solubilized in high concentrations of sodium stearate. *Cereal Chem.* **56**:90–94.

Wieser, H., Seilmeier, W., and Beitz, H.-D. 1980. Vergleichende Untersuchungen über partielle Aminosäuresequenzen von Prolaminen und Glutelinen verschiedener Getreidearten. I. Proteinfraktionierung nach Osborne. *Z. Lebensm. Unters. Forsch.* **170**:17–26.

Wieser, H., Seilmeier, W., and Belitz, H. D. 1987. Vergleichende Untersuchungen über partielle Aminosäuresequenzen von Prolaminen und Glutelinen versheidener Getreidearten. VII. Aminosäuresequenzen von Prolaminpeptiden. *Z. Lebensm. Unters. Forsch.* **184**:366–373.

Wu, Y. V., Cluskey, J. E., and Sexson, K. R. 1967. Effect of ionic strength on the molecular weight and conformation of wheat gluten proteins in 3 M urea solutions. *Biochem. Biophys. Acta* **133**:83–90.

Wu, Y. V., and Dimler, R. J. 1963. Hydrogen ion equilibria of wheat glutenin and gliadin. *Arch. Biochem. Biophys.* **103**:310–318.

Wu, Y. V., and Dimler, R. J. 1964. Conformational studies of wheat gluten, glutenin, and gliadin in urea solutions at various pH's. *Arch. Biochem. Biophys.* **107**:435–440.

Zentner, H. 1968. Effect of ascorbic acid on wheat gluten. *J. Sci. Food Agric.* **19**:464–467.

Chapter 4

Rheological Properties of Cereal Carbohydrates

Jean-Louis Doublier

Starch and proteins are the main macromolecules in cereal grains. Other non-starch carbohydrates occur in cereal grains in low amounts, particularly other pentosans, $(1\rightarrow3)$ $(1\rightarrow4)$ β-glucans, and low molecular weight carbohydrates. This chapter is concerned primarily with carbohydrates that (1) originate from cereal grains or flours and (2) exhibit total or partial solubility in water (at room temperature and/or heated). This definition includes starch, water-soluble pentosans, $(1\rightarrow3)$ $(1\rightarrow4)$ β-glucans, and starch hydrolyzates.

All of these carbohydrates are water soluble, with the exception of starch, which is partly soluble. The rheological behavior of macromolecules in solution relies on the solubility of the polymer and on solvent–polymer interactions. In addition, the rheological properties of macromolecular solutions have been shown to be directly related to macromolecular characteristics such as molecular weight and hydrodynamic volume (Launay et al., 1986). Cereal carbohydrates display a number of interesting rheological properties. Water-soluble pentosans and $(1\rightarrow3)$ $(1\rightarrow4)$ β-glucans possess thickening properties similar to those of other hydrocolloids (Wood, 1984; Hoseney, 1984). Moreover, wheat flour pentosans yield gels by means of an oxidative mechanism (Neukom, 1976; Hoseney, 1984), a phenomenon that has been investigated due to its potential importance in baking. Starch pastes exhibit very specific rheological properties that depend upon a large number of parameters. Due to its importance, the rheology of starch pastes and gels has been the object of numerous investigations (Myers and Knauss, 1965; Schulz, 1974; De Willigen, 1976; Zobel, 1984). However, despite a great deal of effort, particularly in recent years, this subject remains poorly understood. Starch hydrolysates, and particularly the viscosity of glucose syrups, are also of great importance in food technology.

The first, and main, part of this chapter focuses on starch systems. The non-starch carbohydrates are treated in the second part of the chapter; coverage on these carbohydrates is less extensive, since the literature on these molecules is not abundant. In all cases, special attention is paid to the use of the basic rheological methods, that is, viscometry and viscoelastometry, required to obtain fundamental rheological information on the systems and hence on their structures.

STARCH SYSTEMS

Starch properties arise from the unique behavior of its components in aqueous systems. Starch is composed of two types of α-D-glucose polymers: amylose, which is the lower molecular linear component, and amylopectin, which is both large in size and highly branched. Amylose accounts for 20–35% of normal starch, depending upon the botanical origin. In normal cereal starches, this content is remarkably constant at 27–28%. One exception to this is rice (9% to >25%) (Juliano, 1984).

When dispersed in cold water, starch undergoes a limited swelling, but since absorption of water is low (~30%), there is no discernible rheological effect. With sufficient heat a starch–water system or a starch-containing food material (dough, for example) undergoes a series of dramatic changes referred to as gelatinization and pasting. These changes occur beyond the gelatinization temperature (60°C in starch systems containing excess water). Gelatinization temperature is not a constant, but greatly depends upon the characteristics (water content, dissolved solutes) of the medium. It is the changes that occur after gelatinization that result in the development of interesting rheological properties. This chapter does not give an extensive description of the gelatinization and pasting process. The reader can refer to the many reviews dealing with these phenomena (Leach, 1965; Collison, 1968; Sterling, 1978; Ollku and Rha, 1978; Blanshard, 1979; Dengate, 1984).

Rheology of Starch Pastes

Preparation of Starch Pastes

Most of the parameters listed in Table 4-1 are, presumably, affected by the way the starch dispersions are pasted. It therefore is not surprising that a range of rheological behaviors can be reported. Special attention should be paid to the question of the pasting procedure itself. Many parameters must be carefully controlled, including initial temperature, rate of temperature increase, highest cooking temperature, stirring rate, cooking time, pH, and presence of solutes (Radley, 1976).

Several relatively simple methods for cooking starch exist. One possibility consists of the use of a viscograph as the cooking device. This instrument,

Table 14-1. Factors that Might Influence the Rheological Behavior of Starch Pastes and Gels

Dispersed phase = swollen starch granules
Concentration
Granule size and size distribution
Shape of granules
Swelling pattern of the starch
Granule rigidity (e.g., crystallinity) and deformability
Continuous phase
Viscoelasticity of the phase
Amount and type of amylose/amylopectin that has leached from the granules
Entanglements
Interactions between the components
Granule–granule contact
Granule–amylose/amylopectin interactions
Granule–amylose/amylopectin–granule interactions
The surface of the starch granules

Source: From Eliasson (1986) with permission from Food and Nutrition Press.

devised to produce starch pastes under standardized conditions, ensures reproducibility of the preparation of the paste. The possibilities of varying pasting parameters such as the rate of heating or agitation are, however, quite limited with the viscograph. An alternative technique is the use of a rotary evaporator and water bath to prepare pastes. This method allows control of the heating and stirring rates (Evans and Haisman, 1979). Both of these techniques are, however, far removed from industrial conditions. A more versatile method is preparation of the starch dispersion in a flask equipped with a stirrer (Nedonchelle, 1968; Schultz, 1974; Wong and Lelievre, 1981; Doublier et al., 1987a; Eliasson et al., 1988). The vessel is immersed in a water bath (either set at the upper temperature limit or heated at a constant rate). Such a procedure is described as an AFNOR standard method (AFNOR, 1966) in which the following conditions are specified: stirring rate, 750 rpm; upper temperature, 95°C; holding at this maximum, 10 minutes; total pasting time, 40 minutes. This procedure can be easily modified for varying stirring and heating conditions (Schutz, 1974; Doublier et al., 1987a).

Viscosity Measurements—Flow Behavior

The need for absolute measurements of the viscosity of starch pastes is widely recognized. Pioneering work was published in the 1920s and 1930s based on viscosity measurements obtained by using a capillary device (Farrow and Lowe, 1923; Farrow et al., 1928; Richardson and Waite, 1933). Only recently have commercial rotational viscometers come into use for starch characterization. The delay in using commercial rotational viscometers can be ascribed to methodological reasons related to viscosity measurements, as well as to the practical problems (previously described) in the preparation of starch pastes. Starch pastes are shear sensitive and hence do not exhibit a simple flow behavior. Taking viscosity measurements in rotational instruments involves high-shear conditions (or environments), so shear-sensitive materials must be measured by following specific procedures in order that results be reproducible. Earlier work dealt mostly with these practical problems (Schutz, 1974; Djaković and Dokić, 1972).

The best method currently available to measure the flow properties of starch pastes is based on the use of a rotational viscometer with coaxial cylinders or cone and plate geometry. Measurements must be taken at elevated temperatures, preferably above 60°C, to avoid interference by retrogradation. The shear rate range must be as wide as possible, that is, at least two decades (10 s^{-1}–10^3 s^{-1}). Plotting flow curves from increasing and then decreasing shear rates allows an estimate of the thixotropic properties (shear sensitivity) of the paste. Starch pastes display a shear thinning behavior; this shear thinning is illustrated in Figure 4-1 for different types of starches (Evans and Haisman, 1979) as a plot of apparent viscosity η versus shear rate $\dot{\gamma}$. A thixotropic loop is generally exhibited only in the first shear rate cycle (Nedonchelle, 1968; Djaković and Dokić, 1972). After the first cycle is completed, therefore, an equilibrium state is reached and steady pointwise measurements can be taken (Schutz, 1974; Doublier, 1981).

Within the limited shear rate range of the previous example, flow curves can be described by the power law relationship in Equation 4-1.

$$\sigma = k\,\dot{\gamma}^n \quad \text{or} \quad \eta_a = k\,\dot{\gamma}^{n-1} \tag{4-1}$$

where

σ = shear stress
η_a = apparent viscosity
$\dot{\gamma}$ = shear rate
n = shear thinning index ($n = 1$ for a Newtonian fluid)
k = consistency index corresponding to the apparent viscosity at 1 s^{-1}.

A shear rate of 1 s^{-1} is not always accessible experimentally. Because of this, another equivalent relationship Equation 4-2 is often preferred.

$$\sigma = \sigma_1(\dot{\gamma}/\dot{\gamma}_1)^n \quad \text{or} \quad \eta_a = \eta_1(\dot{\gamma}/\dot{\gamma}_1)^{n-1} \tag{4-2}$$

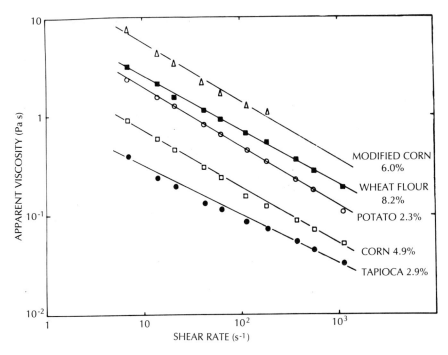

Figure 4–1. Flow curves (in logarithmic scales) of starch suspensions (temperature: 60°C). (*From Evans and Haisman, 1979, with permission of Food and Nutrition Press, Inc.*)

This empirical equation was proposed as a description of starch paste viscosity as early as 1923 (Farrow and Lowe, 1923). While it has been employed by several authors (Farrow et al., 1928; Richardson and Waite, 1933; Nedonchelle, 1968; Schutz, 1970; Cruz et al., 1976; Evans and Haisman, 1979; Doublier, 1981; Wong and Lelievre, 1982a), the relationship is not obeyed at shear rates below 10 s^{-1}. This behavior is illustrated in Figure 4-2, a logarithmic plot of shear stress versus shear rate (Evans and Haisman, 1979). Shear stress appears to be constant at low shear rates, a tendency that has been reported for different starches and pasting conditions (Lancaster et al., 1966; Nedonchelle, 1968; Wong and Lelievre, 1982b). This is usually interpreted as the existence of a yield stress, a behavior typical of suspensions. Such curves have been successfully described by using the Herschel–Bulkley relationship (Eq. 4-3) (Lancaster et al., 1966; Kubota et al., 1978, 1979; Evans and Haisman, 1979; Wong and Lelievre, 1982b).

$$\sigma = k \, \dot{\gamma}^n + \sigma_o \qquad (4\text{--}3)$$

where σ_o = yield stress.

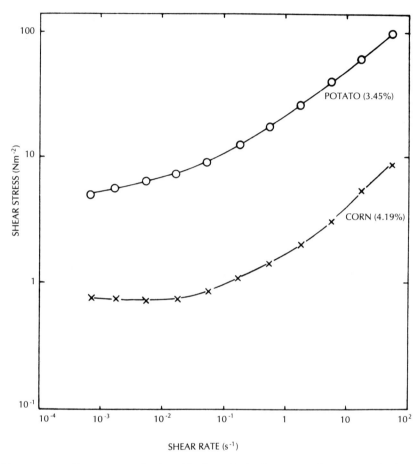

Figure 4–2. Flow curves (in logarithmic scales) of potato and corn starch pastes (temperature: 60°C). (*From Evans and Haisman, 1979, with permission of Food and Nutrition Press, Inc.*)

The rheological significance of the yield stress is that a minimum amount of stress is necessary to initiate flow. The material behaves as a solid below this yield stress. It should be understood, however, that these relations are empirical and, therefore, of limited value in the interpretation of rheological data on a structural basis. In fact, the parameters k and n do not have physical meaning at the molecular level. On the other hand, the yield value σ_o is related to the structure of the material. Unfortunately, σ_o is often determined by extrapolation, so we cannot be sure that its actual value has been measured. This problem was pointed out by Bagley and Christianson (Bagley and Christianson, 1983; Christianson and Bagley, 1984), who emphasized the need to make measurements at very low shear rates. Such a requirement is seldom fulfilled, because σ_o very often cannot be determined due to a lack in the sensitivity of the measuring system.

Although most viscosity measurements are obtained via destructive methods, they can reveal part of the information needed for rheological characterization of starch pastes. Viscosity measurements can provide

1. an estimate of the shear sensitivity of the paste,
2. an overall description of the flow properties of the pastes from a plot of the curves in normal coordinates, and
3. a description of the shear thinning or "pseudoplastic" behavior of the pastes over two or three decades of shear rates by the use of a logarithmic plot.

Viscoelastic Behavior

The fact that viscosity measurements, particularly destructive methods, have limitations in the rheological description of materials as complex as starch pastes prompts the question as to the existence of viscoelastic properties. Earlier work by Myers and co-workers (Myers et al., 1962; Myers and Knauss, 1964, 1965) was based on dynamic measurements with concentric cylinder geometry. The testing instrument was a rotational viscometer modified to operate in an oscillatory mode. The workers used a resonance method that allowed measurements of the shear viscosity η and the shear elasticity G at the resonance frequency. This approach was not very successful, principally due to the complexity of measurements involved. Analysis was made more difficult by the fact that the viscoelastic parameters of interest were frequency dependent and the resonance frequency varied depending upon the paste being studied.

The development and increasing use of commercial oscillatory rheometers since the late 1970s has made rheological measurements more effective. These instruments allow a description of the frequency and strain dependence of the dynamic viscoelastic parameters η', η'', G', G'', and tan δ over a frequency range of two or three decades (Evans and Haisman, 1979; Wong and Lelievre, 1981; Eliasson, 1986; Lindahl and Eliasson, 1986; Eliasson et al., 1988). The illustration given in Figure 4-3 is for a 6.6% (wt/wt) wheat starch dispersion measured at 25°C. The G' and G'' values increase slightly with frequency, but G' is much higher than G'', demonstrating the system is elastic. Note that only two of the three parameters G', G'', and η' need to be measured to determine the dynamic properties, because the parameters are related to each other as follows:

$$\eta' = G''/\omega$$
$$\eta'' = G'/\omega$$
$$\tan \delta = G''/G'$$

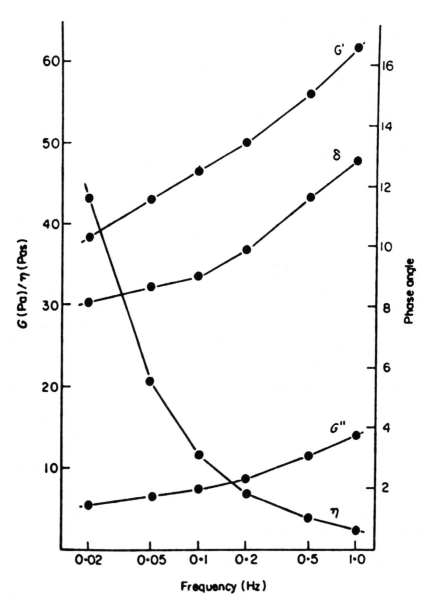

Figure 4–3. The frequency dependence of the storage modulus, G', the loss modulus, G'', the dynamic modulus, η', and the phase angle, δ, for a 6.6% wheat starch paste (temperature: 25°C). (*From Lindahl and Eliasson, 1986, with permission of SCI*)

with

G' = dynamic shear storage modulus
G'' = dynamic shear loss modulus
η' = dynamic viscosity or the in phase component of complex viscosity
η'' = out-of-phase component of complex viscosity
δ = phase lag angle

The coefficient η and the storage modulus G cited in the work by Myers and co-workers (1962) are equivalent to the G' and η' moduli defined above.

The major commercially available instruments that can be used to obtain measurements of these values include the Weissenberg rheogoniometer (Carrimed Ltd., Dorking, U.K.), the Rheometrics mechanical spectrometer (Rheometrics Inc., Piscataway, NJ), the Bohlin rheometer (Bohlin Rheologi, Lund, Sweden), and the Carrimed rheometer (Carrimed Ltd., Dorking, U.K.). None of these rheometers are marketed specifically for the characterization of starch. In fact, work performed in this field remains limited.

When measuring the viscosity of starch pastes, the occurrence of retrogradation during measurements should be avoided. It therefore is preferable to perform measurements at elevated temperatures (60 or 70°C). However, most of the data reported to date were obtained at 25 or 30°C, and within a short period of time ($\sim$ 2 hours) after cooling to room temperature. The investigators (Wong and Lelievre, 1981; Lindahl and Eliasson, 1986; Eliasson, 1986) had to assume that, within this time scale, measurements were not influenced by retrogradation. This assumption was not verified.

Figure 4-4 is an example of variations in G' and G'' during the heating (1.5°C/min) of 10% wheat and corn starch suspensions (Eliasson, 1986). The curves are characterized by initial peaks in both moduli close to the gelatinization temperature, reported for excess water systems, followed by a decrease. A final peak occurred at about 90–95°C. These patterns are comparable to the standard plots obtained with viscographs (Leach, 1965; Ollku and Rha, 1978; Zobel, 1984; Dengate, 1984). They provide, however, additional information regarding the viscoelasticity of the pastes. For wheat starch, G' and G'' were of the same order across the entire temperature range; for corn starch, G' was an order of magnitude higher than G'' at all temperatures above 72°C. These results suggest that corn starch dispersions are significantly more viscoelastic than wheat starch dispersions of comparable concentrations.

Effects of Main Parameters

The ultimate goal of the rheological characterization of starch pastes is to interpret their properties on a structural molecular basis. The factors, listed in

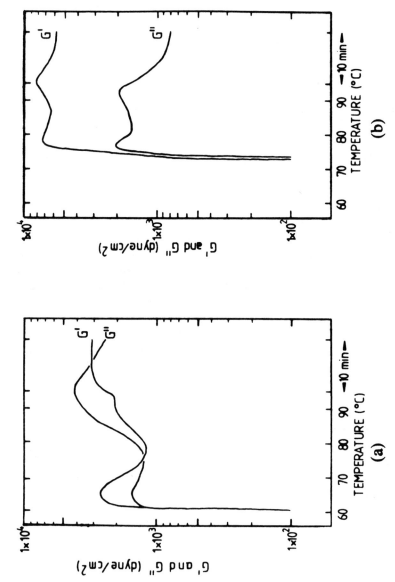

Figure 4–4. Storage modulus, *G'*, and loss modulus, *G''*, as a function of temperature for 10% starch suspensions; (*a*) wheat starch; (*b*) corn starch. (*From Eliasson, 1986, with permission of Food and Nutrition Press, Inc.*)

Table 4-1, are known to influence the rheology of pastes and gels. All of them depend, to a greater or a lesser degree, on a number of other parameters. Some are extrinsic parameters such as the upper heating temperature, the heating rate, the stirring rate, and the presence of solutes. Others, such as the botanical origin of the starch itself, are intrinsic.

Measurement Temperature. Provided that no practical or instrumental problems are encountered, the measurement temperature can be selected between 20°C and 90°C. The highest temperatures are controlled less easily and require, in addition, that water evaporation be controlled. Modifications in paste structure due to retrogradation are not likely to occur at these temperatures, however, and that makes measurements easier to perform. In contrast, low-temperature measurements must be carried out rapidly, so as to avoid retrogradation and subsequent modification of the rheology of the system.

In work on dilute (3–7%, wt/wt) wheat starch pastes, flow curves obtained between 20 and 70°C displayed shapes similar to each other. Thixotropic loops were, however, more pronounced over the range of 25–30°C than above 50°C (Doublier, 1981); these results confirmed the above statements regarding low-temperature measurements. Plots of ln K versus $1/T$ did not exhibit the discontinuity expected if the structure of the dispersion was modified (as by retrogradation). These data were obtained from rapid (within about 30 minutes) measurements after the end of the preparation procedure, and the data clearly indicate that measurements can be carried out whatever the temperature provided they are performed rapidly to avoid retrogradation. The consistency index K of the power law relationship (Equation 4-1) was also reported by other authors (Cruz et al., 1976; Kubota et al., 1979) to obey an Arrhenius-type equation, whereas n was almost independent of temperature. The value of ΔH_a, the activation energy of flow, ranged from 12 kJ mole^{-1} to 30 kJ mole^{-1}, depending on concentration and type of starch. Bagley and Christianson (1983) reported that the viscosity curves of concentrated (10 to 15%, w/w) wheat starch pastes measured at 23°C and 60°C, respectively, differed both quantitatively (in the values of the viscosities) and qualitatively (in the shape of the curves). Moreover, a yield stress was apparent at 23°C but not at 60°C. This result suggests that structure formation occurred during cooling and seemed to be an indication of retrogradation. This conclusion is in an apparent disagreement with the data reported above (Doublier, 1981). The main difference here is the concentration of the systems under examination. Starch concentration is thus an important parameter to account for when choosing the measurement temperature. Specifically, the higher paste concentration makes measurements at low temperatures more difficult.

Preparation Procedures. It is well known (Schutz, 1974; Zobel, 1984) that, as composite systems, the mechanical properties of starch pastes depend, to a large extent, on their method of preparation. Figures 4-5 and 4-6 illustrate the effect of the heating rate and the agitation imposed during heating (Doublier, Llamas, and Le Meur, 1987). Four different procedures were compared with a combination of two heating and two stirring rates. The upper temperature was 96°C. Stirring was at either 200 rev/min (referred to as the low stirring rate, LS) or 750 rev/min (referred to as the high stirring rate, HS). The first heating rate (1°C/min, low heating rate, LR) was comparable to that employed in the Brabender Viscograph. The second heating rate (high heating rate, HR) was achieved by instantaneously heating the starch dispersion up to 96°C. In actuality, the increase from 30 to 90°C required 10 minutes; 96°C was reached after an additional 5 minutes. The figures show clearly that the flow properties estimated from the overall shape of the flow curve (i.e., the apparent viscosity at a given shear rate $\dot{\gamma}$ and the thixotropic loop) are strongly dependent upon the preparation procedure. The results obtained for wheat starch were surprising (Doublier, 1981; Doublier, Llamas, and Le Meur, 1987). Pastes prepared at 1°C/min were much less viscous than those prepared at the high heating rate (compare LR–LS with HR–LS in Fig. 4-5). Similar observations were reported by Wong and Lelievre (1981) in a study of the viscoelasticity of wheat starch pastes. This effect was not found with corn starch (Fig. 4-6). These data make it clear that the effects of each of these preparation parameters differ depending upon the type of starch being analyzed. This implies that it is impossible to accurately predict the performance of a starch prepared under other conditions without actually carrying out the appropriate experiment.

The cooking temperature is another important parameter in determining the rheological properties of pastes. Bagley and Christianson (Bagley and Christianson, 1982, 1983; Christianson and Bagley, 1983, 1984) investigated highly concentrated (10–20%) normal corn starch and wheat starch dispersions that had been heated to temperatures between 60 and 80°C. Flow curves for fully cooked pastes (above 70°C) exhibited a shear thinning behavior with a yield stress. As was expected, the higher the cooking temperature, the greater the shear thinning shown by the flow curve and the higher the yield stress. Because the cooking temperatures were below those of the second swelling solubilization step (85–90°C), the investigators suggested that the observed behaviors were due to suspensions of close-packed swollen granules in the absence of solubles. Kubota and colleagues (1978) studied much less concentrated rice starch suspensions (6%) cooked between 70 and 85°C. Values of K from the power law equation increased with increasing pasting temperature. Unfortunately, no systematic studies have been performed on other cereal starches in dilute concentrations. This omission may be explained by the fact that, at low concentrations, cereal starch dispersions are very thin when the cooking temperature is less than 85°C. Settling of the granules is difficult to avoid and seriously compromises the validity of any measurement.

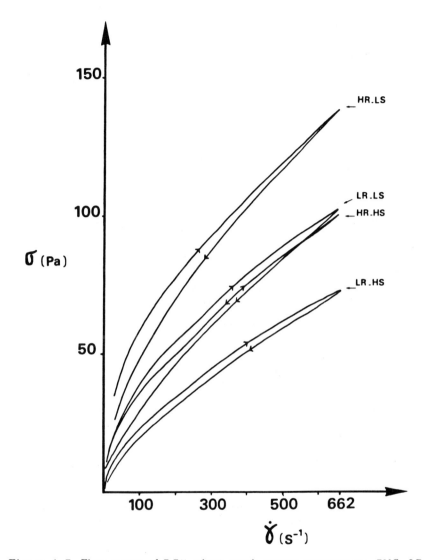

Figure 4–5. Flow curves of 7.7% wheat starch pastes; temperature: 70°C; LR: Low heating rate; HR: high heating rate; LS: low-shear conditions; HS: high-shear conditions (*From Doublier, Llamas, and LeMeur, 1987, with permission of Elsevier Applied Science Publishers, Ltd.*)

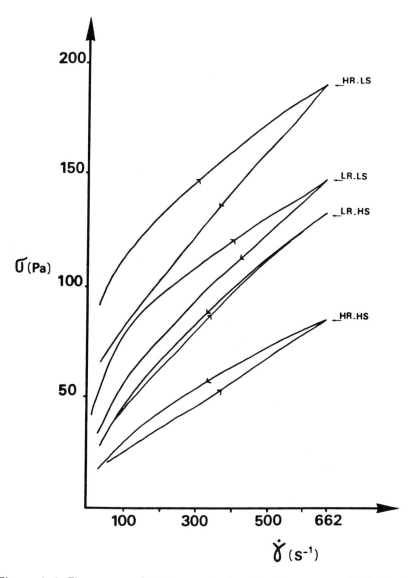

Figure 4–6. Flow curves of 7.7% corn starch pastes; temperature: 70°C. Pasting conditions, high heating rate, low shear. 1 = first cycle, 2 = second cycle. (*From Doublier, Llamas, and Le Meur, 1987, with permission of Elsevier Applied Science Publishers, Ltd.*)

Botanical Origin. Wheat starch and corn starch have been compared extensively by using the Brabender Viscograph or other empirical devices. It is known that, at equivalent concentrations, pastes of corn starch yield much higher viscosities than do wheat starch. Comparison of Figures 4-5 and 4-6 also shows that, by using the same pasting procedure, corn starch pastes are more viscous than wheat starch pastes. However, roughly comparable flow curves were obtained with HR–LS conditions for wheat starch (Fig. 4-5) and LR–LS conditions for corn starch (Fig. 4-6). The data previously reported from viscoelastic measurements (Fig. 4-4) also show that the G' and G'' of corn starch pastes are much higher than those for wheat starch under comparable conditions. Similarly, Lindahl and Eliasson (1986) reported the G' for corn starch to be twice that for wheat starch. In contrast, Bagley and Christianson (1983) reported that yield stresses of concentrated corn starch suspensions ($>10\%$) heated to between 60°C and 80°C were significantly lower than those of wheat starch pastes prepared and measured under equivalent conditions. The presence of a yield stress is the result of long-range structural interactions and can be related to the structural cohesion of the unstressed paste. A higher yield stress can be explained by stronger granule/granule interactions, which would be broken down under shearing. These types of interactions are not expected to influence the apparent viscosities of the pastes or suspensions (Christianson and Bagley, 1984), which are related to the volumes occupied by swollen granules and soluble material. There is, therefore, no contradiction between the behavior at low shear rates, as estimated by the yield stress, and the behavior at high shear rates, as determined with the apparent viscosity at a given shear rate.

Figure 4-7 compares oat, corn, and wheat starch pastes at a relatively low concentration, 3.4% (Doublier, Paton, and Llamas, 1987). Oat starch was found to exhibit a flow behavior greatly different from that of the two other cereal starches. Moreover, as was shown by the presence of a thixotropic loop, oat starch pastes were highly sensitive to shearing. The swelling–solubility curves (Doublier, Paton, and Llamas, 1987) shown in Figure 4-8 correspond to the same starches pasted under conditions similar to those shown in Figure 4-7. It is clearly seen that wheat, corn, and oat starch are in increasing order for swelling characteristics. Also, corn and wheat starches exhibit comparable solubilities that are slightly lower than those for oat starch. These patterns suggest a relation between the swelling and solubility of starches and the rheological behavior of the pastes. This point will be discussed later. These spectacular differences suggest that the structure of oat starch pastes differs from those of the other two pastes. The viscoelastic behaviors of cereal starches from other crops (barley, triticale, rye) have been compared to those for wheat and corn (Lindahl and Eliasson, 1986). Barley and triticale produced pastes comparable to those of wheat, while rye starch was much less viscoelastic. No interpretation has been proposed yet

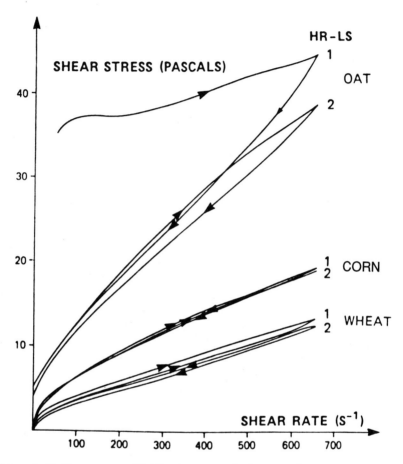

Figure 4–7. Flow curves of 3.4% oat, corn, and wheat starch pastes; temperature: 70°C. Pasting conditions, high heating rate, low shear. 1 = first cycle, 2 = second cycle. (*From Doublier, Paton, and Llamas, 1987, with permission of AACC, Inc.*)

to explain such large differences between normal cereal starches. Attempts to relate rheological properties to the structure and content of internal lipids showed limited success (Doublier, Paton, and Llamas, 1987). A more realistic explanation would have to take into account a number of parameters, such as granular crystallinity, the nature of the lipids, the location of the lipids, and amylose and amylopectin inside the granule. For a given botanical origin, variations in starch paste properties according to the grain variety were also reported. These variations were studied, for instance, for wheat starch (Wong and Lelievre, 1981, 1982b), and rheological properties were shown to vary as a function of two parameters, the volume occupied by the swollen starch granules and also the size distribution of the granules. Figure 4-9 shows a comparison between a normal corn starch, a waxy corn starch, an acetylated

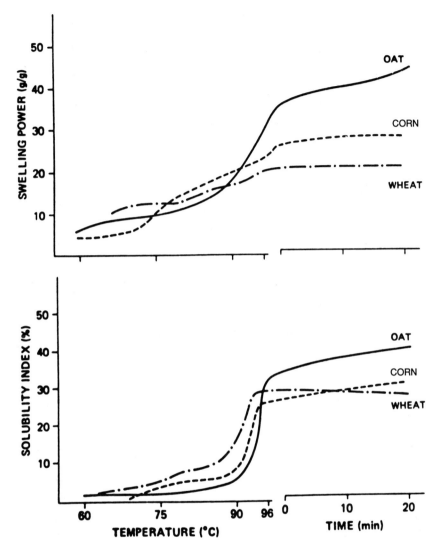

Figure 4–8. Swelling and solubility patterns for oat, corn, and wheat starches cooked under high heating rate and low shear conditions. (*From Doublier, Paton, and Llamas, 1987, with permission of AACC, Inc.*)

high amylose starch, and a cross-linked waxy starch (Eliasson et al., 1988). The G' values for waxy starch were two orders of magnitude lower than those for normal starch. Cross-linking this waxy starch produced behavior comparable to that of the normal starch. This behavior apparently resulted from increased integrity as a consequence of the granule's structure. On the other hand, the chemically modified high-amylose starch was made slightly

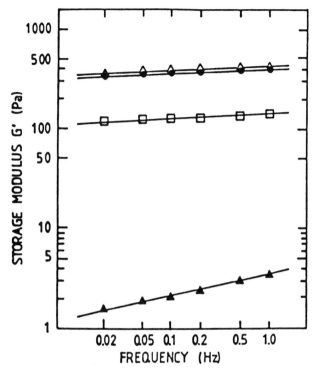

Figure 4–9. G' as a function of frequency for 7.5% corn starch pastes; temperature: 25°C; (●) normal; (▲) waxy; (△) Cross-linked waxy; (□) acetylated high amylose. (*From Eliasson et al., 1988, with permission of Verlag Chemie, Gmbh*)

water soluble and displayed a higher consistency than an unmodified high-amylose starch.

Concentration. Starch pastes can be described in terms of a number of parameters, including η_1 from Equation 4-2, which has been reported to vary as a power law p of concentration (Doublier, 1981; Doublier, Llamas, and Le Meur, 1987):

$$\eta_1 = k' \, C^p \qquad (4-4)$$

with p ranging between 3 and 4.2, depending upon many parameters. This relationship is illustrated in Figure 4-10 for corn starch pastes. The η_1 value is defined there as the apparent viscosity at a shear rate 95 s^{-1} and is plotted as a function of concentration in logarithmic coordinates. Values of the shear thinning index n were reported to vary from 1 (Newtonian behavior) to a limiting value of 0.5 as concentration increased (Evans and Haisman, 1979; Doublier, 1981; Doublier, Llamas, and Le Meur, 1987). The yield stress

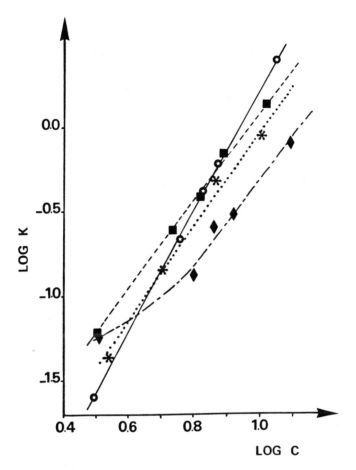

Figure 4–10. Variations of the apparent viscosity at 95 s^{-1} as a function of concentration (in logarithmic coordinates); corn starch pastes; temperature: 70°C; (◉) LR-LS conditions: (*) LR-HS conditions; (■) HR-LS conditions; (♦) HR-HS conditions. (*From Doublier, Llamas, and Le Meur, 1987, with permission of Elsevier Applied Science Publishers, Ltd.*)

also was reported to vary with concentration according to the relationship in Equation 4-5 (Evans and Haisman, 1979; Wong and Lelievre, 1982b; Bagley and Christianson, 1983):

$$\sigma_0 = A \, (C - C_o)^q \tag{4-5}$$

with C_o = concentration at which particles are just close packed, and A = a parameter including the effects of swelling capacity and volume distribution of starch granules (Wong and Lelievre, 1982b). The value of C_o, which is directly related to the swelling capacity of the granules, varies according to the botanical origin of the starches and the grain variety used. The exponent q was shown to range between 1.3 and 2. The fact that G' varies as a function of concentration is illustrated in Figure 4-11 (Evans and Haisman, 1979). For corn starch, a linear variation was exhibited above a critical value, C_o, which was the same as in Equation 4-5. A slightly stronger concentration dependence was reported (Equation 4-6) for wheat starch (Wong and Lelievre, 1981):

$$G' = B (C - C_o)^{Kg} \qquad (4\text{–}6)$$

with $Kg = 1.4$ and C_o as in Equation 4-5. B is similar to A in Equation 4-5. G'' and η' were also shown to exhibit a similar concentration dependence.

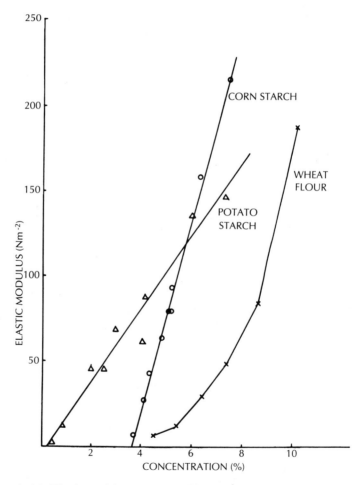

Figure 4–11. Elastic modulus, G', at 0.2 Hz as a function of concentration. (*From Evans and Haisman, 1979, with permission of Food and Nutrition Press, Inc.*)

Highly concentrated starch pastes (>10%) cannot be measured if they are pasted at temperatures above 85°C, the temperature of the second swelling–solubilization stage. The viscosity is too high for the paste to be transferred from the cooking device to the rheometer. Figure 4-12 shows variations of apparent viscosity as a function of starch concentration for different cooking temperatures (<75°C) (Bagley and Christianson, 1983). The effect of this parameter is clearly illustrated as the viscosity increases above a specific

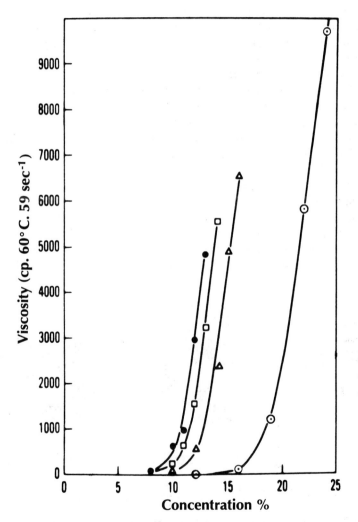

Figure 4–12. Variations of the apparent viscosity at 59 s^{-1} for wheat starch dispersions after 75 minutes at 60°C (⊙), 65°C (△), 70°C (□), and 75°C (●). (*From Bagley and Christianson, 1982, with permission of Food and Nutrition Press, Inc.*)

concentration, which can be also related to the degree of occupancy of the available free volume.

Structure–Properties Relationships

Interpretation of rheological measurements of starch pastes requires a structural description of the systems under investigation. Relationships between starch functionality and the supermolecular order of the starch granules have been presented recently (Blanshard, 1987; Guilbot and Mercier, 1985). It was suggested that, although most granule crystallinity arises from amylopectin, amylose would create a stable structure within the granule, mainly by way of the amylose/lipid complexes. Relationships between granule structure and pasting behavior, however, still are little understood. The main events that occur during the pasting process are related to the swelling–solubilization of starch. These phenomena can easily be studied by sedimentation methods (Leach et al., 1959; Leach, 1965) and microscopic observations (Miller et al., 1973; Bowler et al., 1980; Christianson et al., 1982; Ghiasi et al., 1982; Williams and Bowler, 1982; Varriano-Marston et al., 1985).

Morphological changes taking place during the pasting process can be described by using light microscopy combined with scanning electron microscopy (Miller et al., 1973; Bowler et al., 1980; Williams and Bowler, 1982). An illustration of changes taking place in lenticular wheat starch is given in Figure 4-13. Between 60°C and 80°C (first swelling step), starch granules are slightly swollen and distorted. A clear distinction was reported to exist

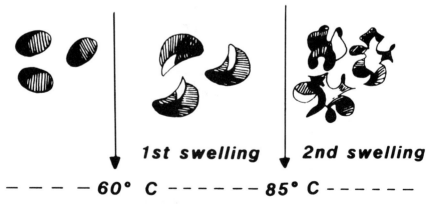

Figure 4–13. Schematic representation of morphological changes occurring during a pasting process for a lenticular wheat starch granule (*From Bowler et al., 1980, with permission of Verlag Chemie Gmbh*)

between cereal starches from the tribe triticae (wheat, rye, and triticale) and the others, maize and oat, particularly (Williams and Bowler, 1982). The first ones exhibited a radial swelling in one plane, resulting in the production of flattened discs. In contrast, with maize and oat starches, the expansion occurred along the three axes, resulting in swollen granules that appeared to be enlarged versions of their unswollen precursors. It is noteworthy that the first swelling step takes place within a narrow temperature range corresponding to the starch's gelatinization temperature. Thus, for maize starch, Christianson and colleagues (Christianson et al., 1982; Christianson and Bagley, 1983) observed that the first swelling took place between 65 and 67°C. At 65°C, starch granules seemed to be unaffected, whereas at 67°C they became swollen and more distorted. Changes taking place in the second swelling stage, within the temperature range 80–85°C, are more complex. Cereal starches of the first group (wheat, rye, triticale) display a tangential swelling in the same plane as the first swelling step. This swelling yields a characteristic convoluted geometrical structure (Bowler et al., 1980; Williams and Bowler, 1982). For the second group (corn, oat, and also certain legume starches), swelling is more dramatic and occurs along the same three axes as in the first step. In spite of these differences, many authors have reported that individual starch granules are still observed at 97°C (Collison and Elton, 1961; Miller et al., 1973; Bowler et al., 1980; Williams and Bowler, 1982). All of these changes take place rapidly, resulting in swollen granules that are quite flexible and hence more shear sensitive.

Microscopic observation can be usefully complemented by sedimentation experiments on dilute starch suspensions (Montgomery and Senti, 1958; Leach et al., 1959; Leach, 1965; Doublier, 1981; Doublier, Llamas, and Le Meur, 1987). The two-step swelling–solubility pattern is easily shown by this method and, moreover, the composition of the solubilized and swollen phases can be determined as well. It has been found that amylose is solubilized before amylopectin. This solubilization occurs mainly at the second step, beyond 85°C (Doublier, 1981; Ghiasi et al., 1982), and seems to be related to the dissociation of the amylose/lipid complexes. This is suggested by differential scanning calorimetry (DSC) measurements (Kugimiya and Donovan, 1980; Bulpin et al., 1982; Stute and Konieczny-Janda, 1983; Eliasson and Krog, 1985). In some cases, a small amount of amylose can remain inside starch granules after pasting is completed (Doublier, 1981). This residual amylose limits the extent of swelling and the shear sensitivity of the granules. As soon as all of the amylose is leached from the granule, part of the amylopectin is susceptible to being released, provided a strong enough shearing force is applied (Doublier, 1981; Ghiasi et al., 1982).

All of these observations provide a basis for describing the structure of starch pastes and their rheological behavior. For normal cereal starches, granules remain individual after heating to 95°C. The paste resulting from the heating process is a suspension of swollen particles composed of amylopectin molecules trapped inside the granules. The continuous phase is a solution of amylose that may also contain amylopectin. Hence, the rheological behavior of such a composite system depends upon diverse parameters (see Table 4-1). Most of these parameters themselves depend upon the botanical origin and the preparation procedure. Several authors (De Willigen, 1976; Ollku and Rha, 1978; Evans and Haisman, 1979; Wong and Lelievre, 1981) ascribe a major role to the dispersed phase in determining overall properties and postulated that the continuous phase was of minor importance. In contrast, the microscopic observations of Miller and co-workers (1973) showed that a large amount of soluble material must be exuded from swollen granules to develop viscosity. This was interpreted as evidence of the major importance of the soluble material. In effect, solubilization is concomitant with the second swelling step, which suggests that both phenomena are required to fully develop a paste's viscosity. Actually, the rheology of a suspension is governed by:

1. the volume fraction occupied by the dispersed phase,
2. the rheological characteristics and the deformability of swollen particles,
3. the rheology of the continuous phase, and
4. the interactions between the dispersed phase and the continuous phase.

As a rough approximation, the apparent viscosity of a suspension at a given shear rate obeys a relation similar to:

$$\eta = \eta_s f(\Phi) \qquad (4\text{--}7)$$

where η_s is the viscosity of the continuous phase and $f(\Phi)$ is a function of the volume fraction of the dispersed phase. For a starch paste, η_s would be a function of the macromolecular composition and the concentration. The $f(\Phi)$ function will conform to the usual relations found for ideal suspensions, which derive from the well-known Einstein equation:

$$f(\Phi) = 1 + 2.5\,\Phi \qquad (4\text{--}8)$$

Highly concentrated wheat and corn starch pastes that were cooked below 85°C were described on such a basis. Bagley and colleagues (Bagley and Christianson, 1982; Christianson et al., 1982; Bagley and Christianson, 1983; Christianson and Bagley, 1983) assumed that, owing to the absence of soluble material in pastes cooked below 85°C, viscosity would be governed primarily by the volume fraction of swollen particles. Here, η_s corresponds to the viscosity of water. Figure 4-14 shows data from Figure 4-11 replotted by using the expression, $C\,Q$, as the variable. Q is the swelling capacity of the starch

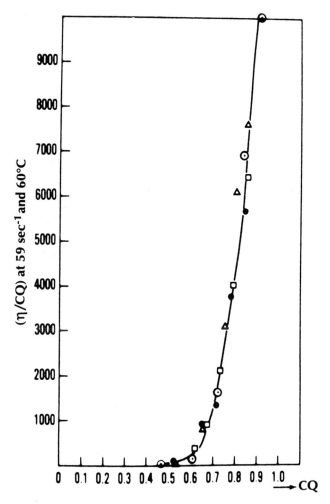

Figure 4–14. Data of Figure 4-3 replotted as $\eta_{sp}/C\ Q$ versus $C\ Q$ (C: concentration; Q: swelling index). (*From Bagley and Christianson, 1982, with permission of Food and Nutrition Press, Inc.*)

at the cooking temperature and C is the concentration. $C\ Q$ was a measure of the volume of swollen starch granules and was used as an estimate of Φ. The master curve generated in Figure 4-13 confirmed the validity of Bagley's treatment of the data. The viscosity increase was shown to be noticeable at values of Φ above about 0.7. In line with this approach, it was suggested that the $f(\Phi)$ function would have an exponential form:

$$f(\Phi) = e\ ([\eta]\Phi) \tag{4–9}$$

This was verified for Φ ranging from 0.5 to 1 (Bagley et al., 1983). The [η] value would vary slightly with shear rate, and a general expression for the relative viscosity ($\eta_r = \eta/\eta_s$) over the shear rate tested range was proposed as:

$$\eta_r = e^{b1\,\Phi}\,\dot{\gamma}^{b2} \qquad (4\text{--}10)$$

with $b_2 < 0$.

For pastes cooked above 90°C, the amount of soluble material can vary from 30% to 80%, depending upon starch and preparation procedure (Doublier, 1981, 1987; Doublier, Llamas, and Le Meur, 1987; Doublier, Paton, and Llamas, 1987). Here the contribution of the continuous phase cannot be neglected and the above treatment is no longer valid. Several investigators (Evans and Haisman, 1979; Wong and Lelievre, 1981, 1982b) attempted to interpret their viscometric or viscoelastic data solely on the basis of swelling capacity. The success of their data treatment was limited and interpretations were only qualitative. Extending Bagley's treatment to take into account the contribution of soluble macromolecules within the continuous phase provided an interesting basis for the interpretation of flow properties. In the model, the volume fraction Φ of the dispersed phase and concentration C_s of the continuous phase were estimated (Doublier, Llamas, and Le Meur, 1987; Doublier, Paton, and Llamas, 1987) from:

$$\Phi = (1 - S/100)\, C\, Q \qquad (4\text{--}11)$$

$$C_s = (1/(1 - \Phi))\, C\, S/100 \qquad (4\text{--}12)$$

where S is the solubility index expressed in percent, C is starch concentration in g/g, and Q is swelling index in g/g. Equation 4-11 is an extension of the expression proposed by Bagley as $\Phi = C\,Q$, where it is assumed that the sediment recovered in swelling–solubilization experiments is filled by swollen particles. This is a reasonable assumption if particles are highly deformable and can easily be close packed. Furthermore, it is evident that Φ cannot be higher than unity, and the proposed treatment therefore is limited to a range of concentrations depending on S and Q. Equation 4-12 is valid provided that $\Phi < 1$. The expression $(1 - \Phi)$ defines the volume available to soluble material, and the expression $C\,S/100$ defines the amount of solubles at a given concentration. Analysis of these expressions shows that Φ is a linear function of concentration, whereas C_s varies hyperbolically with an asymptote for $\Phi = 1$. The concentration at this limit, C_y, was defined as a yield concentration and was shown (Doublier, Llamas, and Le Meur, 1987) to vary from 5% to 28% for corn starch and from 6% to 10% for wheat starch according to pasting procedures. The higher the solubility, the higher the C_y value. Three concentration domains were distinguished on this basis, depending upon the volume filled by swollen particles:

1. a dilute and semidilute regime with Φ less than 0.7
2. a concentrated regime with Φ between 0.7 and 0.9
3. a highly concentrated regime where Φ should be higher than 0.9

Concentrations corresponding to each domain depend upon the type of starch and the pasting procedure. In the dilute domain, the overall viscosity was shown to be very low and directly related to Φ. In the concentrated regime, the contribution of the continuous phase tended to predominate. The C_s value increased dramatically when Φ approached 0.9 because of the limited volume available to soluble material. The upper limit of regime 2 was indeed related to the yield concentration C_y. The highly concentrated domain could not be described as easily as the two others. The Φ value cannot reach unity because of the presence of soluble material. Granule swelling is reduced due to the limited amount of water available and solubility is restricted. It was suggested (Doublier, Llamas, and Le Meur, 1987) that within this domain, the deformability of swollen particles in the dispersed phase should also be taken into account. This parameter would be inversely related to the starch concentration inside the swollen particles, as estimated by $1/Q$.

This treatment provided only a qualitative structural description of starch pastes. It did not provide a complete description of the relationships between rheology and paste structure. It did, however, prove to be successful for comparing the properties of different types of cereal (wheat, corn, oat) (Doublier, Llamas, and Le Meur, 1987; Doublier, Paton, and Llamas, 1987) and legume starches (pea, Vicia faba) (Doublier, 1987). It provides a good basis for describing the structure of starch pastes in relation to the type of starch and the pasting procedure.

Rheology of Starch Gels

Starch gelation results, as starch pastes are cooled, from the reorganization (retrogradation) of the starch molecules. Gelation is a time-dependent process. From the above description of the structure of starch pastes, a cereal starch gel can be "regarded as a composite in which swollen granules are embedded in and reinforce a gel matrix" (Ring, 1985; Miles, Morris, and Ring, 1985b). The main parameters that influence the rheology of starch gels are related to three factors: the rheological characteristics of the matrix gel (primarily amylose), the volume fraction of the swollen particles, and the deformability of the particles. The gel matrix corresponds to the continuous phase of the pastes described previously. Its rheological characteristics are therefore dependent on the amylose concentration within this phase. Pioneering work (Ott and Hester, 1965) on amylose/waxy corn starch mixtures demonstrated that gel formation is related primarily to the amylose content of the continuous phase and the degree of granule swelling. Swollen granules were found to act as fillers in the amylose network. This observation was later confirmed in more detailed studies (Ring, 1985; Miles, Morris, Orford, and Ring, 1985).

Rheological investigations of starch gels have been performed mainly on highly concentrated gels (>10% starch). Experimental protocols have been diverse and included shear measurements in a U-tube (Ring, 1985; Ellis and Ring, 1985), shear oscillatory measurements (Wong and Lelievre, 1982a; Ellis and Ring, 1985; Miles, Morris, and Ring, 1985a), uniaxial compression (Bechtel, 1950; Cluskey et al., 1959; Krüsi and Neukom, 1984; Christianson et al., 1985, 1986), and shear creep-recovery experiments (Collison and Elton, 1961). Dynamic measurements are particularly useful in monitoring development of the network with time. For instance, recent work by Miles and colleagues (Miles, Morris, Orford, and Ring, 1985; Miles, Morris, and Ring, 1985a) utilized the Rank Brothers' Pulse Shearometer, in which the shear modulus is determined at a high frequency (200 Hz) through the application of a shear wave. This frequency is much higher than in usual dynamic measurements (<10 Hz). Comparison of data obtained with the pulse shearometer to measurements with a U-tube yielded good agreement, however, confirming the validity of high-frequency measurements. Examples are given in Figures 4-15 and 4-16 of two starch gels (concentrations, 10% and 20%) and two amylose gels (concentrations, 2.4% and 3.2%) (Miles, Morris, Orford, and Ring, 1985). Starch gels were obtained from starch pastes after rapid quenching at 26°C. A rapid increase in G' was observed for the 20% starch concentration. This increase was slower at 10%. A constant value of G' was reached after 150 minutes in both cases. Amylose gels were produced by quenching amylose solutions at 26°C. Here, the development of G' was rapid and reached a constant value after 100 minutes. Thus, variations in G' are comparable for amylose and starch gels. Figure 4-16 shows the long-term development of the same gels. The G' value for the amylose gels remained constant over this time, while G' values for the starch gels increased slowly with time. Heating the 20% starch gel to 95°C and then cooling it showed that long-term gelation was thermo-reversible (Miles, Morris, Orford, and Ring, 1985). A similar observation previously had been reported for more concentrated (>50%) wheat starch gels (Cluskey et al., 1959). The slow long-term process was ascribed to amylopectin crystallization, which occurred at a much slower rate than amylose gelation (Miles, Morris, Orford, and Ring, 1985, Ring et al., 1987).

The effects of gel concentration in the range 6–30% for wheat starch and corn starch have been investigated (Ring, 1985; Miles, Morris, Orford, and Ring, 1985) in shear experiments using the Ward and Saunders's

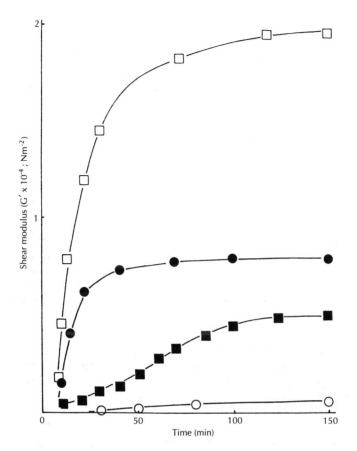

Figure 4–15. Development of the shear modulus, G', with time for 10% (■) and 20% (□) pea starch gels, and 2.4% (○) and 3.2% (●) amylose gels. (*From Miles, Morris, Orford, and Ring, 1985, with permission of Applied Science Publishers, Ltd.*)

U-tube device (Saunders and Ward, 1954). At concentrations (C_o) above 6%, variation of the shear modulus (G) with concentration was linear. Such a linear dependence contrasts with the concentration dependence of amylose gels, which was reported to be at the seventh power (Ellis and Ring, 1985). Below C_o, G could not be obtained due to a lack in the sensitivity of the measuring system. In these experiments, no difference was observed between wheat and corn starch gels (Miles, Morris, and Ring, 1985b). In another comparative study of 10% wheat, corn, and rice starch gels, no difference was found between wheat and corn starch gels, while rice starch resulted in a gel with a much lower rigidity modulus (Christianson et al., 1986). In these cases, data for corn and wheat starch gels did not depend on the pasting temperature

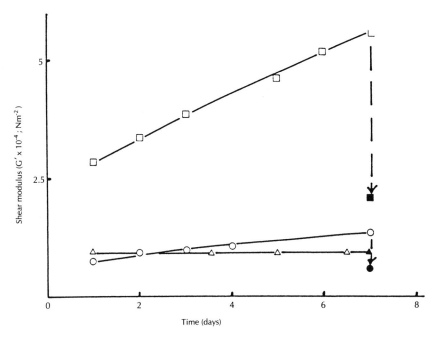

Figure 4–16. Long-time development of the shear modulus, G', for 10% (○) and 20% (□) starch gels and a 3.2% (△) amylose gel at 26° C. The dotted line indicates changes after heating to 90°C. and cooling. (*From Miles, Morris, Orford, and Ring, 1985, with permission of Applied Science Publishers Ltd.*)

(80°C, 94°C, or 121°C). This contrasts with rice starch, in which the rigidity decreased as pasting temperature increased.

Collison and Elton (1961) mentioned the dominant role of pasting temperature on the shear modulus of a 17% corn starch gel. A maximum was found at 90°C. It was suggested that below 90°C, the more "influential factor was adhesion of granules," while loss of granule integrity at temperatures above 90°C became the main phenomenon. It was also reported that the Young's modulus (E) of corn and wheat starch gels varies strongly with gel preparation conditions, specifically with the rate of heating and agitation (Doublier, Llamas, and Le Meur, 1987). Important differences in gel rigidity have also been found between wheat and corn starch. More rigid gels were obtained with wheat than with corn starch. This result was shown to be related to the rigidity of swollen particles. Hence, controversial results seem to arise from the comparison of data obtained by different authors. As for the properties of the starch pastes referred to earlier, it should be kept in mind that dispersed systems are so complex that a range of properties can be found, depending on a diversity of factors.

In a study on wheat starch gels of low concentration (<7%), it was shown that the Avrami equation could be employed successfully (Wong and Lelievre, 1982a):

$$\frac{G_e' - G_t'}{G'_e - G'_o} = \exp(-kt^a) \qquad (4\text{--}13)$$

Analysis of changes in G' with time showed that the process should be separated in two steps, with $a = 1$ in the first period and $a = 2$ in the second one. An a value of 1 is interpreted in the Avrami analysis as the result of a nucleation process. Highly concentrated starch gels (>50%) have been investigated with the aim of studying bread staling by following the same approach (Krüsi and Neukom, 1984). In this case, Avrami exponents were lower than 1, a result that can not be given a physical explanation. The Avrami description is mainly applicable to the crystallization of synthetic polymers, assuming the presence of only one macromolecular species. Since starch gelation involves two separate polysaccharides, the Avrami treatment must be regarded at best as an empirical means to describe the kinetics of starch gelation.

Gelation of amylose has been investigated by using various physical techniques, including light scattering, turbidity measurements, dilatometry, rheological measurements, and X-ray diffraction (Ellis and Ring, 1985; Miles, Morris, and Ring, 1985a). Opaque elastic gels were obtained by cooling amylose solutions. A weak X-ray diffraction pattern of the B-type developed, confirming a partial crystallization of amylose molecules in the gel. It was suggested (Miles, Morris, and Ring, 1985a) that amylose gelation arose from a phase separation into a polymer-rich phase and a polymer-deficient phase. The gel resulted from the crystallization of amylose within the polymer-rich region, and it was found to be thermally irreversible at temperatures below 100°C. It was postulated that amylose gelation would dominate the short-term development of the gel structure (as in Fig. 4-15). Because amylopectin molecules are concentrated within the swollen granules, it is within the granules that crystallization and gelation of amylopectin take place. It was shown (Ring et al., 1987) that gelation of amylopectin is a thermoreversible process at temperatures below 100°C. Hence, the long-term change in modulus as illustrated in Figure 4-16 for a 20% starch gel can be related to gelation of amylopectin within the granule.

The above descriptions assume an ideal situation in which amylose and amylopectin are totally separated during the pasting process. It was shown previously that other situations can occur with mixtures of amylose and amylopectin present either within the continuous phase or within the dispersed phase. Kalichevsky and Ring (1987) found that amylose and amylopectin, although both α-glucans, exhibit thermodynamic incompatibility in aqueous solution. This incompatibility results in the immiscibility of both components and encourages a phase separation if they are present in the same phase. As a result, retrogradation and gelation of amylose should be influenced by the presence of amylopectin or vice-versa. This is the case in most practical situations. The implications of such effects should thus be taken into account in studies of starch gelation.

OTHER CEREAL CARBOHYDRATES

Cereal $\beta(1{\rightarrow}4)$ $(1{\rightarrow}3)$-D-glucans

Cereal $\beta(1{\rightarrow}4)$ $(1{\rightarrow}3)$-D-glucans are present in most cereal grains but occur in larger amounts in barley and oat (Preece and Hobkirk, 1953; Wood, 1984). For the sake of convenience these components will be referred to as β-glucans. Details on the chemical structure of these high molecular weight components can be found in a review by Wood (1984). Briefly, they are linear chains with $(1{\rightarrow}3)$ and $(1{\rightarrow}4)$-β-glucosidic linkages in the average ratio 30:70. This average ratio would correspond to cellotriosyl and cellotetraosyl $(1{\rightarrow}4)$ units linked to each other via $(1{\rightarrow}3)$ bonds. The regularity of this structure, however, is far from being confirmed, and there is some evidence of the presence of longer $(1{\rightarrow}4)$-β-linked backbones (five to eleven units). The high solubility of β-glucans as compared with cellulose (which has only β linkages and is insoluble in water) results from the presence of $(1{\rightarrow}3)$ β bonds, which introduce irregularity into the structure. Molecular weights of β-glucans have been determined in some studies (Forrest and Wainwright, 1977; Woodward et al., 1983). Values ranging from 3×10^5 to 4×10^7 have been reported. It is difficult to compare data due to the diversity of the methods, however. Barley β-glucans have reportedly caused problems in the brewing process when they were present in high amounts (usually 0.5%). These problems were related to the high viscosity of such polysaccharides in solution. Oat β-glucans were also shown to yield high-viscosity solutions, and it has been suggested that they could be used as thickening agents in food formulations (Wood, 1984; Autio et al., 1987). Very few reports on the rheology of these molecules exist, however. Relative viscosities of solutions at 0.5% concentration reported in the literature are summarized in Table 4-2 (Wood et al., 1978; Wood, 1984). All viscosities were obtained by using capillary viscometers. Relative viscosities were very low, on the order of 2–5, except for data reported by Clarke and Stone (1966), which were two orders of magnitude higher. Relative viscosities higher than 1,000 were also found (Wood et al., 1978), but the authors reported difficulties in the accurate determination of such high viscosities with capillary viscometers. The accuracy of the determinations is therefore questionable. Such large differences in relative viscosities probably account for large differences in calculated molecular weights. It must be

Table 4-2. Aqueous Viscosities of Oat and Barley β-glucans (concentration: 0.5%)

Sample	Temp. (°C)	Relative Viscosity [a]	References
Oats	25	1.1	Preece and MacKenzie (1952)
	25	2.4	Preece and Hobkirk (1953)
	25	262	Clarke and Stone (1966)
Barley	25	2.9	Preece and Hobkirk (1953)
	25	3.11	Preece (1957)
	30	3.0	Meredith et al. (1951)
	25	295	Clarke and Stone (1966)
	20	2–7	Scott (1972)
	20	3	Morgan et al. (1983)

Source: From Wood (1984) with permission of AACC.

[a] $\eta_{solution}/\eta_{water}$

emphasized that these data may result from measurements at different shear rates. An oat gum sample, composed of about 80% β-glucan, was shown to display non-Newtonian behavior at concentrations above 0.2% (Autio et al., 1987). Comparison of the data in Table 4-2 may therefore be irrelevant, since the experimental shear rates were not specified and could vary considerably.

The flow properties of oat gums have been described by using coaxial cylinder viscometers (Wood, 1984; Autio et al., 1987). Examples for oat gum in solution in water (concentration 1%), in sucrose solution, and in 1 M NaCl are given in Figure 4-17 (Wood, 1984). Comparison with a medium-viscosity hydroxyethylcellulose illustrates the high viscosity of the oat gum sample. Shear thinning behavior is evident. It was also reported that no thixotropy was evident over an increasing and decreasing shear rate cycle. Comparable observations were reported by Autio and co-workers (1987). In both series of data, flow curves were described by using the power law equation (Eq. 4-1). The power law parameters were compared to data for guar gum found in the literature (Elfak et al., 1979). It must be emphasized that the shear rate ranges employed for the viscosity measurements were different: 200–1,600 s^{-1} and 20–200 s^{-1}, respectively, for oat gums (Wood, 1984; Autio et al., 1987) and 14–1,000 s^{-1} for the galactomannan solution (Elfak et al., 1979). A comparison of data in terms of the power law parameters k and n is thus incorrect, because they depend on the shear rate range. The main conclusion that can be drawn from these results is that oat gum, and hence β-glucans, exhibit strong shear thinning behavior and no thixotropy. They can rival other polysaccharides, galactomannans for instance, with respect to their thickening properties.

A better description of flow properties would require measurements over a large range of shear rates, as has been performed extensively for other hydrocolloids (see Morris and Ross-Murphy, 1981; Launay et al., 1986). The authors of this chapter carried out this type of measurement for an oat gum

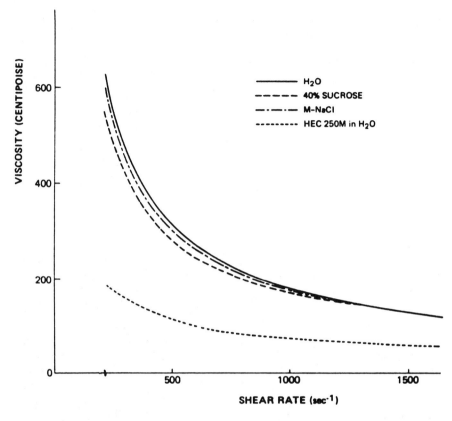

Figure 4–17. Flow curves (in normal coordinates) of solutions of oat gum (1%) in water, 40% sucrose, and 1 M NaCl; comparison with hydroxyethylcellulose (1%) in water; temperature: 25°C. (*From Wood, 1984, with permission of AACC, Inc.*)

sample ($[\eta]$ = 5.6 dl/g) over a shear rate range of 10^{-2}–10^{2} s^{-1} and for concentrations ranging between 0.1% and 2% (Doublier, unpublished). As for the other polysaccharides, non-Newtonian behavior was observed at a concentration of about 0.4%. For concentrations higher than this, the flow curves displayed shear thinning behavior. A limiting Newtonian viscosity η_0 at low shear rates was found. The η_0 variations with concentration were linear at low concentrations (below 0.3%) and were very steep at concentrations above 0.5%. By using the so-called reduced concentration $C[\eta]$, as classically performed for polysaccharides in solution (Launay et al., 1986), the η_0-$C[\eta]$ curve was shown to be close to the curve for galactomannans (Robinson et al., 1982). The viscoelastic behavior of oat gum in solution also resembles that displayed by galactomannans. This resemblance is illustrated in Figure 4-18 for solutions with comparable reduced concentrations ($C[\eta]$ = 11.2 and 13 for oat gum and guar gum, respectively). The G' and G'' variations with frequency are typical

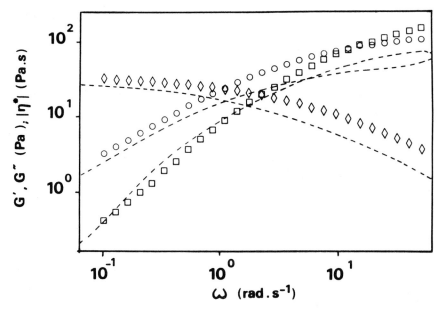

Figure 4–18. Frequency dependence of storage modulus, G' (□), loss modulus, G'' (○), and complex viscosity modulus, $|\eta|^*$ (◇), of an oat gum solution (2%, $c[\eta]$ = 11.2); comparison with a guar gum solution (1.5%, $c[\eta]$ = 13) (dashed lines). (*From J.-L. Doublier, unpublished*)

of polysaccharide solutions where no gelation is exhibited (Morris and Ross-Murphy, 1981). The overall conclusion from this investigation confirms, as in the above-mentioned studies, that the flow and viscoelastic properties of β-glucans are comparable to those of other polysaccharides used as thickening agents. These properties are known to be directly related to the molecular weights of the macromolecules and to depend strongly upon concentration above the so-called entanglement concentration (Morris and Ross–Murphy, 1981; Robinson et al., 1982; Launay et al., 1986). A more systematic study is required, however, to elucidate more details on the possible specific properties of β-glucans as compared with those of other polysaccharides.

Water-Soluble Pentosans

The amount of published data related to the properties in aqueous solution of isolated water-soluble pentosans from cereals is small. These components are mainly composed of arabinoxylans and arabinogalactans (Neukom, 1976; Hoseney, 1984). Arabinoxylans are high molecular weight linear polysaccharides. The main chain of this polysaccharide is composed of (1-4)-β-D-xylose units; single α-L-arabinose side groups are attached to the main chain. The presence of these short side groups causes arabinoxylans to be water soluble, and reportedly to form clear solutions as do the other water-soluble hydrocolloids. No rheological work on arabinoxylans has been published to date, however. Arabinogalactans are, on another hand, highly branched and do not develop interesting rheological properties in aqueous media.

Arabinoxylans have been more specifically investigated for their gelling ability upon the action of oxidative agents (Neukom, 1976; Hoseney, 1984). This ability has been related to the presence of small amounts of ferulic acid, which is esterified to the largest part of the arabinoxylan fraction (Neukom, 1976; Yeh et al., 1980). The mechanism of this oxidative gelation has been thoroughly investigated (Neukom, 1976; Yeh et al., 1980; Hoseney and Faubion, 1981; Hoseney, 1984), since it is of importance to the baking industry. It has been suggested that this gelation process would result from cross-linking of arabinoxylan chains through the ferulic acid residues (Neukom, 1976). Another interpretation would involve a covalent binding of protein with the arabinoxylan chain via a ferulic acid group (Hoseney and Faubion, 1981). Most of the rheological studies have been dealing with the implications of oxidative gelation on the rheology of dough (Hoseney and Faubion, 1981; Jackson and Hoseney, 1986), which is beyond the scope of this chapter.

Some investigations have been reported on the behavior of arabinoxylans in dilute aqueous solution. Molecular weight values were reported to be slightly lower than those of β-glucans (Forrest and Wainwright, 1977). Based on a chromatographic method and by using a similar "universal calibration," two relationships of $[\eta]$ and $\overline{M}_w$ were found for an arabinoxylan sample from rye flour, depending upon the molecular weight range (Anger et al., 1986):

$$[\eta] = 3.47 \times 10^{-3} \overline{M}_w^{0.98} \qquad \text{for } \overline{M}_w < 2 \times 10^5 \qquad (4\text{--}14)$$

$$[\eta] = 29.5 \overline{M}_w^{0.23} \qquad \text{for } \overline{M}_w > 2 \times 10^5 \qquad (4\text{--}15)$$

The exponent b in Equation 4-14 is not too far from typical values for linear polysaccharides (Launay et al., 1986) ($b = 0.8$). The exponent in the second equation ($b = 0.23$) seems more surprising, since it is on the order of values reported for high molecular weight branched polysaccharides (for instance, $b = 0.28$ for branched dextrans) (Granath, 1958). These data can not be discussed further, however, and a confirmation of this series of measurements is required. The only viscosity measurements available for arabinoxylans in dilute solution are too sparse to consider here.

Starch Hydrolysates

The term starch hydrolysates refers to starch products obtained from extensive hydrolysis of starch through an acid, acid/enzymic, or enzymic reaction. Starch hydrolysates are classified according to the degree of conversion on the basis of the dextrose equivalent (DE<100); that is, the higher the DE, the more hydrolyzed the sample. When the DE is lower than 20, products are referred to as maltodextrins or low-converted hydrolysates. Values higher than 20 define starch syrups. The average molecular weights of these compounds are always low, and they do not display complex rheological properties in solution. Highly concentrated solutions can be prepared that exhibit Newtonian behavior. Viscosity is of major importance in food uses, however.

The Hoeppler Viscometer was reported to be convenient over wide ranges of concentrations (20–80%), temperatures (15–82°C) and DE (17–100) (Erickson et al., 1986). Data for acid or enzyme hydrolyzed corn starches within these limits are available (Cakebread, 1971a; Murray and Luft, 1973; Kearsley and Birch, 1977; Jackson, 1985a; Erickson et al., 1986). Illustrations of the dependence of viscosity on concentration and DE are given in Figures 4-19 and 4-20 (Murray and Luft, 1973; Erickson et al., 1986). Figure 4-19 also includes maltodextrins (DE < 20). It is clear that the viscosity variation with DE is stronger within the range of DE 5–20 than above 20. It is also within this range that the molecular weight varies dramatically. It must be emphasized that the DE is an average parameter that cannot reflect the width of molecular weight distribution, which is known to vary depending upon the hydrolysis method. For example, a 42% DE syrup containing 5% higher molecular weight oligosaccharides (>25,000) was shown to yield a solution 50% more viscous than a syrup with only 2% high molecular weight oligomers (Jackson, 1985b).

Glucose syrups and maltodextrins are currently employed in many food applications (confectionery, preserves, fillings, frozen desserts, beverages, etc.) (MacDonald, 1984). In each situation, viscosity is one of the main parameters that must be controlled. High-boiling sweets is probably the field in which viscosity is of particular importance; it is at the basis of graining (sucrose crystallization), which must be either prevented or carefully adjusted. In producing high-boiling sweets, glucose syrups are employed as mixtures, with sucrose at a high total solids content (usually > 90%). Using a rotational viscometer (Rotovisko), Völker (1969) measured the viscosity of sucrose/glucose mixtures (100/50 to 100/100 ratios) at temperatures ranging between 110 and 147°C and with a moisture content of 2%. Viscosities ranged between 300 Pa·s and 10 Pa·s, depending on temperature, sucrose/glucose syrup ratio, DE, and method of preparing the glucose syrup. This last parameter was reported to play a major role at this very high solids content, in relation to the above-mentioned content of higher molecular weight oligosaccharides. These data were discussed further in the context of "graining control" during the glass transition that occurs on cooling (Cakebread, 1971b). Viscosity data obtained at high temperatures were related to the final viscosity of the "glassy state," estimated to be as high as 10^{11} Pa·s. A relationship with the glass transition temperature (T_g) and water activity was also reported. Problems associated with viscosity measurements in this field are not well resolved. They

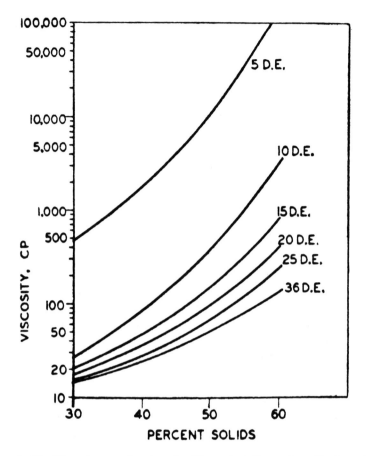

Figure 4–19. Viscosity as a function of solids content (in semilogarithmic scales) of maltodextrins (DE < 20) and glucose syrups; temperature: 27°C. (*From Murray and Luft, 1973, with permission of IFT*)

are different from those associated with macromolecular systems. Viscosities higher than 10^4 Pa·s are poorly measured with conventional viscometers and, to date, the only estimates available for more viscous glucose syrup solutions have been made through indirect means—by extrapolation, for example.

CONCLUSION

This chapter discusses relationships between rheological properties and the molecular structure of aqueous systems formed by cereal carbohydrates. Although empirical methods have been used more often, no review of the data obtained by these means was attempted, because rheological parameters

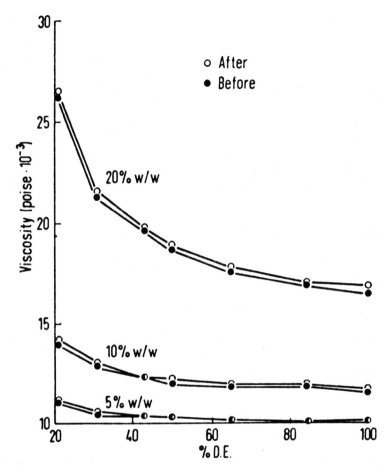

Figure 4–20. Viscosity variations of glucose syrups as a function of DE before (●) and after (○) hydrogenation; temperature: 25°C. (*From Kearsley and Birch, 1977, with permission of Verlag Chemie, Gmbh*)

in these instances are not well defined. A wide range of rheological behaviors is shown by these carbohydrates, from the Newtonian behavior of glucose syrups to the quasi-solid behavior of highly concentrated starch gels. Intermediate situations exist in the cases of macromolecular solutions such as those of β-glucans and suspensions such as starch pastes. Each system must be characterized in an appropriate way.

The description of starch systems (pastes and gels) remains qualitative. Viscosity and viscoelastic measurements of pastes allowed interesting bases for interpreting differences arising from botanical origins and preparation conditions; that is, the volume fractions of the dispersed phases and the compositions of the continuous phases. Moreover, such a description provides a good basis for describing starch gels as composites. Many points must be considered in detail within these descriptions, however. The results discussed above were limited to native starches. Chemically modified starches should be investigated in the same way. Similarly, the rheology of starch in the presence of other food components (e.g., monoacetylated lipids, hydrocolloids, and proteins) has to be described on a structural basis.

β-glucans are likely to exhibit behavior similar to those of other food thickening polysaccharides such as galactomannans. The methodology usually employed to describe the rheology of polysaccharides should be used more extensively. The results must be complemented by those of studies involving the effects of molecular weight (known to be the chief factor determining the rheology of such noninteracting macromolecular systems). The rheology of water-soluble pentosans is believed to be comparable to that of β-glucans, but this remains to be confirmed. In the cases of these two polysaccharides, a problem arises from their limited availability. Since they are not produced on an industrial scale, the large amounts of material required for rheological measurements are difficult to obtain.

Problems inherent in the rheological characterization of glucose hydrolysates are completely different. Low-concentration solutions are easily measured with commercially available instruments. The main difficulties are encountered with highly concentrated systems. These systems should be characterized by methods developed for use with synthetic polymers to investigate the glass transition.

REFERENCES

AFNOR, 1966. *Amidons et fécules. Détermination des coefficients de viscosité des empois.* Norme N.F.V. 03-604.

Anger, H., Dörfer, J., and Berth, G. 1986. Untersuchungen zur molmasse und grenzviskosität von arabinoxylan (pentosan) aus roggen (secale cereale) zur aufstellung der Mark-Houwink-Beziehung. *Die Nahrung* **30**:205–208.

Autio, K., Myllymäki, O. and Mälkki, Y. 1987. Flow properties of solutions of oat β-glucans. *J. Food Sci.* **52**:1364–1366.

Bagley, E. B., and Christianson, D. D. 1982. Swelling capacity of starch and its relationship to suspension viscosity. Effect of cooking time, temperature and concentration. *J. Text. Stud.* **13**:115–126.

Bagley, E. B., and Christianson, D. D. 1983. Yield stresses in cooked wheat starch dispersions. *Staerke* **35**:81–86.

Bagley, E. B., Christianson, D. D., and Beckwith, A. C. 1983. A test of the Arrhenius viscosity volume fraction relationship for concentrated dispersions of deformable particules. *J. Rheol.* **27**:503–507.

Bechtel, W. G. 1950. Measurement of properties of corn starch gels. *J. Colloid Sci.* **5**:260–270.

Blanshard, J. M. V. 1979. Physical aspects of starch gelatinization. In *Polysaccharides in Foods*, ed. J. M. V. Blanshard and J. R. Mitchell, pp. 139–152. London: Butterworths.

Blanshard, J. M. V. 1987. Starch granule and function. In *Starch: Properties and Potential*. Critical Reports on Applied Chemistry, Vol. 13, ed. T. Galliard, pp. 16–54. New York: Wiley.

Bowler, P., Williams, M. R., and Angold, R. E. 1980. A hypothesis for the morphological changes which occur on heating lenticular wheat starch in water. *Staerke* 32(6):186–189.

Bulpin, P. V., Welsh, E. J., and Morris, E. R. 1982. Physical characterization of amylose–fatty complexes in starch granules and in solution. *Staerke* 34:335–339.

Cakebread, S. H. 1971a. Physical properties of confectionery ingredients. Viscosity of carbohydrate solutions. *Confectionery Prod.* 37:662–665.

Cakebread, H. H. 1971b. Physical properties of confectionery ingredients. Viscosity and high boiling mixtures of high solids content at high temperature. *Confectionery Prod.* 37:705–709.

Christianson, D. D., and Bagley, E. G. 1983. Apparent viscosities of dispersion of swollen cornstarch granules. *Cereal Chem.* 60(2):116–121.

Christianson, D. D., and Bagley, E. B. 1984. Yield stresses in dispersions of swollen, deformable corn starch granules. *Cereal Chem.* 61:500–503.

Christianson, D. D., Baker, F. L., Loffredo, A. R. and Bagley, E. B. 1982. Correlation of microscopic structure of corn starch granules with rheological properties of cooked pastes. *Food Microstruct.* 1:13–24.

Christianson, D. D., Casiraghi, E. M., and Bagley, E. B. 1985. Uniaxial compression of bonded and lubricated gels. *J. Rheol.* 29:671–684.

Christianson, D. D., Casiraghi, E. M., and Bagley, E. B. 1986. Deformation and fracture of wheat, corn and rice starch gels in lubricated and bonded uniaxial compression. *Carbohydr. Polymers* 6:335–348.

Clarke, A. E., and Stone, B. A. 1966. Enzymic hydrolysis of barley and other β-glucans by a $\beta(1\rightarrow4)$-glucan hydrolase. *Biochem. J.* 99:582–588.

Cluskey, J. E., Taylor, N. W., and Senti, F. R. 1959. Relations of the rigidity of flour, starch, and gluten gels to bread staling. *Cereal Chem.* 36:236–246.

Collison, R. 1968. Swelling and gelation of starch. In *Starch and Its Derivatives*, ed. J. A. Radley, pp. 168–193. London: Chapman and Hall, Ltd.

Collison, R., and Elton, G. A. H. 1961. Some factors which influence the rheological properties of starch gels. *Staerke* 13:164–173.

Cruz, A., Russel, W. B., and Ollis, D. F. 1976. Shear viscosity of native and enzyme hydrolyzed amioca starch pastes. *AICHE J.* 22(5):832–840.

De Willigen, A. H. A. 1976. The rheology of starch. In *Starch and Its Derivatives*, ed. J. A. Radley, pp. 168–193. London: Chapman and Hall, Ltd.

Dengate, H. N. 1984. Swelling pasting and gelling of wheat starch. In *Advanced Cereal Science Technology*, Vol 6, ed. Y. Pomeranz, pp. 49–82. St. Paul, MN: AACC.

Djaković, L., and Dokić, P. 1972. Die rheologische charakterierung der stärkegele. *Staerke* 24:195–201.

Doublier, J. L. 1981. Rheological studies on starch. Flow behaviour of wheat starch pastes. *Staerke* 33(12):415–420.

Doublier, J. L. 1987. A rheological comparison of wheat, maize, faba bean and smooth pea starches. *J. Cereal Sci.* **5**:247–262.

Doublier, J. L., Llamas, G., and Le Meur, M. 1987. A rheological investigation of cereal starch pastes and gels. Effect of pasting procedures. *Carbohydr. Polymers* **7**:251–275.

Doublier, J. L., Paton, D., and Llamas, G. 1987. A rheological investigation of oat starch pastes. *Cereal Chem.* **64**(1):21–26.

Elfak, A. M., Pass, G., and Phillips, G. O. 1979. The effect of shear rate on the viscosity of solutions of guar gum and locust bean gum. *J. Sci. Food Agric.* **30**:439–444.

Eliasson, A. C. 1986. Viscoelastic behaviour during the gelatinization of starch. I. Comparison of wheat, maize, potato and waxy-barley starches. *J. Text. Stud.* **17**:253–265.

Eliasson, A. C., and Krog, N. 1985. Physical properties of amylose–monoglycerides complexes. *J. Cereal Sci.* **3**:239–248.

Eliasson, A. C., Finstad, H., and Ljunger, G. 1988. A study of starch–lipid interactions for some native and modified maize starches. *Staerke* **40**(3):95–100.

Ellis, H. S., and Ring, S. G. 1985. A study of some factors influencing amylose gelation. *Carbohydr. Polymers* **5**:201–213.

Erickson, E. R., Bernsten, R. A., and Eliason, M. A. 1986. Viscosity of corn syrup. *J. Chem. Eng. Data* **11**:485–488.

Evans, I. D., and Haisman, D. R. 1979. Rheology of gelatinised starch suspensions. *J. Text. Stud.* **10**:347–370.

Farrow, F. D., and Lowe, G. M. 1923. The flow of starch paste through capillary tubes. *J. Text. Inst.* **14**:414–440.

Farrow, F. D., Lowe, G. M. and Neale, S. M. 1928. The flow of starch pastes. Flow at high and low rates of shear. *J. Text. Inst.* **19**:18–31.

Forrest, I. S., and Wainwright, T. 1977. The mode of binding of β-glucans and pentosans in barley endosperm cell walls. *J. Inst. Brew.* **83**:279–286.

Ghiasi, K., Hoseney, R. C., and Varriano-Marston, E. 1982. Gelatinization of wheat starch. I. Excess-water systems. *Cereal Chem.* **59**(2):81–85.

Granath, K. A. 1958. Solution properties of branched dextrans. *J. Colloid Sci.* **13**:308–328.

Guilbot, A., and Mercier, C. 1985. Starch. In *The Polysaccharides*, Vol. 3, ed. G. O. Aspinall, pp. 201–282. New York: Wiley.

Hoseney, R. C. 1984. Functional properties of pentosans in baked foods. *Food Technol.* **38**:114–116.

Hoseney, R. C., and Faubion, J. M. 1981. A mechanism for the oxidative gelation of wheat flour water-soluble pentosans. *Cereal Chem.* **58**:421–424.

Jackson, B. 1985a. Application of glucose syrups in modern confectionery manufacture. I. *Confectionery Manufacture Marketing* **22**:12–14.

Jackson, B. 1985b. Glucose syrups. II. *Confectionery Manufacture Marketing* **22**:30–34.

Jackson, G. M., and Hoseney, R. C. 1986. Fate of ferulic acid in overmixed wheat flour doughs: partial characterization of a cystein-ferulic acid adduct. *J. Cereal Sci.* **4**:87–95.

Juliano, B. O. 1984. Rice starch: production, properties and uses. In *Starch: Chemistry and Technology*, 2nd ed., ed. R. L. Whistler, J. N. BeMiller, and E. F. Paschall, pp. 507–528. New York: Academic Press.

Kalichevsky, M. T., and Ring, S. G. 1987. Incompatibility of amylose and amylopectin in aqueous solution. *Carbohydr. Res.* **162**:323–328.

Kearsley, M. W., and Birch, G. G. 1977. Production and physicochemical properties of hydrogenated glucose syrups. *Staerke* **29**:425–429.

Krüsi, H., and Neukom, H. 1984. Untersuchunger über die retrogradation der stärke in konzentrienten weizenstärkegelen. Teil 1. *Staerke* **36**:40–45.

Kubota, K., Hosokawa, Y., Suzuki, K., and Hosaka, H. 1978. Determination of viscometric constants in empirical flow equations of heated starch solutions. *J. Fac. Fish Animal Husb. Hiroshima Univ.* **17**:1–15.

Kubota, K., Hosokawa, Y., Suzuki, K., and Hosaka, H. 1979. Studies of the gelatinization rate of rice and potato starches. *J. Food Sci.* **44**:1394–1397.

Kugimiya, M., and Donovan, J. W. 1980. Phase transitions of amylose–lipid complexes in starches: a calorimetric study. *Staerke* **32**:265–270.

Lancaster, E. B., Conway, H. F., and Schwab, F. 1966. Power-law rheology of alkaline starch pastes. *Cereal Chem.* **43**:637–643.

Launay, B., Doublier, J.-L., and Cuvelier, G. 1986. Flow properties of aqueous solutions and dispersions of polysaccharides. In *Functional Properties of Food Macromolecules*, ed. J. R. Mitchell and D. A. Ledward, pp. 1–78. London: Elsevier.

Leach, H. W. 1965. Gelatinization of starch. In *Starch: Chemistry and Technology. I. Fundamental Aspects*, ed. R. L. Whistler and E. F. Paschall, pp. 289–307. New York: Academic Press.

Leach, H. W., McCowen, L. D., and Schoch, T. J. 1959. Structure of the starch granule. 1. Swelling and solubility patterns of various starches. *Cereal Chem.* **36**:534–544.

Lindahl, L., and Eliasson, A. C. 1986. Effect of wheat proteins on the viscoelastic properties of starch gels. *J. Sci. Food Agric.* **37**:1125–1132.

MacDonald, M. 1984. Uses of glucose syrups in the food industry. In *Glucose Syrups: Science and Technology*, ed. S. Z. Dziedzic and M. W. Kearsley, pp. 247–263. London: Elsevier.

Meredith, W. O. S., Base, E. J., and Anderson, J. A. 1951. Some characteristics of barley, malt and wort gums. *Cereal Chem.* **28**:177–188.

Miles, M. J., Morris, V. J., and Ring, S. G. 1985a. Gelation of amylose. *Carbohydr. Res.* **135**:257–269.

Miles, S. G., Morris, V. J., and Ring, S. G. 1985b. Recent observations on starch retrogradation. In *New Approaches to Research on Cereal Carbohydrates*, ed. R. D. Hill and L. Munck, pp. 109–114. Amsterdam: Elsevier.

Miles, M. J., Morris, V. J., Orford, P. D., and Ring, S. G. 1985b. The roles of amylose and amylopectin in the gelation and retrogradation of starch. *Carbohydr. Res.* **135**:271–281.

Miller, B., Derby, R. I., and Trimbo, H. B. 1973. A pictorial explanation for the increase on viscosity of a heated wheat starch–water suspension. *Cereal Chem.* **50**:271–280.

Montgomery, E. M., and Senti, F. R. 1958. Separation of amylose from amylopectin by an extraction–sedimentation procedure. *J. Polymer Sci.* **28**:1–9.

Morgan, A. G., Gill, A. A., and Smith, D. B. 1983. Some barley grain and green malt properties and their influence on malt hot water extract. I. β-glucan, β-glucan solubilase and endo-β-glucanase. *J. Inst. Brew.* **89**:283–291.

Morris, E. R., and Ross-Murphy, S. B. 1981. Chain flexibility of polysaccharides and glycoproteins from viscosity measurements. *Tech. Carbohydr. Metabolism* **B310**:1–46.

Murray, D. G., and Luft, L. T. 1973. Low DE corn starch hydrolyzates. *Food Technol.* **27**:32–40.

Myers, R. R., Knauss, C. J., and Hoffman, R. D. 1962. Dynamic rheology of modified starches. *J. Appl. Polymer Sci.* **6**:659–666.

Myers, R. R., and Knauss, C. J. 1964. Dynamic mechanical properties. Determination of viscosity and rigidity of starch pastes. In *Methods in Carbohydrate Chemistry*, Vol. 4, ed. R. L. Whistler, and M. L. Wolfrom, pp. 128–133, New York: Academic Press.

Myers, R. R., and Knauss, C. J. 1965. Mechanical properties of starch pastes. In *Starch: Chemistry and Technology*, Vol 1., ed. R. L. Whistler and E. F. Paschall, pp. 393–407. New York: Academic Press.

Nedonchelle, Y. 1968. Sur la rhéologie des solutions concentrées de carbohydrates macromoléculaires. Dissertation. Université de Strasbourg.

Neukom, H. 1976. Chemistry and properties of the non-starchy polysaccharides (NSP) of wheat flour. *Lebensmitt. Wiss. Technol.* **9**:143–148.

Ollku, J., and Rha, C. K. 1978. Gelatinization of starch and wheat flour starch. *Food Chem.* **3**:293–317.

Ott, M., and Hester, E. E. 1965. Gel formation as related to concentration of amylose and degree of starch swelling. *Cereal Chem.* **42**:476–484.

Preece, I. A., and MacKenzie, K. G. 1952. Non-starch polysaccharides of cereal grains. II. Distribution of water-soluble gum-like materials in cereals. *J. Inst. Brew.* **58**:457–464.

Preece, I. A., and Hobkirk, R. 1953. Non-starchy polysaccharides of cereal grains. III. Higher molecular gums of common cereals. *J. Inst. Brew.* **59**:385–392.

Preece, I. A. 1957. Malting relationship of barley polysaccharides. *Wallersteim Lab. Commun.* **20**:147–161.

Radley, J. A. 1976. Physical methods of characterising starch. In *Examination and Analysis of Starch and Starch Products*, ed. J. A. Radley, pp. 91–131. London: Elsevier.

Richardson, W. A., and Waite, R. 1933. The flow of starch pastes. The effects of soaps and other electrolytes on the apparent viscosity of starch pastes. *J. Text. Inst.* **24**:383–416.

Ring, S. G. 1985. Some studies on starch gelation. *Staerke* **37**:80–83.

Ring, S. G., Colonna, P., I'anson, K. J., Kalichevsky, M. T., Miles, M. J., Morris, V. J., and Ring, S. G. 1987. The gelation and crystallisation of amylopectin. *Carbohydr. Res.* **162**:277–293.

Robinson, G., Ross-Murphy, S. B., and Morris, E. R. 1982. Viscosity–molecular weight relationships, intrinsic chain flexibility, and dynamic solution properties of guar galactomannans. *Carbohydr. Res.* **107**:17–32.

Saunders, R. P., and Ward, A. G. 1954. An absolute method for the rigidity modulus of gelation gels. *Proceedings, 2nd International Congress Rheol. Oxford*, ed. V. C. G. Harrison, pp. 284–290. Pergamon.

Schutz, R. A. 1970. De la rhéologie de systèmes aqueux à base de gomme. *Staerke* **22**:116–125.

Schutz, R. A. 1974. *Die rheologie auf dem stärkegebeit*. Berlin: Verlag Paul Parey.

Scott, R. W. 1972. The viscosity of worts in relation to their content of β-glucan. *J. Inst. Brew.* **78**:179–186.

Sterling, C. 1978. Textural qualities and molecular structure of starch products. *J. Text. Stud.* **9**:225–255.

Stute, R., and Konieczny-Janda, G. 1983. DSC—untersuchungen an stärken. Teill II. Untersuchungen an Stärke-lipid-complexen. *Staerke* **35**:340–347.

Varriano-Marston, E., Zeleznak, K., and Nowota, A. 1985. Structural characteristics of gelatinized starch. *Staerke* **37**(10):326–329.

Völker, H. H. 1969. Die eingluss von glukosesirupen auf die beschaffenheit von hartkaramelmassen. *Zucker Suesswarenwirtschaft* **22**:114–118.

Williams, M. R., and Bowler, P. 1982. Starch gelatinization: a morphological study of triticeae and other starches. *Staerke* **34**(7):221–223.

Wong, R. B. K., and Lelievre, J. 1981. Viscoelastic behaviour of wheat starch pastes. *Rheol. Acta* **20**:299–307.

Wong, R. B., and Lelievre, J. 1982a. Effects of storage on dynamic rheological properties of wheat starch pastes. *Staerke* **34**:231–233.

Wong, R. B. K., and Lelievre, J. 1982b. Rheological characteristics of wheat starch pastes measured under steady shear conditions. *J. Appl. Polymer Sci.* **27**:1433–1440.

Wood, P. J., Siddiqui, I. R., and Paton, D. 1978. Extraction of high-viscosity gums from oats. *Cereal Chem.* **55**:1038–1049.

Wood, P. J. 1984. Physicochemical properties and technological and nutritional significance of cereal β-glucans. In *Cereal Polysaccharides in Technology and Nutrition*, ed. V. F. Rasper, pp. 52–57. St. Paul, MN: AACC.

Woodward, J. R., Phillips, D. R., and Fincher, G. B. 1983. Water soluble (1-3) (1-4)-β-D-glucans from barley (*Hordeum vulgare*) endosperm. I. Physicochemical properties. *Carbohydr. Polymers* **3**:143–156.

Yeh, Y. F., Hoseney, R. C., Lineback, D. R. 1980. Changes in wheat flour pentosans as a result of dough mixing and oxidation. *Cereal Chem.* **57**:144–148.

Zobel, H. F. 1984. Gelatinization of starch and mechanical properties of starch pastes. In *Starch Chemistry and Technology*, 2d ed., ed. J. A. Radley, pp. 49–82. St. Paul, MN: AACC.

Chapter 5

Influences of the Glassy and Rubbery States on the Thermal, Mechanical, and Structural Properties of Doughs and Baked Products

Harry Levine and Louise Slade

The technological importance of the glass transition in amorphous polymers and the characteristic temperature at which it occurs (the glass transition temperature, Tg) is well known as a key aspect of synthetic polymer science (Ferry, 1980; Rowland, 1980; Sears and Darby, 1982). Eisenberg (1984) has stated that "the glass transition is perhaps the most important single parameter which one needs to know before one can decide on the application of the many noncrystalline [synthetic] polymers that are now available." Especially in the last several years, a growing number of food scientists have followed the compelling lead of the synthetic polymers field by increasingly recognizing the practical significance of the glass transition as a physicochemical event that can govern food processing, product properties, quality, and stability (Slade, 1984; Slade and Levine, 1984a, 1984b, 1987a, 1987b, 1988a–e; Franks, 1985a, 1985b; Blanshard, 1986, 1987, 1988; Levine and Slade, 1986, 1987, 1988a–f; Blanshard and Franks, 1987; Schenz, 1987; Slade et al., 1988; Simatos and Karel, 1988; Karel and Langer, 1988; Marsh and Blanshard, 1988).

This recognition has gone hand in hand with an increasing awareness of the inherent nonequilibrium nature of all "real world" food products and processes, as represented by the microcosm of the bakery industry, in which amorphous polymeric carbohydrates and proteins are major functional components (Slade,

Material in this chapter is taken from previous reports and reviews by the authors, and is updated and reproduced here with permission of the various publishers (see references list) holding copyrights.

1984; Slade and Levine, 1984a, 1984b, 1987a, 1988a–d; Blanshard, 1986, 1987, 1988; Levine and Slade, 1987, 1988e; Slade et al., 1988). Thermal and thermomechanical analysis methods have been shown to be particularly well suited for studying such nonequilibrium systems, in order to define structure–activity relationships from measurements of the thermal and mechanical properties of synthetic amorphous polymers (Turi, 1981). Differential scanning calorimetry (DSC) and dynamic mechanical analysis (DMA) have become established methods for characterizing the kinetic transition from the glassy solid to the rubbery liquid state. This transition occurs at Tg in completely amorphous and partially crystalline, synthetic and natural polymer systems (Fuzek, 1980), including many food materials (Ma and Harwalkar, 1988). The central focus of a polymer science approach to thermal analysis studies of structure–function relationships in food systems (Levine and Slade, 1987, 1988e) emphasizes the insights gained by an appreciation of the fundamental similarities between synthetic amorphous polymers and glass-forming aqueous food materials with regard to their thermal, mechanical, and structural properties. Based on this approach, DSC results have been used to demonstrate the validity of the concept that product quality and stability often depend on the maintenance of food systems (including baked goods) in kinetically metastable, dynamically constrained, time-dependent glassy and/or rubbery states rather than equilibrium thermodynamic phases. These nonequilibrium physical states determine the time-dependent thermomechanical, rheological, and textural properties of food ingredients and products (Slade, 1984; Slade and Levine, 1984a, 1984b, 1987a, 1987b, 1988a–e; Levine and Slade, 1986, 1988a–f; Slade et al., 1988). Bloksma and Bushuk (1988) have pointed out the time and temperature dependence of both the rheological behavior of doughs and the textural characteristics of baked products, which depend directly upon their thermal, mechanical, and structural properties. It is precisely in that context that this chapter, which describes the influences of glassy and rubbery states on the thermal, mechanical, and structural properties of doughs and baked products, fits within the basic theory section of this book.

Plasticization, and its modulation of the temperature of the glass transition, is another key technological aspect of synthetic polymer science (Sears and Darby, 1982). In that field, the classical definition of a plasticizer is "a material incorporated in a polymer to increase the polymer's workability, flexibility, or extensibility" (Sears and Darby, 1982). Characteristically, the Tg of an undiluted polymer is much higher than that of a typical low molecular weight (LMW), glass-forming diluent. As the diluent concentration of a solution increases, Tg decreases monotonically, because the average molecular weight (MW) of the homogeneous polymer–plasticizer mixture decreases, and its free volume increases (Ferry, 1980). A polymer science approach to the thermal analysis of foods in general, and dough and baked product systems (model and real) in particular, involves recognition of the critical role of water as an effective plasticizer of amorphous polymeric, oligomeric, and monomeric food materials (Slade, 1984; Slade and Levine, 1984a, 1984b; Karel, 1985, 1986; Levine and Slade, 1987; Slade et al., 1988). Sears and Darby (1982) have stated unequivocally that "water is the most ubiquitous plasticizer in our world." Karel (1985) has noted that "water is the most important . . . plasticizer for hydrophilic food components." It has become well documented, in large part through DSC studies, that plasticization by water results in a depression of the Tg (and of the elastic modulus and melt viscosity) of completely amorphous and partially crystalline food ingredients. This Tg depression may be advantageous or disadvantageous to product processing, functional properties, and storage stability. Recently, interest has increased in the importance of the effect of water as a plasticizer of many different food materials and other biopolymers (Lillford, 1988; Simatos and Karel, 1988), including the following: starch (Kainuma and French, 1972; van den Berg, 1981, 1986; van den Berg and Bruin, 1981; Slade, 1984; Slade and Levine, 1984a, 1984b, 1987a, 1988b–d; Maurice et al., 1985; Biliaderis et al., 1985; Biliaderis, Page, Maurice, and Juliano, 1986; Biliaderis, Page, and Maurice, 1986a, 1986b; Ablett et al., 1986; Blanshard, 1986, 1987, 1988; Yost and Hoseney, 1986; Levine and Slade, 1987, 1988e; Zeleznak and Hoseney, 1987; Marsh and Blanshard, 1988; Zobel, 1988), gluten (Slade, 1984; Hoseney et al., 1986; Doescher, Hoseney, and Milliken, 1987; Levine and Slade, 1987, 1988e; Slade et al., 1988), starch hydrolysis products (SHPs) (Levine and Slade, 1986, 1988a,c,e,f; Slade and Levine, 1988d; Karel and Langer, 1988), LMW sugars and polyhydric alcohols (Schenz et al., 1984; Chan et al., 1986; Levine and Slade, 1988a–f; Slade and Levine, 1988a, 1988e; Quinquenet et al., 1988), gelatin (Jolley, 1970; Yannas, 1972; Marshall and Petrie, 1980; Slade and Levine, 1984a, 1984b, 1987b; Tomka, 1986; Levine and Slade, 1987), collagen (Batzer and Kreibich, 1981), elastin (Kakivaya and Hoeve, 1975; Hoeve and Hoeve, 1978; Scandola et al., 1981; Atkins, 1987), lysozyme and other enzymes (Bone and Pethig, 1982, 1985; Poole and Finney, 1983a, 1983b; Finney and Poole, 1984; Morozov and Gevorkian, 1985), and the semicrystalline cellulose (Salmen and Back, 1977) and amorphous hemicelluloses and lignin (Kelley et al., 1987) components of wood.

Atkins (1987) has succinctly stated the important observation that "water acts as a plasticizer, dropping the Tg of most biological materials from about 200°C [for anhydrous polymers, e.g., starch and gluten] to about -10°C or so [under physiological conditions of water content], without which they would be glassy" (in their native, in vivo state). The latter Tg of about -10°C is in fact characteristic of high molecular weight (HMW) biopolymers with moisture contents of about 30% or more, which corresponds to physiological conditions. This characteristic Tg has been reported for many polymeric carbohydrates and proteins, including starch, gluten, gelatin (Slade, 1984; Slade and Levine, 1987a, 1987b, 1988b–d; Levine and Slade, 1987, 1988c, 1988e; Slade et al., 1988), hemicelluloses (Kelley et al., 1987), and elastin (Atkins, 1987). Elastin epitomizes a case in which this subzero Tg is critical to healthy physiological function. Elastin exists as a completely amorphous, water-plasticized, covalently cross-linked (via disulfide bonds), network-forming polymer system whose viscoelastic properties have been likened to those of wheat gluten (Edwards et al., 1987; Ablett et al., 1988). In its role as a major fibrous structural protein of skin, ligaments, and arteries, elastin exists in vivo as a rubbery liquid that demonstrates classical rubberlike elasticity (Ferry, 1980) only as long as its Tg remains well below 0°C, due to a water content of 0.40 g/g protein. In contrast, in the pathologic state of arteriosclerosis ("hardening of the arteries"), elastin becomes a glassy solid at body temperature due to a decrease in water content to 0.17 g/g and a corresponding increase in Tg to 40°C (Hoeve and Hoeve, 1978). The importance of water as a plasticizer of major amorphous polymeric and monomeric components in doughs and baked products can be inferred from a review by Bloksma and Bushuk (1988). These authors noted that water plays a key role in (1) the formation of dough, (2) determining the rheological properties of dough, and (3) determining the texture of baked products. Bloksma and Bushuk pointed out that increasing the water content of dough (thereby increasing the extent of plasticization of dough polymers such as damaged starch, gluten, and pentosans, as well as monomeric sugars) results in decreasing viscosity and modulus. These parameter changes lead to effects on the rheological properties of dough, including decreasing stiffness, increasing cohesion, increasing extensibility, and decreasing resistance to mixing. In this context, plasticization by water near room temperature can be advantageous for processibility, for example, in sheeting of cracker dough, while plasticization by water at elevated temperatures may be disadvantageous to product quality, for example, in excessive gelatinization of starch during the baking of cracker dough. In other situations, for example, in the storage of dual-texture baked goods with a crisp component, plasticization due to moisture uptake or internal migration can be detrimental to product textural stability.

A unified conceptual approach to research on food polymer systems, which is based on established principles translated from synthetic polymer science, has enhanced our qualitative understanding of structure–function relationships in food ingredients and products, including baked goods systems (Slade, 1984; Slade and Levine, 1984a, 1984b, 1987a, 1987b, 1988a–e; Levine and Slade, 1986, 1987, 1988a–f; Slade et al., 1988). Ablett and co-workers (1986) have advocated a related "materials science approach" to studies of the influence of water on the mechanical behavior of dough and batter before, during, and after baking. Similarly, in a review of structure–property relationships in starch, Zobel (1988) cited concepts used to characterize synthetic polymers and advocated this approach to provide an increased understanding of the amorphous state and its role in determining the physical properties of native and gelled starches. A central theme of our so-called "food polymer science" approach focuses on the effect of water as a plasticizer on the glass transition and resulting diffusion-limited behavior of water-soluble or water-miscible (collectively referred to as water-compatible) and water-sensitive amorphous materials or amorphous regions of partially crystalline materials (Levine and Slade, 1987; Slade et al., 1988). Plasticization on a molecular level leads to increased intermolecular space or free volume, decreased local viscosity, and a concomitant increased mobility (Ferry, 1980). Plasticization implies intimate mixing, such that a plasticizer is homogeneously blended in a polymer, or a polymer in a plasticizer. Note that a true solvent, capable of cooperative dissolution of the ordered crystalline state and having high thermodynamic compatibility and miscibility at all proportions, is also always a plasticizer, but a plasticizer is not always a solvent (Sears and Darby, 1982). Water-compatible food polymers such as starch, gluten, and gelatin, for which water is an efficient plasticizer but not necessarily a good solvent, exhibit essentially the same physicochemical responses to plasticization by water as do many water-compatible synthetic polymers and many readily soluble monomeric and oligomeric carbohydrates (Levine and Slade, 1987). This fact demonstrates two underlying precepts of the food polymer science approach: (1) synthetic amorphous polymers and glass-forming aqueous food materials are fundamentally similar in behavior, and (2) food ingredients can be viewed generically as members of homologous families of completely amorphous or partially crystalline polymers, oligomers, and monomers, soluble in and/or plasticized by water. The series from glucose through the malto-oligosaccharides to the amylose and amylopectin components of starch exemplifies such a homologous polymer family.

On a theoretical basis of established structure–property relationships for synthetic polymers, the functional properties of food materials during processing and product storage can be successfully explained and can often be predicted (Levine and Slade, 1987; Slade et al., 1988). The discipline of food polymer science has developed to unify structural aspects of foods, conceptualized as completely amorphous or partially crystalline polymer systems (the latter typically based on the classical "fringed micelle" morphological model [Flory, 1953; Wunderlich, 1973; Billmeyer, 1984]), with functional aspects, described in terms of the integrated concepts of "water dynamics" and "glass dynamics." Through this unification, the appropriate kinetic description of the nonequilibrium thermomechanical behavior of food systems such as doughs and baked goods has been illustrated in the context of a "dynamics map," shown in Figure 5-1. This map was derived from a generic solute–solvent state diagram (MacKenzie, 1977; Franks et al., 1977), which in turn was based on a more familiar equilibrium phase diagram of temperature versus composition. The dynamics map, like the "supplemented state diagram" (MacKenzie, 1977), is complicated by an attempt to represent aspects of both equilibrium and nonequilibrium thermodynamics in a single figure. The primary distinction at

MOBILITY TRANSFORMATION MAP

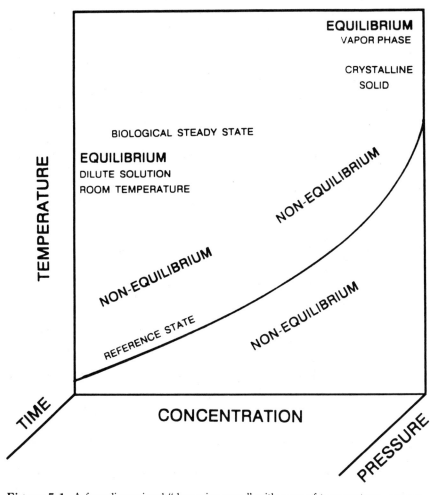

Figure 5-1. A four-dimensional "dynamics map," with axes of temperature, concentration, time, and pressure, which can be used to describe mobility transformations in nonequilibrium glassy and rubbery systems. (*From Slade and Levine, 1988a, with permission.*)

atmospheric pressure is that the equilibrium regions are completely described as shown in two dimensions of temperature and composition, with no time dependence, while the nonequilibrium regions emphatically require the third dimension of time. The established principle of time–temperature superpositioning (Sichina, 1988) has been extended to define "mobility transformations" in terms of the critical variables of time, temperature, and moisture content (with pressure as another variable of potential technological importance). The dynamics map has been used (Slade and Levine, 1988a) to describe mobility transformations in water-compatible food polymer systems which exist in kinetically metastable glassy and rubbery states (Lillford, 1988) that are always subject to conditionally beneficial or detrimental plasticization by water (Levine and Slade, 1987; Slade et al., 1988). For example, the kinetics of starch gelatinization can be explained in terms of mobility transformations by locating on the dynamics map the alternative pathways of complementary plasticization by heat and moisture, as discussed later with regard to Figure 5-26. The map domains of moisture content and temperature, traditionally described with only limited success by using concepts such as "water activity" (A_w), "bound water," water vapor sorption isotherms, and sorption hysteresis, have been treated alternatively as aspects of water dynamics (Levine and Slade, 1987). This concept has provided an innovative perspective on the moisture management and structural stabilization of "intermediate moisture food" systems (Slade and Levine, 1988e) and the cryostabilization of frozen, freezer-stored, and freeze-dried aqueous glass-forming food materials (Levine and Slade, 1986, 1988c, 1988e).

Glass dynamics deals with the temperature dependence of relationships among composition, structure, thermomechanical properties, and functional behavior. Glass dynamics has been used to describe a unifying concept for interpreting "collapse" phenomena; for example, shrinkage due to loss of entrapped leavening gases during storage of pre-proofed frozen bread dough (Levine and Slade, 1986, 1988c, 1988e, 1988f). Collapse phenomena in completely amorphous or partially crystalline food systems (MacKenzie, 1975; To and Flink, 1978; Flink, 1983; Karel and Flink, 1983; Karel, 1985) are diffusion-limited consequences of a material-specific structural and/or mechanical relaxation process. The microscopic and macroscopic manifestations of these consequences occur in real time at a temperature about 20°C above that of an underlying molecular state transformation (Slade and Levine, 1988a). This transformation from a kinetically metastable amorphous solid to an unstable amorphous liquid occurs at Tg (Levine and Slade, 1986). The critical effect of plasticization (leading to increased free volume and mobility in the dynamically constrained glass) by water on Tg is a key aspect of collapse and its mechanism. It is interesting to note that an official definition of starch gelatinization as a collapse process has recently been proposed by a committee of distinguished starch scientists (Atwell et al., 1988).

A general physicochemical mechanism for collapse has been described (Levine and Slade, 1986), based on the occurrence of a material-specific structural transition at Tg, followed by viscous flow in the rubbery liquid state (Flink, 1983). The mechanism was derived from Williams–Landel–Ferry (WLF) free volume theory for (synthetic) amorphous polymers (Williams, Landel, and Ferry, 1955; Ferry, 1980). It has been concluded that Tg is identical to the phenomenological transition temperatures observed for structural collapse (*Tc*) and recrystallization (*Tr*). The non-Arrhenius kinetics of collapse and/or recrystallization in the high-viscosity (η) rubbery state are governed by the mobility of the water-plasticized polymer matrix. These kinetics depend on the magnitude of ΔT above *Tg* (Levine and Slade, 1988a), as defined by a temperature-dependent exponential relationship derived from WLF theory. Glass dynamics has proved a useful concept for elucidating the physicochemical mechanisms of structural/mechanical changes involved in various melting and (re)crystallization processes. Such phenomena are observed in many partially crystalline food polymers and processing/storage situations relevant to baked goods, including, for example, the gelatinization and retrogradation of starches (Slade and Levine, 1987a). Glass dynamics has also been used to describe the viscoelastic behavior of amorphous polymeric network-forming proteins such as gluten and elastin (Slade et al., 1988).

The key to a new perspective on concentrated, water-plasticized food polymer systems relates to recognition of the fundamental importance of the dynamics map mentioned above. The major area of the map in Figure 5-1, that is, the area surrounding the reference state in two dimensions and projecting into the third time dimension, represents a nonequilibrium situation corresponding to the temperature–composition region of greatest technological significance for the food industry. The critical feature in the use of this map is identification of the glass transition as the reference state (contour line) on the kinetic map surface, a conclusion based on WLF theory. This line of demarcation (representing the glass curve of *Tg* versus concentration) serves as a basis for describing the nonequilibrium thermomechanical behavior of polymeric materials in glassy and rubbery states, in response to changes in moisture content, temperature, and time (Levine and Slade, 1987; Slade and Levine, 1988a). Mobility is the transcendent principle underlying the definition of the glass transition as the appropriate reference state, because mobility is the key to all transformations in time (or frequency), temperature, and concentration between different relaxation states for a technologically practical system. The interdependent concepts embodied in the dynamics map have provided insights into the relevance to the functional aspects of food systems (including doughs and baked goods) of this glassy reference state (Slade and Levine, 1988a). For example, the kinetics of all diffusion-limited relaxation processes, which are governed by the mobility of a water-plasticized polymer matrix, vary (from Arrhenius to WLF-type) between distinct temperature/structure domains, which are divided by this glass transition. The viscoelastic, rubbery fluid state, for which WLF kinetics apply (Ferry, 1980), represents the most significant domain for the application of water dynamics (Slade and Levine, 1988e). One particular location among the continuum of *Tg* values along the reference glass curve in Figure 5-1 results from the behavior of water as a crystallizing plasticizer and corresponds to an operationally invariant point (called *Tg'*) on a state diagram for any particular solute (Slade and

Levine, 1988a; Levine and Slade, 1988c). Tg' represents the solute-specific subzero Tg of the maximally freeze-concentrated, amorphous solute/unfrozen water (UFW) matrix surrounding the ice crystals in a frozen solution (Franks et al., 1977; Franks, 1982, 1985a, 1985b). This solute-specific location defines the glass that contains the maximum practical amount of plasticizing moisture (called Wg', expressed as grams UFW/gram solute or weight percent [w%] water [Levine and Slade, 1986]) and represents the transition from a concentrated fluid to a kinetically metastable, dynamically constrained solid (Levine and Slade, 1987). This insight has proved pivotal to the characterization of structure–function relationships in completely amorphous and partially crystalline food polymer systems, including baked goods (Slade and Levine, 1987a; Slade et al., 1988). It is critical to recall that Tg' corresponds to the subzero Tg mentioned by Atkins (1987) as being characteristic of water-plasticized, rubbery biopolymers in vivo.

The experimental background section following this introduction provides a brief description of the principles and methods of thermal analysis used to characterize structure–property relationships in typical polymeric food materials. Next, a theoretical background section introduces a conceptual framework of established structure–property principles from the field of synthetic polymer science, which has been applied with equal success to studies of various food polymer systems. In the following section, experimental results on starch are reviewed that serve to illustrate many of these concepts. The final section highlights the physicochemical properties of wheat gluten as a highly amorphous, viscoelastic polymer system. Gluten is a food protein that exhibits unique functional characteristics, and it is certainly the most important protein in the baking industry. Yet, in comparison to such extensively studied proteins as gelatin and collagen, the polymer physicochemical properties of gluten have been sparsely researched, and many questions regarding the structure–function relationships of gluten in doughs and baked products remain to be answered. In the hope that it will help bring to light some of these vital questions and stimulate further research, a hypothesis is reviewed on the specific role of gluten in the baking mechanism for sugar-snap cookies (Slade et al., 1988). This mechanism depends on the capacity of gluten to function as either a thermoplastic or a thermosetting amorphous polymer.

EXPERIMENTAL BACKGROUND

The material in this section on DSC theory and experimental methods is deliberately limited to background information required to provide a context for understanding the DSC results discussed in the sections on starch and gluten. Much of this information was originally derived from DSC studies of synthetic polymers, and the discussion draws on the fundamental analogies between synthetic amorphous polymers and aqueous glass-forming food polymer systems with regard to their thermal and thermomechanical properties. An extensive discussion of DSC theory and practice is beyond the scope of this chapter. The interested reader is referred to previous reviews of DSC in food research by Biliaderis (1983), Lund (1983), Wright (1984), Hoseney (1984), and Donovan (1985), which provide descriptions of instrumentation and methods plus illustrative DSC results and interpretations, and to a general review of DSC methods by Richardson (1978). The reader is also referred to the excellent book *Thermal Characterization of Polymeric Materials* edited by Turi (1981), and another, *Thermal Analysis of Foods*, edited by Ma and Harwalkar (1988).

Characterization of Structure–Thermal Property Relationships in Food Polymers by Differential Scanning Calorimetry

All materials, synthetic and natural alike, when observed in their physical solid state in the appropriate temperature range, exist in one of several possible structural forms: completely crystalline, semicrystalline, partially crystalline, or completely amorphous. Table sugar is a familiar example of an essentially completely crystalline food material. The term semicrystalline usually denotes materials of greater than 50% crystallinity (Flory, 1953; Wunderlich, 1973), such as cellulose (Slade et al., 1988). The term partially crystalline refers to materials of much less than 50% crystallinity, such as native wheat starch (Levine and Slade, 1987). Completely amorphous, homogeneous polymers manifest a single, "quasi-second-order" transition from a metastable glassy solid to an unstable rubbery liquid at a characteristic Tg. In contrast, partially crystalline, pure polymers show two characteristic transitions: one at Tg for the amorphous component, and the other at Tm for the crystalline component. Tm represents a first-order transition from a crystalline solid to an amorphous liquid, and always occurs at a higher temperature than Tg for homopolymers (Wunderlich, 1981). Both Tg and Tm are measurable as thermal or thermomechanical transitions by various instrumental methods, including DSC, DMA, differential thermal analysis (DTA), and thermomechanical analysis (TMA) (Fuzek, 1980).

For both completely amorphous and partially crystalline polymer systems, the glass transition is manifested as a discontinuous change in heat capacity, a second-order transition. The magnitude of the change at Tg reflects the free volume contribution to heat capacity, which is supplementary to the more familiar vibrational contribution (Wunderlich, 1976). The diagnostic step change is observed as an endothermic baseline shift in a DSC heat flow curve (Wunderlich, 1981). This shift is illustrated in all three of the idealized thermograms shown in Figure 5-2. The usual convention is to denote as Tg the inflection point of the step change in heat capacity, which corresponds to a peak maximum in the negative first derivative of the heat flow curve. This convention is illustrated in parts A and B of Figure 5-3 (Levine and Slade, 1986), which show typical low-temperature DSC thermograms for frozen aqueous carbohydrate solutions. For partially crystalline materials, continued heating beyond Tg results in the appearance of an endothermic peak (signifying a first-order transition) in the heat flow curve (Figure 5-2 curve b), corresponding to Tm. Specifically for the frozen solutions exemplified in Figure 5-3, the Tm that immediately follows Tg' is that of pure ice. Melting of a crystalline solid involves cooperative dissociation of an ordered molecular structure. Dissociation of the lattice components is generally concomitant with an increase in volume of up to 15% (Wunderlich, 1980). (The ice lattice is a well-known exception to this generalization, in that ice decreases in

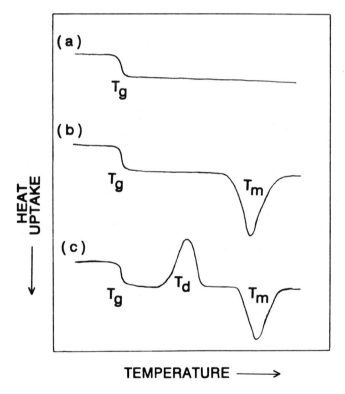

TEMPERATURE ⟶

Figure 5-2. Idealized DSC heat flow curves showing (a) Tg for a completely amorphous material; (b) Tg and Tm for a partially crystalline material; (c) Td between Tg and Tm for a completely amorphous but crystallizable material, during rewarming following melting and rapid cooling. (*From Slade et al., 1988, with permission.*)

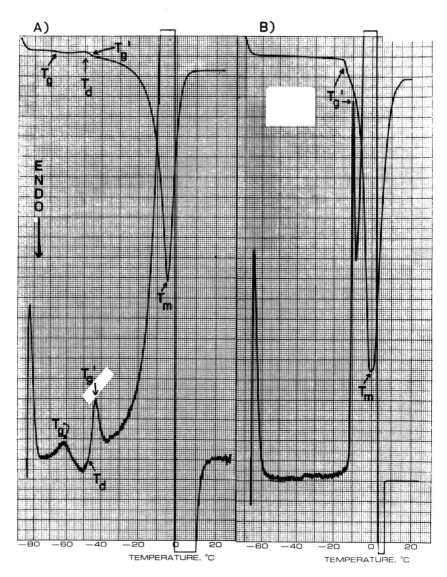

Figure 5-3. Du Pont 990 DSC thermograms for 20 w% solutions of (A) glucose, and (B) Star Dri 10 (10 DE) maltodextrin. In each, the heat flow curve begins at the top (endothermic down), and the analog derivative trace (endothermic up and zeroed to the temperature axis) at the bottom. (*From Levine and Slade, 1986, with permission.*)

volume by 8.3% on melting to liquid water at $Tm = 0°C$.) "Softening" of a glass at Tg involves cooperative relaxation of a molecular assembly that lacks "long-range" order. Relaxation of the immobilized units always involves an increase in volume, of typically less than 1%, as the system transforms to a liquid state characterized by greater free volume, lower local viscosity, and greater degrees of translational and rotational freedom (Ferry, 1980). Thus, both Tg and Tm denote endothermic events, because heat input is required to raise the energy level of a solid to that of the higher energy, generally higher entropy liquid state (Wunderlich, 1981). But Tm is defined on the basis of energetics (the dependence of changes on free energy), as the temperature where the free energies of the coexisting solid and liquid states are equal. In contrast, Tg is defined on the basis of kinetics (the dependence of changes on temperature), as the temperature where the cooperative relaxation rate of the constrained matrix matches the operational observation time.

Once melted, crystalline or partially crystalline materials can often be immobilized in a completely amorphous solid form by sufficiently rapid cooling of the melt from $T > Tm$ to $T < Tg$, which "locks in" the disordered molecular structure of the liquid state (Wunderlich, 1981). In the case of many concentrated, glass-forming, aqueous solutions, fast cooling to subzero temperatures can lead to complete vitrification of the original liquid solution, without any freezing of water or eutectic crystallization of solute. The initial solution concentration is captured and the original homogeneous distribution of the solute molecules is preserved in a glass with $Tg \neq Tg'$, rather than allowing the solute concentration to increase due to disproportionation (freeze concentration) during crystallization of ice (Franks, 1982). This rapid-cooling technique permits analysis of Tg for metastable glasses, including quench-cooled aqueous glasses. Subsequent rewarming of such a completely amorphous but crystallizable system to the operationally defined devitrification temperature (Td, where $Tg < Td < Tm$ [Luyet, 1960]) would allow crystallization to occur, as manifested by an exothermic peak in a DSC heat flow curve (Fig. 5-2 curve *c*). This phenomenon of devitrification during warming is illustrated in Figure 5-3a for the case of a vitrified aqueous solution, in which the solvent can readily crystallize in the timeframe of the experiment, but the solute cannot. Here, the devitrification exotherm is evidence for additional crystallization of ice from a partially vitrified glucose solution. In contrast to rapid cooling, slower cooling can permit complete crystallization (or more extensive partial crystallization) to occur during cooling from an undercooled melt or solution (Wunderlich, 1976). In every case, crystallization from the undercooled, metastable liquid state, at $T < Tm$, whether it occurs during cooling of a melt or heating from a glass to a rubber, is an exothermic event, opposite in a thermodynamic sense from the process of crystalline melting, since it always involves a first-order transition from a disordered liquid to an ordered solid state (Wunderlich, 1973).

Low-Temperature DSC:
The Physicochemical Significance
of Tg' and Wg'

As is illustrated by a comparison of Figures 5-2 and 5-3, higher-moisture food polymer solutions and gels exhibit sequential thermal events, during warming after freezing, which are analogous to those of low-moisture or dry, partially crystalline polymers. A low-temperature DSC method (Levine and Slade, 1986, 1988c, 1988e, 1988f) has been used to measure two thermomechanical properties characteristic of individual noncrystallizing solutes: Tg' and Wg'. Wg' is the amount of unfrozen water in the dynamically constrained solid of extremely high viscosity that is formed on cooling to $T < Tg'$ (Franks, 1985b). This water in the homogeneous, freeze-concentrated solute–water glass is rendered "unfreezable," on a practical timescale, by immobilization with the solute. Wg', as a measure of the composition of this glass at Tg', can also be expressed in terms of Cg', as weight percent solute. Wg' is calculated from the measured area (enthalpy) under the ice melting endotherm of a DSC thermogram. By calibration with pure water, this measurement yields the weight of ice in a maximally frozen sample. The difference between the weight of ice and the known weight of total water in an initial solution is the weight of UFW in the glass at Tg', per unit weight of solute. This DSC procedure is one of several methods used routinely in the food industry to determine the so-called "water binding capacity" of a solute (Labuza, 1985).

As is illustrated by the idealized state diagram in Figure 5-4, the matrix surrounding the ice crystals in a maximally frozen solution or gel is a super-saturated solution of all the solute in the fraction of water remaining unfrozen. This matrix exists as a glass of constant composition at any temperature below Tg', but as a rubbery fluid of lower concentration at higher temperatures between Tg' and Tm of ice. The maximally freeze-concentrated glass at Tg' is one particular glass on the continuous glass curve for any specific solute–water system. In other words, $Tg'–Cg'$ is the solute–characteristic point on the reference contour line of the dynamics map in Figure 5-1. This point is of special technological significance because of the behavior of water as a crystallizing plasticizer (Levine and Slade, 1988c, 1988e). Marsh and Blanshard (1988) have documented the technological importance of freeze-concentration and the practical implication of the description of water as a readily crystallizable plasticizer, characterized by a high Tm/Tg ratio of about 2 (Slade and Levine, 1988a). A theoretical calculation (Marsh and Blanshard, 1988) of the Tg of a typically dilute (i.e., 50%) wheat starch gel fell well below the measured value of about $-5°C$ to $-7°C$ for Tg' (Slade, 1984; Slade and Levine, 1984a, 1984b), because the theoretical calculation did not account for the formation of ice and freeze-concentration that occur below about $-3°C$. Recognition of the practical limitation of water as a plasticizer of water-compatible solutes,

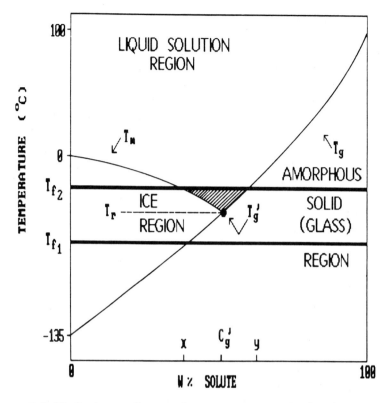

Figure 5-4. Idealized state diagram of temperature vs. w% solute for an aqueous solution of a hypothetical small carbohydrate (representing a model frozen food system), illustrating the critical relationship between *Tg'* and freezer temperature (*Tf*), and the resulting impact on the physical state of the freeze-concentrated amorphous matrix. (*From Levine and Slade, 1988c, with permission.*)

due to the phase separation of ice, reconciled the difference between theoretical and measured values of *Tg* (Marsh and Blanshard, 1988). Moreover, the theoretical calculations supported the measured value of about 27% water (Slade, 1984; Slade and Levine, 1984a, 1984b) for *Wg'*, the maximum practical water content of an aqueous wheat starch glass. The calculated water content of the wheat starch glass with a *Tg* of about −7°C is about 28% (Marsh and Blanshard, 1988).

The *Tg'*−*Cg'* point in Figure 5-4 represents the end of ice formation in real time. Crystallization on cooling of water, which would be readily crystallizable if pure, essentially ceases at an incomplete extent, due to the immobility imposed by the vitrification of the glass-forming solute–water blend of composition *Cg'*−*Wg'*. At solute concentrations near and above the eutectic composition, melting of the metastable solution is described by a nonequilibrium extension of the thermodynamically defined equilibrium liquidus curve. The shape of the nonequilibrium extension of the liquidus curve is kinetically determined by the underlying glass curve. Thus, for a typical solute that does not readily undergo eutectic crystallization, Tg' does not represent the intersection of the equilibrium liquidus curve with the glass curve, but rather corresponds to the intersection of the nonequilibrium extension of the liquidus curve and the kinetically determined supersaturated glass curve (Slade and Levine, 1988a). As such, Franks (1982) has described Tg' as a quasi-invariant point in the state diagram, in terms of its characteristic temperature (Tg') and composition (Cg' or Wg') for any particular solute. The $Tg'-Cg'$ point represents a focal point of Figure 5-4, in that it can be reached by adding water to a dry polymer, as well as by freeze-concentrating a dilute solution (Slade and Levine, 1988a). In the former case, which is relevant to the situation mentioned earlier for water-plasticized biopolymers in vivo, Tg decreases to Tg', as the water content increases to Wg'. Any further increase in water content, followed by crystallization of the excess plasticizer upon slow cooling, results once again in the characteristic Tg' and composition for the non-ice portion. Thus, Tg' and Wg', as physicochemically invariant, although kinetically determined, thermal properties of aqueous glass-forming food polymer systems, are key conceptual features of the dynamics map.

THEORETICAL BACKGROUND FROM PHYSICAL POLYMER SCIENCE—SYNTHETIC POLYMER STRUCTURE–PROPERTY PRINCIPLES

"Fringed Micelle" Structural Model for Partially Crystalline Polymers

The "fringed micelle" model, originally developed to describe the morphology of partially crystalline synthetic polymers, is illustrated in Figure 5-5. It is particularly useful for conceptualizing a three-dimensional network composed of microcrystallites (with crystalline melting temperature, Tm) which cross-link amorphous regions (with glass transition temperature, Tg) of flexible-coil chain segments (Jolley, 1970). The model is especially applicable to synthetic polymers that crystallize from an undercooled melt or concentrated solution to produce a metastable network of relatively low percent crystallinity. Typically, such polymers contain small crystalline regions of only about 100Å dimensions (Wunderlich, 1973). Thus, the model has also often been used to describe the partially crystalline structure of aqueous gels of biopolymers such as starch and gelatin (Jolley, 1970; Wunderlich, 1973; Slade, 1984; Slade and Levine, 1984a, 1984b, 1987a, 1987b, 1988b; Guilbot and Godon, 1984; Kuge

Figure 5-5. "Fringed micelle" model of the crystalline–amorphous structure of partially crystalline polymers. (*From Slade and Levine, 1987b, with permission.*)

and Kitamura, 1985; Zobel, 1988), in which the amorphous regions contain plasticizing water and the microcrystalline regions, which serve as physical junction zones, are crystalline hydrates. The model has also been used to conceptualize the partially crystalline morphology of frozen aqueous food polymer systems, in which case ice crystals represent the "micelles" dispersed in a continuous amorphous matrix (the "fringe") of solute-UFW (Levine and Slade, 1987). An important feature of the model, as applied to HMW polymer systems such as native granular as well as retrograded starches, concerns the interconnections between crystalline and amorphous regions. A single long polymer chain can have helical (or other ordered) segments located within one or more microcrystallites, which are covalently linked to flexible-coil segments in one or more amorphous regions (Jolley, 1970). Moreover, in the amorphous regions, chain segments may experience random intermolecular "entanglement couplings" (Graessley, 1984), which are topological interactions rather than covalent or noncovalent chemical bonds (Mitchell, 1980).

Thus, in terms of their thermomechanical behavior in response to plasticization by water and/or heat, the crystalline and amorphous phases are neither independent of each other nor homogeneous (Wunderlich, 1981).

Crystallization/Gelation Mechanism for Partially Crystalline Polymers

A classical description of crystallization as a three-step mechanism has been widely used for partially crystalline synthetic polymers crystallized, from the melt or concentrated solution, by undercooling from $T > Tm$ to $Tg < T < Tm$ (Jolley, 1970; Wunderlich, 1976). The mechanism is conceptually compatible with the "fringed micelle" model (Hiltner and Baer, 1986). It involves the following sequential steps, which apply universally to all crystallizable substances (Wunderlich, 1976): (1) nucleation (homogeneous), formation of critical nuclei; (2) propagation, growth of crystals from nuclei by intermolecular association; and (3) maturation, crystal perfection (by annealing of metastable microcrystallites) and/or continued slow growth (via "Ostwald ripening"). Within this universal description, flexible macromolecules are distinguished from small molecules by the possibility of nucleation by intramolecular initiation of ordered (e.g., helical) chain segments and propagation by association of chain segments for the high polymers (Wunderlich, 1976).

Thermoreversible gelation of a number of crystallizable synthetic homopolymers and copolymers from concentrated solution has been reported to occur by this crystallization mechanism (Boyer et al., 1985; Hiltner and Baer, 1986; Mandelkern, 1986; Domszy et al., 1986). In contrast, a different gelation mechanism, not involving crystallization and concomitant thermoreversibility, pertains to polymers in solution that remain completely amorphous in the gel state. Such high polymers are distinguished from oligomers by their capacity for intermolecular entanglement coupling, resulting in the formation of rubberlike viscoelastic random networks (called gels, in accord with Flory's [1974] nomenclature for disordered three-dimensional networks formed by physical aggregation) above a critical polymer concentration (Ferry, 1980). Examples of food polymers that can form such amorphous entanglement gels include gluten in unoriented wheat flour dough, sodium caseinate in imitation mozzarella cheese, and casein in real cheese (Slade et al., 1988). As has been summarized by Mitchell (1980), "entanglement coupling is seen in most high MW polymer systems. Entanglements [in completely amorphous gels] behave as crosslinks with short lifetimes. They are believed to be topological in origin rather than involving chemical bonds." It is important to note that hydrogen bonding need not be invoked to explain the viscoelastic behavior of completely amorphous gels formed from solutions of entangling polysaccharides or proteins (Slade et al., 1988).

The gelation-via-crystallization process (described as a nucleation-limited growth process [Domszy et al., 1986]) produces a metastable three-dimensional network (Mandelkern, 1986) cross-linked by "fringed micellar" (Hiltner and Baer, 1986) or chain-folded lamellar (Domszy et al., 1986) microcrystalline junction zones composed of intermolecularly associated helical chain segments (Boyer et al., 1985). Such partially crystalline gel networks may also contain random interchain entanglements in their amorphous regions (Domszy et al., 1986). The nonequilibrium nature of the process is manifested by "well known aging phenomena" (Hiltner and Baer, 1986) (i.e., maturation), attributed to time-dependent crystallization processes that occur after the initial gelation. The thermoreversibility of such gels is explained in terms of a crystallization (on undercooling) $<->$ melting (on heating to $T > Tm$) process (Domszy et al., 1986). Only recently has it been recognized that for synthetic polymer–organic diluent systems (e.g., polystyrene–toluene), such gels are not glasses (Blum and Nagara, 1986) ("gelation is not the glass transition of highly plasticized polymer" [Hiltner and Baer, 1986]) but partially crystalline rubbers (Boyer et al., 1985), in which the mobility of the diluent (in terms of rotational and translational motion) is not significantly restricted by the gel structure (Blum and Nagara, 1986). Similarly, water is highly mobile as a diluent in starch and gelatin gels, and amounts greater than Wg' freeze readily at subzero temperatures. The temperature of gelation ($Tgel$) is above Tg (Blum and Nagara, 1986), in the rubbery fluid range up to about 100°C above Tg. $Tgel$ is related to the flow relaxation temperature Tfr, which is observed in flow relaxation of rigid amorphous entangled polymers (Boyer et al., 1985), and to Tm observed in melts of partially crystalline polymers (Hiltner and Baer, 1986). The basis for the MW dependence of $Tgel$ has been identified (Boyer et al., 1985) as an isoviscous state (which may include the existence of interchain entanglements) of $\eta\,gel/\eta g = 10^5/10^{12} = 1/10^7$, where ηg at $Tg \simeq 10^{12}$ Pa s (Ferry, 1980).

The distinction among these transition temperatures becomes especially important for elucidating how the morphology and structure of food polymer systems such as doughs and baked products relate to their thermal and mechanical behavior. This distinction is a particularly important consideration when experimental methods involve very different timeframes (e.g., mechanical measurements during compression tests or over prolonged storage; thermal analysis at scanning rates varying over four orders of magnitude; relaxation times from experiments at acoustic, microwave, or nuclear magnetic resonance [NMR] frequencies) and sample preparation histories (i.e., temperature, concentration, time) (Slade et al., 1988). In the case of morphologically homogeneous, amorphous molecular solids, Tg corresponds to the limiting relaxation temperature for mobile polymer backbone chain segments. In the case of morphologically heterogeneous, supramolecular networks, the effective network Tg corresponds to the Tfr transition above Tg for flow relaxation (Boyer et al., 1985) of the network. For example, the ratio of Tfr/Tg varies with MW from 1.02 to 1.20 for polystyrene above its entanglement MW (Keinath and Boyer, 1981). Tfr defines an isoviscous state of 10^5 Pa s for entanglement networks (corresponding to $Tgel$ for partially crystalline networks) (Boyer et al., 1985). $Tgel$ of a partially crystalline network would always be observed at or above Tfr ($\equiv$ network Tg) of an entanglement network; both transitions occur above Tg, with an analogous influence of MW and plasticizing water. As an example, the effective network Tg responsible for mechanical firmness of freshly baked bread would be near room temperature for low extents of network formation, well above room temperature for mature networks, and equivalent to $Tgel$ near 60°C for staled bread, even though the underlying Tg for segmental motion, responsible for the predominant second-order thermal transition, remains below 0°C at Tg' (Slade et al., 1988).

Curiously, it has been well established for a much longer time (Domszy et al., 1986) that the same three-step polymer crystallization mechanism describes the gelation mechanism for the classic gelling system, gelatin–water (Jolley, 1970; Wunderlich, 1980). The fact that the resulting partially crystalline gels (Marshall and Petrie, 1980) can be modeled by the "fringed micelle" structure is also widely recognized (Jolley, 1970; Wunderlich, 1976; Slade and Levine, 1984a, 1984b, 1987b). However, while the same facts are true with regard to the aqueous gelation of starch (i.e., retrogradation, a thermoreversible gelation-via-crystallization process which follows gelatinization and "pasting" of partially crystalline native granular starch–water mixtures [Atwell et al., 1988]), and despite the established importance of gelatinization to the rheological properties of high-moisture wheat flour doughs during baking (Bloksma, 1986; Bloksma and Bushuk, 1988) and of retrogradation to the time-dependent texture of fresh-baked versus aged breads (Hoseney, 1986), recognition of starch (or pure amylose or amylopectin) retrogradation as a thermoreversible polymer crystallization process has been much more recent and less widespread (Ring, 1985a, 1985b; Miles et al., 1985a; Blanshard, 1986; Ablett et al., 1986; Biliaderis, Page, Maurice, and Juliano, 1986; Biliaderis, Page, and Maurice, 1986a, 1986b; Ring and Orford, 1986; Ring et al., 1987; Russell, 1987b; l'Anson et al., 1988; Mestres et al., 1988; Zobel, 1988). Blanshard (1988; Marsh and Blanshard, 1988) has recently applied synthetic polymer crystallization theory to investigate the kinetics of starch recrystallization and thereby gain insight into the time-dependent textural changes (i.e., staling due to firming) that occur in baked products such as bread. Similarly, Zeleznak and Hoseney (1987b) have applied principles of polymer crystallization to the interpretation of results on annealing of retrograded starch during aging of bread stored at superambient temperatures. Many of the persuasive early insights in this area have resulted from studies by Slade and co-workers, using the food polymer science approach to investigate structure–property relationships in starch (Slade, 1984; Slade and Levine, 1984a, 1984b, 1987a, 1988b; Maurice et al., 1985; Biliaderis et al., 1985;

Biliaderis, Page, Maurice, and Juliano, 1986; Biliaderis, Page, and Maurice, 1986a, 1986b).

Slade and co-workers (Slade, 1984; Slade and Levine, 1984a, 1984b, 1987a, 1988b; Maurice et al., 1985; Biliaderis et al., 1985; Biliaderis, Page, Maurice, and Juliano, 1986; Biliaderis, Page, and Maurice, 1986a, 1986b) have used DSC results to demonstrate that native granular starches, both normal and waxy, exhibit nonequilibrium melting (Wunderlich, 1981), annealing, and recrystallization behavior characteristic of a kinetically metastable, water-plasticized, partially crystalline polymer system with a small extent of crystallinity. This group has stressed the significance of the conclusion, in which others have concurred (Reid and Charoenrein, 1985; Blanshard, 1986, 1987, 1988; Chungcharoen and Lund, 1987; Burros et al., 1987; Paton, 1987; Russell, 1987a; Zobel, 1988; Zobel et al., 1988), that gelatinization is a nonequilibrium polymer melting process. Gelatinization actually represents a continuum of relaxation processes (underlying a structural collapse) that occurs (at $T > Tg$) while starch is heated in the presence of plasticizing water and in which crystallite melting is indirectly controlled by the dynamically constrained, continuous amorphous surroundings. That is, melting of microcrystallites, which are hydrated clusters of amylopectin branches (Whistler and Daniel, 1984; French, 1984), is controlled by prerequisite plasticization ("softening" above Tg) of flexibly coiled, possibly entangled chain segments in the interconnected amorphous regions of the native granule, for which the local structure is conceptualized according to the "fringed micelle" model. Such nonequilibrium melting in metastable, partially crystalline polymer network systems, in which the crystalline and amorphous phases are neither independent of each other nor homogeneous, is an established concept for synthetic polymers (Wunderlich, 1981). Slade and co-workers (Slade, 1984; Slade and Levine, 1984a, 1984b, 1987a, 1988b; Maurice et al., 1985; Biliaderis et al., 1985; Biliaderis, Page, and Maurice, 1986a, 1986b; Biliaderis, Page, Maurice, and Juliano, 1986) have suggested, and others have agreed (Burros et al., 1987; Paton, 1987; Russell, 1987a; Mestres et al., 1988) that previous attempts (e.g., Lelievre, 1976; Donovan, 1979; Biliaderis et al., 1980) to use the Flory-Huggins thermodynamic treatment to interpret the effect of water content on the Tm observed during gelatinization of native starch have failed to provide a mechanistic model, because Flory-Huggins theory (Flory, 1953) applies only to melting of polymers in the presence of diluent under the conditions of the equilibrium portion of the solidus curve.

An interesting and graphic illustration of the concept of nonequilibrium melting in partially crystalline synthetic polymer systems has been described by Wunderlich (personal communication). This illustration is detailed here to help the reader better understand the applicability of this concept to the gelatinization of native granular starch. Wunderlich related the case of a synthetic block copolymer produced from comonomers A and B. Monomer A was readily crystallizable and capable of producing a HMW crystalline homopolymer of relatively low "equilibrium" *Tm*. In contrast, monomer B was not crystallizable and produced a completely amorphous, HMW homopolymer with a *Tg* much higher than the "equilibrium" *Tm* of homopolymer A. When a minor amount of A and a major amount of B were copolymerized to produce a linear block polymer (with runs of repeat A covalently backbone-bonded to runs of repeat B to yield a molecular structure of the type -BBBBB-AAAA-BBBBBBB-AAA-BBBBBB-), the resulting product was partially crystalline. Because the A and B domains were covalently linked, macroscopic phase separation upon crystallization of A was prevented, and microcrystalline "micelles" of A blocks remained dispersed in a three-dimensional amorphous network of B block "fringes." When the melting behavior of this block copolymer was analyzed by DSC, the melting transition of the crystalline A domains was observed at a temperature *above* the *Tg* of the amorphous B domains. The A domains were kinetically constrained against melting (by dissociation and concomitant volume expansion) at their "equilibrium" *Tm* by the surrounding continuous glassy matrix of B. The A domains were only free to melt (at a nonequilibrium *Tm* ≫ "equilibrium" *Tm*) *after* the B domains transformed from glassy solid to rubbery liquid at their *Tg*.

Retrogradation has been described as a nonequilibrium (i.e., time/temperature/moisture dependent) polymer recrystallization process in completely amorphous (in the case of waxy starches) starch–water melts (Slade, 1984; Slade and Levine, 1984a, 1984b, 1987a, 1988b). In normal starches, retrogradation has been confirmed to involve both fast crystallization of amylose and slow recrystallization of amylopectin (Miles et al., 1985a; Ring and Orford, 1986; Ring et al., 1987; Russell, 1987b; l'Anson et al., 1988; Mestres et al., 1988). Amylopectin recrystallization has been described as a nucleation-limited growth process that occurs, at $T > Tg$, in the mobile, viscoelastic, "fringed micelle" gel network plasticized by water, and which is thermally reversible at $T > Tm$ (Slade, 1984; Slade and Levine, 1984a, 1984b, 1987a, 1988b). This description has also been confirmed for both amylopectin (Miles et al., 1985a; Ring et al., 1987) and amylose (l'Anson et al., 1988). The aging effects typically observed in starch gels and baked bread have been attributed (as in synthetic polymer–organic diluent gels) to time-dependent crystallization processes (i.e., maturation), primarily involving amylopectin, that occur after initial gelation (Ring et al., 1987; Paton, 1987; Russell, 1987b; Marsh and Blanshard, 1988; Mestres et al., 1988). With regard to these effects, Slade (1984) has reported that

> analysis of results [of measurements of the extent of recrystallization versus time after gelatinization] by the classical Avrami equation may provide a convenient means to represent empirical data from retrogradation experiments [Jankowski and Rha, 1986a; Russell, 1987b; Marsh and Blanshard, 1988], but some published theoretical interpretations [e.g., Kulp and Ponte 1981] have been misleading.

Complications, due to the nonequilibrium nature of starch recrystallization via the three-step mechanism, limit the theoretical utility of the Avrami parameters, which were originally derived to describe crystallization under conditions far above the glass curve and where details about nucleation events and constant linear growth rates were readily measurable (Wunderlich, 1976). Others have agreed with this conclusion (Miles et al., 1985a) and have pointed out that such an Avrami analysis allows no insight into crystal morphology (Russell, 1987b) and provides no clear mechanistic information (Blanshard, 1988). Furthermore, the Avrami theory gives no indication of the temperature dependence of the rate of crystallization (Marsh and Blanshard, 1988).

It should be recalled that the same three-step crystallization mechanism also applies to LMW compounds (Wunderlich, 1976; Franks, 1982), such as concentrated aqueous solutions and melts of LMW carbohydrates (Slade and Levine, 1988a, 1988e), and to recrystallization processes in frozen systems of water-compatible food materials (Levine and Slade, 1987, 1988c, 1988f).

Crystallization Kinetics for Partially Crystalline Polymers

The classical theory of crystallization kinetics, applied to partially crystalline synthetic polymers (Wunderlich, 1976), is illustrated in Figure 5-6. This theory has also been shown to describe the kinetics of starch retrogradation (Slade, 1984; Slade and Levine, 1984a, 1984b, 1987a; Zeleznak and Hoseney, 1987a; Blanshard, 1988; Marsh and Blanshard, 1988) and gelatin gelation (Flory and Weaver, 1960; Jolley, 1970; Slade and Levine, 1984a, 1984b, 1987b; Domszy et al., 1986). Figure 5-6 shows the dependence of crystallization rate on temperature within the range $Tg < T < Tm$, and emphasizes the fact that gelation via crystallization can occur only in the rubbery (undercooled liquid) state, between the temperature limits defined by Tg and Tm (Levine and Slade, 1987; Marsh and Blanshard, 1988). These limits, for gels recrystallized from HMW gelatin solutions of concentrations up to about 65 w% gelatin (i.e., $W > Wg' \simeq 35$ w% water), are about $-12°C$ ($= Tg'$) and 37°C, respectively. For B-type starch (or purified amylopectin) gels recrystallized from homogeneous and completely amorphous gelatinized sols or pastes containing $\gtrsim 27$ w% water ($= Wg'$), they are about $-5°C$ ($= Tg'$) and 60°C, respectively (Slade, 1984; Slade and Levine, 1984a, 1984b). In gelatinized potato starch:water mixtures (1:1 w:w), retrogradation has been demonstrated at single storage temperatures between 5°C and 50°C (Nakazawa et al., 1985). In retrograding potato and wheat starch gels, low-temperature storage (at 5°C and 4°C, respectively) results in recrystallization to lower Tm, less symmetrically perfect polymorphs than those produced by storage at room temperature (Nakazawa et al., 1985; Jankowski and Rha, 1986a).

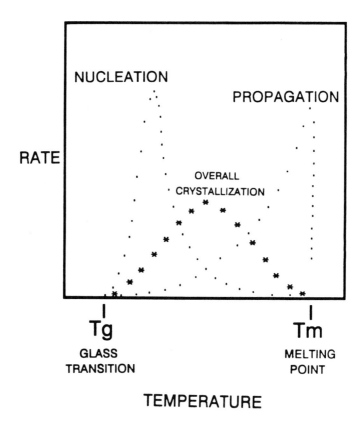

Figure 5-6. Crystallization kinetics of partially crystalline polymers, expressed in terms of crystallization rate as a function of temperature. (*Adapted from Jolley, 1970; Wunderlich, 1976; Baro et al., 1977.*)

Conversely, a higher crystallization temperature generally favors the formation of the higher Tm, more stable A-type, rather than B-type, starch polymorph (Gidley, 1987; Gidley and Bulpin, 1987). For amylopectins from waxy maize and other botanical sources, thermoreversible gelation via crystallization from concentrated (> 10 w% solute) aqueous solution has been observed after long-term storage at 1–5°C (Ring, 1985a; Ring and Orford, 1986; Ring et al., 1987). In baked bread, low-, intermediate-, and high-temperature storage (4°C, 25°C, and 40°C, respectively) results in starch recrystallization manifested by corresponding lower-, intermediate-, and higher-Tm staling endotherms (Zeleznak and Hoseney, 1987b). In a 50% wheat starch gel, the extent of crystallization increases with decreasing storage temperature in the range 2–37°C (i.e., displays a negative temperature dependence), and the rate of recrystallization to the B-form is more rapid at 2°C than at 37°C (Marsh and

Blanshard, 1988). In contrast to the familiar *Tm* of about 60°C for thermore-versible B-type amylopectin gels with excess moisture stored at room tem-perature (and for stale bread) (Slade and Levine, 1987a), the corresponding *Tm* for thermoreversible V-type amylose gels is well above 100°C (Biliaderis et al., 1985; Ring et al., 1987), owing in part to the much higher weight-average degree of polymerization ($\overline{DP}w$) of the amylose chain segments (i.e., $\overline{DP}w \simeq 50$ vs. $\simeq 15$ for amylopectin [Ring et al., 1987]) comprising the micro-crystalline junction zones. Analogously, the familiar *Tm* well above 100°C for various V-type lipid-amylose crystalline complexes (Biliaderis et al., 1985) is much higher than the corresponding *Tm* of about 70°C reported for a lipid–amylopectin crystalline complex (Slade and Levine, 1987a). These findings are fully consistent with the established relationship between increasing chain length (and MW) and increasing *Tm* within homologous families of partially crystalline synthetic polymers (Wunderlich, 1980; Billmeyer, 1984).

As has been illustrated by Figure 5-6 and the results on the tempera-ture dependence of starch recrystallization, the rate of crystallization would be practically negligible at $T < Tg$, because nucleation is a liquid-state phe-nomenon (i.e., in part, a transport process through a viscous medium [Blan-shard, 1988; Marsh and Blanshard, 1988]) that requires translational and ori-entational mobility; such mobility is virtually disallowed (over realistic times) in a mechanical solid of $\eta \gtrsim 10^{12}$ Pa s (Franks, 1982). The temperature of homogeneous nucleation (*Th*) can be estimated from the ratio of *Th*/*Tm* (°K), which is typically near 0.8 for partially crystalline synthetic polymers as well as small molecules, with a reported range of 0.78–0.85 (Walton, 1969; Wun-derlich, 1976). The rate of propagation goes essentially to zero below *Tg*, because propagation is a diffusion-limited process (Baro et al., 1977) for which practical rates also require the liquid state. At $T > Tm$, the rate of overall crystallization also goes to zero, because, intuitively, one realizes that crystals can neither nucleate nor propagate at any temperature at which they would be melted instantaneously.

Figure 5-6 illustrates the complex temperature dependence of the over-all crystallization rate and of the rates of the separate mechanistic steps of nucleation and propagation. According to classical nucleation theory, the nucleation rate is zero at *Tm* and increases rapidly with decreasing tempera-ture (and increasing extent of undercooling [$Tm - T$]) over a relatively nar-row temperature interval, which for undiluted synthetic polymers begins at an undercooling of 30–100°C (Wunderlich, 1976). Within this temperature region, the nucleation rate shows a large negative temperature coefficient (Jolley, 1970; Marsh and Blanshard, 1988). At still lower temperatures (and greater extents of undercooling), where nucleation relies on transport and depends on local viscosity, the nucleation rate decreases with decreasing temperature and increasing local viscosity, to near zero at *Tg* (Wunderlich, 1976; Marsh and Blanshard, 1988). In contrast, the propagation rate increases rapidly with increasing temperature, from a near-zero rate at Tg, and shows a large positive temperature coefficient over nearly the entire rubbery range, until it drops precipitously to a zero rate at Tm (Jolley, 1970; Wunderlich, 1976; Marsh and Blanshard, 1988). The fact that the nucleation and propagation rates show temperature coefficients of opposite signs in the temperature region of intermediate undercooling has been explained (Marsh and Blanshard, 1988) by pointing out that "when the temperature has been lowered sufficiently to allow the formation of [critical] nuclei [whose size decreases with decreasing temperature (Franks, 1982; Blanshard, 1988)], the [local] viscosity is already so high that it prevents growth of crystalline material" (Baro et al., 1977). The maturation rate for nonequilibrium crystallization processes, like the propagation rate, increases with increasing temperature, up to the maximum Tm of the most mature crystals (Slade, 1984).

As is shown by the symmetrical curve in Figure 5-6, the overall crystallization rate (i.e., the resultant rate of both the nucleation and propagation processes), at a single holding temperature, reaches a maximum at a temperature about midway between Tg and Tm, and approaches zero at Tg and Tm (Jolley, 1970; Wunderlich, 1976; Slade, 1984; Slade and Levine, 1984a, 1984b; Blanshard, 1988; Marsh and Blanshard, 1988). Identification of the location of the temperature of maximum crystallization rate has been described (Wunderlich, 1976) in terms of a universal empirical relationship (based on two underlying concepts) for the crystallization kinetics of synthetic high polymers. The first concept identifies a model polymer (e.g., a readily crystallizable elastomer with $Tg = 200$ K and $Tm = 400$ K) as one for which the temperature dependence of polymer melt viscosity is described by WLF kinetics (Wunderlich, 1976). The same concept has been shown to be applicable to describe the nonequilibrium thermomechanical relaxation behavior of "typical" and "atypical" food carbohydrates in aqueous glassy and rubbery states (Slade and Levine, 1988a). The second concept empirically defines a reduced temperature, based on Tg and Tm for typical polymers, as $(T - Tg + 50 \text{ K})/(Tm - Tg + 50 \text{ K})$ (Wunderlich, 1976). An analogous reduced temperature scale, based on Tg' and Tm, has been shown to describe the rotational mobility (i.e., dielectric relaxation behavior) of concentrated aqueous sugar solutions in the supra-glassy fluid state (Slade and Levine, 1988a). For all synthetic high polymers analyzed, the temperature position of the maximum crystallization rate, on a universal master curve like the one shown in Figure 5-6, occurs at about 0.6 of the reduced temperature scale (Wunderlich, 1976). Low-MW synthetic compounds have been fitted to a similar curve, but with a different position for the maximum crystallization rate, at about 0.8 of the reduced temperature scale (Wunderlich, 1976). Based on this empirical relationship for synthetic high polymers, the calculated single holding temperature for maximum crystallization rate would be about 300°K for the model elastomer (in fact, exactly midway between Tg and Tm), -3°C for a gelatin gel with $\gtrsim 35$ w% water (a temperature made inaccessible, without detriment to product quality, due to unavoidable ice formation), 14°C for a typical B-type starch (or amylopectin) gel with $\geq$ 27 w% water, and 70°C for a V-type amylose gel (based on Tm = 153°C [Ring et al., 1987]). It is interesting to note that the calculated value of about 14°C for B-type starch is similar to (1) the empirically determined subambient temperature for the maximum rate of starch recrystallization and concomitant crumb firming during aging, reported in an excellent study of the kinetics of bread staling by Guilbot and Godon (1984), but not previously explained on the basis of the polymer crystallization kinetics theory described above; and (2) the temperature of about 5°C calculated from Lauritzen-Hoffman polymer crystallization kinetics theory by Marsh and Blanshard (1988) for a 50% wheat starch gel. The fact that these subambient temperatures are much closer to the operative Tg (i.e. Tg') than to Tm (unlike the situation depicted by the symmetrical shape of the crystallization rate curve in Figure 5-6, which typifies the behavior of many synthetic polymers) clearly indicates that the crystallization process for B-type starch (or pure amylopectin) is strongly nucleation-limited (Slade, 1984; Slade and Levine, 1987a; Marsh and Blanshard, 1988).

In contrast to the maximum crystallization rate achievable at a single temperature, Ferry (1948) showed that the rate of gelation for gelatin can be further increased, while the phenomenon of steadily increasing gel maturation over an extended storage time can be eliminated, by a two-step temperature cycling gelation protocol that capitalizes on the crystallization kinetics defined in Figure 5-6. Ferry showed that a short period for fast nucleation at 0°C (a temperature above Tg' and near the peak of the nucleation rate curve), followed by another short period for fast crystal growth at a temperature just below Tm, produced a gelatin gel of maximum and unchanging gel strength in the shortest possible time overall. Recently, Slade has shown that a similar temperature-cycling protocol can be used to maximize the rate of starch recrystallization in freshly gelatinized starch–water mixtures with at least 27 w% water (Slade, 1984; Slade and Levine, 1987a), resulting in a patented process for the accelerated staling of starch-based food products (Slade et al., 1987). Zeleznak and Hoseney (1987b) subsequently adopted this protocol in their study of the temperature dependence of bread staling.

Viscoelastic Properties of Amorphous and Partially Crystalline Polymers—Glass-Rubber Transition Behavior—Effect of Molecular Weight on Tg

In the absence of a plasticizer, the viscoelastic properties of completely amorphous and partially crystalline polymers depend critically on temperature relative to Tg of the undiluted polymer. These properties include, e.g., polymer specific volume, V, as illustrated in Figure 5-7 for glassy, partially crystalline, and crystalline polymers. From free volume theory, Tg is defined as the temperature at which the slope changes (due to a discontinuity in the thermal expansion coefficient) in the V versus temperature plot for a glass and for the glassy regions of a partially crystalline polymer as shown in Figure 5-7 (Ferry, 1980). In contrast, V shows a characteristic discontinuity at *Tm*, typically increasing up to about 15% for a crystal and for the crystalline regions of a partially crystalline polymer (Wunderlich, 1980). Glassy and partially crystalline polymers also manifest viscosity versus temperature behavior as illustrated in Figure 5-8. This figure provides operational definitions of a glass and of the glass transition, based on mechanical properties, in terms of microscopic viscosity as a mechanical relaxation process. Figure 5-8 shows that, as the temperature is lowered from that of the low-viscosity liquid state above *Tm*, where familiar Arrhenius kinetics apply, through a temperature range from *Tm* to *Tg*, a completely different, very non-Arrhenius, form of the kinetics, that is, WLF kinetics, becomes operative (Slade and Levine, 1988a). Then, at a temperature where mobility becomes limiting, a state transition occurs, typically manifested as a change in viscosity, modulus, or mechanical relaxation rate of three orders of magnitude (Eisenberg, 1984; Bair, 1985). A "mechanical" glass transition can be defined by combinations of temperature and deformation frequency for which sufficiently large numbers of mobile units (e.g., small molecules or backbone chain segments of a macromolecule) become cooperatively immobilized during a time comparable to the experimental period (Buchanan and Walters, 1977; Morozov and Gevorkian, 1985), such that the material becomes a mechanical solid capable of supporting its own weight against flow. Arrhenius kinetics become operative once again in the glassy solid, but the rates of all diffusion-limited processes are much lower in this high-viscosity solid state than in the liquid state (Levine and Slade, 1987). In fact, the difference in average relaxation times between the two Arrhenius regimes is typically more than fourteen orders of magnitude (Slade and Levine, 1988a). The microscopic viscosity equals about 10^{11} to 10^{14} Pa s ($= 10^{12}$ to 10^{15} Poise) at *Tg* (Soesanto and Williams, 1981; Downton

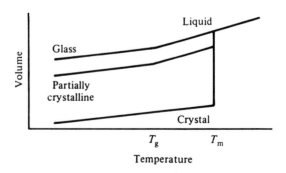

Figure 5-7. Specific volume as a function of temperature for glassy, crystalline, and partially crystalline polymers. (*From Levine and Slade, 1987, with permission.*)

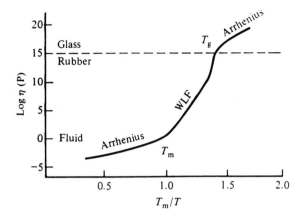

Figure 5-8. Viscosity as a function of reduced temperature (Tm/T) for glassy and partially crystalline polymers. (*From Levine and Slade, 1987, with permission.*)

et al., 1982; Franks, 1982), which represents the intersection of the curve of log η versus reduced temperature, Tm/T, in Figure 5-8 with the boundary between the glassy solid and rubbery liquid states. Figures 5-7 and 5-8 further illustrate the basis for the operational definition of the glass transition in amorphous polymers as a second-order thermodynamic transition (Wunderlich, 1981): This transition represents a temperature-, time- (or frequency-), and composition-dependent, material-specific change in nonequilibrium physical state from an amorphous mechanical solid to a viscoelastic liquid (Petrie, 1975; Ferry, 1980).

For many partially crystalline synthetic polymers, in the absence of diluent, Tg equals 0.5 to 0.8 of the value of Tm in °K (Wunderlich, 1980; Batzer and Kreibich, 1981; Franks, 1982). For highly symmetrical pure polymers, oligomers, and monomers, the value of the ratio Tg/Tm is often close to 0.5, while for highly unsymmetrical ones, the ratio is often greater than 0.8 (Brydson, 1972; Wunderlich, 1980). For food materials such as monomeric fructose, anomalously high values of Tg result in ratios of $Tg/Tm \gg 0.8$, which have been found to correlate with anomalous relaxation behavior due to contributions of excess free volume or decreased local effective viscosity (Slade, 1984; Slade and Levine, 1984a, 1984b, 1987a, 1987b, 1988a–e; Levine and Slade, 1987, 1988a; Slade et al., 1988). Anomalously high values of apparent Tg/Tm ratio have been observed for water-plasticized, partially crystalline food polymer systems such as native granular starches and gelatin, but in these cases, anomalously low free volume results from relative dehydration of the amorphous regions, compared to the crystalline hydrate regions (Slade and Levine, 1987a, 1987b, 1988b).

For pure synthetic polymers in the absence of diluent, Tg is also known to vary with MW in a characteristic and theoretically predictable fashion, which has a significant impact on resulting mechanical and rheological properties. For a homologous series of amorphous linear polymers, Tg increases with increasing number–average MW ($\overline{\text{Mn}}$), due to decreasing free volume contributed by chain ends, up to a plateau limit for the region of entanglement coupling in rubberlike viscoelastic random networks (typically at $\overline{\text{Mn}} = 1.25 \times 10^3$–$10^5$ daltons (Graessley, 1984)), then levels off with further increases in $\overline{\text{Mn}}$ (Billmeyer, 1984; Ferry, 1980). Below the entanglement $\overline{\text{Mn}}$ limit, there is a theoretical linear relationship between increasing Tg and decreasing inverse $\overline{\text{Mn}}$ (Sperling, 1986). For polymers with constant values of $\overline{\text{Mn}}$, Tg increases with increasing weight-average MW ($\overline{\text{Mw}}$), due to increasing local effective viscosity (Slade and Levine, 1988a, 1988e; Levine and Slade, 1988b). This contribution of local effective viscosity is reported to be especially important when different MWs in the low range are being compared (Ferry, 1980). The difference in three-dimensional morphology and resultant mechanical and rheological properties between a collection of nonentangling, LMW polymer chains and a network of entangling, HMW, randomly coiled polymer chains can be imagined as analogous to the difference between masses of elbow macaroni and spaghetti. For synthetic polymers, the $\overline{\text{Mn}}$ at the boundary of the entanglement plateau often corresponds to about 600 backbone chain atoms (Sperling, 1986). Since there are typically about 20–50 backbone chain atoms in each polymer segmental unit involved in the cooperative translational motions at Tg (Brydson, 1972), entangling high polymers are those with at least about 12–30 segmental units per chain. Figure 5-9 illustrates the characteristic dependence of Tg on $\overline{\text{Mn}}$ (expressed in terms of log DP [degree of polymerization]) for several homologous series of synthetic amorphous polymers. In this semi-log plot, the Tg values for each polymer reveal three distinguishable intersecting linear regions: Region III) a steeply-rising region for nonentangling small oligomers; Region II) an intermediate region for nonentangling low polymers; and Region I) the horizontal plateau region for entangling high polymers (Shalaby, 1981). From extensive data from the literature for a variety of synthetic polymers, it has been concluded that this three-region behavior is a general feature of such Tg versus log $\overline{\text{Mn}}$ plots, and demonstrated that the data in the nonentanglement Regions II and III show the theoretically predicted linear relationship between Tg and inverse $\overline{\text{Mn}}$ (Shalaby, 1981).

Low-temperature DSC results for the characteristic Tg' values of individual carbohydrate and protein solutes have recently shown that Tg' is a function of MW for both homologous and quasi-homologous families of water-compatible monomers, oligomers, and polymers, which include many food ingredients commonly used in baked products (Levine and Slade, 1986, 1987; Slade et al., 1988). However, the glass at Tg' is not that of the pure, undiluted polymer, and so there is no theoretical basis for assuming that this Tg of the freeze-concentrated glass should depend on the MW of the dry polymer. Yet, if the relative shapes of the polymer–diluent glass curves are similar within a

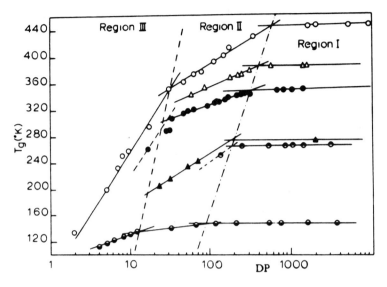

Figure 5-9. Plot of Tg as a function of log DP (degree of polymerization) (a measure of $\overline{Mn}$), for poly(alpha-methylstyrene) (open circles); poly(methylmethacrylate) (open triangles); poly(vinyl chloride) (solid circles); isotactic polypropylene (solid triangles); atactic polypropylene (circles, top half solid); and poly(dimethylsiloxane) (circles, bottom half solid). (*From Sperling, 1986, with permission.*)

polymer series, increases in MW lead to proportional increases in both dry Tg and Tg' (Slade and Levine, 1988a). Thus, it has been shown, for two extensive series of polyhydroxy compounds (PHCs), that the linear relationship between Tg and inverse MW of the solute does apply to the particular Tg' of the solute–unfrozen water glass (Levine and Slade, 1986, 1988b, 1988f). Moreover, it has been demonstrated that carbohydrate polymers such as SHPs and LMW PHCs such as sugars, polyhydric alcohols, and glycosides show exactly the same characteristic Tg versus $\overline{Mn}$ behavior as described above for synthetic amorphous polymers (Levine and Slade, 1986, 1987, 1988a–f). Tg' values for an extensive series of commercial SHPs (of polydisperse MWs, in the range from 180 for monodisperse glucose itself to about 60,000 for a 360-DP polymer, characterized in terms of dextrose equivalent (DE) value, where DE $= 18,016/\overline{Mn}$) have demonstrated their classical behavior as a homologous family of amorphous glucose oligomers and polymers. The plot of Tg' versus $\overline{Mn}$ in Figure 5-10 clearly exhibits the same three-region behavior as shown in Figure 5-9: (I) the plateau region indicative of the capability for entanglement coupling by high polymeric SHPs of DE ≤ 6 and $Tg' \geq -8°C$; (II) the intermediate region of nonentangling low polymeric SHPs of $6 < DE < 20$; and (III) the steeply rising region of nonentangling, small SHP oligomers of DE > 20. The plot of Tg' versus $1/\overline{Mn}$ in the inset of Figure 5-10, with a linear correlation coefficient $r = -0.98$, demonstrates the theoretically predicted linear relationship for all the SHPs in Regions II and III, with DE values greater than 6. The plateau region evident in Figure 5-10 has identified a lower limit of $\overline{Mn} \simeq 3,000$ ($\overline{DPn} \simeq 18$) for entanglement leading to viscoelastic network formation by such polymeric SHPs in the freeze-concentrated glass formed at Tg' and Cg'. This $\overline{Mn}$ is within the typical range of 1,250–19,000 for minimum entanglement MWs of many synthetic amorphous linear high polymers (Graessley, 1984). The corresponding $\overline{DPn}$ of about 18 is within the range of 12–30 segmental units in an entangling high polymer chain, thus suggesting that the glucose repeat in the glucan chain (with a total of 23 atoms/hexose ring) may represent the mobile backbone unit involved in cooperative solute motions at Tg'. The entanglement capability has been suggested to correlate well with various functional attributes (see labels on plateau region in Figure 5-10) of low-DE SHPs, including a predicted and subsequently demonstrated (Levine and Slade, 1986, 1988f) ability to form thermoreversible, partially crystalline gels from aqueous solution (Richter et al., 1976a, 1976b; Braudo et al., 1979, 1984; Bulpin et al., 1984; Reuther et al., 1984; Lenchin et al., 1985; Miles et al., 1985a; Ellis and Ring, 1985). It has been suggested (Levine and Slade, 1987) that SHP gelation occurs by a mechanism involving crystallization plus entanglement in concentrated solutions undercooled to $T <$ Tm, as described in the section entitled Crystallization/Gelation Mechanism for Partially Crystalline Polymers.

In contrast to the commercial SHPs, a large series of quasi-homologous, monodisperse (i.e., MW $= \overline{Mn} = \overline{Mw}$) PHCs with known MWs in the range 62–1,153, including a homologous set of malto-oligosaccharides up to DP 7, has been found to manifest Tg' values below the Tg' limit defined by SHPs for entanglement and the onset of viscoelastic rheological properties and to be incapable of gelling from solution (Levine and Slade, 1987, 1988b, 1988f). The plot of Tg' versus MW in Figure 5-11, drawn conventionally as a smooth curve through all the points (Billmeyer, 1984), can easily be visualized to represent two intersecting linear regions (III for MW < 300 and II for $300 <$ MW $< 1,200$). From the fair linearity of the Tg' versus 1/MW plot ($r = -0.93$) for all the data in the inset of Figure 5-11 (and from the better linearity of the corresponding plot [$r = -0.99$] for the series of malto-oligosaccharides [Levine and Slade, 1987], which exemplify the theoretical glass-forming behavior characteristic of a homologous family of nonentangling, linear, monodisperse oligomers [Ferry 1980]), it has been concluded that these diverse LMW sugars, polyols, and glycosides show no evidence of entanglement in the freeze-concentrated glass at Tg'. For these PHCs, none larger than a heptamer of MW 1,153, the main plot in Figure 5-11 shows that Region I, representing the entanglement plateau where Tg remains constant with increasing MW, has not been reached, in accord with the MW (and corresponding DP) range cited above as the lower limit for polymer entanglement.

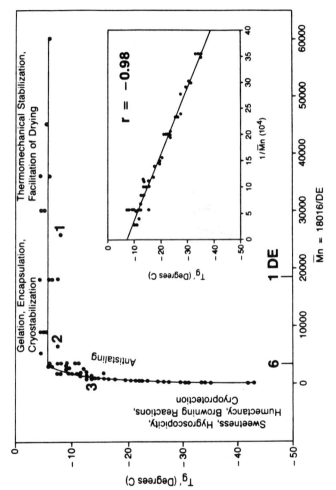

Figure 5-10. Variation of the glass transition temperature, Tg', for maximally frozen 20 w% solutions against $\overline{M}n$ (expressed as a function of DE) for commercial SHPs. DE values are indicated by numbers marked above the x-axis. Areas of specific functional attributes, corresponding to three regions of the diagram, are labeled. *Inset:* plot of Tg' versus $1/\overline{M}n$ ($\times$ 10,000) for SHPs with $\overline{M}n$ values below entanglement limit, illustrating the theoretically predicted linear dependence. (*From Levine and Slade, 1988e, with permission.*)

189

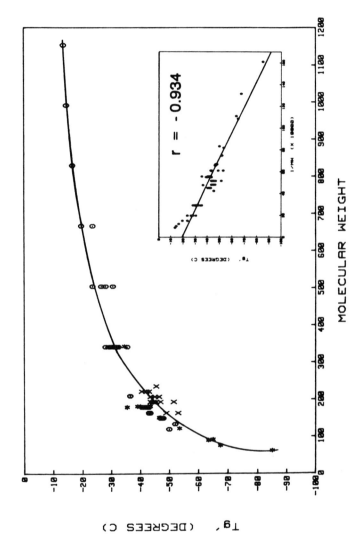

Figure 5-11. Variation of the glass transition temperature, Tg', for maximally frozen 20 w% solutions against MW for a collection of sugars (o), glycosides (x), and polyols (*). *Inset:* plot of Tg' versus 1/MW ($\times$ 10,000), illustrating the theoretically predicted linear dependence. (*From Levine and Slade, 1988f, with permission.*)

Effects of Water Plasticization on Thermomechanical Properties of Polymers — the State Diagram — *Tg* Curves

Water as a plasticizer is known to affect both the *Tg* and *Tm* of partially crystalline polymers (Slade, 1984; Slade and Levine, 1984a, 1984b, 1987a, 1987b; Levine and Slade, 1986, 1987; Slade et al., 1988). Water is a "mobility enhancer," in that its low MW leads to a large increase in mobility, due to increased free volume and decreased local viscosity (Ferry, 1980), as moisture content is increased from that of a dry solute to a solution (Slade and Levine, 1988e). The direct plasticizing effect of increasing moisture content at constant temperature is equivalent to the effect of increasing temperature at constant moisture and leads to increased segmental mobility of chains in amorphous regions of glassy and partially crystalline polymers, allowing in turn a primary structural relaxation transition at decreased *Tg* (Rowland, 1980; Sears and Darby, 1982; Flink, 1983). State diagrams illustrating the extent of this *Tg*-depressing effect (some of which are shown here, as listed below) have been reported for a variety of synthetic and natural water-compatible polymers and monomers, including the following: nylon 6 (see Fig. 5-16), poly(vinyl pyrrolidone) (PVP) (MacKenzie and Rasmussen, 1972; Franks et al., 1977; see Fig. 5-13), starch (see Fig. 5-12; Fig. 5-14A; and Fig. 5-14B), elastin (Kakivaya and Hoeve, 1975), collagen (see Fig. 5-16), gelatin (Marshall and Petrie, 1980; Borchard et al., 1980; Reutner et al., 1985; Slade and Levine, 1987b), gluten (see Fig. 5-15), and LMW carbohydrates such as sorbitol (Quinquenet, 1988), sucrose, glucose (Luyet and Rasmussen, 1968; MacKenzie, 1977; Franks, 1982; Schenz et al., 1984; Blanshard and Franks, 1987), and fructose (see Fig. 5-13). In these diagrams, the smooth "glass" curves of *Tg* versus weight percent water (or solute) show the dramatic effect of water on *Tg*, especially at low moisture contents (i.e., $\leq$ 10 w% water). In this region, *Tg* generally decreases by about 5–10°C/w% water, from the neighborhood of 200°C for the dry polymer (as mentioned by Atkins [1987]). For example, Figure 5-12 depicts the amylopectin of freshly gelatinized starch as a typical water-compatible, completely amorphous polymer, which manifests a *Tg* curve from about 125°C for pure "dry" starch to about −135°C, the *Tg* of pure amorphous solid water (Mayer, 1988). Figure 5-12 shows the *Tg* of starch decreasing about 6°C/w% water for the first 10 w% moisture, in good agreement with another published glass curve for starch (calculated from free volume theory [Blanshard, 1988]) shown in Figure 5-14A. Similarly, the glass curve for amorphous gluten in Figure 5-15 shows a decrease in *Tg* from greater than 160°C at less than 1 w% water to 15°C at 16 w% water, a depression of about 10°C/w% water in this moisture range. As is shown in Figures 5-12 through 5-16, similar values of the extent of plasticization at low moisture, in the range of about 5–10°C/w% water, have been found to apply widely to water-compatible glassy and partially crystalline polymers,

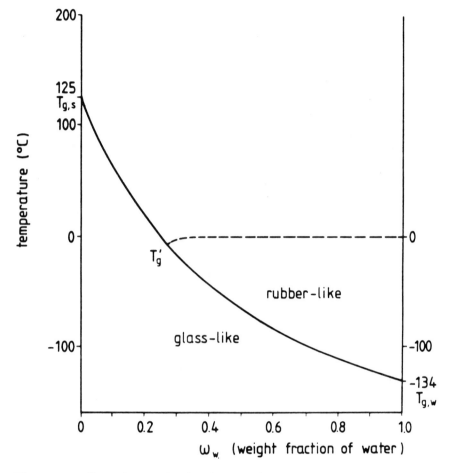

Figure 5-12. State diagram, showing the approximate *Tg* temperatures as a function of mass fraction, for a gelatinized starch–water system. (*From van den Berg, 1986, with permission.*)

oligomers, and monomers, including many food ingredients and components of doughs and baked products (Levine and Slade, 1987; Slade et al., 1988).

Mechanism of Water Plasticization — the Myth of "Bound" Water

According to the prevailing view in the literature on synthetic polymers, the predominant contribution to the mechanism of plasticization of water-compatible glassy polymers by water derives from a free volume effect (Ellis

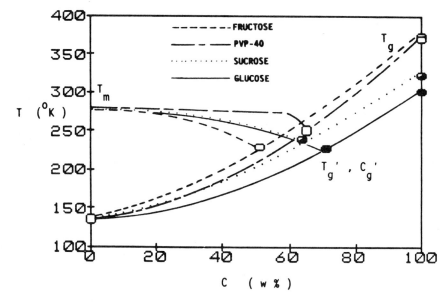

Figure 5-13. Solute–water state diagrams of temperature versus concentration for fructose, glucose, sucrose, and PVP-40 (poly(vinyl pyrrolidone), MW = 40,000), which illustrate the effect of water plasticization on experimentally measured glass curves, and the location of the invariant point of intersection of the glass curve and the nonequilibrium portion of the liquidus curve at Tg' and Cg', for each solute. (*From Slade and Levine, 1988a, with permission.*)

et al., 1984; Jin et al., 1984; Ellis, 1988). Free volume theory (Ferry, 1980) provides the general concept that free volume is proportional to inverse $\overline{Mn}$, so that the presence of a plasticizing diluent of low MW leads to increased free volume, allowing increased backbone chain segmental mobility. The increased mobility is manifested as a decreased Tg of the binary polymer–diluent glass (Bone and Pethig, 1982; Sears and Darby, 1982; Marsh and Blanshard, 1988). For synthetic polymers, it is well known that the ability of a diluent to depress Tg decreases with increasing diluent MW (Boyer et al., 1985), as predicted by free volume theory. Recent reports have demonstrated that the effectiveness of water as a plasticizer of synthetic polymers primarily reflects its low molar mass (Jin et al., 1984; Ellis et al., 1984; Ellis, 1988). These workers have discounted older concepts of specific interactions, such as disruptive water–polymer hydrogen bonding in polymer hydrogen-bonded networks, or plasticizing molecules becoming "firmly bound" to polar sites along a polymer chain, in explaining the plasticizing ability of water. Although hydrogen bonding certainly affects solubility parameters and contributes to the compatibility of polymer–water blends (Sears and Darby, 1982), it has been convincingly shown that polymer flexibility does not depend on specific hydrogen bonding to backbone polar groups (Buchanan and Walters, 1977). Rather, the relative size of the mobile segment of the linear backbone (Sperling, 1986), and thus the relative $\overline{M}w$ of its blend with water, governs the magnitude of plasticization and so determines Tg (Buchanan and Walters, 1977). To negate the older arguments for site-specific hydrogen bonding, NMR results have been cited which clearly indicate that water molecules in polymers with polar sites have a large degree of mobility (Ellis et al., 1984; Jin et al., 1984). In this context, mobility is defined in terms of translational and rotational degrees of freedom for molecular diffusion on a timescale of experimental measurements. Franks (1982, 1983a, 1983b, 1985a–c, 1986) has advocated a similar view and presented similar evidence to try to dispel the popular (Labuza, 1985) but outdated (Simatos and Karel, 1988) myths about "bound" water and "water binding capacity" in glass-forming food polymers or LMW materials. For example, proton NMR has been used to test the accessibility of water with reduced mobility in the crystalline regions of retrograded wheat starch gels. Such gels are partially crystalline, with B-type hydrated crystalline regions in which water molecules constitute an integral structural part of the crystal unit cell (French, 1984; Imberty and Perez, 1988). Results of NMR studies have shown that all the water in such a starch gel can be freely exchanged with D_2O (Wynne-Jones and Blanshard, 1986). Most recently, Ellis (1988) has reported results of a comprehensive DSC study which show that several diverse synthetic "amorphous polyamides in pure and blended form exhibit a monotonic depression of Tg as a function of water content," and which "lend further credence to the simple and straightforward plasticizing action of water in polar polymers irrespective of their chemical and physical constitution." These results have helped to confirm the conclusions that (1) the behavior of hydrophilic polymers with aqueous diluents is precisely the same as that of nonpolar synthetic polymers with organic diluents, and (2) water-compatible food polymers such as starch and gluten, for which water is an efficient plasticizer but not necessarily a good solvent, exhibit the same physicochemical responses to plasticization as do many water-compatible synthetic polymers (Levine and Slade, 1987; Slade et al., 1988). The excellent agreement between the measured value of Tg' (Slade, 1984; Slade and Levine, 1984a, 1984b) and the theoretical value recently calculated from free volume theory (Blanshard, 1988; Marsh and Blanshard, 1988) for an aqueous wheat starch gel with $\geq 27\%$ moisture lends further support to these conclusions.

Figure 5-14. (A) Variation of Tg with the volume fraction of starch as calculated by free volume theory. (*From Blanshard, 1988, with permission.*) (B) Plot of the response of Tg to sample moisture in native and pregelatinized wheat starch. (*From Zeleznak and Hoseney, 1987, with permission.*)

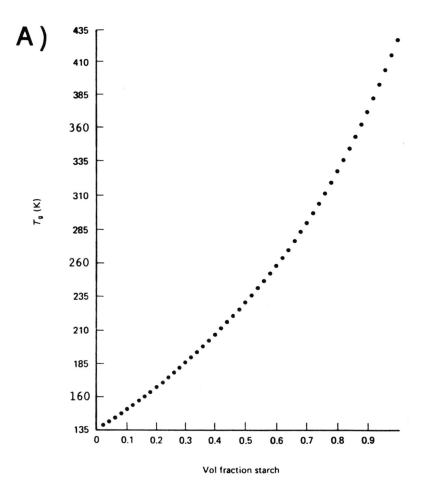

A)

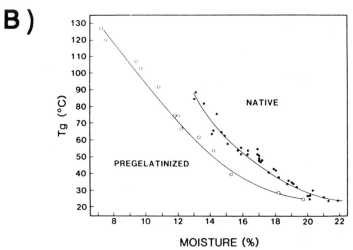

B)

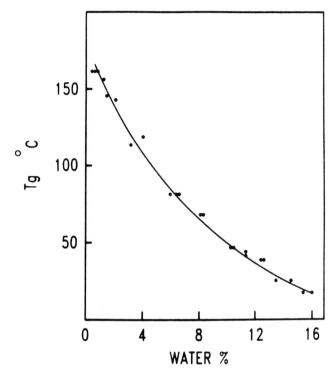

Figure 5-15. Change in *Tg* as a function of moisture for a hand-washed and lyophilized wheat gluten. (*From Hoseney et al., 1986, with permission.*)

To put this modern concept of water plasticization in a more familiar context of the older, more traditional literature on "bound" and "unfreezable" water and on water sorption by food polymers at low moisture (reviewed in detail elsewhere [Levine and Slade, 1987; Slade et al., 1988]), the earliest-sorbed water fraction is most strongly plasticizing, is always said to be "unfreezable" in a practical timeframe, and is often referred to as "bound." The later-sorbed water fraction is said to be freezable; is referred to as "free," "mobile," or "loosely bound"; and is either weakly plasticizing or nonplasticizing, depending on the degree of water compatibility of the specific polymer. The degree of water compatibility relates to the ability of water to depress *Tg* to *Tg'*, and to the magnitude of *Wg'* (Levine and Slade, 1987; Slade and Levine, 1988e). Regardless of context, a key fact about the "freezability" of water relates to the homogeneous nucleation process for ice (Franks, 1987). Even at temperatures as low as −40°C, a minimum on the order of 200 water molecules must associate within a domain of about 40Å in order to form a critical nucleus that will grow spontaneously into an ice crystal (Franks,

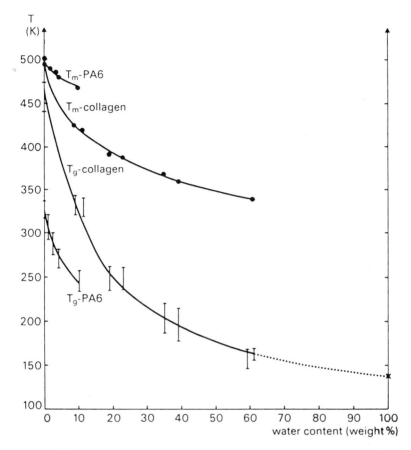

Figure 5-16. Comparison of the influence of water on the *Tg* and *Tm* of a natural (collagen) and a synthetic polyamide (PA6 = nylon 6). (*From Batzer and Kreibich, 1981, with permission.*)

1982). Thus, within any food material at low moisture, clusters of water molecules of lower density than about 200 molecules/40Å cannot freeze under any circumstances (Slade et al., 1988).

The solute-specific, invariant quantity of unfrozen water captured in the glass which forms at Tg', defined earlier as Wg' (Levine and Slade, 1986), is traditionally referred to by many food scientists and technologists as one measure of "bound" water (Labuza, 1985). However, "bound" water, with regard to either frozen or room temperature food systems, including doughs and baked products, is a misnomer that has persisted for at least the last 30 years, despite constant debate (Kuprianoff, 1958; Levine and Slade, 1986, 1987, 1988a–f; Lillford, 1988; Simatos and Karel, 1988; Slade et al., 1988) and ever more convincing arguments that the concepts of "bound" water, "water binding," and "water binding capacity" of a solute are incorrect, inappropriate, and misleading rather than helpful (Franks, 1982, 1983a, 1983b, 1985a, 1985b, 1985c, 1986). The concept of "bound" water originated in large part from a fundamental misconception that discrete "free" and "bound" physical states of water in food materials (or "free," "loosely bound," and "tightly bound" states) could provide a valid representation of water molecules in a solution at ambient temperature. Actually, at $T > Tg'$, water molecules in a solution exist within a single physical state (i.e., liquid) characterized not by any kind of static geometry but rather by a dynamic continuum of degrees of hindered instantaneous mobility (Slade et al., 1988). In this liquid solution state, individual water molecules are only transitorily hydrogen bonded to individual polar sites on the solute (Franks, 1983a, 1983b).

As has been explained by Slade and Levine (Slade and Levine, 1988a; Levine and Slade, 1988e), the solute-specific value of Wg' is the maximum amount of water that can exist with a solute in a spatially homogeneous, compatible blend that, in the rubbery state, exhibits long-range cooperative relaxation behavior described by WLF kinetics, but not long-range lattice order. Further dilution beyond Wg' results in a loss of cooperative mobility and the onset of short-range fluid mechanics, described by Arrhenius kinetics. Thus, expression of Wg' as a water/solute number ratio (i.e., a "notional hydration number" [Franks, 1983a]) actually represents the technologically practical maximum limit for the amount of water that can act as a plasticizer of a particular solute (Franks, 1985b; Levine and Slade, 1987), rather than the amount of water that is "bound" to or whose dynamics are governed by that solute. Part of the reason for the persistence of the concept of "bound" water in such concentrated solute systems, despite convincing evidence of its invalidity, relates to a conclusion inadvisedly extrapolated from findings for very dilute solutions. The addition of a few isolated solute molecules to pure water already causes a profound effect on the self-diffusion properties in the solution. The hindered diffusion of water molecules instantaneously in the vicinity of individual solute molecules is construed as the effect of "viscous drag"; these less-mobile water molecules are visualized to be "pulled along" with the solute during flow. But it has been demonstrated repeatedly (Franks, 1983a, 1983b) that the less-mobile water molecules are freely exchangeable with all of the water in the solution, leading to the inescapeable consensus view that the water is not bound to the solute. On the other hand, in describing dilute solutions, no one has ever suggested that the solute molecules are "bound" to water molecules. When the situation is reversed, adding a few water molecules to an anhydrous solute profoundly changes the viscoelastic properties of the solute via water plasticization, which increases the free volume and decreases the local viscosity (Levine and Slade, 1988e). Why then, in light of this evidence of a dramatic increase in the mobility of the solute, have many found it so easy to jump to the conclusion that these water molecules must be "bound" to solute molecules?

It is only recently becoming more widely acknowledged and accepted that the so-called "bound" water corresponding to Wg' is not energetically bound in any equilibrium thermodynamic sense (Blanshard and Franks, 1987; Levine and Slade, 1987, 1988e; Lillford, 1988; Simatos and Karel, 1988; Slade et al., 1988). Rather, it is simply kinetically immobilized, due to the extremely high local viscosity of the metastable glass at Tg', and thus dynamically constrained from the translational and rotational diffusion required for ice crystal growth (Biros et al., 1979; Pouchly et al., 1979; Franks, 1982, 1986). This does not mean that no solute–water hydrogen bonds exist in the glass at Tg', only that such hydrogen bonds are the normal consequence of dissolution of a solute in water rather than the cause of the kinetic immobility that renders this water "unfreezable" in real time. The stabilizing free energy of such solute–water hydrogen bonds is no greater than for water–water hydrogen bonds in ice (Franks, 1983b, 1985c). Analogously, for model solutions of small sugars at room temperature, results of NMR and dielectric relaxation measurements have shown that "the residence time of a given water molecule at a solvation site [i.e., a hydroxyl group on a sugar] is extremely short, less than 1 ns" (Franks, 1983a, 1983b). Furthermore, results from studies of synthetic polymers (Jin et al., 1984) and polymeric carbohydrate and protein gels (Wynne-Jones and Blanshard, 1986; Lillford, 1988) alike have demonstrated conclusively that water molecules said to be "bound" to polar groups on such polymeric solutes are in fact highly mobile (especially compared to the mobility of water in ice [Derbyshire, 1982]) and able to exchange freely and rapidly, likewise on an NMR timescale, with other (so-called "free" or "bulk") water molecules and D_2O. Other studies have concluded that "bound" water has thermally labile hydrogen bonds (Pouchly et al., 1979; Biros et al., 1979), shows cooperative molecular mobility (Hoeve, 1980), has a heat capacity approximately equal to that of liquid water rather than ice (Hoeve and Hoeve, 1978; Hoeve, 1980; Pouchly, 1979), and has some capability to dissolve salts (Burghoff and Pusch, 1980).

It has been concluded that "in the past, too much emphasis has been given to 'water binding'" (Simatos and Karel, 1988). In fact, the typical observation of two relaxation peaks (ascribed, following traditional dogma, to "free" and "bound" water) for all biological tissues and solutions that have been examined in dielectric experiments (Mashimo et al., 1987) is entirely consistent with and exactly analogous to the behavior of synthetic polymers with their non-aqueous, nonhydrogen bonding organic plasticizers (Sears and Darby, 1982). The traditional point of view on the "structuring" effect of solutes on water, which helped give rise to the myth of "bound" water, is rightfully being replaced by a new perspective and emphasis on the mobilizing effect of water acting as a plasticizer on solutes. This new perspective has led to a deeper qualitative understanding of structure–function relationships in aqueous food polymer systems, including doughs and baked products (Slade, 1984; Slade and Levine, 1984a, 1984b, 1987a, 1987b, 1988a–e; Levine and Slade, 1986, 1987, 1988a–f; Slade et al., 1988).

Effects of Water Plasticization
on Partially Crystalline Polymers

In partially crystalline polymers, water plasticization occurs only in the amorphous regions (Starkweather, 1980; van den Berg, 1981; Gaeta et al., 1982; Jin et al., 1984). In linear synthetic polymers with anhydrous crystalline regions and a relatively low capacity for water in the amorphous regions (e.g., nylons [Starkweather, 1980]), the percent crystallinity affects Tg, such that increasing percent crystallinity generally leads to increasing Tg (Jin et al., 1984). This behavior is due primarily to the stiffening or "antiplasticizing" effect of disperse microcrystalline cross-links, which leads to decreased mobility of the chain segments in the interconnected amorphous regions (Gaeta et al., 1982). The same effect is produced by covalent cross-links (Jin et al., 1984), which, when produced by radiation, occur only in the amorphous regions (Ellis et al., 1984). In polymers with anhydrous crystalline regions, only the amorphous regions are accessible to penetration and therefore to plasticization by moisture (Ellis et al., 1984; Jin et al., 1984).

Similar phenomena are observed in partially crystalline polymers with hydrated crystalline regions, such as starch (Maurice et al., 1985), gelatin (Marshall and Petrie, 1980), and collagen (Batzer and Kreibich, 1981). For example, in native starches, hydrolysis by aqueous acid ("acid etching") or enzymes, at $T < Tm$, can occur initially only in amorphous regions (French, 1984). Similarly, acid etching of retrograded starch progresses in amorphous regions, leading to increased relative crystallinity (or even increased absolute crystallinity, by crystal growth) of the residue (French, 1984). Dehumidification of granular starch proceeds most readily from initially mobile amorphous regions, leading to nonuniform moisture distribution (van den Berg, 1981). In partially gelatinized starches, dyeability by a pigment increases with increasing amorphous content (Kuge and Kitamura, 1985). The effective Tg, which immediately precedes and thereby determines the temperature of gelatinization ($Tgelat$) in native starch, depends on the extent and type (B versus A versus V polymorphs) of crystallinity in the granule (but not on amylose content), and on total moisture content and moisture distribution (Slade, 1984; Slade and Levine, 1984a, 1984b, 1987a, 1988b; Maurice et al., 1985). For normal and waxy starches, $Tgelat$ increases with increasing percent crystallinity (Juliano, 1982), an indirect effect due to the disproportionation of mobile short branches of amylopectin from amorphous regions to microcrystalline "micelles," thereby increasing the average MW and effective Tg of the residual amorphous constituents (Slade and Levine, 1988b), because these branches are unavailable to serve as "internal" plasticizers (Sears and Darby, 1982). Two other related phenomena are observed as a result of the nonuniform moisture distribution in situations of overall low moisture content for polymers with hydrated crystalline regions: (1) atypically high Tg/Tm (in K) ratios much greater than 0.80 but, of course, less than 1.0 (Wunderlich, 1980; Soesanto and Williams, 1981), in contrast to the characteristic range of 0.5–0.8 for many partially crystalline synthetic polymers (Brydson, 1972), as is illustrated in Figure 5-16 by collagen (Tg/Tm = 0.93) versus nylon 6 (Tg/Tm = 0.67); and (2) a pronounced apparent depressing effect of water on Tm (Lelievre, 1976; Zobel et al., 1988) as well as Tg, such that both Tg and Tm decrease with increasing moisture content, as is also illustrated in Figure 5-16 for collagen and nylon 6.

The fact that water plasticization occurs only in the amorphous regions of partially crystalline, water-compatible polymers is critical to the explanation of how these metastable amorphous regions control the nonequilibrium melting behavior of the crystalline regions. As has been mentioned, the concept of nonequilibrium melting established for partially crystalline synthetic polymers (Wunderlich, 1981) has been applied to biopolymer systems such as native starch and gelatin (Slade, 1984; Slade and Levine, 1984a, 1984b, 1987a, 1987b, 1988a, 1988b, 1988e), in order to describe the mechanical relaxation process that occurs as a consequence of a dynamic heat/moisture/time treatment (Blanshard, 1979; Lund, 1984). The existence of contiguous microcrystalline and amorphous regions (e.g., in native starch, the crystallizable short branches and backbone segments with their branch points, respectively, of amylopectin molecules) covalently linked through individual polymer chains creates a "fringed micelle" network. Relative dehydration of the amorphous regions to an initial low overall moisture content leads to the kinetically stable condition in which the effective Tg is higher than the "equilibrium" Tm of the hydrated crystalline regions. Consequently, the effective Tm (Slade and Levine, 1988b) is elevated and observed only after softening of the amorphous regions at Tg. Added water acts directly to plasticize the continuous glassy regions, leading to the kinetically metastable condition in which their effective Tg is depressed. Thus, the "fringe" becomes an unstable rubber at $T > Tg$, allowing sufficient mobility and swelling by thermal expansion and water uptake for the interconnected microcrystallites, embedded in the "fringed micelle" network, to melt (by dissociation, with concomitant volume expansion) on heating to a less kinetically constrained Tm only slightly above the depressed Tg. For such a melting process, use of the Flory-Huggins thermodynamic treatment to interpret the effect of water content on Tm has no theoretical basis (Lelievre, 1976; Wunderlich, 1981; Alfonso and Russell, 1986), because while water as a plasticizer does directly affect the Tg and indirectly affect the

Tm of polymers such as starch and gelatin, the effect on Tm is not the direct effect experienced in equilibrium melting (i.e., dissolution) along the solidus curve (Slade, 1984; Slade and Levine, 1984a, 1984b, 1987a, 1987b, 1988b, 1988e). In contrast to the case of native starch, where initial "as is" moisture is limiting, in a case where excess moisture exists, such as a retrograded wheat starch gel with $\gtrsim$ 27 w% water (Wg'), where the amorphous matrix would be fully plasticized and ambient temperature would be above Tg (i.e., $Tg' \simeq -5°C$), the fully hydrated and matured crystalline junctions would show the actual, lower (and closer to equilibrium) Tm of $\simeq$ 60°C for retrograded B-type starch (Slade, 1984; Slade and Levine, 1984a, 1984b, 1987a). Note the analogy between the description of nonequilibrium melting in native granular starch and the case described earlier of nonequilibrium melting in a synthetic, partially crystalline block copolymer. In this context, it is interesting that Wunderlich (1980) defines branched polymers as a special case of copolymers, using the example of a synthetic polymer with crystallizable branches.

Another interesting case for which the concept of nonequilibrium melting can be used to explain the melting behavior observed in a starch polymer–water system involves acid-hydrolyzed starches. The "lintnerization" process (French, 1984; Blanshard, 1987), as applied to native granular starches (Buleon et al., 1987) or retrograded starch gels (Mestres et al., 1988), involves hydrolysis with aqueous hydrochloric or sulfuric acid, initially in the accessible amorphous regions. As acid etching of the amorphous "fringes" surrounding the microcrystalline "micelles" progresses and the three-dimensional network is broken down, the apparent crystallinity of the isolated residue increases (even in the absence of possible crystal growth or perfection during hydrolysis [Buleon et al., 1987]). This residue is composed mainly of the more resistant (i.e., not depolymerized during hydrolysis [Buleon et al., 1987]), complexed amylose–lipid and retrograded amylose microcrystallites with decreasing amounts of contiguous, covalently attached amorphous material. However, even though the relative percent crystallinity of the residue increases with increasing extent of lintnerization, the measured Tm of the highly crystalline, complexed amylose–lipid lintner product can be observed to *decrease* (Mestres et al., 1988), relative to the Tm of that crystalline component in the native granule or retrograded gel prior to lintnerization. In the lintnerized normal corn starch gels studied by Mestres and co-workers (1988), this measured decrease in Tm of the crystalline complexed amylose–lipid was an unexpected and unexplained observation said to "merit further examination." The authors of this chapter suggest that the explanation lies in the fact that the measured Tm, prior to lintnerization, is the effective Tm for the nonequilibrium melting process allowed only above the effective Tg of the kinetically constrained amorphous regions surrounding the crystalline complexed amylose–lipid, which can be seen for wheat starch as the step change in heat capacity contributed by free volume (Wunderlich, 1981) that precedes the endotherm above 100°C for both native granular starch (see Fig. 5-22a) and mature retrograded starch gel (see Fig. 5-22d). In contrast, the apparent Tm of the lintnerized product, substantially freed from the immobility previously imposed by the surrounding amorphous network, is a lower effective Tm closer to the "equilibrium" Tm of the complexed amylose–lipid microcrystals.

Effects of Water Plasticization on Viscoelastic Properties of Polymers in the Rubbery State: the Domain of WLF Theory, the WLF Equation, and WLF Kinetics

At temperatures above the effective Tg, plasticization by water affects the viscoelastic, thermomechanical, electrical, guest/host diffusion, and gas permeability properties of completely amorphous and partially crystalline polymer systems to an extent mirrored in its effect on Tg (Levine and Slade, 1987). In the rubbery range above Tg (typically from Tg to $Tg + 100°C$), the dependence of viscoelastic properties on temperature is successfully predicted (Cowie, 1973) by the WLF equation, an empirical equation whose form was originally derived from free volume theory (Williams et al., 1955; Ferry, 1980). The WLF equation (Williams et al., 1955; Soesanto and Williams, 1981) can be written as:

$$\log_{10} \left| \frac{\eta}{\rho T} \middle/ \frac{\eta g}{\rho g \, Tg} \right| = -\frac{C1(T - Tg)}{C2 + (T - Tg)} \tag{5–1}$$

where η is the viscosity or other diffusion-controlled relaxation process, ρ the density, and $C1$ and $C2$ are "universal constants" (17.44 and 51.6, respectively, as extracted from experimental data on numerous synthetic polymers [Williams et al., 1955]). Sperling (1986) has made the salient remark that "for a generation of [synthetic] polymer scientists and rheologists, the WLF equation has provided a mainstay both in utility and theory." This equation describes the kinetic nature of the glass transition, and has been shown to be applicable to any glass-forming polymer, oligomer, or monomer (Ferry, 1980), including the following: molten glucose (Williams et al., 1955), amorphous glucose–water mixtures (Chan et al., 1986), and concentrated solutions of mixed sugars (Soesanto and Williams, 1981) as examples of interest for foods such as baked products. The equation defines mobility in terms of the non-Arrhenius temperature dependence of the rate of any diffusion-controlled relaxation process occurring at a temperature T compared to the rate of the relaxation at the reference temperature Tg, shown here in terms of log η related usefully to ΔT, where $\Delta T = T - Tg$. The WLF equation is valid in the temperature range of the rubbery or undercooled liquid state and is based on the assumptions that polymer free volume increases linearly with increasing temperature above Tg (as illustrated implicitly in Fig. 5-7) and that segmental or mobile unit viscosity, in turn, decreases rapidly with increasing free volume (as illustrated implicitly in Fig. 5-8) (Ferry, 1980). Thus, the greater the ΔT, the faster a system is able to move (due to increased free volume and decreased mobile unit viscosity); so the greater the mobility, and the shorter the relaxation time. In essence, the WLF equation and resulting master curve of log ($\eta/\eta g$) versus $T - Tg$ (Williams et al., 1955; Soesanto and Williams, 1981) represent a mobility transformation, described in terms of a time–temperature superposition (Slade and Levine, 1988a). Such WLF plots typically show a five orders of magnitude change in viscosity (or in the rates of other relaxation processes) over a 20°C interval near Tg (Soesanto and Williams, 1981; Franks, 1985b), which is characteristic of WLF behavior in the rubbery fluid range (Slade and Levine, 1988a).

In the context of the utility of the WLF equation, the underlying basis of the principle of time–temperature superpositioning is the equivalence between time (or frequency) and temperature as they affect the molecular relaxation processes that influence the viscoelastic behavior (i.e., the dual characteristics of viscous liquids and elastic solids) of polymeric materials and glass-forming small molecules (Ferry, 1980; Sichina, 1988). This principle is illustrated in Figure 5-17, which shows a master curve of the modulus as a function of temperature or frequency for a typical partially crystalline synthetic high polymer (Graessley, 1984). Figure 5-17 has been used to describe the viscoelastic behavior of such materials, as exemplified by a kinetically metastable gelatin gel in an undercooled liquid state, in the context of WLF theory (Borchard et al., 1980). At $T > Tg$, gelatin gels manifest a characteristic rubberlike elasticity (Tomka et al., 1975), due to the existence of a network of entangled, randomly coiled chains (Yannas, 1972). With increasing temperature, a gelatin gel traverses the five regions of viscoelastic behavior characteristic of synthetic, partially crystalline polymers (Yannas, 1972), as illustrated in Figure 5-17: (1) at $T < Tg$, vitrified glass; (2) at $T = Tg$, glass transition to leathery region; (3,4) at $Tg < T < Tm$, rubbery plateau to rubbery flow; and (5) at $T > Tm$, viscous liquid flow. It is interesting to note that at $Tg < T < Tm$, a gelatin gel is freely permeable to the diffusion of dispersed dyes and molecules as large as hemoglobin (Slade et al., 1988); only at $T < Tg$ is such dye diffusion greatly inhibited (Wesson et al., 1982).

The WLF equation is not intended for use much below Tg (i.e., in the glassy solid state) or in the very low-viscosity liquid state ($\eta < 10$ Pa s [Soesanto and Williams, 1981]), typically 100°C or more above Tg, where Arrhenius kinetics apply (Ferry, 1980). For partially crystalline polymers, the breadth of the temperature range of the rubbery domain of WLF behavior corresponds to the temperature interval between Tg and Tm (Wunderlich, 1976; Ferry, 1980; Franks, 1982), as illustrated in Figure 5-8. An analysis (Slade and Levine,

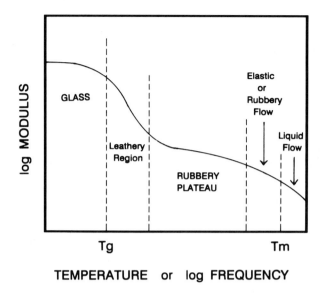

Figure 5-17. Master curve of the modulus as a function of temperature or frequency illustrating the five regions of viscoelastic behavior characteristic of partially crystalline synthetic polymers.

1988a) of the variation of the size of this temperature interval with the Tg/Tm ratio of representational synthetic polymers and glass-forming, LMW carbohydrates has illustrated the origin of the usual 100°C range for a typical, "well-behaved" synthetic elastomer of $Tg = 200$ K, $Tm = 300$ K, and $Tg/Tm = 0.67$. In contrast, the analysis has demonstrated WLF domains of (1) 200°C for a highly symmetrical polymer of $Tg = 200$ K, $Tm = 400$ K, and $Tg/Tm = 0.5$; (2) only 50°C for a highly unsymmetrical polymer of $Tg = 200$ K, $Tm = 250$ K, and $Tg/Tm = 0.8$; and (3) less than 50°C for an "atypical, poorly behaved" polymer of $Tg/Tm > 0.9$. The synthetic polymer cited as the classic example of category (2) behavior, which has been attributed to anomalously large free volume at Tg, is bisphenol polycarbonate, with $Tg/Tm = 0.85$ (Brydson, 1972). Category (3) behavior has been reported to be exemplified by food materials such as native starch and gelatin (due to nonuniform distribution of moisture in amorphous and crystalline regions) and fructose and galactose (due to anomalous translational free volume) (Slade and Levine, 1988a). This analysis has pointed out the critical significance of anomalous values of the Tg/Tm ratio close to 1.0 on the mobility, resultant relaxation behavior, and consequent technological process control for such nonequilibrium food polymer systems in their supra-glassy fluid state above Tg (Slade and Levine, 1988a), in terms of the WLF kinetics of various translational diffusion-limited, mechanical/structural relaxation processes, such as gelatinization, annealing, and recrystallization of starch (Slade and Levine, 1988b).

Description of time-/temperature-dependent behavior by the WLF equation requires selection of the appropriate reference Tg for any particular glass-forming material (of a given MW and extent of plasticization [Williams et al., 1955; Soesanto and Williams, 1981; Chan et al., 1986]), be it Tg for a low-moisture system or Tg' for a frozen system. Tg for a typical undercooled liquid is defined in terms of an isofree volume state of limiting free volume and also approximately as an isoviscosity state somewhere in the range from 10^{11}–10^{14} Pa s (Soesanto and Williams, 1981; Franks, 1982). This isoviscosity state refers to local, not macroscopic, viscosity. This fact constitutes a critical conceptual distinction (Slade and Levine, 1988a), because the glass transition is a cooperative transition (Wunderlich, 1981) resulting from local cooperative constraints on mobility, and Tg represents a thermomechanical property controlled by the local segmental or small molecule, rather than macroscopic, environment of a polymer. Cooperative constraints of local viscosity and free volume on translational diffusion determine the temperature at which the glass transition is manifested, as a dramatic increase in relaxation times compared to the experimental timeframe. As has been mentioned, a Tg/Tm ratio close to 1.0 is accounted for by an anomalously large free volume requirement for diffusion (Brydson, 1972). When the free volume requirement is so large, a glass transition (i.e., vitrification of the rubbery fluid) on cooling can actually occur even when the local viscosity of the system is relatively low. Thus, instead of the typical "firmness" for a glass ($\simeq 10^{12}$ Pa s (Ferry, 1980)), such a glass (e.g., of fructose or galactose) may manifest a $\eta g \ll 10^{12}$ Pa s (Soesanto and Williams, 1981; Slade and Levine, 1988a).

Based on WLF theory, a generalized physicochemical mechanism for mechanical/structural relaxation processes ("collapse" phenomena) in glass-forming food systems, including doughs and baked products, has been described as follows (Levine and Slade, 1986). When the ambient temperature (during processing or storage) becomes greater than Tg, the free volume of the system increases, leading to increased cooperative mobility in the environment of polymer chain segments or small molecules in the amorphous regions of both glassy and partially crystalline systems. Consequently, the viscosity of the dynamically constrained solid falls to below ηg (Downton et al., 1982), which permits viscous liquid flow (Flink, 1983). In this rubbery state, translational diffusion can occur on a practical timescale, and diffusion-controlled relaxations (which also frequently exhibit a prerequisite for nucleation) are free to proceed with rates defined by the WLF equation (i.e., rates that increase logarithmically in a nonlinear fashion with increasing ΔT). This generalization is made possible by the fact that there are two paths by which the ambient temperature can become greater than Tg (Levine and Slade, 1988e). Heating (plasticization by heat) elevates the ambient temperature to above Tg for both diluent-free (bone dry) and diluted (water-containing) systems. Increasing the moisture content (plasticization by water) depresses the Tg to below the ambient temperature for diluted systems only.

WLF kinetics differ from Arrhenius kinetics in three important ways (Levine and Slade, 1988e):

1. The reference temperature, Tg, is a solute-specific property whose location depends on the linear DP and stereochemistry of mobile units for WLF kinetics, but the reference temperature is arbitrarily taken to be 0 K for Arrhenius kinetics and is not explicitly designated.
2. The exponential dependence of relaxation rates on temperature is itself temperature dependent for WLF kinetics, but the exponential dependence of relaxation rates is a linear function of temperature for Arrhenius kinetics.
3. The magnitude of the temperature dependence of WLF kinetics near Tg is dramatically greater than that of Arrhenius kinetics below Tg or very far above Tg.

The impact of WLF behavior on the kinetics of diffusion-controlled relaxation processes in water-plasticized, rubbery food polymer systems has been conceptually illustrated by the idealized curve shown in Figure 5-18 (Levine and Slade, 1988e). Relative relaxation rates, calculated from the WLF equation with its universal numerical constants, demonstrate the nonlinear logarithmic relationship: for $\Delta T = 0, 3, 7, 11,$ and $21°C$, corresponding relative rates would be $1, 10, 10^2, 10^3,$ and 10^5, respectively. These rates illustrate the five orders of magnitude change over a 20°C interval above Tg, which is

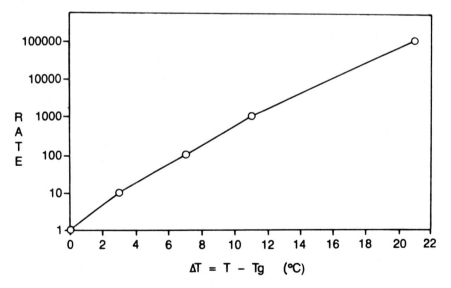

Figure 5-18. Variation of the rate of a diffusion-controlled relaxation process against $\Delta T = T - Tg$, as defined by the WLF equation with its "universal" constants of $C1 = 17.44$ and $C2 = 51.6$. (*From Levine and Slade, 1988e, with permission.*)

typically shown by WLF plots. These rates are dramatically different from the rates defined by the familiar Q_{10} rule of Arrhenius kinetics for dilute solutions. For Arrhenius behavior above Tm, a factor of ten change in relaxation rate would require a 33°C change in temperature, in comparison to a 3°C change for WLF behavior near Tg of a partially crystalline polymer of $Tg/Tm = 0.67$ (Slade and Levine, 1988a). Another general example of WLF-governed relaxation behavior was illustrated earlier by the crystallization kinetics diagram in Figure 5-6. Recrystallization is a diffusion-limited process (Baro et al., 1977). The propagation step in the recrystallization mechanism approaches a zero rate at $T < Tg$ for an amorphous but crystallizable polymer (Wunderlich, 1976) (e.g., freshly gelatinized starch), initially quenched from the melt to a kinetically metastable solid state. Due to immobility in the glass, migratory diffusion of large main-chain segments (e.g., of amylose, or of crystallizable branches of amylopectin) required for crystal growth would be inhibited over realistic times. However, the propagation rate increases exponentially with increasing ΔT above Tg (up to Tm) (Marsh and Blanshard, 1988), due to the mobility allowed in the rubbery state. (Note the similar shapes of the WLF curve in Fig. 5-18 and the propagation rate curve in Fig. 5-6.) Thus, a recrystallization transition from unstable (i.e., undercooled) amorphous liquid to (partially) crystalline solid may occur at $T > Tg$ (White and Cakebread, 1966; Phillips et al., 1986; Karel, 1986), with a rate defined by the WLF equation.

As was mentioned in the section Crystallization Kinetics for Partially Crystalline Polymers, on a timescale of technological significance, crystallization can occur only within the WLF rubbery domain (Wunderlich, 1976). In the process of crystallization for a polymer that is completely amorphous and unseeded, homogeneous nucleation in an undercooled melt is the first mechanistic stage, which must precede crystal growth. The necessary extent of undercooling in K from Tm to Th is a universal property of crystallizable materials. Just as Tg is related to Tm by the ratio $Tg/Tm \simeq 0.67$, with a range of 0.5–1.0, for essentially all molecular glass-formers, including high polymers and small molecules, Th is related to Tm by the ratio $Th/Tm \simeq 0.8$, with a narrow range of 0.78–0.85 for essentially all crystallizable substances (Walton, 1969; Wunderlich, 1976). The relationships among Th, Tm, and Tg for representational small sugars in water is illustrated in Figure 5-19, which shows schematic state diagrams for three different situations that can govern

--→

Figure 5-19. Schematic state diagrams for representational small sugars with Tg/Tm ratios of (A) 0.67, (B) about 0.75, and (C) > 0.9. The diagrams illustrate the locations of the solute–water Tm and Tg curves relative to the curve of estimated Th, and emphasize how the influence of the glass transition on the homogeneous nucleation of solute from undercooled concentrated solutions differs according to the location of Th within or outside the WLF region between Tm and Tg. (*From Slade and Levine, 1988a, with permission.*)

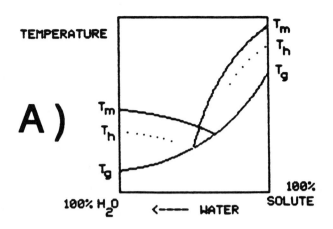

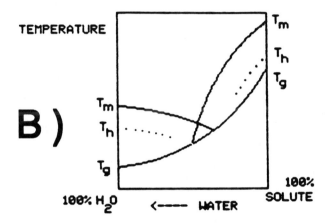

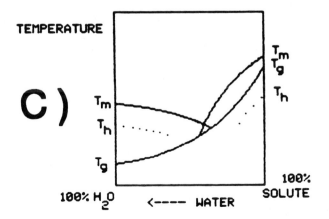

homogeneous nucleation. The situations result from different values of the Tg/Tm ratio, which reflects the magnitude of the metastable WLF region in which crystallization can occur. In each case, Th was located according to the typical ratio of Th to Tm. These stylized state diagrams highlight the different ways in which Th and Tg can be related, depending on the Tg/Tm ratio for a particular solute, and thus reveal how the relationship between Th and Tg determines and allows prediction of the stability against recrystallization of concentrated or supersaturated aqueous solutions of specific sugars (Levine and Slade, 1987). In the first case (Fig. 5-19A), for a sugar with a typical Tg/Tm ratio of 0.67, homogeneous nucleation of the solute would be very efficient, because Th lies well above Tg. Therefore, upon undercooling this concentrated solution from $T > Tm$, homogeneous nucleation would occur at Th within the liquid zone, before vitrification could immobilize the system at Tg. Common examples of sugars whose actual state diagrams resemble the one in Figure 5-19A, and which are known to crystallize readily from concentrated aqueous solution, include xylose ($Tg/Tm = 0.66$) and glucose ($Tg/Tm = 0.70$), and to a lesser extent, sucrose ($Tg/Tm = 0.70$) (Slade and Levine, 1988a). In the second case (Fig. 5-19B), for a sugar with a higher Tg/Tm ratio $\simeq 0.75$, homogeneous nucleation would be retarded, because Th falls much closer to Tg in the more viscous fluid region where transport properties can become a significant limiting factor on nucleation in nonequilibrium systems (Leubner, 1987). Ribose ($Tg/Tm = 0.73$) is an example of a sugar with a state diagram resembling Figure 5-19B, whose nucleation would be so retarded. In the last case (Fig. 5-19C), for a sugar with an anomalously high Tg/Tm ratio close to 1.0, homogeneous nucleation would be prevented on a practical timescale, because Th actually lies below Tg. Thus, on undercooling a concentrated solution, vitrification would occur first, thereby immobilizing the system and preventing the possibility of solute nucleation at Th. Fructose ($Tg/Tm = 0.94$), which is well known to be almost impossible to crystallize from aqueous solution without preseeding or precipitating with non-solvent, exemplifies the state diagram in Figure 5-19C and the nucleation inhibition behavior predicted from it (Slade and Levine, 1988a, 1988e). Once again, the explanation for this behavior derives from the WLF kinetics governing the rubbery domain near Tg, where a 20°C temperature interval is equivalent to a range of five orders of magnitude in relaxation rates. Hence, within practical timeframes, the immobility imposed by the glassy domain can have an all-or-nothing effect on homogeneous nucleation and crystal growth (Slade and Levine, 1988a).

One of the most critical messages to be distilled at this point is that the structure–property relationships of water-compatible food polymer systems are dictated by a moisture–temperature–time superposition (Starkweather, 1980; Flink, 1983; Levine and Slade, 1986). Referring to the idealized state diagram in Figure 5-4 as a conceptual mobility map (which reflects the "real world" cases illustrated in Figs. 5-12 through 5-16), one sees that the Tg curve represents a boundary between nonequilibrium glassy and rubbery physical states in which various diffusion-controlled processes (e.g., mechanical and structural relaxations) either can occur over realistic times (at $T > Tg$ and $W > Wg'$, the domain of "water dynamics," or $T > Tg$ and $W < Wg'$, the domain of "glass dynamics" [Levine and Slade, 1987]) or cannot occur over realistic times (at $T < Tg$, in the glassy state) (Levine and Slade, 1986). The WLF equation defines the kinetics of molecular-level relaxation processes, which will occur in practical timeframes only in the rubbery state above the effective Tg, in terms of an exponential, but non-Arrhenius, function of ΔT above this boundary condition.

Cryostabilization Technology— Collapse Processes

Cryostabilization Technology

Upon a foundation of pioneering studies (from the field of cryobiology) of the low-temperature thermal properties of frozen aqueous model systems relevant to the cryopreservation of biopolymers, by Luyet (1939, 1960; Luyet and Rasmussen, 1968; Rasmussen and Luyet, 1969), MacKenzie (1977; MacKenzie and Rasmussen, 1972), Rasmussen, and Franks (1982, 1985a, 1985b; Franks et al., 1977), an extensive "cryostabilization technology" database of DSC results for Tg' and Wg' values of carbohydrate and protein food ingredients has been built (Levine and Slade, 1986, 1987, 1988a–f; Slade and Levine, 1988a, 1988e; Slade et al., 1988). Cryostabilization is a concept introduced to describe a practical industrial technology for the storage stabilization of frozen, freezer-stored, and freeze-dried foods (Levine and Slade, 1986, 1988c, 1988e). Cryostabilization provides a means of protecting products, stored for long periods at typical freezer temperature ($Tf = -18°C$), from deleterious changes in texture (e.g., "grain growth" of ice, solute crystallization), structure (e.g., "collapse," shrinkage), and chemical composition (e.g., enzymatic activity, oxidative reactions such as fat rancidity, flavor/color degradation). Such changes are exacerbated in many typical fabricated foods whose formulas are dominated by LMW carbohydrates of characteristically low Tg' and high Wg'. The key to this protection, and resulting improvement in product quality and storage stability, lies in controlling the structural state, by controlling the physicochemical and thermomechanical properties, of the freeze-concentrated amorphous matrix surrounding the ice crystals in a frozen system.

The importance of the glassy state of the maximally freeze-concentrated solute/UFW matrix and the special technological significance of its particular Tg, Tg', relative to Tf, have been described in the context of solute–water state diagrams such as the idealized one shown in Figure 5-4 (Levine and Slade, 1988c, 1988e). Figure 5-4 has been used to illustrate the fundamental physicochemical basis of cryostabilization and to explain why Tg' is the essence of the conceptual framework of this technology. If the amorphous matrix surrounding the ice crystals in a frozen food system is maintained as a mechanical solid (i.e., at $Tf1 < Tg'$ and $\eta > \eta g$), then diffusion-controlled processes that typically result in reduced quality and stability can be virtually prevented, or at least greatly inhibited. This physical situation has been illustrated by scanning electron microscopy (SEM) photomicrographs of frozen model solutions (Franks, 1980, in Levine and Slade, 1988c), which show small, discrete ice crystals embedded and immobilized in a continuous amorphous matrix of freeze-concentrated solute/UFW that exists as a glassy solid at $T < Tg'$. The situation has been described by analogy to an unyielding block of window glass with captured air bubbles. In contrast, storage stability is reduced if a natural food material is improperly stored at too high a Tf, or if a fabricated product is improperly formulated, so that the matrix is allowed to exist as a rubbery liquid at $Tf2 > Tg'$ (see Fig. 5-4), in and through which translational diffusion is free to occur. Thus, the Tg' glass has been viewed as the manifestation of a kinetic barrier to any diffusion-controlled process (Franks, 1985b; Levine and Slade, 1986), including further ice formation (within the experimental timeframe), despite the continued presence of UFW at all temperatures below Tg'. Analogously, Ellis (1988) has remarked on the ability of a continuous glassy matrix of synthetic polyamide to act as a barrier to rapid moisture loss during DSC heating of a water-plasticized, partially crystalline glassy polyamide sample. The "high activation energy" of this kinetic barrier to relaxation processes has been recognized as the extreme temperature dependence that governs the decrease in free volume and increase in local viscosity near Tg. This perspective on the glass at $Tg'-Cg'$ as a mechanical barrier has provided a long-sought (Guegov, 1981) theoretical explanation of how undercooled water can persist (over a realistic time period [Franks, 1986]) in a solution in the presence of ice crystals (Levine and Slade, 1988c, 1988e). Recognizing these facts, and relating them to the conceptual framework described by Figure 5-4, one can appreciate why the temperature of this glass transition is so important to aspects of frozen food technology involving freezer-storage stability, freeze-concentration, and freeze-drying (Franks, 1982, 1985b; Blanshard and Franks, 1987), all of which are subject to various recrystallization and collapse phenomena at $T > Tg'$ (Levine and Slade, 1986, 1988b, 1988c, 1988e, 1988f).

The optimum Tf for a natural material or optimum formula for a fabricated product is dictated by the characteristic Tg' of the specific solute(s)/UFW matrix composition. Tg' has been shown to be governed in turn by the $\overline{M}w$ of the particular combination of water-compatible solids in a complex food system (Franks, 1985b; Slade and Levine, 1988a; Levine and Slade, 1986,

1987). Moreover, the dynamic behavior of rubbery frozen food products during storage above Tg' has been demonstrated to be dramatically temperature dependent, and the rates of diffusion-controlled deterioration processes to be quantitatively determined by the ΔT between Tf and Tg' (Levine and Slade, 1986, 1987, 1988c, 1988e). Thus, the successful practice of the principles of cryostabilization technology has often been shown to rely on the critical stabilizing role of polymeric carbohydrates and proteins in preventing collapse in frozen food systems (Levine and Slade, 1986, 1988b–f). [Furthermore, the same stabilization principles have been shown to apply equally well to low-moisture foods (Levine and Slade, 1988a).] Based on the established fact that Tg' is a function of MW for both homologous and quasi-homologous families of water-compatible polymers, examples of how the selection and use of appropriate ingredients in a fabricated product (such as gluten supplementation in frozen bread dough) allow the food technologist to manipulate the composite Tg', and thus deliberately formulate to elevate Tg' relative to Tf and so enhance product stability, have been described (Levine and Slade, 1986, 1987, 1988a–f; Slade et al., 1988).

The Tg' and Wg' results for a broad range of food ingredients have permitted the definition of a cryostabilization technology "spectrum," illustrated schematically by the hatched band in Figure 5-20 (Levine and Slade, 1988c). Insights gained from Figure 5-20 have led to the identification of "polymeric cryostabilizers" as a class of common water-compatible food ingredients (including low-DE SHPs and proteins) with a characteristic combination of high Tg' and low Wg' values (Levine and Slade, 1986; Slade et al., 1988). Differential scanning calorimetry investigations of polymeric cryostabilizers have provided an understanding of their stabilizing function, via their influence on the structural state of a complex amorphous matrix, as conceptualized in Figure 5-4. This function derives from their HMW, and the resulting elevating effect of such materials on the composition-dependent Tg' (and corresponding depressing effect on Wg') of a complex frozen system. Increased Tg' leads to decreased ΔT (relative to Tf), which in turn results in decreased rates of change during storage, and so to increased stability of hard-frozen products. The efficacy of polymeric cryostabilizers has also been explained in terms of the breadth of the temperature range for the WLF region, that is, the magnitude of the rubbery domain between Tg' and the Tm of ice (Levine and Slade, 1987). This ΔT ($= Tm - Tg'$) decreases with increasing solute MW, from $> 50°C$ for very low MW polyols and amino acids (with $Tg' < -50°C$) to $< 5°C$ for very low-DE SHPs (with $Tg' > -5°C$). Thus, as this ΔT decreases, the magnitude of the corresponding temperature region of freezer-storage instability (between Tf and a lower Tg') likewise decreases, so that even at higher freezer temperatures, the probability of acceptable product stability would be improved (Levine and Slade, 1988e). In contrast to cryostabilizers, LMW solutes (e.g., sugars, polyols, glycosides, and amino

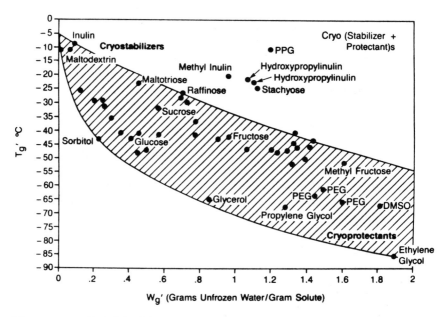

Figure 5-20. Variation of the glass transition temperature, Tg', for maximally-frozen 20 w% solutions against Wg', the composition of the glass at Tg', in g unfrozen water/g solute, for a series of water-compatible carbohydrates, including many compounds represented in Figs. 5-9 and 5-10, illustrating the cryostabilization "spectrum" from monomeric cryoprotectants to polymeric cryostabilizers. [Solutes lying outside the hatched area of the "spectrum" (toward the upper right corner) exhibit properties of both cryoprotectants and cryostabilizers.] (*From Levine and Slade, 1988c, with permission.*)

acids [Slade et al., 1988]) with a characteristic combination of low Tg' and high Wg' values have been predicted and subsequently demonstrated to have utility as "monomeric cryoprotectants" (see Fig. 5-20) in freezer-stored food products with a desirable soft-frozen texture, but undesirably poor stability (Levine and Slade, 1986, 1987, 1988c, 1988e).

The elucidation of the structure–property relationships of food cryostabilizers and cryoprotectants, illustrated by the results in Figure 5-20, has also revealed the underlying physicochemical basis of fundamental (and intuitive) correlations between the critical functional attributes of storage stability and texture of frozen foods (Levine and Slade, 1988c, 1988e). As an essentially universal rule, for both complex products and model systems of single or multiple solutes, higher Tg' and lower Wg' values have been shown to go hand in hand with and be predictive of harder-frozen texture and increased storage stability at a given Tf. Conversely, lower Tg' and higher Wg' values go hand in hand with and are predictive of softer-frozen texture and decreased storage stability. These intrinsic correlations, which have been found to apply to various types of frozen foods (Cole et al., 1983, 1984; Slade et al., 1988), have been recognized as underlying precepts of cryostabilization technology, as illustrated by the "spectrum" shown in Figure 5-20, which derives from the effect of solute MW on Tg' and Wg'. A second contribution to textural hardness of products characterized by high Tg' values derives from the elevated modulus of the amorphous matrix and is directly related to the mechanical stabilization against diffusion-controlled processes such as ice crystal growth (Levine and Slade, 1988e).

Collapse Processes

Extensive studies of the thermal and thermomechanical properties of concentrated aqueous solutions of small carbohydrates at subzero temperatures have established that Tr, the microscopically observed temperature of irruptive ice recrystallization in such glass-forming systems of LMW sugars and polyols (Franks, 1982; Franks et al., 1977; Luyet, 1960; Rasmussen and Luyet, 1969; MacKenzie and Rasmussen, 1972; MacKenzie, 1977; Forsyth and MacFarlane, 1986; Thom and Matthes, 1986), coincides with the solute-specific Tg', measured by thermal or thermomechanical analysis (Levine and Slade, 1986, 1988b, 1988f; Franks, 1985b; Blanshard and Franks, 1987; Reid, 1985; Schenz, 1987). It has also been recognized that ice recrystallization is but one of many possible manifestations (referred to as collapse phenomena) of the dynamically controlled behavior of aqueous glasses and rubbers, which exist at subzero temperatures in kinetically constrained, metastable states rather than equilibrium phases (Levine and Slade, 1986, 1987, 1988a–f; Slade and Levine, 1988a, 1988e; Karel, 1985, 1986; Simatos and Karel, 1988). Generic use of the term "rubber" in this context describes all glass-forming liquids at $Tg < T < Tm$, including both molecular rubbers (viscous liquids) of LMW compounds and viscoelastic network-forming rubbers of entangling high polymers (Ferry, 1980).

As has been discussed, collapse phenomena have been defined as translational diffusion-limited, material-specific, structural and/or mechanical relaxation processes that occur far from equilibrium and are therefore governed by the critical system variables of moisture content, temperature, and time (Tsourouflis et al., 1976). Collapse processes occur at a characteristic collapse transition temperature, Tc (MacKenzie, 1975; To and Flink, 1978; Franks, 1982), which can coincide, on a comparable timescale (e.g., during storage), with an underlying Tg, or can be manifested in real time (e.g., during processing) at a temperature about 20°C above Tg (i.e., at a relaxation rate 10^5 faster than at Tg) (Levine and Slade, 1986; Slade and Levine, 1988a). At the relaxation temperature, moisture content is the critical determinant of collapse and its consequences (Karel and Flink, 1983), through the plasticizing effect of water on Tg (Cakebread, 1969) of the amorphous regions in both glassy and partially crystalline food polymer systems. The kinetics of collapse processes have been shown to be governed by the free volume, local viscosity, and resultant mobility of the water-plasticized glass or rubber (Slade and Levine, 1988a). In the glassy solid, Arrhenius kinetics apply, whereas in the rubbery liquid, rates of all translational diffusion-controlled relaxation processes, including structural collapse and recrystallization, increase according to WLF kinetics with increasing ΔT above Tg (Levine and Slade, 1986). For example, DSC results have been used to demonstrate that, during storage of frozen desserts such as ice cream and novelty products, ice crystal growth rates increase dramatically with increasing $\Delta T = Tf - Tg'$, in accord with WLF rather than Arrhenius kinetics in the rubbery region (Levine and Slade, 1987, 1988c, 1988e). In a related vein, Chan and co-workers (1986) have pointed out that the dielectric relaxation behavior of amorphous glucose plasticized by water is "remarkably similar" to that of synthetic amorphous polymers in glassy and rubbery states. These investigators have shown that the rates of this mechanical relaxation process (which depends on rotational rather than translational mobility) follow the WLF equation for glucose–water mixtures in their rubbery state above Tg, but follow the Arrhenius equation for water-plasticized glucose glasses below Tg. As is discussed later in the section on gluten as a polymeric cryostabilizer, the storage stability of frozen bread dough has been interpreted in terms of a mechanical relaxation process, in which the rate of structural deterioration (via collapse processes facilitated by water plasticization of the rubbery amorphous polymer matrix) is governed by WLF kinetics (Slade et al., 1988; Levine and Slade, 1988e).

A comparison of literature values of subzero collapse transition temperatures, for a variety of water-soluble monomers and polymers, has established the fundamental identity of Tg' with the minimum onset temperatures observed for various structural collapse and recrystallization processes (Levine and Slade, 1986, 1988a–f), for both model solutions and real systems of foods, as well as pharmaceuticals and biologicals (Franks, 1982; Pikal et al., 1983; Pikal, 1985; Reid, 1985). An extensive list of collapse processes, all of which are governed by Tg' of frozen systems (or by a higher Tg pertaining to low-moisture systems processed or stored at $T > 0°C$), and which involve potentially detrimental plasticization by water, has been identified and elucidated (Levine and Slade, 1986, 1987, 1988a–f). Table 5-1 shows an abbreviated list of high- and low-temperature collapse processes that are directly relevant to doughs during baking and baked products during storage. This table and previous versions of it emphasize how the Tg values relevant to high- and low-temperature collapse processes are systematically related through the corresponding product moisture contents, thus illustrating how this interpretation of collapse phenomena has been generalized to include

Table 5-1. Collapse Processes in Baked Goods Systems—
Mechanical/Structural Relaxations Governed by Tg and
Dependent upon Plasticization by Water

Processing and/or Storage at $T > 0°C$
- Gelatinization of native granular starches, $\geq Tg$
- Sugar-snap cookie spreading (so-called "setting") during baking, $\geq Tg$
- Structural collapse during baking of high-ratio cake batter formulated
  with unchlorinated wheat flour or with reconstituted flour
  containing waxy corn starch in place of wheat starch (due to lack
  of development of leached amylose network Tg), $\geq Tg'$
- Recrystallization of amorphous sugars in ("dual texture") cookies at the
  end of baking vs. during storage, $\geq Tg$
- "Melting" (i.e., rubbery flow) of bakery icings and glazes (hygroscopic
  mixed sugar glasses) due to moisture uptake during storage, $\geq Tg$
- Staling due to starch retrogradation via recrystallization in breads and
  other high-moisture, lean baked products during storage, $\geq Tg'$

Processing and/or Storage at $T < 0°C$
- Structural collapse or shrinkage due to loss of entrapped leavening
  gases during frozen storage, $\geq Tg'$
- Ice recrystallization ("grain growth"), $\geq Tr = Tg'$
- Staling due to starch retrogradation via recrystallization in breads and
  other high-moisture, lean baked products during freezer storage, $\geq Tg'$

Source: Adapted from Levine and Slade (1986).

both high-temperature/low-moisture and low-temperature/high-moisture food
products and processes (Levine and Slade, 1986, 1988f). In all cases, a
partially or completely amorphous system in the mechanical solid state at
$T < Tg$ and $\eta > \eta g$ would be stable against collapse, within the period of
experimental measurements of Tg, Tc, and/or Tr. Increased moisture content
(and concomitant plasticization) would lead to decreased stability and shelf-life,
at any particular storage temperature (Karel and Flink, 1983). The various
phenomenological threshold temperatures for a given collapse process all
correspond to the particular Tg' or other Tg relevant to the solute(s) system
and its content of plasticizing water (Levine and Slade, 1986, 1988f).

The examination of the relationship between cryostabilization technology
and collapse processes, and the interpretation of the latter within a conceptual context of glass dynamics, has resulted in a conclusion and explanation
of the fundamental equivalence of Tg, Tc, and Tr, and their dependence on
solute MW and concentration (Levine and Slade, 1986, 1987). This new inter-
pretive approach has provided a better qualitative understanding of empirical
results previously observed but not explained on a basis of polymer structure–
property principles (To and Flink, 1978). This approach has also provided
insights to the traditional countermeasures empirically employed to inhibit
collapse processes (Levine and Slade, 1988f). In practice, collapse (and all

its different manifestations) can be prevented, and food product quality and stability can be maintained, by the following measures:

1. storage at temperatures below or sufficiently near Tg (White and Cakebread, 1966)
2. deliberate formulation to increase Tc (i.e., Tg) to a temperature above or sufficiently near the processing or storage temperature, by increasing the composite $\overline{M}w$ of the water-compatible solids in a product mixture— this has often been accomplished by adding polymeric (cryo) stabilizers such as low-DE SHPs (or other polymeric carbohydrates, proteins, or cellulose and polysaccharide gums) to formulations dominated by LMW solutes (i.e., cryoprotectants) such as sugars and/or polyols (White and Cakebread, 1966; Cakebread, 1969; Tsourouflis et al., 1976; To and Flink, 1978; Downton et al., 1982; Karel and Flink, 1983)
3. in hygroscopic glassy solids and other low-moisture amorphous food systems especially prone to the detrimental effects of plasticization by water (including various forms of "candy" glasses [White and Cakebread, 1966; Cakebread, 1969]):
   a. reduction of the residual moisture content to $\leq 3\%$ during processing
   b. packaging in superior moisture barrier film or foil to prevent moisture pickup during storage
   c. avoidance of high-temperature/high-humidity ($\gtrsim 20\%$ RH) conditions during storage (White and Cakebread, 1966; Cakebread, 1969; Flink, 1983).

One example relevant to baked products, which illustrates the above countermeasures, involves the use of low-DE maltodextrins as stabilizers of bakery icings and glazes. As is described in Table 5-1, such icings and glazes are hygroscopic mixed sugar glasses that are prone to a collapse process referred to as "melting," due to softening and viscous liquid flow above Tg as a consequence of plasticization by absorbed moisture. The incorporation of low-DE maltodextrin in such a sugar blend raises the composite $\overline{M}w$, thus decreasing the hygroscopicity and increasing the Tg of the amorphous matrix relative to a given storage temperature (Levine and Slade, 1986).

Gelation Mechanism and Viscoelastic Properties of Thermosetting Amorphous Polymers—Effect of Tg on Rubbery Thermosets

The classical thermosetting process for synthetic amorphous polymers, for example, in the production of rigid epoxy resins and flexible natural or synthetic rubbers, is typically described by a time–temperature–transformation

(TTT) reaction diagram (Prime, 1981). As is illustrated in Figure 5-21 (Sperling, 1986), this TTT diagram has been "used to provide an intellectual framework for understanding and comparing the cure and glass transition properties of thermosetting systems." As has been reviewed by Prime (1981), thermosetting polymers become infusible and insoluble due to chemical cross-linking reactions during curing, which produce a three-dimensional network with essentially infinite MW. Curing of amorphous polymers at a curing temperature (*Tcure*), somewhat above the initial molecular *Tg* of the polymer reactant (*Tgo*), allows very slow conversion with concomitant elevation of the molecular *Tg*, until the molecular *Tg* approaches *Tcure* and the system vitrifies. The molecular *Tg* depends only on the extent of conversion and not on *Tcure* (Bair, 1985). At a sufficiently higher *Tcure*, gelation occurs before vitrification, at a characteristic extent of conversion. At this system-specific degree of cure, the partially cured polymer thermosets by undergoing a sudden and thermally irreversible transformation from a viscous liquid (*Tcure* $\gg$ *Tgo* of the polymer reactant) to a nonflowable, elastic network (*Tcure* > the gel point, initial *gelTg*), which rapidly approaches infinite macroscopic viscosity as *gelTg* approaches *Tcure*, and the system vitrifies. Gelation, per se, does not retard the rate of the curing reaction (which is chemically, rather than diffusion, controlled as long as *Tcure* is well above the effective instantaneous network *Tg*), and is detectable by the inability of large gas bubbles to rise in a thermosetting mass (Prime, 1981). In Figure 5-21, the S-shaped curing curve between *Tgo* and *Tg*∞ (the network *Tg* of the fully cured system) results because the reaction rate increases with increasing temperature (Sperling, 1986). Thus the gel point, initial *gelTg*, is the critical minimum value of *gelTg*, which represents the temperature at which gelation and vitrification (of the elastic gel to a glass) occur simultaneously. As has been described, at temperatures between *gelTg* and *Tg*∞, the thermosetting mass first reacts chemically to form a cross-linked rubbery network, then vitrifies as the continuously increasing network *Tg* becomes coincident with *Tcure*, essentially quenching the curing reaction (now diffusion controlled) before completion, but giving the illusion that the product is completely cured (Sperling, 1986). Once the network *Tg* exceeds *Tcure*, the diffusion of reactants becomes so restricted that the reaction rate decreases by more than three orders of magnitude (Bair, 1985). The resulting incompletely cured network would have a *Tg* denoted as *ultTg*, where *gelTg* < *ultTg* < *Tg*∞ (Prime, 1981). However, since vitrification is thermally reversible, curing can be extended by a post-cure heat treatment of the rubbery thermoset at T > *ultTg*, or rapidly completed by a post-cure heat treatment at T > *Tg*∞. It is important to note that the TTT diagram in Figure 5-21 refers to the curing and vitrification kinetics of a dry polymer system, in the absence of a plasticizer or solvent. Plasticizer sorption or dilution has a profound effect on the timeframe of the TTT diagram, due to depression of all operative *Tg* values and concomitant acceleration of the curing reactions at a given *Tcure*. The role of water as a

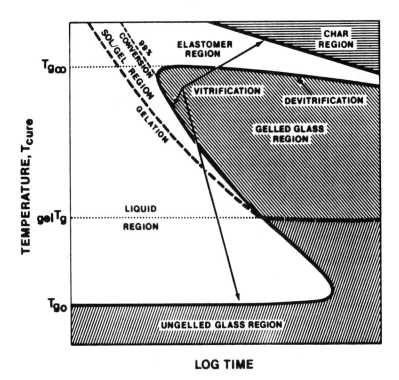

Figure 5-21. The thermosetting process for amorphous polymers, as illustrated by a time–temperature–transformation reaction diagram. (*From Sperling, 1986, with permission.*)

plasticizer is perhaps even more critical for the storage stability of both the original reaction mixture and the final cured product. Moisture sorption or loss has been shown to be a major factor in the aging behavior of synthetic thermosets (Prime, 1981). Plasticization of the thermoset network depresses the network Tg and modulus, whereas loss of plasticizer leads to toughness or brittleness.

The TTT diagram in Figure 5-21 illustrates the four material states encountered during curing: liquid, rubber (elastomer), ungelled glass, and gelled glass. The time axis signifies measured times to gelation and to vitrification as a function of cure temperature (Prime, 1981; Sperling, 1986), and underlines the importance of the kinetics of curing reactions in thermosettable polymers. In a DSC experiment, for example, if a polymer sample (of $Tg = Tgo$) is heated to $T > gelTg$ in an initial scan, but then is immediately cooled to $T < Tgo$ and rescanned, the extent of curing achieved during the first scan may be undetectably small (Slade et al., 1988). Consequently, Tgo in the rescan may appear unchanged, and the sample may behave as a thermoplastic polymer. In contrast, if the same sample is heated to $T \gg gelTg$, or to $T > gelTg$ and then held for a longer time before cooling and rescanning, the curing reaction can proceed. In this case, the rescan would not show Tgo, but rather a new network $Tg > Tgo$, and the sample would appear to be thermoset (Slade et al., 1988). An actual example of this contrasting behavior has been reported for curing of an epoxy–amine system in the DSC (Prime, 1981). On holding at $Tcure$ a few degrees above $gelTg$, about an hour was required to achieve the characteristic extent of cure to thermoset. When $Tcure$ was more than 80°C above $gelTg$, the same extent of cure was achieved in about 3 minutes, so that thermosetting could occur within the timeframe of a dynamic DSC experiment. Indeed, in the latter case, because $Tcure$ was more than 20°C above $Tg\infty$, complete curing was achieved within 10 minutes.

The apparent parallels between the established behavioral characteristics of thermosetting synthetic amorphous polymers, and the empirically known structure–function relationships of wheat gluten in various baking applications (and post-baking heat treatments, e.g., microwave "refreshening") have been described (Slade et al., 1988) and will be reviewed in the section, Behavior of Gluten as an Amorphous Thermosetting Polymer.

Water Vapor Sorption by Glassy and Rubbery Polymers

Water sorption, and its beneficial or detrimental effects, can play a critical role in bakery processes and products; that is, before, during, and after baking. Absorption of liquid water by the amorphous polymeric carbohydrates and proteins in flours, leading to their kinetically controlled plasticization ("hydration") during dough mixing, is obviously a key element of dough formation and an important determinant of the rheological properties and mechanical behavior of doughs (Ablett et al., 1986; Bloksma and Bushuk, 1988). Nonequilibrium evaporative desorption of water vapor (Lillford, 1988) is a crucial aspect of the baking process. Time- and temperature-dependent ad-/desorption of water vapor during ambient storage is potentially an important aspect of the stability and shelf life of baked products. The above facts justify the relevance of the following discussion of water vapor sorption by glassy and rubbery polymers.

It has been stated (Bryan, 1987a) that the reasons for the well-known irreversibility and hysteresis shown by water vapor sorption isotherms of biopolymers, including various food proteins and carbohydrates (Watt, 1980; D'Arcy and Watt, 1981; van den Berg, 1981), are not established. Bryan (1987a) has suggested that observed "irregularities reflect changes in the conformation and/or dynamic behavior of the biopolymer molecule." He has noted that "recent work on water–protein interactions (Finney and Poole, 1984; Bone and Pethig, 1985) is compatible with the occurrence of small conformational changes and increased flexibility (perhaps a loosening of the protein structure) as more water is added to a protein" in the solid state. Lillford (1988) has expressed a similar opinion, crediting (as others had previously [Levine and Slade, 1986]) the onset of enzymatic activity in low-moisture, amorphous lysozyme powders to plasticization by sorbed water, leading to sufficient segmental mobility for diffusion-limited enzyme–substrate interactions to occur. With regard to water–protein sorption hysteresis, Bryan (1987b) has suggested that hysteresis "could result from slow, incomplete conformational changes occurring upon addition of water" to proteins in the solid state, and its subsequent removal. These changes could result from incomplete "intermolecular phase annealing" or "phase changes," and the resulting hysteresis "might be related to the physical state and prior history of the sample" (Bryan, 1987b). Again, Lillford (1988) has expressed similar sentiments, pointing out (again as others had previously [Levine and Slade, 1987]) that starch (van den Berg, 1981) and other hydrophilic polymers plasticized by water "are far from inert in the adsorption process," should not be modeled as "an immobile and unaffected substrate," manifest a "nonequilibrium desorption process responsible for hysteresis," and probably are "usually in a metastable kinetic state . . . where the interaction of polymers and water cannot be treated in terms of equilibrium thermodynamics." Bryan (1987b) has also noted the fact that freeze-dried protein samples to be used as substrates in sorption experiments "might form an amorphous [solid] phase," a possibility overlooked by Lioutas and co-workers (1986) in their discussion of the water sorption behavior of lysozyme.

In a recent review that included this subject (Levine and Slade, 1987), an explanation (subsequently endorsed by Lillford [1988]) for the sorption behavior described by Bryan (1987a, 1987b) has been offered. Water vapor absorption in food systems can be treated as a diffusion-controlled transport process involving structural relaxations in glassy and rubbery polymers. The kinetics of diffusion rates associated with adsorption leading to absorption (D'Arcy and Watt, 1981) and with sorption-desorption hysteresis (Watt 1980) depend, in part, on the ever-changing structural state of a polymer relative to its Tg and on the extent of plasticization of the polymer by water (Levine and Slade, 1987). For sorption by synthetic amorphous polymers at $T \ll Tg$, it has been suggested that classical Fickian diffusion of LMW plasticizing sorbate, which may appear to be time-independent (Berens and Hopfenberg, 1980) and to show Arrhenius-type temperature dependence, may actually be an indication of extremely slow and inconspicuous relaxation in a kinetically metastable glassy polymer moving toward its equilibrium state (Neogi, 1983). In the temperature range near but below Tg to 5–10°C above Tg (the latter called the leathery region), observations of anomalous, non-Fickian or viscoelastic

(Roussis, 1981; Durning and Tabor, 1986), time-dependent, cooperative diffusion have been suggested to indicate that the glassy state is relaxing more rapidly to the rubber (Neogi, 1983), or in other words, that the "polymer relaxation time matches the sorption time scale" (Durning and Tabor, 1986). These sorption situations also reflect the fact that water plasticization of glassy polymers leads to increasing permeability of the substrate to gases and vapors, due to increasing segmental mobility of the polymer as Tg decreases relative to the constant sorption temperature Ts (Levine and Slade, 1987). In the rubbery state far above Tg, diffusion and relaxation rates increase sharply, as does polymer free volume (which is known to cause a dramatic increase in permeability to gases and vapors [Karel, 1986]). At the point where WLF-governed temperature dependence begins to approach Arrhenius temperature dependence (Hoeve, 1980), Fickian diffusion behavior again applies (Gaeta et al., 1982; Neogi, 1983).

The dependence of nonequilibrium sorption behavior on Tg relative to Ts, which has been described, has been illustrated by the results of a study by Wolf and co-workers (1985). These investigators studied the kinetics of water uptake, as a function of environmental relative humidity (RH) at room temperature, in a low-moisture, amorphous food polymer system representative of other amorphous or partially crystalline substrates, including baked products. In the system studied, the sorption behavior reflected both adsorption of water vapor and absorption of condensed liquid water via a diffusion-controlled transport process (i.e., a mechanical relaxation process governed by the mobility of the substrate matrix). The sorption results were said to "reveal that the time which is required to reach equilibrium conditions needs special attention," and that "equilibration times of at least 14 days might be recommended" (Wolf et al., 1985). It has been suggested (Slade and Levine, 1988a) that these results in fact demonstrated that times which are orders of magnitude greater than 14 days would be required to approach equilibrium conditions. In such sorption experiments, a low-moisture, amorphous food system may be in an extremely low mobility, "stationary" solid state so far from equilibrium that it can be easily confused with equilibrium. Results of the study showed that, even in 25 days, there was no change in water uptake when the environmental RH was 11%. The essentially immobile solid sample remained far from equilibrium in a low-moisture, negligibly plasticized "apparent steady state." In sharp contrast, in less than 25 days, there was a dramatic change in water uptake by the same substrate material when the RH was 90%. Under these conditions, the higher-moisture sample was significantly plasticized and exhibited sufficient mobility to allow a more rapid approach toward a still-higher-moisture and not-yet-achieved equilibrium condition. The fundamental trend of increasing sorption rate with increasing environmental RH evidenced by these results has suggested a mechanistic correlation between increasing mobility and increasing rate of relaxation of the substrate–water system toward its unique final state of equilibrium (Slade and Levine, 1988a). This correlation reflects the sequential relationship between increasing water uptake, increasing plasticization, increasing free volume, and decreasing local viscosity, which result in decreasing Tg. Viewed in isolation on a practical timescale, the unchanging nature of the low-moisture substrate at low RH would prevent an observer from recognizing this nonequilibrium situation of extremely slow mechanical relaxation.

From countless studies of so-called "equilibrium" water vapor sorption and sorption isotherms for completely amorphous or partially crystalline, water-compatible polymers, two general characteristics have become widely acknowledged (Levine and Slade, 1987). One is that such experiments do not usually represent a true thermodynamic equilibrium situation, since the polymer substrate is changing structurally—and slowly, during sorption experiments that are often, and are sometimes recognized to be (Bizot et al., 1985; Wolf et al., 1985) much too short—due to plasticization by sorbed water (Watt, 1980; Roussis, 1981; van den Berg, 1981; Franks, 1982; Neogi, 1983). Second, since Tg decreases during sorption, such experiments are not even isothermal with respect to the ΔT-governed viscoelastic properties of the polymer, because ΔT ($= Ts - Tg$) changes dynamically over the sorption time course (Watt, 1980; Neogi, 1983; Levine and Slade, 1987). Consequently, both the extent of sorption and the mobility of sorbed molecules generally increase with increasing plasticization by water (Hopfenberg et al., 1981; Apicella and Hopfenberg, 1982).

As has been reviewed in detail elsewhere (Levine and Slade, 1987), the basic premises that underlie the interpretation of the water sorption characteristics of polymeric food carbohydrates and proteins include the following: (1) the sorption properties of a "dry" polymer depend on its initial structural state and thermodynamic compatibility with water (Roussis, 1981); (2) in both completely amorphous and partially crystalline polymers, only the amorphous regions preferentially absorb water (Starkweather and Barkley, 1981); and (3) the shape of a particular sorption isotherm depends critically on the relationship between Ts and both the initial Tg of the "dry" polymer and the Tg of the water-plasticized polymer during the sorption experiment (Pace and Datyner, 1981). Several interesting illustrations of the last point have been described. For example, for sorption experiments done at a series of temperatures bracketing Tg of the water-plasticized polymer, the classic sigmoidal shape of the isotherm, at $Ts < Tg$, flattens suddenly, at $Ts > Tg$, to one characteristic of solution sorption by a rubbery polymer (Moy and Karasz, 1980). For sorption at a single temperature, initially below Tg of the "dry" polymer but subsequently above the "wet" Tg, which is depressed due to water uptake and plasticization, the isotherm can show a sudden change in slope due to the structural transition at Tg (Pouchly et al., 1979; Pace and Datyner, 1981). Similar dynamic behavior has been observed in cases in which an amorphous but crystallizable substrate (e.g., amorphous gelatin [Slade and Levine, 1987b]), initially in the form of a low-moisture glass, recrystallizes from the rubbery state after plasticization due to sorption of sufficient moisture, which allows the glass transition to occur, either deliberately during short-term sorption experiments or inadvertently during long-term product storage (Flink, 1983; Karel, 1985).

Sorption–desorption hysteresis has been called "the outstanding unexplained problem in sorption studies" (Watt, 1980). Many completely amorphous and partially crystalline polymers, both synthetic and natural, that swell slowly and irreversibly during water sorption show marked hysteresis between their absorption and desorption isotherms (D'Arcy and Watt, 1981). In partially crystalline biopolymers (e.g., native starch), hysteresis increases with increasing percent crystallinity (van den Berg, 1981). Buleon and colleagues (1987) have attributed the sorption hysteresis observed in acid-hydrolyzed (lintnerized) native starches to their metastable semicrystalline structure, by analogy to synthetic elastomers with crystalline domains, which "are more prone to develop hysteresis in association with internal stresses than other [completely amorphous] polymers, in which relaxation processes occur with ease." Water-soluble polymers such as PVP have been reported to show hysteresis at moisture contents less than Wg', which corresponds to the unfreezable fraction of total sorbed water (Franks, 1982).

The extent of hysteresis at a particular Ts depends on the rate of the swelling/deswelling relaxation process in a polymer substrate (Berens and Hopfenberg, 1980; Watt, 1980; D'Arcy and Watt, 1981; Pace and Datyner, 1981; Neogi, 1983), and thus has been suggested to depend on Tg and the magnitude of ΔT, during both the sorption and desorption experiments, in a manner consistent with WLF free volume theory (Levine and Slade, 1987). Unfortunately, as has been frequently noted (Bizot et al., 1985; Lillford, 1988), hysteresis cannot be explained by any thermodynamic treatment that applies only to, or any model originally derived for, reversible equilibrium states. Hysteresis is not an intrinsic feature of a sorbing polymer, but depends on experimental conditions (D'Arcy and Watt, 1981). The often anomalous nature of sorption–desorption that has been associated with hysteresis, including non-Fickian diffusion behavior, is due to dynamic plasticization of a glassy or partially crystalline polymer by water during a sorption experiment (Roussis, 1981; van den Berg, 1981; Neogi, 1983). It has been concluded that hysteresis characteristically results from a moisture-/temperature-/time-dependent, slow, nonequilibrium, swelling-related conformational change (involving a structural relaxation, and in some cases even a subsequent phase change), which is facilitated by increasing free volume and segmental mobility in a polymer that is being plasticized dynamically during sorption (D'Arcy and Watt, 1981; Bizot et al., 1985; Lillford, 1988). For example, Buleon and co-workers (1987) have suggested that the irreversible sorption behavior of lintnerized starches is due to a water-plasticized conformational change, leading to a "structural hysteresis" that develops in a temperature/moisture content domain corresponding to a partially rubbery, partially crystalline system. In such systems, true thermodynamic equilibration can take weeks, months, or even years to be achieved (Berens and Hopfenberg, 1980; Apicella and Hopfenberg, 1982).

Sorption hysteresis has also been observed in molten polymers at temperatures well above their Tg; for example, in concentrated molten synthetic polymer–organic solvent solutions (Bonner and Prausnitz, 1974). It has been argued that such hysteresis cannot be explained simply by linking this behavior to nonequilibrium effects imposed by the properties of the glassy solid state (Bizot et al., 1985). It has been demonstrated, however, that the correlated parameters of local viscosity and polymer Tg/Tm ratio do in fact critically influence the mobility of supra-glassy liquids (i.e., low-viscosity liquids well above their rubbery domain), such as molten polymers, even at temperatures 100°C or more above Tg (Slade and Levine, 1988a). It has been suggested that this influence of local viscosity and polymer Tg/Tm ratio carries over to the nonequilibrium behavior, including sorption hysteresis, of supra-glassy molten polymers (Slade and Levine, 1988a). The hysteresis between water vapor ad-/absorption and liquid water desorption in native starch has been reported to result from desorption that remains nonequilibrated even after two years, versus adsorption that achieves and remains in "well-defined equilibrium" states over the same period (Bizot et al., 1985). An alternative explanation for the observed hysteresis has been suggested (Slade and Levine, 1988a), whereby both limbs of the isotherm reflect the persistence of nonequilibrium states. The desorption limb represents the behavior of supra-glassy, partially crystalline starch drying slowly and irreversibly to a partially crystalline glassy state (Slade and Levine, 1987a, 1988b) different from the original native state. In contrast, the ad-/absorption limb represents the behavior of partially crystalline glassy native starch undergoing an extremely slow, water-plasticized relaxation process (van den Berg, 1981), which remains very far from equilibrium even after two years, to a supra-glassy, partially crystalline state, that is, a "pseudo steady state" easily mistaken for equilibrium. As has been mentioned, this same "pseudo steady state" behavior has been observed for water sorption by another glassy food polymer system (Wolf et al., 1985).

A complete mechanistic understanding of the water sorption process for a water-plasticizable polymer, and the resulting predictive capability that this would provide, would require definition of the dependence of sorption behavior on the independent variables of moisture content, temperature, time, and polymer MW (and dry Tg), and the dependent variable of the structural state of a polymer, in terms of its Tg and ΔT, during sorption. Unfortunately, no single currently available sorption isotherm equation, model, or theory is capable of such a complete description (Levine and Slade, 1987).

STARCH AS A PARTIALLY CRYSTALLINE
POLYMER SYSTEM PLASTICIZED BY WATER

Starch is a water-compatible biopolymer system, which, in terms of abundance and utility, represents an important food polymer, particularly as a major component of flours in doughs and baked products. Starch also serves as an outstanding model system to illustrate the conceptual food polymer science approach to the study of structure–property relationships of food molecules, which are treated as homologous systems of polymers, oligomers, and monomers with their plasticizers and solvents (Levine and Slade, 1987; Slade and Levine, 1987a).

Normal starch is not a homopolymer, but a mixture of two glucose polymers, predominantly linear amylose (of MW 10^5–10^6) and highly branched amylopectin (of MW 10^8–10^9), typically in a weight ratio of about 25:75 amylose:amylopectin, arranged in a supramolecular structure in the form of a layered granule (Whistler and Daniel, 1984). Even mutant starch, such as waxy corn starch containing only amylopectin, is not best described as a homopolymer of glucose, but rather as a special type of block copolymer in which backbone segments with their branch points (e.g., intercrystallite segments of B2 and B3 chains [Hizukuri, 1986]) exist in amorphous domains and the crystallizable branches exist in microcrystalline domains (Wunderlich, 1980; Whistler and Daniel, 1984; French, 1984). These amylopectin branches also account for the characteristic low extent of crystallinity, about 15–45% (Zobel, 1984, 1988; Blanshard, 1986), in normal granular starches (French, 1984). The native granule of normal starch is not a polymer spherulite (Wunderlich, 1973), but rather a layered structure in which space-filling is achieved by amylopectin branching with radial alignment of the branches in crystalline regions (French, 1984). Each of the glucose polymers may be partially crystalline (French, 1984), and their amorphous components may each manifest a distinctive Tg (at a temperature dependent on MW as well as local moisture content) characteristic of a predominant amorphous domain (Slade and Levine, 1987a).

Despite the microscopic and macroscopic structural complexity of starch, many workers since 1980 have usefully discussed the physicochemical effect of water, acting as a plasticizer of the amorphous regions in the native granule, on the Tg of starch (van den Berg, 1981, 1986; van den Berg and Bruin, 1981; Slade, 1984; Slade and Levine, 1984a, 1984b, 1987a, 1988b–e; Maurice et al., 1985; Biliaderis et al., 1985, Biliaderis, Page, and Maurice, 1986a, 1986b; Biliaderis, Page, Maurice, and Juliano, 1986; Blanshard, 1986, 1987, 1988; Ablett et al., 1986; Yost and Hoseney, 1986; Zeleznak and Hoseney, 1987; Levine and Slade, 1987; Chungcharoen and Lund, 1987; Marsh and Blanshard, 1988; Biliaderis and Galloway, 1989). Many of these studies have employed DSC to demonstrate that native granular starches, both normal and waxy, exhibit the nonequilibrium melting ("gelatinization"), annealing, and recrystallization ("retrogradation") behavior characteristic of a kinetically metastable, water-plasticized, partially crystalline polymer system with a small extent of crystallinity. This behavior is known to be capable of influencing the baking process (Blanshard, 1986, 1987, 1988), through the critical effect of gelatinization on the rheological properties of high-moisture doughs during baking (Bloksma, 1986; Bloksma and Bushuk, 1988). This behavior can also influence the quality and storage stability of breads and related high-moisture baked products, through the major effect of retrogradation on product shelf-life (Guilbot and Godon, 1984; Hoseney, 1986; Slade and Levine, 1987a). Starch gelatinization during baking is recognized to be a dynamic heat/moisture/time treatment (Blanshard, 1979; Lund, 1984). For native, freshly gelatinized, and retrograded starches alike, thermal analysis by DSC has revealed the crucial role of water as a plasticizer of both completely amorphous and partially crystalline starch systems and the importance of the glass transition as a physicochemical event that can govern starch properties, processing, and stability. In this section, recent DSC studies on the following topics are reviewed: gelatinization as a nonequilibrium melting/mechanical relaxation process, retrogradation as a recrystallization process, the effect of sugars on gelatinization and retrogradation, the mechanism of gelation/recrystallization, and the kinetics of recrystallization and its acceleration. All are evaluated from the perspective of the influences of nonequilibrium glassy and rubbery states in starch on the time-dependent thermal, mechanical, and structural properties of high-moisture doughs and baked products.

Thermal Analysis of Nonequilibrium Melting, Annealing, and Recrystallization Behavior in Native Granular Starch–Water Model Systems

Figure 5-22 shows DSC (Perkin-Elmer DSC-2C) heat flow curves of commercially isolated Aytex P wheat starch, with an initial as is moisture content of 10 w%, mixed with an equal weight of added water to achieve a final moisture content of 55 w% after gelatinization. Because the initial rate of plasticization at room temperature is near zero, the added water remains predominantly outside the granules, and the native starch with no pretreatment (curve *a*) demonstrated a major glass transition and subsequent superimposed crystalline transition(s) in the temperature range 50–90°C, which comprise the events of gelatinization (initial swelling) and pasting (second-stage swelling) of a starch granule (Zobel, 1984; Atwell et al., 1988). A minor glass transition and subsequent melting of crystalline amylose-lipid complex was seen in the temperature range 90–115°C (Biliaderis et al., 1985; Biliaderis, Page,

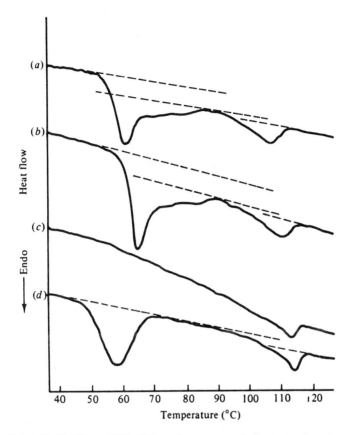

Figure 5-22. Perkin-Elmer DSC-2C heat flow curves of wheat starch:water mixtures (45:55 by weight): (a) native; (b) native, after 55 days at 25°C; (c) immediate rescan after gelatinization of sample in (a); (d) sample in c, after 55 days at 25°C. Dashed lines represent extrapolated baselines. (*From Slade and Levine, 1987a, with permission.*)

and Maurice, 1986a). The glass transition of the amorphous regions of amylopectin was observed at the leading edge of the first melting peak, between about 50°C and 60°C (Slade, 1984; Slade and Levine, 1984a, 1984b). This near superposition of the glass transition on the crystalline melt, due to the inhomogeneity of moisture contents inside and outside the granules and to heating rate, was revealed by the expected characteristic shift in heat capacity (diagnostic of a glass transition [Wunderlich, 1981; Yost and Hoseney, 1986]) shown by the extrapolated baselines (dashed lines) in Figure 5-22, curve *a*. The important insight into the thermal properties of starch represented by the identification of this *Tg* was subsequently corroborated by DSC results for granular rice starches (Biliaderis, Page, Maurice, and Juliano, 1986;

Chungcharoen and Lund, 1987). These results likewise demonstrated that melting of the microcrystallites is governed by the requirement for previous softening of the glassy regions of amylopectin.

When native wheat starch was allowed to anneal at 55 w% sample moisture (initially 10 w% inside the granules and 100 w% outside) for 55 days at 25°C (Fig. 5-22, curve b), the transition temperatures and extent of crystallinity increased. However, the characteristic baseline shift indicative of a preceding glass transition was still evident, demonstrating that melting of the microcrystallites was still governed by the requirement for previous softening of the glassy regions of amylopectin. Annealing is used in this context to describe a crystal growth/perfection process, in a metastable, partially crystalline polymer system (Wunderlich, 1976) such as native starch (Kuge and Kitamura, 1985). Annealing is ordinarily carried out at a temperature Ta in the rubbery range above Tg but below Tm, typically at an optimal $Ta = 0.75$–$0.88\ Tm$ (K) (Brydson, 1972), for polymers with Tg/Tm ratios of 0.5–0.8. In contrast, recrystallization is a process that occurs in a crystallizable but completely amorphous metastable polymer at $Tg < Tr < Tm$ (Wunderlich, 1976).

After gelatinization during heating to 130°C, and quench-cooling to 25°C, an immediate rescan of wheat starch at 55 w% moisture (Fig. 5-22, curve c) showed no transitions in the temperature range 30–100°C. However, when this completely amorphous (i.e., no remaining amylopectin microcrystals) sample was allowed to recrystallize (at uniformly distributed 55 w% moisture) for 55 days at 25°C (Fig. 5-22, curve d), it showed a major Tm at about 60°C (as a symmetrical endothermic peak, with essentially no baseline shift), which was not immediately preceded by a Tg. This Tm is well known to characterize the melting transition observed in retrograded wheat starch gels with excess moisture (which are partially crystalline and contain hydrated B-type starch crystals) and in staled bread and other high-moisture wheat starch–based baked goods (Kulp and Ponte, 1981; Longton and LeGrys, 1981; Fearn and Russell, 1982; Russell, 1983; Lund, 1983; Eliasson, 1985; Hoseney, 1986; Russell, 1987b).

Results of complementary low-temperature DSC analysis (Slade, 1984; Slade and Levine, 1984a, 1984b, 1987a), shown in Figure 5-23, revealed why no Tg was observed in the sample of freshly gelatinized (thus completely amorphous amylopectin) wheat starch rescanned from room temperature to 130°C (Fig. 5-22, curve c), or immediately before Tm in the sample of recrystallized starch (Fig. 5-22, curve d). Native wheat starch, at 55 w% total sample moisture, showed only a Tm of ice at the instrumental sensitivity used for the thermogram (curve a) in Figure 5-23. In contrast, a gelatinized sample (curve b), at the same water content and instrumental sensitivity, showed a prominent (and reversible) glass transition of fully plasticized amorphous starch at about -5°C, preceding and superimposed on the ice melt. As has been mentioned, this Tg is actually Tg' for gelatinized (but not hydrolyzed) wheat starch

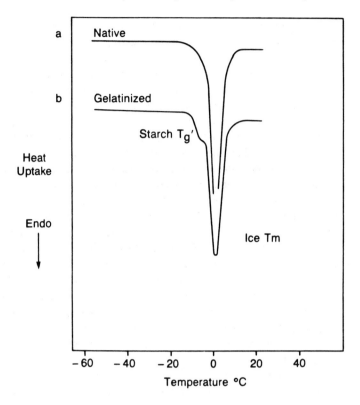

Figure 5-23. Du Pont 990 DSC heat flow curves of wheat starch:water mixtures (45:55 w:w): (a) native granular; (b) immediate rescan after gelatinization of sample in (a), which reveals a prominent Tg' at −5°C, preceding the Tm of ice. (*From Slade and Levine, 1988b, with permission.*)

in excess moisture (Levine and Slade, 1986, 1987), defined by $Wg' \gtrsim 27$ w% water (i.e., about 0.37 g UFW/g starch [Kuge and Kitamura, 1985]), as is illustrated by the state diagram in Figure 5-12. (It should be noted in passing that this characteristic subzero Tg' of the maximally freeze-concentrated gelatinized starch/UFW matrix was recently incorrectly identified, in a DSC study of starch retrogradation by Kwak and Winter [1988], as a crystallization exotherm signifying a retrogradation marker.) For the same instrumental sensitivity settings, Tg' was not detectable in Figure 5-23, curve *a*, because the cooperative, controlling majority of the amorphous regions of partially crystalline native starch prior to gelatinization show a much higher Tg, indicative of a much lower local effective moisture content (Slade and Levine, 1987a) and the absence of contributions from short amylopectin branches that are sequestered in crystalline regions.

A critical conclusion of these earlier studies was that knowledge of total sample moisture alone cannot reveal the instantaneously operative extent of plasticization of amorphous regions of a starch granule (Slade and Levine, 1987a). Amorphous regions of a native granule are only partially plasticized by excess water in a sample at room temperature, so that softening of the glassy matrix must occur (observed at about 50–60°C during heating at 10°C/min in the DSC) before microcrystallites can melt. (This situation of partial and dynamically changing plasticization at room temperature also explains why slow annealing was possible for the native starch sample in curve *b* of Fig. 5-22.) After gelatinization, the homogeneously amorphous starch is fully and uniformly plasticized (by uniformly distributed water) at 55 w% moisture, and the metastable amorphous matrix exists at room temperature as a mobile, viscoelastic rubber in which diffusion-controlled recrystallization, governed by WLF rather than Arrhenius kinetics (Levine and Slade, 1987), can proceed with rates proportional to $\Delta T \simeq 30$°C above Tg. Another insight revealed by these DSC results concerns the dynamic effects on starch caused by the DSC measurement itself (Slade and Levine, 1987a). During a DSC heating scan, effective plasticizer (water) content increases dynamically from the initial 6–10 w% in a native sample before heating to the final 55 w% at the end of melting, and this kinetically constrained moisture uptake leads to dynamic swelling of starch granules at temperatures above Tg, which is not reversible on cooling. The same behavior is manifested in volume expansion measurements on starch–water systems performed by TMA (Maurice et al., 1985; Biliaderis, Page, Maurice, and Juliano, 1986). The major contribution to the experimentally observed increase in volume above Tg is a typical polymer swelling process, characteristic of compatible polymer–diluent systems (Sears and Darby, 1982), which is linear with the amount of water taken up. Thermal expansion of amorphous starch is also allowed above Tg, but it represents only a minor contribution to the observed volume increase, that is, about 0.1%/K for typical polymers (Ferry, 1980). Thus, the predominant mechanism, swelling, is indirectly related to the role of water as a plasticizer of starch, but the minor mechanism is directly related.

Once it had been established (Slade, 1984; Slade and Levine, 1984a, 1984b) that the thermal behavior of native wheat starch at 55 w% total moisture in the temperature range 50–100°C represents the superposition of a second-order glass transition followed by a first-order crystalline melting transition, it was shown (Fig. 5-24) that plasticization of amorphous regions by water can be accelerated without melting crystalline regions. In Figure 5-24, waxy corn starch was used as a model system to study amylopectin in the absence of amylose. Like wheat starch, native waxy corn starch (curve *a*) exhibited nonequilibrium melting of a partially crystalline glassy polymer, with a requisite glass transition (signified by a baseline shift) preceding multiple crystalline transitions in the temperature range 50–100°C, when total sample moisture was 55 w%. When this sample was annealed for 15 minutes at 70°C (Fig. 5-24, curve *b*), total excess heat uptake below the baseline was reduced by about 25% and the temperature range of the multiple transitions was shifted

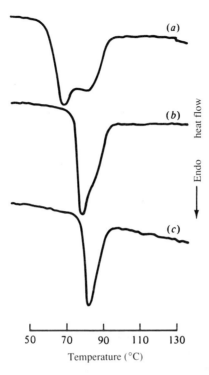

Figure 5-24. Du Pont 1090 DSC heat flow curves of waxy maize starch:water mixtures (45:55 by weight): (a) native; (b) native, after 15 min at 70°C; (c) native, after 30 min at 70°C. (*From Maurice et al., 1985, with permission.*)

upward and became narrower, but the glass transition immediately before the crystalline melt was still evidenced by the characteristic baseline shift. A similar result had been seen (Fig. 5-22, curve *b*) for wheat starch annealed at 25°C for 55 days, and also had been reported for potato starch annealed at 50°C for 24 hours (Welsh et al., 1982). (Note the apparently exponential dependence on ΔT, dictated by WLF kinetics, of rates for different annealing conditions.) Similar consequences of annealing by heat/moisture treatment have been observed during cooking of whole wheat grains (Jankowski and Rha, 1986a). In contrast, when native waxy corn starch was annealed for 30 minutes at 70°C (Fig. 5-24, curve *c*), total excess heat capacity below the baseline was reduced by 50% and represented only the enthalpy of the first-order crystalline melting transition. This conclusion was confirmed by the symmetry of the endotherm and the absence of an obvious baseline shift associated with it. The baseline shift was not observed in the temperature range 40–130°C, because the glass transition preceding the crystal melt had been depressed to $Tg' < 0$°C, due to complete plasticization of the amorphous regions by water (Slade, 1984; Slade and Levine, 1984a, 1984b; Maurice et al., 1985).

The amorphous regions of a starch granule represent a continuous phase, and the covalently attached microcrystalline branches of amylopectin plus discrete amylose–lipid crystallites represent a discontinuous phase. For each polymer class (distinguished by an arbitrarily small range of linear DP), water added outside the granule acts to depress the Tg of the continuous amorphous regions, thus permitting sufficient mobility for the metastable crystallites to melt on heating to a Tm above Tg (Slade and Levine, 1984a, 1984b). The effect of changing the amount of added water on the thermal behavior of native rice starch can be seen in Figure 5-25. At the "as is" 10 w% moisture content, with no added water (curve a), the glass transition of amylopectin occurred at temperatures above 100°C and multiple crystalline melting transitions occurred at temperatures above 150°C. Similar profiles would have been seen at increasing total sample moisture contents up to about 30 w% (i.e., Wg'), with the initial glass transition occurring at decreasing temperatures. At moisture contents higher than about 30 w% (Fig. 5-25, curves b–f), the initial glass transition occurred at about the same temperature, and the subsequent cooperative events occurred at lower and narrower temperature ranges as water content was increased. The moisture content that is sufficient to completely plasticize starch after gelatinization is about 27 w% (i.e., Wg') (Kuge and Kitamura, 1985). However, unlike gelatin gels, which can be dried to different moisture contents so that water is uniformly distributed throughout the amorphous regions (Slade and Levine, 1984a, 1984b), native starch starts out typically at 6–10 w% moisture, and the added water is outside the granule (and so is initially nonplasticizing), prior to moisture uptake and swelling during DSC heating.

The results and conclusions indicated by Figure 5-25 (Slade and Levine, 1984a, 1984b; Maurice et al., 1985) were confirmed by subsequent DSC results of Biliaderis and co-workers (Biliaderis, Page, Maurice, and Juliano, 1986) on several other varietal rice starches. In addition to several composite thermograms similar in appearance to Figure 5-25, these investigators presented a graph of Tg versus starch concentration (weight percent) for partially crystalline native starch:water mixtures; the graph started at about 240°C for the dry sample, and decreased with increasing moisture to 68°C at about 30 w% total moisture ("as is" plus added). Beyond this moisture content (i.e., for further additions of [initially nonplasticizing] water outside the granules), and before gelatinization during DSC heating, the initial Tg appeared to remain constant. This graph of the effect of water on the dynamically measured value of Tg for partially crystalline native starch (Biliaderis, Page, Maurice, and Juliano, 1986) should not be confused with the Tg curve in a state diagram for a homogeneous starch–water system; that is, completely amorphous gelatinized starch–water, as shown in Figure 5-12 (van den Berg, 1986). The latter illustrates the smooth glass curve that connects the Tg of dry starch

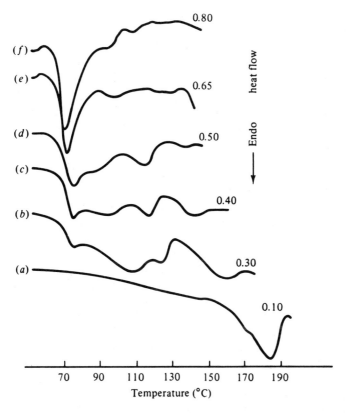

Figure 5-25. Du Pont 1090 DSC heat flow curves of native rice starch at various water contents: (a) starch with "as is" moisture of 10 w%; (b–f) starch with moisture added to water weight fractions indicated. (*From Maurice et al., 1985, with permission.*)

with the Tg of amorphous water, and passes through the Tg' ($-5°C$) and Wg' (27 w% water) for gelatinized starch (Slade, 1984; Slade and Levine, 1984a, 1984b).

Temperature Location of the Glass Transition Associated with Gelatinization of A-Type Cereal Starches

Zobel (1988) has recently remarked that "glass transition is an expression familiar to the polymer chemist but still somewhat foreign to starch chemists" and that "while recognized as a factor in starch characterization, glass transition temperatures have been, and to a degree still are, somewhat elusive."

While the existence of a glass transition and measurable Tg in partially crystalline native granular starches has only recently become established by DSC studies reported in the past five years (Slade, 1984; Slade and Levine, 1984a, 1984b, 1987a, 1988b; Maurice et al., 1985; Biliaderis, Page, Maurice, and Juliano, 1986; Levine and Slade, 1987), and is therefore still in the process of becoming more widely recognized and accepted (Blanshard, 1986, 1987, 1988; Chungcharoen and Lund, 1987; Russell, 1987a; Marsh and Blanshard, 1988; Zobel, 1988; Zobel et al., 1988), the temperature location of the glass transition associated with gelatinization of native starch has become a point of contention in the current literature (Yost and Hoseney, 1986; Zeleznak and Hoseney, 1987a; Slade and Levine, 1988a).

Two reports (Yost and Hoseney, 1986; Zeleznak and Hoseney, 1987a) have explored the validity of the model in which the thermal behavior of native starch at 55 w% total moisture in the temperature range 50–100°C represents the superposition of a second-order glass transition followed by a first-order crystalline melting transition. Yost and Hoseney (1986) presented DSC results for gelatinization and annealing by heat/moisture treatment of wheat starch in water at 50 w% total sample moisture content. These investigators concluded that annealing occurred (in samples previously held for 24 hours at room temperature) at temperatures 3–8°C below the gelatinization Tm for wheat starch, but not at lower temperatures. These results did not contradict previous DSC results and conclusions (Slade, 1984; Slade and Levine, 1984a, 1984b, 1987a; Maurice et al., 1985) about the relative locations of Tg and Tm for gelatinization of starch, especially in light of existing knowledge about annealing of metastable, partially crystalline synthetic polymers (Brydson, 1972; Wunderlich, 1976). As has been mentioned, annealing occurs at $Tg < Ta < Tm$, typically at $Ta = 0.75$–$0.88\ Tm$ (K), for polymers with Tg/Tm ratios of 0.5–0.8. In this metastable rubbery domain defined by WLF theory, annealing is another diffusion-controlled, nonequilibrium process for which rate is governed by WLF, rather than Arrhenius, kinetics (Nakazawa et al., 1984; Levine and Slade, 1987; Slade and Levine, 1988b). As has been demonstrated for various native granular starches under conditions of limited moisture (Nakazawa et al., 1984; Slade and Levine, 1984a, 1984b, 1987a; Kuge and Kitamura, 1985; Maurice et al., 1985; Yost and Hoseney, 1986; Krueger et al., 1987; Blanshard, 1987), the time required to achieve a measurable and comparable (in a reasonable and similar experimental timeframe) extent of annealing is shortest at Ta just below Tm (greatest ΔT above Tg) and longest at Ta just above Tg (smallest ΔT). The minimum value of the Tg/Tm ratio for wheat starch (at a uniformly distributed excess moisture content $\gtrsim 27$ w%) is about 0.80 (i.e., $Tg'/Tm = 268/333K$); this ratio increases with decreasing moisture content to an anomalously high value greater than 0.9 (Slade and Levine, 1988b). This anomalous situation corresponds to conditions of the nonequilibrium gelatinization or annealing of native starch upon heating in the presence of water added to 50 w%. Consequently, the temperature range that encompasses the effective locations of Tg, Ta, and Tm for native starch heated with excess added water is quite narrow, a conclusion that is also suggested by the DSC results of Nakazawa and colleagues (1984). Zeleznak and Hoseney (1987a) investigated the Tg of both native and prege-latinized wheat starches as a function of moisture content, and concluded, in seeming conflict with the annealing results previously reported (Yost and Hoseney, 1986), that their findings "contradicted the suggestion that Tg imme-diately precedes melting in starch." In both papers, Hoseney and co-workers based their argument in large part on the failure to observe a glass tran-sition (in the form of a discontinuous change in heat capacity) in a rescan after native starch was gelatinized in excess added moisture in the DSC. Unfortunately, their DSC measurements were not extended below 0°C, and so the prominent Tg at $Tg' \simeq -5°C$ for gelatinized starch (Fig. 5-23) was not observed.

In an effort to resolve any questions raised by the conclusions of Yost and Hoseney (1986) and Zeleznak and Hoseney (1987a), a subsequent DSC study of representative A-type cereal starches (Slade and Levine, 1988b) verified and further quantified the temperature location of the effective glass transition that immediately precedes the nonequilibrium melting transition of amylopectin microcrystallites and thereby controls the melting process asso-ciated with gelatinization. For that study, native granular wheat and waxy corn starches were heated at 10°C/min in the presence of water added to 55 w% total moisture to facilitate temporal resolution of the thermal events. The experimental DSC protocol represented a novel and critically discriminating extension of procedures previously recommended (Blanshard, 1987) and used (Yost and Hoseney, 1986; Russell, 1987a) to analyze starch gelatinization. Definitive DSC experiments, involving partial initial heating scans to inter-mediate temperatures in the range from 30°C to 130°C, followed by quench cooling and immediate complete rescans, revealed the operational location of Tg for wheat starch (above 54°C and completed by 63°C) and waxy corn starch (above 63°C and completed by 71.5°C). Corresponding effective "end of melting" temperatures (Tm) for the nonequilibrium melting transition of annealed amylopectin microcrystallites in normal wheat (about 92°C) and waxy corn (about 95°C) starches were also identified. Results of this study made it possible to achieve a deconvolution of the contributions of amylopectin and amylose to the nonequilibrium melting behavior of native granular starches, through DSC analyses of normal wheat and waxy corn starches. These results were also used to demonstrate the kinetically controlled relationship (based on the dynamics of plasticization by water) between the operative Ta, at which a nonequilibrium process of annealing can occur in native granular starches subjected to various heat/moisture/time treatments, and the effective Tg and Tm that are relevant to gelatinization and which bracket Ta, thereby confirm-ing that the location of Ta 3–8°C below Tm (Yost and Hoseney, 1986) is not inconsistent with a Tg immediately preceding melting.

The study demonstrated that the effective Tg associated with gelatinization of native granular starch, most readily resolved by heating in excess added moisture at nearly equal weights of starch and water, depends on the instantaneously operative conditions of moisture content, temperature, and time (Slade and Levine, 1988b). This finding helped to eliminate any potential confusion over the absence of a single, "absolute" value of Tg for starch (or, for that matter, for any other amorphous material), by illustrating the established fact (Ferry, 1980; Petrie, 1975) that the operational designation of any specific Tg value is only relevant to the instantaneously-operative conditions of its measurement. This point can be illustrated conceptually by the schematic state diagram for the amorphous regions of native granular cereal starch in Figure 5-26, which is used here in the context of a dynamics map to describe the process of gelatinization. Figure 5-26 traces the route followed during a DSC experiment, in terms of the following path of temperature–moisture content loci:

1. initial heating of native starch (N) at "as is" moisture from room temperature, through the instantaneously operative Tg, into the rubbery region (whereupon the rates of moisture uptake and swelling increase dramatically with increasing temperature [Blanshard, 1987]), through the instantaneously operative Tm (not shown, but located along the dotted line between Tg and 100°C), to G, representing the gelatinized sample with a final uniform moisture content at the end of the first heating scan (130°C)
2. cooling to $T <$ the effective Tg of the gelatinized sample (has become Tg'), which is accompanied by freezing and freeze-concentration to $W = Wg'$
3. reheating, from $T < Tg'$, through Tg' (whereupon ice melting and starch dilution begin), back up to G at the end of the second heating scan.

The concept implicit in Figure 5-26 is that the glass transition represents a rate-limiting stage of a mechanical relaxation process, for which the spectrum of relaxation rates depends on the instantaneous magnitude of the free volume and/or local viscosity, which in turn depends on the relative values of experimental moisture content compared to the moisture content (Wg) of the operative glass, experimental temperature compared to the instantaneous Tg, and experimental timeframe compared to the instantaneous relaxation rate (Slade and Levine, 1988b). Such a relaxation process, described by WLF theory, underlies various functional aspects of starch in high-moisture doughs and baked products, such as gelatinization, crystallite melting, annealing, and recrystallization. The academic distinction between the kinetic control of gelatinization with this operational Tg as the reference temperature versus the energetic control through Tm might be dismissed as a semantic issue were it not for its overriding technological importance. Control through Tm would dictate that the gelatinization temperature and mechanical behavior of a starch system during the events of gelatinization depend solely on the identity of the

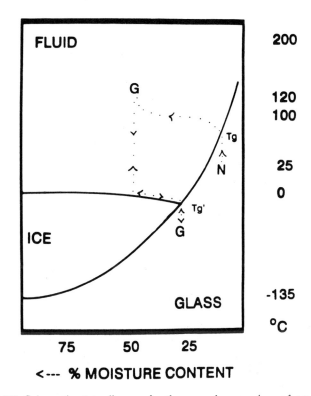

Figure 5-26. Schematic state diagram for the amorphous regions of granular starch, on which is traced the temperature–moisture content path followed during (1) initial heating, to $T >$ the instantaneously operative Tg (and subsequent instantaneously operative Tm), to gelatinize (G) native (N) starch, (2) subsequent cooling, to $T <$ the effective Tg (i.e., Tg') of gelatinized starch, and (3) reheating to $T > Tg'$.

starch crystalline polymorph, without regard to the previous history of the sample. Practical experience related to the aging of cereal grains prior to industrial processing, variations in the wet-milling process to isolate starches, and other heat/moisture treatments clearly confirms the role of sample history and the kinetic control of gelatinization through the path-determined operational location of the starch Tg (Slade and Levine, 1988c).

The composite diagram of DSC heat flow curves for Aytex P wheat starch, shown in Figure 5-27, demonstrated conclusively that initial heating to at least 92°C was required (for a heating rate of 10°C/min, to a final sample moisture content of 50 w%) to complete the nonequilibrium melting process associated with gelatinization and pasting of native wheat starch. The familiar $Tm \simeq 115$°C for an amylose–lipid crystalline complex (Biliaderis et al., 1985) was evident in every curve in Figure 5-27. Partial initial scanning to temperatures greater than or equal to 72°C but less than 92°C resulted in only partial melting,

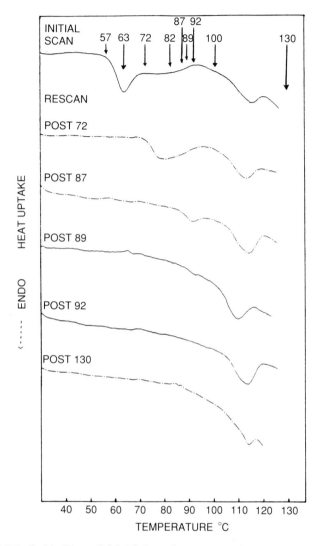

Figure 5-27. Perkin-Elmer DSC-2C heat flow curves of wheat starch:water mixtures (50:50 w:w). *Top*, initial scan at 10°C/min for native starch; *Others*, rescans at 10°C/min, immediately following partial initial scanning, at 10°C/min, from 20°C to maximum temperatures indicated and then immediate cooling, at nominal instrument rate of 320°C/min, to 20°C. (*From Slade and Levine, 1988b, with permission.*)

as evidenced in the rescans by a remnant of the melting profile, compared to the thermal profile of the complete melting process shown at the top of Figure 5-27. This remnant decreased in area with increasing maximum temperature (in the range 72–92°C) of the partial initial scan, but disappeared completely (yielding a featureless thermogram, as evidenced by a flat baseline, from 30–100°C) only after initial heating to 92°C or higher. From these preliminary DSC measurements, it was concluded that 92°C represents the effective *Tm* at the "*end* of melting" (Lelievre, 1976; Nakazawa et al., 1984) for native wheat starch heated at 10°C/min with 50 w% total moisture. The temperature at which melting begins was deduced from the DSC results shown in Figure 5-28, which also revealed the temperature location of the effective glass transition that must precede the onset of this nonequilibrium melting process for Aytex P with water added to 55 w% total moisture. On the timescale of the DSC measurement, these two temperatures were essentially identical. Implicit in the results shown in Figure 5-27 is the fact that, at temperatures within the range from the effective *Tg* to the end-of-melting *Tm*, the extent of gelatinization is temperature dependent. As has been mentioned, this condition prohibits the application of Arrhenius kinetics to model the gelatinization process (Reid and Charoenrein, 1985; Burros et al., 1987) and emphasizes the applicability of WLF kinetics (Slade, 1984; Slade and Levine, 1984a, 1984b, 1987a, 1988b).

The composite diagram of DSC heat flow curves for wheat starch with 55 w% water in Figure 5-28 shows the complete nonequilibrium melting process as an initial scan in curve *A*; the partial initial scans, with their end-of-scan temperatures indicated, as solid lines in parts *B–K*; and the complete rescans as dashed lines in parts *B–L*. For parts *B–L*, the Thermal Analysis Data Station (TADS) computer on the Perkin-Elmer DSC-2C was instructed to display simultaneously the initial scans and the rescans, and it was allowed to confirm that the same instrumental baseline response of heat uptake (in mcal/s) at the instrumentally "equilibrated" starting temperature of 20°C occurred in both scans. This data processing step evidenced successful experimental execution by the near-perfect superposition of the 20–50°C baseline portions of the initial scans and corresponding rescan segments in parts *B*, *C*, and *D*, and the 90–100°C baseline portions of the scans and rescans in parts *J*, *K*, and *L*. This step was critical to the subsequent identification of the effective *Tg* that governs gelatinization of commercially isolated native cereal starch in excess moisture during heating from room temperature to 100°C. The results in Figure 5-28 were used to deduce the location of the effective *Tg* as the temperature by which there had occurred, in the initial scan, a characteristic and diagnostic baseline shift in heat capacity, during the timescale of the experimental measurements, for heating at 10°C/min. As an example, within the temperature range 30–100°C in part *L*, comparison of the 30–60°C baseline portions of the initial scan and the rescan demonstrated such a diagnostic difference in heat capacity, thereby documenting that a glass transition had occurred during the initial scan. Because of the previous occurrence of the change in heat capacity, the featureless (at $T < 100°C$) rescan was superimposed on the initial scan only after the latter returned to baseline after the end of the endothermic melting process at $Tm \simeq 92°C$. As expected, this effective end-of-melting Tm was the same as that shown in Figure 5-27, since both 50 w% and 55 w% water represent conditions of large excess moisture for gelatinized starch. This effective end-of-melting Tm was also suggested by Figure 5-28I, for which the initial thermal profile, upon heating to 89°C, stopped just short of a return to baseline. The reason for the apparent absence of a Tg in the rescan of part L has been explained by reference to Figure 5-23. After complete gelatinization upon heating to 130°C and 55 w% final moisture content, $Tg = Tg' \simeq -5°C$.

The effective Tg preceding and controlling the nonequilibrium melting process associated with gelatinization was identified as that minimum narrow temperature span in the initial scan, below which the change in heat capacity had not yet occurred (as reflected by superimposed baselines for scan and rescan), but at and above which it already had (as reflected by a displacement of baselines, at $T < Tg$, between scan and rescan) (Slade and Levine, 1988b). It can be seen in Figure 5-28 parts B, C, and D that the scans and corresponding rescan segments were essentially identical up to 54°C, and the heat capacity change had not yet occurred before the rescans, because initial heating to 37°C, 47°C, or 54°C had not yet reached the uniform requirements of time, temperature, and moisture for cooperative relaxation at Tg. In contrast, in parts $E–K$, the scan and rescan baselines, at $T < Tg$, were displaced, because initial heating to $T \geq 63°C$ had allowed the amorphous regions of the native granules to undergo a glass transition. By the convention described above, the effective "end of softening" Tg preceding crystallite melting was thus identified as greater than 54°C and less than 63°C, as illustrated in Figure 5-28E. This upper limit for Tg corresponds to the temperature at the "peak minimum" in the characteristic DSC thermal profile for wheat starch gelatinization shown in Figure 5-28A. The narrow 55–63°C temperature span of the effective Tg occurs along the leading edge of the "gelatinization endotherm" (Slade, 1984; Slade and Levine, 1984a, 1984b). Figure 5-28E represents a temporal and thermal deconvolution of the melting transition of microcrystalline regions from the preceding glass transition of amorphous regions of water-plasticized starch. Nakazawa and co-workers (1984) alluded to a similar differentiation

$\longrightarrow$

Figure 5-28. Perkin-Elmer DSC-2C heat flow curves of wheat starch:water mixtures (45:55 w:w): (A) initial scan at 10°C/min for native starch; (B–L) solid lines = partial initial scans for native starch, at 10°C/min, from 20°C to maximum temperatures indicated; dashed lines = rescans at 10°C/min, immediately following partial initial scanning and then immediate cooling, at nominal 320°C/min, to 20°C. (*From Slade and Levine, 1988b, with permission.*)

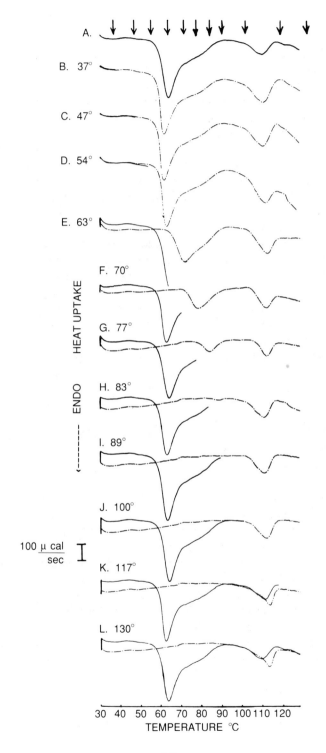

A.

B. 37°

C. 47°

D. 54°

E. 63°

F. 70°

G. 77°

H. 83°

I. 89°

J. 100°

K. 117°

L. 130°

HEAT UPTAKE

ENDO

$100 \dfrac{\mu \ cal}{sec}$

30 40 50 60 70 80 90 100 110 120
TEMPERATURE °C

between the mobile amorphous regions and immobile crystalline regions with regard to the timeframe of their DSC results for normal rice starches analyzed at 50 w% total moisture, but they implausibly suggested that the elevated Tm observed in annealed starches was due to increased stability in the amorphous regions. Rather, for this case of starch in excess moisture, annealing allows a relaxation in the amorphous regions from a more to a less kinetically metastable state, while the crystalline regions perfect from a less metastable state to a more stable state with a higher Tm (Wunderlich, 1981). The rescan of Figure 5-28, part E, exhibits the following features, compared to the typical appearance of curve A below 100°C; a more symmetrical melting endotherm with essentially no baseline shift, an onset temperature (essentially coincident with the initial effective Tg) of 63°C, a peak minimum of 70°C, and an effective end-of-melting Tm of 92°C. The essentially undetectable baseline shift in heat capacity associated with the crystalline melting transition at the instrumental settings that allowed ready demonstration of the large change in heat capacity associated with the glass transition was expected, as explained by Wunderlich (1981, Figs. 5-13, 5-16), due to the free volume contribution to heat capacity. The appearance of the thermal profile in the region of the glass transition is analogous in shape to an endothermic hysteresis peak, a common characteristic manifested by partially crystalline polymers (Wunderlich, 1981). An endothermic hysteresis peak is indicative of some jump, during the sample history, in temperature, plasticizer content, or pressure at a rate exceeding the relaxation rate of the appropriate process. This hysteresis peak is observed during subsequent DSC analysis as a "stress relief" via "enthalpic relaxation" (Wunderlich, 1981). The apparent enthalpic relaxation of starch (Zeleznak and Hoseney, 1987a), with a peak minimum at 63°C, is superimposed on the universal step change in heat capacity. One can further imagine summing the glass and melting transitions, superimposed on one another in the temperature range 50°C to 100°C, by adding together the scan and rescan in Figure 5-28, part E, thus reconstituting, with no discernible loss of total heat below the baseline, the characteristic DSC thermal profile (curve A) for wheat starch gelatinization in excess water.

Figure 5-29 (Slade and Levine, 1988b) contains the analogous composite diagram of DSC heat flow curves for waxy corn starch with 55 w% total moisture. As in Figure 5-28, Figure 5-29 shows the complete nonequilibrium melting process as an initial scan in curve A; the partial initial scans, with their maximum temperatures indicated, as solid lines in parts B–G; and the complete rescans as dashed lines in parts B–H. In contrast to Figure 5-28, amylose–lipid melting transitions above 100°C are absent for this essentially amylose-free starch. Based on the same analysis and logic described for Figure 5-28, and equally successful superposition of the instrumental baseline response of initial and rescans (e.g., Fig. 5-29, part B, in the temperature range from 30°C to 60°C and H from 95°C to 130°C), the following results

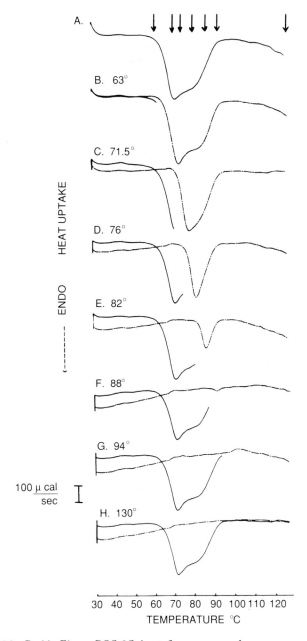

Figure 5-29. Perkin-Elmer DSC-2C heat flow curves of waxy corn starch:water mixtures (45:55 w:w): (A) initial scan at 10°C/min for native starch; (B–H) solid lines = partial initial scans for native starch, at 10°C/min, from 20°C to maximum temperatures indicated; dashed lines = rescans at 10°C/min, immediately following partial initial scanning and then immediate cooling, at nominal 320°C/min, to 20°C. (*From Slade and Levine, 1988b, with permission.*)

were obtained from Figure 5-29. Parts *C–H* manifested displaced baselines, in the temperature range 30°C to ≲ 65°C, for the initial scans and the rescans, while part *B* showed superimposed baselines in the same temperature range. Accordingly, the effective end-of-softening Tg preceding crystallite melting was identified as greater than 63 and less than 71.5°C from Figure 5-29*C*. As for wheat starch in Figure 5-28, this upper limit for Tg corresponds to the temperature at the "peak minimum" in the typical DSC thermogram for waxy corn starch gelatinization shown in Figure 5-29*A*. The narrow temperature span of this Tg occurred within the range 64–71.5°C, along the leading edge of the endotherm. Part *H* revealed an effective end-of-melting Tm ≃ 95°C, where the initial scan returned to baseline after gelatinization. This observation was corroborated in part *G*, where the thermal profile stopped just short of a return to baseline upon initial heating to 94°C. As in Figure 5-28*E*, Figure 5-29*C* illustrates a separation of the melting transition of A-type microcrystallites from the glass transition that must precede it. In this case, however, the melting transition can be unambiguously assigned to the microcrystalline, clustered amylopectin branches, and the glass transition to the contiguous amorphous regions of water-plasticized amylopectin. The rescan of Figure 5-29*C* shows a nearly symmetrical melting endotherm with the following features: onset temperature of 71.5°C, coinciding with Tg; peak minimum of about 78°C; and end-of-melting Tm at 95°C. Waxy corn starch in Figure 5-29*C*, like normal wheat starch in Figure 5-28*E*, showed an undetectably small baseline shift from its leading to trailing end for the isolated melting transition associated with gelatinization in excess moisture. Thus, these DSC results demonstrated conclusively that the change in heat capacity, illustrated in Figures 5-28 and 5-29, was associated entirely with the glass transition immediately preceding the crystalline melting endotherm. In contrast, Yost and Hoseney (1986) also observed such a change in heat capacity (between an initial partial heating scan, to a single intermediate temperature, and a complete rescan) for native wheat starch in 50 w% water, but because these investigators did not observe a Tg (at the subzero Tg') in the DSC rescan, Zeleznak and Hoseney (1987a) suggested instead that it "merely indicates that the heat capacity of a starch–water suspension is lower than that of gelatinized starch." Figures 5-28*E* and 5-29*C* demonstrated the actual explanation for their observation.

Russell (1987a) has published DSC thermograms (showing initial complete heating scans and superimposed immediate complete rescans) for native wheat and waxy corn starches at 57 w% total moisture content. These thermograms are very similar in appearance to Figure 5-28 (curves *A* and *L*) and Figure 5-29 (curves *A* and *H*), respectively. He also observed and recognized the characteristic change in heat capacity (ΔCp) as signifying a glass transition immediately preceding crystallite melting. Because of the "very small ΔCp (about 0.1 J/°C g sample)" associated with this baseline shift, however, Russell's

(1987a) concluding remarks stopped short of a wholehearted endorsement ("it is likely that a glass transition is associated with starch gelatinization") of the concept (Slade, 1984; Slade and Levine, 1984a, 1984b; Maurice et al., 1985), by subsequently referring to "the putative glass transition." With regard to the magnitude of ΔCp for the glass transition exhibited in Figures 5-28 and 5-29, it has been shown (Slade and Levine, 1988c) that the magnitude of the baseline shift for the glass transition of amylopectin in a DSC thermogram (essentially identical to the heat flow curve in Fig. 5-22a) for a 23 mg sample of 47:53 native wheat starch:water (assumed to represent a starch of about 73% amylopectin, about 24% crystallinity, and thus about 6 mg of amorphous amylopectin) is equivalent in magnitude to the ΔCp that would be observed for 6 mg of completely amorphous polystyrene at its Tg.

With the aim of deconvoluting the contributions of amylopectin and amylose to the nonequilibrium melting behavior of native granular wheat and waxy corn starches in 55 w% moisture, the effective values of Tg and end-of-melting Tm were compared: for wheat starch, $Tg \simeq 63°C$ and $Tm \simeq 92°C$, while for waxy corn starch, $Tg \simeq 71.5°C$ and $Tm \simeq 95°C$. These effective end-of-melting Tm values, rather than the corresponding onset or peak values, were chosen for comparison (Slade and Levine, 1988b) because they would represent melting of the largest and/or most perfected microcrystals (Flory, 1953; Wunderlich, 1980), and so would be most relevant to the comparison of Tg/Tm ratios (Lelievre, 1976; Russell, 1987a). For both starches, the Tm values were similar. In contrast, the effective Tg for wheat starch was significantly lower than the value for waxy corn starch.

The values of the ratio of effective Tg/end-of-melting Tm, relevant to gelatinization of these native granular starches in 55 w% water by DSC heating at 10°C/min, were 0.92 for wheat and 0.94 for waxy corn. For water-compatible polymers other than starch, Tg/Tm ratios greater than 0.9 have been attributed to the influence of a metastable supramolecular structure with nonuniform moisture distribution (Batzer and Kreibich, 1981; Levine and Slade, 1987). As has been mentioned, for those carbohydrate–water systems that are spatially homogeneous due to uniform moisture content and compositionally homogeneous with well-defined $\overline{Mn}$ and $\overline{Mw}$, it has been found that an increasing Tg/Tm ratio (above 0.8) correlates with increasingly anomalous relaxation behavior, due to contributions of excess free volume (i.e., high mobility and free volume) or decreased local viscosity in a glass at its Tg and in concentrated solutions above Tg (Slade and Levine, 1988a, 1988e).

The effective values of Tg identified (Slade and Levine 1988b), which are associated with first-stage swelling of native granular starches heated in 55 w% water, do *not* represent the Tg of amorphous regions of native granules at about 10 w% total moisture. Recent evidence has suggested that the value for that operative Tg is greater than 100°C for several different normal and waxy cereal grain starches (Maurice et al., 1985; Biliaderis, Page, Maurice, and Juliano, 1986; Zeleznak and Hoseney, 1987). Nor do they represent the Tg of completely amorphous gelatinized starch at 55 w% moisture, which is actually Tg' of about $-5°C$, as illustrated in Figure 5-12. The values of Tg reported by Slade and Levine (1988b) are manifested by the amorphous regions of native granules *during* the dynamic process of plasticization by heat (increasing at 10°C/min in the temperature range from 20°C to 130°C) and moisture uptake (increasing in the range from 10 w% to 55 w%) and represent particular, intermediate values, within a continuum, that depend on the instantaneous temperature and content of plasticizing water.

As has been discussed, when the experimental history with respect to heating rate, temperature range, and total sample moisture content was the same, the thermal profiles of amylose-containing normal wheat starch (Fig. 5-28) and essentially amylose-free waxy corn starch (Fig. 5-29) did not differ qualitatively below 100°C. The major qualitative difference was the presence of a melting transition above 100°C for crystalline lipid–amylose complex in the initial scans of normal wheat starch (also seen in immediate rescans as a result of recrystallization from the self-seeded melt), and the absence of this transition in the thermal profiles for waxy corn starch. The qualitative similarity of the thermal behavior of normal and waxy starches below 100°C indicated that the thermal profiles represented nonequilibrium melting of microcrystals composed of hydrated clusters of amylopectin branches in both cases, with no significant contributions from amylose (Slade and Levine, 1988b). Thus, the quantitative differences between the values of the operative end-of-softening Tg and end-of-melting Tm for normal wheat versus waxy corn starch should be explained on the basis of structure–property differences in their amylopectin components. As has been stated, sample history (path dependence such as jumps in moisture, temperature, or pressure) is often more important than inherent equilibrium thermodynamic properties, and as important as chemical structure for the explication of structure–property differences in nonequilibrium systems. Moreover, the starch–water system is neither spatially nor molecularly homogeneous, and the greater anomaly in Tg/Tm ratio for waxy corn starch compared to normal wheat starch will also depend highly on contributions of sample history as well as the structural biochemistry of the starch.

For a similar initial operative level of water plasticization in both the normal wheat and waxy corn starch systems, the quantitative differences seen for Tg, nonequilibrium Tm, and Tg/Tm ratio associated with gelatinization and pasting, can be explained by the previously mentioned fact that, for homologous amorphous polymers, Tg increases with increasing average MW (Slade and Levine, 1988b). Significantly lower average MW of the amorphous regions of the starch granule would allow a greater rate of water uptake and greater values for the instantaneous extent of water plasticization at each time point in the DSC experiment. The underlying basis for the difference in operative average MW of the amorphous regions of the native amylopectins has been described. Disproportionation of more mobile branches with lower linear DP to the microcrystalline domains leads to higher average MW in the residual amorphous regions, and consequently to higher effective values of Tg and kinetically constrained Tm (Slade and Levine, 1987a). For this reason, the relative extents of crystallinity, ranging from 15% to 45%, of native starches from various sources and with both A- and B-type diffraction patterns, are inversely related to their gelatinization temperatures (Zobel, 1984; Snyder, 1984). The "high amylose" starches that result from the amylose-extender mutation, and that give misleading blue value determinations of 60% amylose content (Shannon and Garwood, 1984), are an apparent exception to this rule. But even in the case of this so-called high-amylose starch, it is the anomalous amylopectin, with its long, unclustered, noncrystalline branches, that produces the dramatically elevated values of Tg and, indirectly, of Tm; this is true despite the inherently low Tm of (isolated) B-type crystals (Slade and Levine, 1988b). (B-type crystals, isolated to remove kinetic constraints on melting due to amorphous surroundings, would melt at a lower temperature than isolated A-type crystals.) Like the silo-aging process for rice and the high-humidity drying process for potato (Snyder, 1984), the wet-milling process for corn provides an opportunity for annealing of starch (Krueger et al., 1987) and concomitant elevation in the extent of crystallinity, the average MW of residual disproportionated amorphous regions of amylopectin, and gelatinization Tg.

The study of the gelatinization process by Slade and Levine (1988b) dealt exclusively with A-type cereal grain starches rather than B-type tuber and root starches, such as from potatoes. The same was true of earlier studies: (1) by Slade and co-workers (Slade, 1984; Slade and Levine, 1984a, 1984b, 1987a; Maurice et al., 1985; Biliaderis, Page, Maurice, and Juliano, 1986), (2) of Tg and annealing by Yost and Hoseney (1986) and Zeleznak and Hoseney (1987a), and (3) of annealing by Krueger and co-workers (1987) and Nakazawa and co-workers (1984). In addition to possible differences in the extent of crystallinity due to process variations (Snyder, 1984), B-type native granular starches often have higher "as is" moisture contents in both the amorphous and crystalline regions than do A-type starches (i.e., overall, but likewise heterogeneously distributed, moisture contents of about 18–20 w% for B-type versus about 6–10 w% for A-type) (French, 1984; Whistler and Daniel, 1984). Thus, the initial instantaneously operative extent of plasticization of the continuous amorphous matrix, which subsequently governs the nonequilibrium melting of the disperse microcrystalline regions, can be significantly different for B- versus A-type starches (Slade and Levine, 1987a) and can contribute to the observed lower gelatinization temperature of potato starch (Snyder, 1984). However, the generic description of the gelatinization process for cereal starches (Slade, 1984; Slade and Levine, 1984a, 1984b) is still valid for potato starch. Less extensive drying subsequent to starch biosynthesis results in a greater preexisting moisture content in the amorphous regions of commercial potato starch, greater free volume, and depressed effective Tg, and in the crystalline regions, depressed effective end-of-melting Tm. The functional attributes and physical properties, including the extent of crystallinity and X-ray diffraction pattern, of potato starch can be altered by deliberate drying (Donovan, 1979; Snyder, 1984) or heat/moisture treatment (Donovan et al., 1983; Snyder, 1984; Kuge and Kitamura, 1985) to resemble those of cereal starches. As a consequence of preexisting plasticization by water, depressed initial Tg, greater initial mobility, and lower end-of-melting Tm, the entire heating profile of the gelatinization of native potato starch is sharper and narrower than for cereal-like, treated potato starch at the same total moisture content of the sample (moisture content of starch plus added water) and is centered at a lower temperature (Slade and Levine, 1988b).

It cannot be overemphasized that the glass transition in starch, as in any other amorphous or partially crystalline material, represents a rate-limiting stage of a relaxation process (Ferry, 1980) for which the spectrum of relaxation rates depends on the instantaneous magnitude of the free volume and/or local viscosity, which in turn depends on the relative values of experimental moisture compared to the Wg of the operative glass, experimental temperature compared to instantaneous Tg, and experimental timeframe compared to the instantaneous relaxation time (Slade and Levine, 1988b). Thus, when DSC heating rates approach the operative relaxation rates for a measured process, a lower heating rate would result in observation of a lower Tg value. Theoretically, heating at 1°C/min rather than 10°C/min would result in a Tg lower by about 3°C, as calculated from the WLF equation for "well-behaved" polymers with Tg/Tm ratios near 0.67. Tg differences of 3–5°C per order of magnitude are expected over broader ranges of experimental rates (or frequencies) for such well-behaved polymers, since the relaxation spectrum changes gradually from WLF to Arrhenius kinetics over a temperature interval of about 100°C above Tg (Slade and Levine, 1988a). Experimentally, this expectation has been confirmed in the case of polystyrene, for which the value of Tg was lower by 15°C when the heating rate was decreased from 1°/s to 1°/hr, a factor of 3,600 (Wunderlich, 1981). Differential scanning calorimetry experiments with slow heating rates of less than 0.5°C/min, for very dilute aqueous potato starch suspensions of about 2% solids, have allowed a very small scale micro-reversibility, which was misinterpreted as the ability to achieve and maintain equilibrium throughout the gelatinization process (Shiotsubo and Takahashi, 1986). Loss of temporal resolution of the thermal events due to the greatly excess moisture content accounted for part of the apparent micro-reversibility (Slade and Levine, 1988b). Actually, isothermal treatment of aqueous rice starch slurries with excess moisture (50% solids), for more than 100 hours at various temperatures between 40°C and 85°C (equivalent to infinitely slow heating rates), was not sufficient to approach an equilibrium state (Nakazawa et al., 1984), as expected, since the melting of partially crystalline systems is *never* an equilibrium process (Wunderlich, 1981).

The dynamic nature of the glass transition is also reflected in the nonequilibrium annealing process for native starches in the presence of moisture that is insufficient for massive second-stage swelling. For example, recent results for annealing (to measurable, but not necessarily equal, extents, for different granular starches) at different temperatures and times have included the following:

1. waxy corn at 55 w% moisture, 70°C for 10 minutes or 65°C for 30 minutes (Maurice et al., 1985)
2. wheat at 50 w% moisture, 72°C for 30 minutes or room temperature for 24 hours (Yost and Hoseney, 1986)
3. dent corn in excess added water, 60°C for 15 minutes, 55°C for 2 hours, or 50°C for 48 hours (Krueger et al., 1987)
4. normal and waxy rices at 50 w% moisture, at from 85°C for 5 minutes to 40°C for 140 hours (Nakazawa et al., 1984)
5. wheat at 55 w% moisture, 25°C for 55 days (Slade, 1984; Slade and Levine, 1984a, 1984b).

Comparison of these results demonstrates a mobility transformation with respect to time, temperature, and effectively plasticizing moisture content, and has led to the conclusion that significant annealing at lower temperatures for longer times is controlled by a lower operative Tg (resulting from a longer experimental timeframe) than the Tg that precedes crystallite melting by heat/moisture treatment in the DSC (Slade and Levine, 1988b). In other words, the operative Tg relevant to annealing at $Tg < Ta < Tm$ decreases with increasing holding time in excess added moisture at lower temperatures, due to the effect of dynamic plasticization. Under conditions in which the operative Tg has clearly fallen to Tg' at $-5°C$ (and Wg has increased to Wg'), significant annealing has also been observed in retrograded starch gels and baked bread aged at storage temperatures well above room temperature, and so much closer to Tm than Tg (Longton and LeGrys, 1981; Zeleznak and Hoseney, 1987b). The unifying explanation lies in the fact that the progressively resultant events of plasticization, mechanical relaxation above the glass transition, and functional manifestation (including starch gelatinization, crystallite melting, annealing, and recrystallization, which can occur during or after baking of high-moisture doughs) are all dynamic, nonequilibrium processes, the kinetics of which are governed by WLF theory for glass-forming systems (Slade and Levine, 1988b).

Thermal Properties of Three-Component Model Systems: The Antiplasticizing Effect of Added Sugars on Gelatinization of Starch

The description of the effect of water as a plasticizer on native starch, from the perspective of starch as a partially crystalline glassy polymer system, has been extended to the next level of complexity; that is, three-component model systems of native starch, water, plus added sugars (Slade, 1984; Slade and Levine, 1984a, 1984b, 1987a). This extended description is based on a recognition of gelatinization of granular starch in aqueous media as (1) a diffusion-limited, mechanical relaxation process with non-Arrhenius kinetics (Burros et al., 1987), which depend on the mobility of the added plasticizer (Slade and Levine, 1988a, 1988b); and (2) a nonequilibrium melting process (as a consequence of heat/moisture treatment), which becomes cooperative and occurs at a significant rate at a characteristic "gelatinization temperature" ($Tgelat$) corresponding to the instantaneous Tg (i.e., $Tgelat = Tg$) of the water-plasticized amorphous regions of amylopectin (Slade and Levine, 1988b). Gelatinization in concentrated aqueous solutions of common small sugars begins at a higher $Tgelat$ than in water alone, a retardation effect that has been demonstrated to result from "antiplasticization" (as defined for synthetic polymers [Sears and Darby, 1982]) by sugar–water cosolvents, relative to the extent of plasticization by water alone (Slade, 1984; Slade and Levine, 1984a, 1984b, 1987a). Sugar–water, of higher average MW than water alone, results in a smaller depression of starch Tg than does pure water. In fact, isothermal treatment of starch in sugar–water, at a temperature that would result in nonequilibrium melting of amylopectin in water alone, results instead in antiplasticization by annealing and crystallite perfection (Slade and Levine, 1987a).

It has been known empirically for decades that various sugars, including sucrose, fructose, and glucose, raise the temperature of gelatinization of starch in water and delay the increase in viscosity ("pasting"), and that this effect increases with increasing sugar concentration (Zobel, 1984; Lund, 1984; Blanshard, 1987). This effect of sugars on the gelatinization and pasting behavior of native and modified starches is also well known to be important to the processing and properties of real baked products, in that the effect influences the extent of wheat starch gelatinization, and its retardation or even inhibition during baking of high-sugar cookie doughs and cake batters (Blanshard, 1986, 1988). The effect of sugars on $Tgelat$ had been attributed in the past in part to a depression of "Aw" by sugars and in part to an unexplained interaction (called "sugar bridges" by Ghiasi et al., 1983) of sugars with the amorphous areas of starch granules. However, no successful attempt had been made to show how these two aspects of the effect might be related, or to explain the mechanism of elevation of $Tgelat$, prior to the description of the antiplasticizing effect of sugar–water cosolvents (Slade, 1984; Slade and Levine, 1984a, 1984b, 1987a).

The effect of sucrose on *Tgelat* for wheat starch is illustrated in Figure 5-30. For convenience, *Tgelat* was taken as the temperature at the peak of heat uptake, as measured by DSC. Figure 5-30 shows that as the weight of sucrose was increased in a ternary mixture with constant equal weight ratio of starch and water, *Tgelat* increased monotonically for samples up to a 1:1:1 mixture. The DSC heat flow curve of this 1:1:1 mixture (shown in the inset of Fig. 5-30), in which 50 w% sucrose was the added fluid outside the granules, exhibited a glass transition at an effective $Tg > 30°C$ higher than that seen for a 1:0:1 mixture, when the added fluid was water alone. This transition was evidenced by a characteristic baseline shift associated with the free volume contribution to Tg (Wunderlich, 1981) at the leading edge of the gelatinization endotherm. Immediately following and superimposed on the elevated glass transition was a relatively narrow crystalline melting transition, similar to that shown earlier for native starch annealed either for 55 days at 25°C (Fig. 5-22*b*) or for less than 30 minutes at 70°C (Fig. 5-24*b*).

The effect of sucrose on *Tgelat* has been explained, within predictions of the conceptual framework of starch as a partially crystalline glassy polymer,

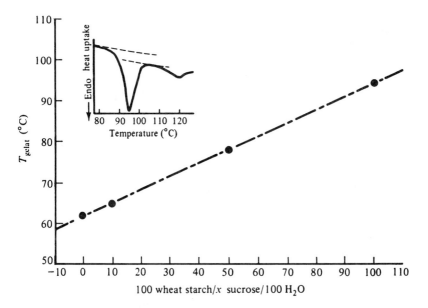

Figure 5-30. Gelatinization temperature, *Tgelat*, as a function of added sucrose content for three-component mixtures of native wheat starch:sucrose:water (100:x:100 parts by weight). *Inset:* DSC heat flow curve of 100:100:100 mixture. (*From Slade and Levine, 1987a, with permission.*)

on the basis of WLF theory (Slade, 1984; Slade and Levine, 1984a, 1984b, 1987a). If a sugar–water solution is viewed as a plasticizing cosolvent, it is evident that such a coplasticizer, of greater average MW than water alone, would be less effective in mobilizing and increasing the free volume of the amorphous fringes in the "fringed micelle" structure of a starch granule. Less effective plasticization would result directly in less depression of Tg, and thus indirectly in less depression of nonequilibrium crystalline melting transition temperatures. In this sense, in comparing the efficiencies of aqueous solvents (including solutions of nonionic solutes such as sugars and polyols) as plasticizers of the glassy regions of native starch, water alone is the best plasticizer, and sugar–water cosolvents are actually *anti*plasticizers relative to water itself. By most effectively depressing the requisite Tg at which gelatinization is initiated, added water results in the lowest $Tgelat$. Increasing concentrations of a given sugar result in increasing average molecular weights and decreasing free volumes of the cosolvent and thus increasing antiplasticization and $Tgelat$ versus water alone.

Of course, increasing the concentration of a given sugar in an aqueous cosolvent also decreases Aw (actually relative vapor pressure, RVP [Levine and Slade, 1987]), but it has been established that the temperature of the effective glass transition of native starch in the presence of added water is independent of total sample moisture above about 30 w% total water content (i.e., Wg') (Slade, 1984; Slade and Levine, 1984a, 1984b, 1987a; Maurice et al., 1985). Moreover, according to WLF theory, the extent of antiplasticization would be expected to increase with increasing $\overline{M}w$ of the cosolvent, within a homologous series of cosolvent components, from monomer to dimer to oligomer to polymer. Yet, in such a case, the RVP of cosolvents at equal weight concentrations would generally increase with increasing cosolvent $\overline{M}w$. For example, for the homologous series glucose, maltose, maltotriose, 10-DE maltodextrin, RVPs of 50 w% solutions increase from 0.85 to 0.95 (for both maltose and maltotriose) to 0.99 (Levine and Slade, 1987; Slade and Levine, 1988e).

Experimental results for the effect of this homologous series of cosolvents on $Tgelat$ of starch have confirmed the prediction based on WLF theory, as illustrated in Figure 5-31 (Slade and Levine, 1984a, 1984b, 1987a), and thus demonstrated that the underlying basis of this effect is not a depression of "Aw" by sugars. Figure 5-31 shows that as the $\overline{M}w$ of the antiplasticizing cosolvent increases, free volume decreases, and the resultant Tg of the cosolvent increases. Consequently, the cosolvent becomes less efficient in depressing the effective Tg at which the gelatinization transitions are initiated in native starch, so $Tgelat$ increases with increasing cosolvent $\overline{M}w$, despite a corresponding increase in cosolvent RVP.

A comparison of the degree of elevation of $Tgelat$ of native wheat starch, for 1:1:1 sugar–water–starch mixtures of a larger and nonhomologous series of

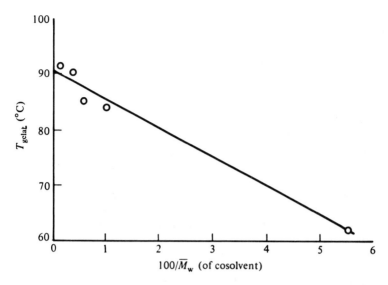

Figure 5-31. Gelatinization temperature, *Tgelat*, as a function of $100/\overline{M}w$ (of cosolvent, water + glucose polymer), for three-component mixtures of native wheat starch:glucose polymer:water (1:1:1 parts by weight) and two-component mixture of native wheat starch:water (1:2 parts by weight). (*From Slade and Levine, 1987a, with permission.*)

sugars, has also been reported recently (Slade and Levine, 1987a). Results of DSC studies showed that *Tgelat* increases in the following order: water alone < galactose < xylose < fructose < mannose < glucose < maltose < lactose < maltotriose < 10-DE maltodextrin < sucrose. It has been suggested that the same structure–property relationships that act as determinants of "water availability" (as opposed to "Aw") in the water dynamics domain of the dynamics map appear to influence the elevation of *Tgelat* (Levine and Slade, 1987; Slade and Levine, 1988e). It has been concluded that no single parameter (e.g., *Tg*, *Wg'*, MW, or *Tm/Tg* ratio) can completely explain the mechanism of elevation of *Tgelat* by sugars, but a combination of these parameters that predicts the contributions of free volume and local viscosity provides useful clues to explain why the elevating effect on *Tgelat* is greater for sucrose than it is for glucose, for which the effect is in turn greater than for fructose (Slade and Levine, 1987a).

   As has been pointed out with regard to the effect of water content in binary starch–water systems, previous attempts to analyze the effect of sugar content in ternary starch–sugar–water systems on the observed *Tm* of starch by the Flory-Huggins equilibrium thermodynamics treatment of melting point depression by diluents (e.g., Lelievre, 1976) have been recognized as unjustified on a rigorous theoretical basis (Slade, 1984; Levine and Slade, 1987; Blanshard, 1987). Aside from theoretical problems introduced by the facts that the interaction parameter chi is concentration dependent and amylopectin microcrystallites are not a monodisperse system, Flory-Huggins theory is appropriate only to describe an equilibrium melting process. Regardless of the amount of water or sugar solution added to native starch, initial melting of a native granule remains a nonequilibrium process, in which melting of the microcrystallites is controlled by previous plasticization of the glassy regions via heat/moisture treatment to T greater than the effective T_g (Slade and Levine, 1987a).

Starch Gelatinization as a Mechanical Relaxation Process Affected by the Mobility of Aqueous Sugar Solutions

In general, mechanical relaxations depend on both translational and rotational mobility (Ferry, 1980). For a typical, well-behaved amorphous polymer–plasticizer system, an increase in free volume (which is related to rotational mobility) would be expected to go hand in hand with a decrease in local viscosity (which is related to translational mobility and reflects the molecular level environment). Depending on the underlying mechanism of a specific mechanical relaxation, however, either the rotational or the translational relaxation time can be the limiting aspect for a particular glass-forming system. The gelatinization of native granular starch in three-component starch–sugar–water model systems has been shown to be a mechanical relaxation process that appears to depend on both translational and rotational diffusion, but can be completely controlled and limited by the translational mobility of aqueous sugar solutions (Slade and Levine, 1988a).

Use of the dynamics map, in the form of the mobility transformation map in Figure 5-1, as a new conceptual approach to the study of nonequilibrium thermomechanical behavior of glass-forming food polymer systems, has facilitated the identification of a discriminating experimental approach and conditions that are capable of separating the effects of translational and rotational mobility on different mechanical relaxation properties, and thus elucidating the underlying basis of the differences in behavior of sugars during starch gelatinization (Slade and Levine, 1988a). The experimental approach developed to analyze the gelatinization process has utilized starch as a "reporter" molecule (probe) to study the relative translational mobilities of aqueous solutions of different sugars. The selection of a molecule to be used as a reporter to probe the local (i.e., molecular level) environment is a critical element of experiments designed to study mechanical relaxation processes. A very low concentration of reporter molecule (e.g., a dye) is often required for translational and rotational diffusion experiments, in order to avoid concentration gradients and perturbation of the local relaxation due to plasticization by the reporter (Huang et al., 1987). Water itself is generally *not* a good candidate for the role of a reporter molecule to study the mobility of aqueous glasses, because water would then be both a functional part of the sample matrix and a reporter in many experiments. For example, in an NMR investigation of the mobility of water in an amorphous polymer, the water concentration (at levels $< Wg'$) cannot be changed without significantly changing the system itself, because of the effect of water as a plasticizer (Slade et al., 1988). Thus, a third molecule would be needed to act as the reporter. High polymeric starch can be used to fill this key role. Experimental results (Slade and Levine, 1988a) have also demonstrated that investigation of the nonequilibrium relaxation behavior of different supra-glassy sugar–water solutions, in the context of the effect of their translational mobility on the diffusion-limited $Tgelat$ of partially crystalline starch, is greatly enhanced by the simultaneous investigation of their rotational mobility, as measured by dielectric relaxation experiments (Tait et al., 1972; Franks et al., 1973; Suggett and Clark, 1976; Suggett et al., 1976; Suggett, 1976).

The response to microwaves in a microwave dielectric dispersion experiment is a rotational response (Suggett and Clark, 1976). The dielectric relaxation time τ for a sugar in aqueous solution is directly related to the rotational diffusion time. Maximum absorbance of electromagnetic radiation by pure water at room temperature occurs at a frequency of about 17 GHz in a microwave dielectric dispersion experiment. Microwave absorption maxima at lower frequencies result when free volume becomes limiting and relaxations occur at lower frequencies due to hindered rotation. For comparison, the commercial frequency used for domestic microwave ovens is 2.45 GHz. In the case of a dilute solution, when free volume is not limiting, the dielectric relaxation time is determined mainly by the intrinsic hydrodynamic volume of the solute (Tait et al., 1972; Franks et al., 1973; Suggett and Clark, 1976). For each sugar solute in water at a given temperature, there is a limiting concentration below which the mobility shows a simple dependence on the average molar volume and above which the free volume limitation would begin to contribute to hindered rotation and increased local viscosity (which is equivalent to macroscopic solution viscosity only for solute MWs below the entanglement limit [Slade and Levine, 1988a]). At 293°K, this limiting concentration has been shown to be about 33 w% for sucrose and about 38 w% for glucose (Soesanto and Williams, 1981). In other words, the hindered mobility characteristic of WLF behavior in the rubbery domain would be observed when $\Delta T \simeq 52$ K and $\Delta W \simeq 31$ w% above the $Tg'-Wg'$ reference state for a sucrose solution and when $\Delta T \simeq 63$ K and $\Delta W \simeq 33$ w% above the $Tg'-Wg'$ reference state for a glucose solution (Slade and Levine, 1988a).

Suggett and Clark (1976) have assessed the rotational diffusion behavior of concentrated aqueous solutions (24.0–33.5 w% solute) of a series of sugars, including the pentoses ribose and xylose, the hexoses glucose and mannose, and the disaccharides sucrose and maltose. These investigators determined dielectric relaxation times from microwave dispersion measurements made over a frequency range from 100 KHz to 35 GHz at 278 K, where these supra-glassy sugar solutions would be expected to exhibit hindered rotation and the WLF behavior that has been mentioned. The effect of the same sugars on starch gelatinization has been assessed from DSC measurements of $Tgelat$ for native granular wheat starch suspensions in 50 w% aqueous sugar solutions (Slade and Levine, 1987a). The relative effects of the different sugar solutions on translational diffusion in the sugar–water–starch suspension have been estimated from these measurements of $Tgelat$, which reflect the relative deficit in the depression of Tg of the amylopectin component of starch by sugar–water compared to water alone (Slade and Levine, 1988a). As has been revealed by a graph of $Tgelat$ versus dielectric relaxation time (in picoseconds) in Figure 5-32, the effects of these sugars on starch gelatinization are highly linearly correlated ($r = 0.97$) with their rotational diffusion times in solution, as measured by dielectric relaxation. It is especially interesting to note that the surprising behavior of glucose and its dimer maltose, which showed very similar rotational diffusion times, was reflected in exactly the same way by their very similar effect on $Tgelat$ for the mechanical relaxation process reported by starch.

It has been suggested that the underlying explanation for this correlation is revealed by the graph in Figure 5-33. This graph shows the fundamental relationship between the measured rotational diffusion times from the dielectric relaxation experiment (Suggett and Clark, 1976) and the calculated relative mobilities of the supra-glassy sugar–water solutions. A mobility transformation map was constructed for each sugar solution, and the relative mobility

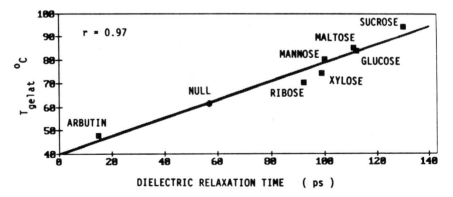

Figure 5-32. Variation of the gelatinization temperature of native wheat starch suspended in 50 w% aqueous sugar solutions with the corresponding dielectric relaxation time measured at 278 K for concentrated aqueous solutions of the same sugars. (*From Slade and Levine, 1988a, with permission.*)

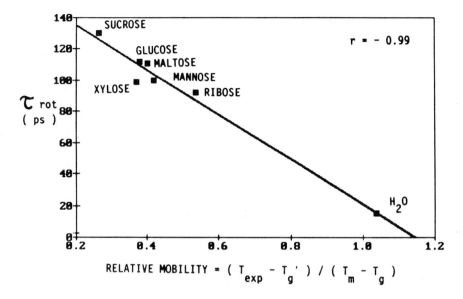

Figure 5-33. Variation of rotational diffusion time measured at *Texp* = 278 K for concentrated aqueous sugar solutions with the corresponding relative mobility parameter of aqueous solutions of the same sugars, calculated from the ratio of *(Texp −Tg')/(Tm−Tg)* for each sugar. (*From Slade and Levine, 1988a, with permission.*)

was estimated from the relative distance between the experimental conditions and the reference glass curve, normalized with respect to the inherent mobility of the sugar. The inherent mobility of a sugar is related to the distance (in units of temperature) required by its dry glass to achieve the mobility of an Arrhenius liquid (Slade and Levine, 1988a). Thus, the relative mobility scale shown in Figure 5-33, calculated from the ratio of $(T - Tg')/(Tm - Tg)$ for each sugar, was defined in terms of the temperature difference between the experimental temperature T (278 K) and Tg' of the freeze-concentrated glass as the reference state, compared to the magnitude of the temperature difference in the WLF domain between Tm and Tg of the dry solute as a measure of inherent mobility. (Note the similarity between this relative mobility scale and the reduced temperature scale, mentioned in the section Crystallization Kinetics for Partially Crystalline Polymers, which defines mobility in the temperature range from Tg to Tm with respect to the kinetics of polymer crystallization, another mechanical relaxation process that depends on both translational and rotational mobility [Wunderlich, 1976; Slade and Levine, 1988a].) Both dielectric relaxation times and translational diffusion coefficients of a broad range of glass-forming systems, including polyvinyl acetate and glucose, have been shown experimentally to follow the WLF equation near Tg (Huang et al., 1987; Matsuoka et al., 1985). The factor $Tm - Tg$ was chosen, in this case as a preferred alternative to the Tg/Tm ratio, for the comparison of mobilities, at $T \gg Tg$, of different sugars with different values of dry Tg (Slade and Levine, 1988a). Figure 5-33 illustrates the excellent linear correlation ($r = 0.99$) between the mobility, expressed in terms of WLF behavior, and the dielectric relaxation behavior of the aqueous sugar solutions in their nonequilibrium, supra-glassy states. As expected, in the absence of anomalous anisotropic requirements of free volume for mobility, both translational and rotational mobility depend correlatively on free volume, but translational motion becomes limiting at a higher temperature and determines Tg (Slade and Levine, 1988a). It should be noted that the calculated translational mobility for xylose was significantly lower than expected, based on its rotational relaxation time. The relatively low value of Tg/Tm for xylose accounted for the low calculated mobility (Slade and Levine, 1988a), which had been previously confirmed experimentally by demonstrating the anomalous ability of xylose to retard the recrystallization of starch (Slade and Levine, 1987a).

Retrogradation/Staling as a Starch Recrystallization Process

Retrogradation of gelatinized starch involves recrystallization of both amylopectin and amylose (Matsukura et al., 1983; Miles et al., 1985a; Russell, 1987b; Mestres et al., 1988; Zobel, 1988). It has been demonstrated for commercial SHPs that the minimum linear chain length required for intermolecular entanglement upon concentration to Cg' corresponds to $\overline{DPn} \approx 18$ and $\overline{Mn} \approx 3,000$ (Levine and Slade, 1986). Sufficiently long linear chain length ($\overline{DPn} \gtrsim 15$–$20$) has also been correlated with intermolecular network formation and thermoreversible gelation of SHPs (Reuther et al., 1984; Slade, 1984; Levine and Slade, 1987, 1988f; Slade and Levine, 1987a) and amylopectin (Ring et al., 1987), and with starch (re)crystallization (Welsh et al., 1982; Miles et al., 1985a; Ring and Orford, 1986; Ring et al., 1987; Mestres et al., 1988). It has been suggested that, in a partially crystalline starch, SHP, or amylopectin gel network, the existence of random interchain entanglements in amorphous regions and "fringed micelle" or chain-folded microcrystalline junction zones (Reuther et al., 1984) each represents a manifestation of sufficiently long chain length (Levine and Slade, 1986). This suggestion is supported by other work (Ellis and Ring, 1985; Miles et al., 1985a, 1985b) showing that amylose gels, which are partially crystalline (Welsh et al., 1982), are formed by cooling solutions of entangled chains. Miles and co-workers (1985b) have stated that amylose gelation requires network formation, and this network formation requires entanglement. These investigators have concluded that "polymer entanglement is important in understanding the gelation of amylose."

Time-dependent gelation of amylose ($\overline{Mw} = 5 \times 10^5$) from dilute aqueous solution ($\gtrsim 1.5$ w% amylose, the critical concentration for entanglement of very high polymer amylose [Miles et al., 1985b]) is said to occur in two stages:

a relatively fast but finite stage due to viscoelastic network formation via entanglement (which is reversible by dilution but not thermoreversible), followed closely by a slower, but continually maturing, crystallization (in a chain-folded- or extended-chain morphology) process (which is thermoreversible above 100°C) (Ellis and Ring, 1985; Miles et al., 1985a, 1985b; Ring, 1985b; Gidley et al., 1986; Ring et al., 1987; l'Anson et al., 1988; Mestres et al., 1988). In contrast, in partially crystalline, thermoreversible (below 100°C), aqueous amylopectin gels, viscoelastic network formation (which is relatively slow and time dependent) is more closely related to the presence of microcrystalline junctions than to entanglements, although entanglement does occur (Ring and Orford, 1986; Ring et al., 1987; Mestres et al., 1988). Since most normal starches are 70–80% amylopectin (Whistler and Daniel, 1984), their gelatinization and retrogradation processes are dominated by the nonequilibrium melting and recrystallization behavior of amylopectin (Slade, 1984; Slade and Levine, 1984a, 1984b, 1987a; Russell, 1987b), although contributions due to amylose can be observed (Matsukura et al., 1983; Jankowski and Rha, 1986a). Generally, the early stages of starch retrogradation are dominated by chain-folded amylose (of DP from about 15 to about 50 and a fold length of about 100Å [Buleon et al., 1984; Ring et al., 1987; Mestres et al., 1988]); the later stages by extended-chain amylopectin (Marsh and Blanshard, 1988) outer branches (of DP about 12–16 [Hizukuri, 1986; Ring et al., 1987]) (Slade, 1984; Slade and Levine, 1987a).

Experimental evidence supporting these conclusions about the thermoreversible gelation mechanism for partially crystalline polymeric gels of starch, amylopectin, and SHPs has come from DSC studies (Slade and Levine, 1987a), the favored technique for evaluating starch retrogradation (Atwell et al., 1988). Analysis of 25 w% SHP gels, set by overnight refrigeration, has revealed a small crystalline melting endotherm with $Tm \simeq 60°C$ (Levine and Slade, 1987), similar to the characteristic melting transition of retrograded B-type wheat starch gels. Similar DSC results have been reported for 20 w% amylopectin (from waxy maize) gels (Ring, 1985a; Ring and Orford, 1986). The small extent of crystallinity in SHP gels can be increased significantly by an alternative two-step temperature-cycling gelation protocol (12 hr at 0°C, followed by 12 hr at 40°C) (Slade, 1984; Slade and Levine, 1987a), adapted from one originally developed by Ferry (1948) for gelatin gels and subsequently applied by Slade and co-workers (1987) to retrograded starch gels. In many fundamental respects, the thermoreversible gelation of concentrated aqueous solutions of polymeric SHPs, amylopectin, and gelatinized starch is analogous to the gelation-via-crystallization of synthetic homopolymer and copolymer–organic diluent systems, as described in the section Crystallization/Gelation Mechanism for Partially Crystalline Polymers (Slade and Levine, 1987a). For the latter partially crystalline gels, the possibly simultaneous presence of random interchain entanglements in amorphous regions (Boyer et al., 1985) and microcrystalline junction zones (Domszy et al., 1986) has been reported. However, controversy exists (Boyer et al., 1985; Domszy et al., 1986) over which of the two conditions (if either alone) might be necessary and sufficient to be primarily responsible for the structure/viscoelastic property relationships of such polymeric systems. This controversy could be resolved by a simple dilution test, which could also be applied to polysaccharide gels; entanglement gels can be dispersed by dilution at room temperature, whereas microcrystalline gels cannot be when room temperature is less than Tm (Levine and Slade, 1987).

The rate and extent of staling in high-moisture ($> Wg' \simeq 27$ w% water), lean (low sugar/fat), wheat starch–based baked products (e.g., breads, rolls, and English muffins) are well known to be correlated with the rate and extent of starch retrogradation during storage (Kulp and Ponte, 1981; Guilbot and Godon, 1984). Retrogradation has been demonstrated to typify a nonequilibrium recrystallization process in a completely amorphous but crystallizable polymer system, which exists (around room temperature) in a kinetically metastable rubbery state and is sensitive to plasticization by water and heat (Slade, 1984; Slade and Levine, 1987a). The rate and extent of starch recrystallization are determined primarily by the mobility of the crystallizable outer branches of amylopectin (Slade, 1984; Ring et al., 1987; Russell, 1987b; Slade and Levine, 1987a; Marsh and Blanshard, 1988). In their retrogradation behavior, the baked crumb of wheat starch–based breads and experimental starch model systems (e.g., elastic amylopectin gels) are known to be analogous (Kulp and Ponte, 1981). If adequate packaging prevents simple moisture loss, the predominant mechanism of staling in bread crumb or concentrated aqueous starch gels is the time-dependent recrystallization of amylopectin (Russell, 1983, 1987b; Miles et al., 1985a; Zeleznak and Hoseney, 1986; Paton, 1987; Marsh and Blanshard, 1988; Siljestrom et al., 1988) from the completely amorphous state of a freshly heated product to the partially crystalline state of a stale product, with concomitant formation of network junction zones, redistribution of moisture, and increased firmness (Slade, 1984). This recrystallization depends strongly on sample history, since both initial heating during baking and subsequent aging during storage are nonequilibrium processes (Slade, 1984; Ring et al., 1987). The local moisture content in the amorphous regions of a native granule determines the effective Tg that precedes melting of the crystalline regions (A-type in wheat) during gelatinization; complete melting of amylopectin crystallites during typical baking eliminates residual seed nuclei available for subsequent recrystallization (Slade, 1984). The extents of swelling and release of protruding and extragranular polymer available for subsequent three-dimensional network formation by recrystallization depend on total moisture content during gelatinization and pasting (Zobel, 1984; Miles et al., 1985a).

Immediately after baking, the amylopectin in the central crumb is completely amorphous, and the gelatinized starch network in white pan bread begins to recrystallize to a partially crystalline structure upon cooling to room temperature (Slade, 1984). Concomitantly, freshly baked bread begins a process of mechanical firming (manifested by increasing modulus) and moisture redistribution, which is perceived sensorily as a loss of "softness and moistness" (Kulp and Ponte, 1981). As has been described, the early stages of these concurrent processes are dominated by amylose: formation of entanglement networks (followed closely by crystallization) by very high MW amylose alone (Miles et al., 1985a, 1985b; l'Anson et al., 1988), and partially crystalline networks or chain-folded crystals by lower MW amylose–lipid complexes. Crystallization of amylose–lipid is favored over retrogradation (Slade and Levine, 1987a). The baking process is insufficient to melt the seeds of pre-existing amylose–lipid crystals, and homogeneous nucleation of new amylose–lipid crystals should occur somewhat above room temperature, while nucleation of retrograded amylose crystals in a high-moisture environment would be most rapid near $-5°C$ (i.e., Tg') (Slade, 1984; Slade and Levine, 1987a).

The later stages of these concurrent processes, and the overall aging of bread, are dominated by recrystallization of amylopectin to a partially crystalline structure containing disperse B-type crystalline regions (e.g., Slade, 1984). The B-type polymorph is a higher moisture crystalline hydrate than A-type starch (French, 1984; Whistler and Daniel, 1984; Wynne-Jones and Blanshard, 1986; Imberty and Perez, 1988). Its recrystallization requires incorporation of water molecules into the crystal lattice, which must occur while starch chain segments are becoming aligned. Thus, this recrystallization necessitates moisture migration within the crumb structure, whereby water (previously homogeneously distributed) must diffuse from the surrounding amorphous matrix and be incorporated in crystalline regions (Slade, 1984). Since crystalline hydrate water can plasticize neither amorphous regions of the starch network nor other networks (glutenin, pentosans) of the baked crumb matrix and cannot be perceived organoleptically (Slade and Levine, 1987a), the overall consequence of this phenomenon is a drier and firmer texture characteristic of stale bread (Kulp and Ponte, 1981). An implicit requirement of starch recrystallization is availability of sufficient moisture, at least locally within the matrix, both for mobilizing long polymer chain segments (by plasticization) and for being incorporated in B-type crystal lattices (Longton and LeGrys, 1981; Slade, 1984; Hoseney, 1986; Russell, 1987b; Slade and Levine, 1987a; Marsh and Blanshard, 1988). For the propagation step of crystallization, plasticization by heat may suffice for growth of A- or V-type crystals, but the negative temperature coefficient of the nucleation step limits the nucleation process to plasticization by water. For gelatinized wheat starch, a moisture content greater than about 27 w% (Wg') represents the minimum requirement for the nucleation process (Slade, 1984), because Wg' establishes sufficient ΔT above the reference Tg' for mobility at typical staling temperatures (Slade and Levine, 1988c) and 27% is the water content of B-type crystals (French, 1984; Whistler and Daniel, 1984; Wynne-Jones and Blanshard, 1986; Imberty and Perez, 1988). In fact, in low-moisture baked goods, native starch granules in dough are not even gelatinized during baking (Blanshard, 1986). Slade (1984) reported, from DSC results for model wheat starch gels, that percent recrystallization of completely amorphous (unseeded) amylopectin at room temperature increases monotonically with increasing percent total moisture in the range 27–50 w% (due to increasingly effective plasticization), then decreases with further increases in moisture up to 90 w% (apparently due to a dilution effect). These results were subsequently confirmed by Zeleznak and Hoseney (1986). Marsh and Blanshard (1988) recently reported that

> No crystallization of [unseeded] starch–water systems which contain < 15% water is predicted to occur when they are stored at temperatures < 40°C. For gels containing 20% moisture, no crystallization should occur at storage temperatures below room temperature. Thus, as is found in practice, no crystallization of starch is predicted when low moisture-content systems are stored below room temperature.

Investigations to relate the concurrent processes of entanglement/network formation, B-type partially crystalline network formation, and moisture migration, in order to assign their roles in the overall process of bread aging and firming, are complicated by the presence of many contributing polymer components with poorly characterized interactions, nonuniform local moisture, and multiple thermal/mechanical relaxation transitions (Slade and Levine, 1987a). Blanshard (1986) has suggested that the textural changes observed at room temperature immediately after baking, before the onset of retrogradation, result from product cooling and transformation "through a continuum of rubbery states," possibly to a glassy state. At a typical moisture content of baked bread (> Wg'), the glassy state occurs at $Tg' \approx -5°C$, which corresponds to the limiting relaxation temperature for mobile polymer segments (Slade, 1984; Zobel, 1988). As has been mentioned, the effective network Tg (for either entanglement or partially crystalline networks) corresponds to the Tfr transition above Tg for flow relaxation, in this case of the crumb matrix (containing networks of starch, gluten, and pentosans) plasticized by water. This effective network Tg for baked bread would be near room temperature for low extents of network formation, and well above room temperature for mature networks; this is true even though the underlying Tg for segmental motion, responsible for the predominant second-order thermal transition, remains below 0°C at Tg' (Slade and Levine, 1987a). An additional complicating feature is introduced by the disproportionation of mobile short branches from water-plasticized amorphous regions upon retrogradation of amylopectin. The average DP of these outer branches (Hizukuri, 1986) is well above the minimum chain length required for recrystallization (Gidley et al., 1986; Ring et al., 1987; Mestres et al., 1988) at the effective local concentration of clustered branches (French, 1984). While 27 w% is the maximum moisture content (French, 1984; Whistler and Daniel, 1984) of B-type crystals, it is the limiting moisture content for segmental relaxation of completely amorphous starch above Tg', and higher moisture would be required to achieve the same segmental mobility in the absence of mobile amylopectin branches (Slade and Levine, 1987a). Because network Tg (Tfr) is higher than segmental Tg (Tg'), an even higher moisture content is required to maintain sufficient ΔT for mobility and softness in staled bread than in freshly baked bread (Slade and Levine, 1988c). Although currently available data are insufficient, it is likely that the use of NMR to investigate changes in water mobility during aging of bread and starch gels (Wynne-Jones and Blanshard, 1986) will be especially complicated by these phenomena. It thus will be necessary to further explore the effects of initial moisture content, storage temperature, branching density, and branch length on the time course of the variation in relaxation times (Slade and Levine, 1987a; Ring et al., 1987; Marsh and Blanshard, 1988). Moreover, NMR relaxation times for water do not reflect contributions from network Tg superimposed on contributions from segmental Tg. It is likely that solid state NMR studies (Schaefer et al., 1987), conducted to examine the changing mobility of starch and crumb matrix networks directly, will provide a more complete understanding of the mechanism of firming during staling (Slade and Levine, 1988c).

Retrogradation/staling has been studied extensively in bread crumb and model starch gels (e.g., wheat, potato, corn, pea) by using DSC, mechanical compression tests, and X-ray crystallography to monitor formation and aging of the recrystallized starch network (e.g., Kulp and Ponte, 1981; Longton and LeGrys, 1981; Fearn and Russell, 1982; Lund, 1983; Russell, 1983, 1987b; Slade, 1984; Eliasson, 1985; Miles et al., 1985a; Ring and Orford, 1986; Hoseney, 1986; Levine and Slade, 1987; Marsh and Blanshard, 1988). A quantitative DSC method has been developed to measure the rate and extent of starch (or pure amylopectin) recrystallization as functions of additional ingredients, time, temperature, and moisture content during gelatinization and storage, in terms of increasing content of retrograded B-type crystalline starch (measured from the area of the characteristic melting endotherm at $Tm \approx 60°C$ associated with crystalline amylopectin) (e.g., Welsh et al., 1982; Ghiasi et al., 1984; Slade, 1984; Slade and Levine, 1984a, 1984b; Nakazawa et al., 1985; Miles et al., 1985a; Ring and Orford, 1986; Jankowski and Rha, 1986a; Zeleznak and Hoseney, 1986, 1987b; Paton, 1987). (These studies have recognized that this so-called "staling endotherm" does not represent the melting of V-type crystalline retrograded amylose, which may also be present.) Typical DSC results are illustrated in Figure 5-34 for two commercial bakery products: part A shows white pan bread immediately after baking (completely amorphous = "fresh") and after 7 days' storage at 25°C (extensively recrystallized amylopectin = "stale"); part B shows progressively increasing amylopectin recrystallization in English muffins after 1, 7, and 13 days' ambient storage after baking.

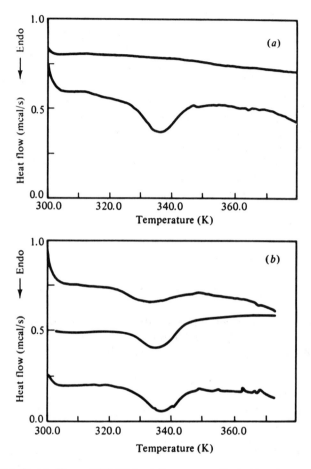

Figure 5-34. Perkin-Elmer DSC-2C heat flow curves illustrating the rate and extent of starch recrystallization in two types of commercial baked goods: (*a*) white bread, immediately after baking (upper curve) and after 1 week at room temperature (lower curve); (*b*) English muffins, on days 1 (upper curve), 7 (middle curve), and 13 (lower curve) after baking. (*From Slade and Levine, 1987a, with permission.*)

Mechanism of Anti-Staling by Sugars

As has been described, sugars, acting as antiplasticizers relative to water alone, raise *Tgelat* and retard pasting of native starch. Sugars are also known to function as anti-staling ingredients in starch-based baked goods. For example, glucose oligomers of DP 3–8 (i.e., nonentangling SHPs [Levine and Slade, 1986]), used as anti-staling additives, have been reported to be effective in inhibiting, and not participating in, starch recrystallization (Krusi and Neukom,

1984). It has been suggested (Slade, 1984; Slade and Levine, 1987a) that these two effects are related as follows. Analogous to the inefficient depression of Tg and consequently Tm of partially crystalline native starch, and therefore $Tgelat$ (relative to the plasticizing action of water alone), sugar solutions produce an elevated Tg of the continuous rubbery local environment of the resulting three-dimensional, completely amorphous, entangled starch gel matrix in a freshly baked product. When Wg' of the sugar solution is similar to that of starch–water alone, as is the case for the homologous family of glucose oligomers, greater Tg of the local environment effects greater network Tg of the starch gel. The elevated Tg of the local environment and network Tg (relative to Tg of the corresponding network plasticized by water alone) control subsequent recrystallization of B-type starch in the undercooled rubbery gel, by controlling the rate of propagation at ΔT above Tg, according to WLF kinetics. For storage at $T >$ network Tg, there is sufficient mobility for devitrification and subsequent formation of crystalline junction zones, resulting in a partially crystalline polymer system that constitutes the retrograded starch gel. Relative to typical storage at ambient temperature, a higher Tg of the local environment and of the network Tg (due to addition of sugars with Wg' similar to starch–water alone) translates to smaller values of ΔT and so a lower rate of propagation of starch recrystallization at the storage temperature. Thus do such sugars act to retard the rate and extent of starch staling during ambient storage. Moreover, WLF theory predicts that a greater MW of a sugar would translate to greater antiplasticization by the sugar solution, and so to a greater anti-staling effect (Slade, 1984).

The situation is somewhat more complicated when Wg' of the sugar solution is much greater than that of starch–water alone. Tg in the local environment of amylopectin branches is still elevated, relative to water alone, but a greater network Tg may be compensated for by increased plasticizing effectiveness of the sugar solution. A systematic study of anti-staling by a large and non-homologous series of common sugars has been reported (Slade and Levine, 1987a). Results of DSC studies compared the degree of elevation of $Tgelat$ of native wheat starch (described in the section Thermal Properties of Three-Component Model Systems) with the degree of inhibition of recrystallization of gelatinized starch, for the same series of sugars. Starch:sugar:water mixtures (1:1:1) were analyzed after complete gelatinization (i.e., heating to $T \gg Tgelat$ in each case, whatever the specific $Tgelat$) followed by 8 days of storage at 25°C. The results showed that the extent of recrystallization decreased in the order fructose > mannose > water alone > galactose > glucose > maltose > sucrose > maltotriose > xylose > lactose > malto-oligosaccharides (enzyme-hydrolyzed, DP > 3). For the glucose homologs within this sugar series, MW and resultant Tg were the apparent primary determinants of anti-staling activity. For the other sugars, however, it was suggested, as it had been with regard to their effect on gelatinization, that "water availability" (in this case during storage rather than during gelatinization), as determined by mobility and free volume and reflected by Wg' (Levine and Slade, 1987), appeared to play a key role in their anti-staling effect. It should be recognized that this suggestion does not contradict the statement by Paton (1987) that "anti-staling agents do not operate [primarily] by a mechanism that alters moisture availability to the starch *during* the baking process, thus affecting retrogradation." In their inhibitory action on starch recrystallization, as in their elevating action on $Tgelat$, sucrose was more effective than glucose, which in turn was more effective than fructose. The fact that fructose–water, relative to water alone, actually accelerated starch staling was a particularly salient finding (Slade and Levine, 1987a), since it supported the previous observation of anomalously large translational diffusion, which promoted mold spore germination in fructose solutions, relative to glucose solutions (Slade and Levine, 1988a).

Mechanism and Kinetics of Recrystallization: Acceleration of Nucleation

Since retrogradation in gelatinized starch (or pure amylopectin)–water systems (at $W > Wg' \simeq 27$ w%) is a nucleation-controlled (Miles et al., 1985a), thermoreversible gelation-via-crystallization process (Ring and Orford, 1986), the crystallization kinetics are described as shown in Figure 5-6, within the temperature range between Tg (i.e., Tg') $\simeq -5°C$ and $Tm \simeq 60°C$ of B-type hydrate crystals (Slade, 1984; Slade and Levine, 1987a; Zeleznak and Hoseney, 1987b; Marsh and Blanshard, 1988). The maximum overall rate of crystallization (i.e., the combined rates of nucleation and propagation), at a *single* storage temperature, would be expected about midway between Tg and Tm (Nakazawa et al., 1985), which for the amylopectin of B-type wheat starch is about room temperature. The observed maximum rate at a temperature in the range 0–10°C for both white bread (Guilbot and Godon, 1984; Zeleznak and Hoseney, 1987a) and 50% wheat starch gel (Marsh and Blanshard, 1988), similar to the value of 14°C calculated from the empirical relationship, $(T - Tg + 50$ K$)/(Tm - Tg + 50$ K$) = 0.6$, for the crystallization kinetics of typical partially crystalline synthetic high polymers (described in the section Crystallization Kinetics for Partially Crystalline Polymers), shows the dominant role of nucleation control (Slade, 1984; Slade and Levine, 1987a). Typical DSC results for the extent of starch staling in freshly baked white bread crumb stored at 25°C for 39 days are illustrated in curve *a* of Figure 5-35.

Starch recrystallization can also be treated as a time-/temperature-/moisture-governed polymer process that can be manipulated (Slade, 1984). For partially crystalline synthetic (Wunderlich, 1976) and food (Slade and Levine, 1987b) polymers in general, and B-type amylopectin in particular (Guilbot and Godon, 1984; Slade, 1984; Miles et al., 1985a; Jankowski and

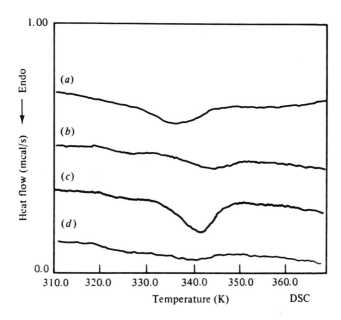

Figure 5-35. Perkin-Elmer DSC-2C heat flow curves of freshly baked white bread crumb staled at different storage temperatures and times: (*a*) 25°C for 39 days; (*b*) 25°C for 42 hr, then 40°C for 2.5 hr; (*c*) 0°C for 42 hr, then 40°C for 2.5 hr; (*d*) −11°C for 42 hr, then 40°C for 2.5 hr. (*From Slade and Levine, 1987a, with permission.*)

Rha, 1986a; Slade and Levine, 1987a; Marsh and Blanshard, 1988), for which the time and temperature of superheating above *Tm* of native amylopectin crystallites during baking is sufficient to eliminate self-seeding upon subsequent cooling (Slade and Levine, 1987a), the rate-limiting step in the crystallization process is nucleation (which is enhanced at lower temperatures) rather than propagation (which is enhanced at higher temperatures). Curves *b–d* in Figure 5-35 illustrate how the rate and extent of staling can be influenced by separating these mechanistic steps and maximizing the nucleation rate. Compared to nucleation and propagation at 25°C (curve *a*), faster propagation at 40°C produced significant staling (curve *b*) in much less time. Initial storage at −11°C (< *Tg′*) inhibited nucleation (except during cooling) and so produced less staling in an equivalent time (curve *d* versus *b*). However, the greatest rate and extent of staling in the shortest time were achieved (curve *c*) by faster nucleation at 0°C (for a long enough time to allow extensive nucleation), followed by faster propagation at 40°C. These results were confirmed in a subsequent study by Zeleznak and Hoseney (1987b).

Figure 5-36 shows DSC results for model wheat starch:water (1:1) mixtures, exposed to different temperature/time storage protocols immediately

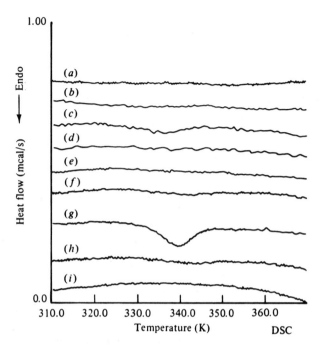

Figure 5-36. Perkin-Elmer DSC-2C heat flow curves of wheat starch:water 1:1 mixtures, stored under different temperature/time conditions immediately following gelatinization: (a) 40°C for 4.5 hr; (b) 25°C for 3.5 hr; (c) 4°C for 3 hr; (d) −23°C for 2.5 hr; (e) −196°C for 1 min, then −23°C for 2 hr; (f) 25°C for 2 hr, then 40°C for 5 hr; (g) 4°C for 2 hr, then 40°C for 6.5 hr; (h) −23°C for 2 hr, then 40°C for 6 hr; (i) −196°C for 1 min, then −23°C for 2 hr, then 40°C for 5.5 hr. (*From Slade and Levine, 1987a, with permission.*)

following gelatinization. This study represented an exploration of the optimum nucleation temperature to produce a maximum rate of recrystallization. For single-temperature storage conditions, the rate of nucleation and thus overall crystallization increased with decreasing temperature (i.e., 40 < 25 < 4°C, as shown in curves a–c, a finding in agreement with subsequent results of Zeleznak and Hoseney [1987b] and Marsh and Blanshard [1988]), as long as the temperature was above Tg'. These results demonstrated the negative temperature coefficient expected for the nucleation rate in a typical polymer crystallization process (Marsh and Blanshard, 1988), as mentioned in the section Crystallization Kinetics for Partially Crystalline Polymers. Freezer storage at $T < Tg'$ inhibited nucleation (curves d, e), and so retarded recrystallization, even when followed by propagation at 40°C. Zobel (1988) reached the same conclusion that Slade had from her experimental results (1984). Zobel remarked that "knowledge about glass transitions appears to offer a way to control starch gel properties in food . . . applications" and that "in the context of bread staling, a rational explanation of why freezing preserves bread from staling is that the bread is held below its *Tg*." It is important to note that, as shown in Table 5-1, starch recrystallization would be inhibited only *while* bread remained at a *Tf* < *Tg′*, but not during cooling to *Tf* (when fast nucleation could occur) or after thawing (when propagation could occur in a previously nucleated matrix) (Slade, 1984; Levine and Slade, 1986). Once again, as for white bread crumb in Figure 5-35, curve *c*, starch gels first held at 4°C to promote rapid nucleation, then at 40°C to allow rapid propagation (Fig. 5-36, curve *g*), showed by far the greatest overall rate and extent of starch staling via amylopectin recrystallization. This finding of an optimum practical nucleation temperature of 4°C (Slade, 1984) to produce a maximum rate of recrystallization (also confirmed by Zeleznak and Hoseney, 1987b) is critically relevant to high-moisture (i.e., starch gelatinized during baking) baked products with coatings, such as chocolate-covered cake donuts. When such a product is run through a refrigerated cooling tunnel (at 40°F) immediately after baking and coating, in order to rapidly solidify ("set") the coating, this process step can result in accelerated starch staling within the crumb matrix during subsequent storage at ambient temperatures, relative to the corresponding rate of staling in the same product not subjected to faster nucleation at subambient temperature via the cooling tunnel treatment.

Figure 5-37 shows DSC results for model wheat starch:water (1:1) mixtures, examined for the effect of increasing nucleation time at 0°C, immediately after gelatinization, and prior to 30 minutes of propagation at 40°C. It was apparent from the trend of steadily increasing endotherm areas in curves *a–f* that the extent of nucleation and overall crystallization increased monotonically with increasing time of nucleation. Moreover, amylopectin recrystallization was already measurable after only 1 hour of total storage, and quite extensive after 5.5 hours. The extraordinary rate and extent of recrystallization were evidenced by comparing the endotherm areas in Figure 5-37 with others in Figures 5-34 through 5-36, all of which were plotted with 1.0 mcal/s full-scale and similar sample weights.

These results (Slade, 1984) were used to design a patented industrial process for the accelerated staling of bread (for stuffing) and other high-moisture, lean, starch-based foods (Slade et al., 1987). By a two-step temperature cycling protocol involving, first, a holding period of several hours at 4°C (to maximize the nucleation rate), followed by a second holding period of several hours at 40°C (to maximize the propagation rate), a much greater overall extent of staling due to amylopectin recrystallization is achieved (versus the same total time spent under constant ambient storage), which is equivalent to staling bread for several days at room temperature (Slade, 1984; Slade and Levine, 1987a).

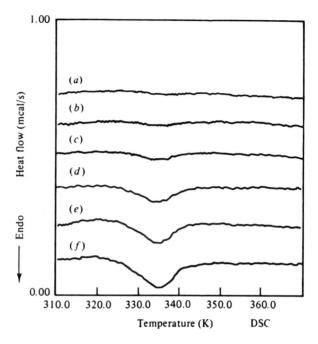

Figure 5-37. Perkin-Elmer DSC-2C heat flow curves of wheat starch:water 1:1 mixtures, nucleated for different times at 0°C, then propagated for 30 min at 40°C, immediately following gelatinization: (a) 10 min; (b) 30 min; (c) 60 min; (d) 180 min; (e) 240 min; (f) 300 min. (*From Slade and Levine, 1987a, with permission.*)

Amylopectin–Lipid Crystalline Complex Formation at Low Moisture

It has long been known that amylose forms a helical complex with lipids; this complex crystallizes readily from water as anhydrous crystals that give rise to V-type X-ray diffraction patterns (Buleon et al., 1984; Jane and Robyt, 1984; Yamamoto et al., 1984; Kowblansky, 1985; Evans, 1986; Zobel, 1988; Biliaderis and Galloway, 1989). The Tm of crystalline amylose–lipid complex is about 110°C (see Fig. 5-22 for typical DSC thermograms) for endogenous lipids of cereals such as wheat, corn, and rice (Biliaderis et al., 1985; Biliaderis, Page, and Maurice, 1986a; Kowblansky, 1985; Eliasson, 1986; Evans, 1986; Slade and Levine, 1987a; Biliaderis and Galloway, 1989). Amylose–lipid complex is more stable than the well-known amylose–iodine complex (Yamamoto et al., 1984), and linear chain–lipid crystals are more thermostable than linear chain–hydrate polymorphs (B-type, $Tm \simeq 60°C$, in retrograded amylose gels, or A-type, $Tm \simeq 85°C$, in low-moisture granular starches) (Slade, 1984; Biliaderis et al., 1985; Biliaderis, Page, and Maurice, 1986a; Slade and Levine, 1987a; Biliaderis and Galloway, 1989). For a given MW of amylose, pure anhydrous amylose crystals are most thermostable of all, with $Tm > 140°C$ (Biliaderis et al., 1985; Biliaderis, Page, and Maurice, 1986a; Ring et al., 1987; Russell, 1987b; Mestres et al., 1988; Biliaderis and Galloway, 1989).

Depending on endogenous lipid content, the amylose (about 20–30% [Whistler and Daniel, 1984]) in a normal native starch granule may exist as a glassy or crystalline, hydrate or amylose–lipid complex (Slade and Levine, 1987a). Upon heating starch for DSC analysis at total sample moisture contents greater than 27 w%, preexisting crystalline amylose–lipid is seen as a melting endotherm at about 110°C. Preexisting glassy amylose, in the presence of but not necessarily pre-complexed with lipid, is evidenced by a crystallization exotherm near 95°C (Slade and Levine, 1987a). For starches with low endogenous lipid contents, addition of exogenous lipid results in a crystallization exotherm on the first heating, and a melting endotherm on the second heating (Biliaderis et al., 1985; Biliaderis, Page, and Maurice, 1986a; Biliaderis and Galloway, 1989). Thus, amorphous amylose, rendered mobile and available via gelatinization and pasting during baking, may (1) complex with endogenous lipid and crystallize, (2) crystallize if previously complexed but restrained from crystallization, or (3) complex and crystallize with exogenous lipid, added as emulsifiers in dough conditioners or shortening. In contrast, even when endogenous lipid content is high or exogenous lipid is added, DSC analysis of waxy starches (which contain essentially no amylose) in the conventional temperature range 20–130°C and the moisture range greater than 30 w% had not (Evans, 1986) until recently (Slade and Levine, 1987a) revealed evidence of crystallization or melting of amylopectin–lipid complexes.

It is known that amylose can be precipitated with butanol but typical amylopectin cannot; this behavior is the basis for the traditional distinction between these two polymers (Buleon et al., 1984; Evans, 1986). The longest accessible linear chain segments in amylopectin are outer branches of DP $\simeq$ 16–20 (Hizukuri, 1986). These branches, which are responsible for the microcrystalline regions of amylopectin, are not long enough to form complexes with iodine or butanol that are stable to dilution at room temperature (Evans, 1986). It had been assumed (Evans, 1986) until recently (Slade and Levine, 1987a) that amylopectin branches are also too short to form stable complexes with lipids, and the absence of DSC transitions (Evans, 1986) had supported this assumption. The most stable complexes of linear amylose with iodine are formed by chains of DP > 40 (Yamamoto et al., 1984). Despite the lack of evidence from DSC and other analytical methods for interactions between amylopectin and lipid, addition of stearoyl lipids (e.g., sodium stearoyl lactylate, SSL), is known to affect the texture and rheology of waxy corn starch samples (Evans, 1986). Similarly, while amylopectin is believed to be capable of forming insoluble complexes (via its outer branches) with surfactants, including monoglycerides such as glycerol monostearate, thereby implicating such monoglycerides in bread shortenings and emulsifiers with a possible role as an anti-staling agent, the nature of such amylopectin complexes remains obscure (Batres and White, 1986). It was believed that amylopectin–lipid complexes may occur, but Tm of the crystals, if determined by the low MW of amylopectin branches, might be well below the DSC temperature range usually examined for starch–lipid complexes (Slade and Levine, 1987a).

Slade used native waxy maize starch, with only "as is" moisture (< 10 w%), as the amylopectin source. A starch:SSL (10:1 w:w) mixture was heated at 120°C for 15 minutes (at 15 pounds pressure), to assure comelting of the reactants. Starch alone, treated the same way, produced the featureless thermogram shown at the top of Figure 5-38, while SSL alone melted at 45–50°C. The rationale for this experimental approach was that, in the presence of only enough moisture (< 10 w%) to permit plasticization and melting of starch at high temperature, but not enough to allow the formation of amylopectin A- or B-type crystal hydrates, the formation of amylopectin–lipid crystalline complex would be possible and favored. The starch:SSL comelt was then nucleated at 4°C for 24 hours, heated to 120°C at 10°C/min and recooled, and then analyzed by DSC. The thermogram (bottom of Fig. 5-38) showed a

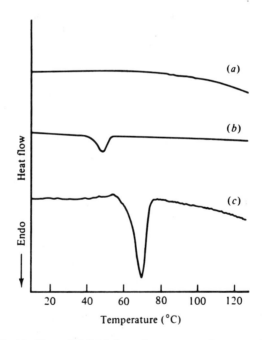

Figure 5-38. Perkin-Elmer DSC-2C heat flow curves of waxy maize starch (< 10 w% water), after heating at 120°C (15 lb pressure) for 15 min (*a*); sodium stearoyl lactylate (SSL) alone, same heat treatment (*b*); and 10:1 (w:w) waxy maize starch:SSL (waxy maize starch < 10 w% water), same heat treatment, followed by 24 hr at 4°C, then heating to 120°C at 10°C/min and recooling, before rescanning (*c*). (*From Slade and Levine, 1987a, with permission.*)

small crystallization exotherm at 55°C, followed by a large and narrow melting endotherm at $Tm \simeq 70°C$. These observations were presented (Slade and Levine, 1987a) as the first DSC evidence of a crystalline amylopectin–lipid complex produced at low moisture. The low Tm, relative to that for amylose–lipid complex, was suggested to indicate a lower-MW complex, formed with the short, crystallizable outer branches of amylopectin.

PHYSICOCHEMICAL PROPERTIES OF WHEAT GLUTEN AS A VISCOELASTIC POLYMER SYSTEM

Unique among cereal grain storage proteins, native ("vital") wheat gluten is a major functional food ingredient, especially as a flour component in baked goods (Bloksma, 1978; Kasarda et al., 1978; Pomeranz, 1978; Magnuson, 1985). The structure–property relationships of gluten have been described in terms of a highly amorphous, multipolymer system, water-plasticizable but not water-soluble, and capable of forming continuous, multidimensional, viscoelastic films and networks (Schofield et al., 1984; Slade, 1984; Hoseney et al., 1986; Levine and Slade, 1987; Doescher, Hoseney, and Milliken, 1987; Edwards et al., 1987). Gluten also appears to function as either a thermoplastic or a thermosetting amorphous polymer in response to the heat/moisture treatment constituted by baking (Slade et al., 1988). The thermosetting behavior, when it occurs at practical heating rates and sample concentrations, is considered to be a typical consequence of irreversible protein denaturation and "heat-set" gelation above a critical temperature (Schofield et al., 1984; Magnuson, 1985; Davies, 1986; Schofield, 1986; Ablett et al., 1988; Blanshard, 1988). The resulting dramatic changes in the mechanical and rheological properties (i.e., large increases in viscosity and elastic modulus at temperatures between about 50°C and 100°C) of bread dough during baking are responsible in part (along with starch gelatinization) for the transformation from a predominantly viscous dough to a predominantly elastic baked crumb (Schofield et al., 1984; Schofield, 1986; Bloksma, 1986; Blanshard, 1988; Bloksma and Bushuk, 1988; Dreese et al., 1988b). The mechanism of such thermo-irreversible "heat-set" gelation involves the formation of permanent networks, due to rheologically effective, intermolecular cross-links composed of covalent disulfide bonds, resulting in a polymerization of the gluten proteins (Bloksma, 1978, 1986; Pomeranz, 1978; Cantor and Schimmel, 1980; Schofield et al., 1984; Ablett et al., 1988; Dreese et al., 1988a). Such completely amorphous, thermoset gel networks (which may also contain entangled polypeptide chains) represent a category separate from the two other types of gels described earlier (i.e., partially crystalline, thermoreversible gels, e.g., gelatin–water, and completely amorphous, topologically reversible "entanglement" gels, e.g., sodium caseinate–water, not heat-treated, which can be dispersed by dilution [Levine and Slade, 1987]). It is important to stress at the outset the distinction between local intermolecular disulfide cross-linking of gliadins (Schofield et al., 1984) and long-range intermolecular disulfide network formation of HMW glutenins (Weegels et al., 1988). "Gluten" defies the usual biological rule of thumb for the genetic design of stable proteins (Cantor and Schimmel, 1980): single polypeptide chains should contain *either* free thiols (cysteine) or disulfides (cystine), but not both in a single gene product. Coexistence of both –SH and –S–S– results in dynamic catalysis of disulfide exchange and a consequent lack of definition of highly probable, predominant subunit conformations in the gluten population (Cantor and Schimmel, 1980). Disulfide exchange, cross-linking, and thermosetting reactions during heat–moisture treatments such as baking result in (1) progressive depletion of the least stable cross-links and catalytically effective thiols and (2) eventual establishment of permanent local cross-links in gliadins, a permanent long-range (cooperative) network by HMW glutenins, and residual catalytically ineffective thiols. Extensibility and viscous flow of the film-forming gliadins in doughs with moisture contents greater than Wg' are diminished by disulfide cross-linking during baking (Schofield et al., 1984); decreased extensibility and increased hardness, but not increased elasticity, result. Elasticity and rigidity of the network-forming HMW glutenins in such doughs are enhanced by disulfide thermosetting during heat treatment (Weegels et al., 1988), which results in an increased number of permanent disulfide junction zones, such that a loss of extensibility *is* accompanied by increased elasticity. Thus, increased firmness results from increase in the loss modulus due to gliadin cross-linking (Schofield et al., 1984) and from increase in the elastic modulus due to HMW glutenin network thermosetting (Weegels et al., 1988). Clearly, simple determination of neither residual thiol content nor the amount of insolubilized protein is sufficient for the interpretation of the mechanism of firming or development of rubbery texture during heat–moisture treatment of gluten doughs, such as microwave baking or reheating of bread. The causative insolubilized proteins should be identified by demonstration of gel electrophoresis bands, high-performance liquid chromatography (HPLC) peaks (Lookhart et al., 1987; Pomeranz et al., 1987; Menkovska et al., 1987; Ng and Bushuk, 1987), or antigens to monoclonal antibodies (Skerritt et al., 1988; Hill et al., 1988), which are present in doughs but are missing in baked products. Missing gliadin subunits could account for increased firmness and decreased extensibility without increased rubberiness, while missing glutenin subunits could account for a dramatic increase in rubberiness with only a moderate increase in firmness. These structure–property relationships of gluten are critical to the processing, product quality, and storage stability of various types of flour-based doughs and finished baked goods (Schofield et al., 1984; Davies, 1986; Schofield, 1986; Blanshard, 1988; Slade et al., 1988).

   Wheat gluten's unique combination of cohesive, viscoelastic, film-forming, thermoplastic or thermosetting, and water-absorbing properties (Schofield et al., 1984; Magnuson, 1985) is derived from the structure of the native proteins (Payne, 1987). Gluten is a complex of interacting proteins, which have been categorized traditionally into two major groups, gliadins and glutenins, based on their solubility and insolubility, respectively, in 70–90% aqueous ethanol (Osborne, 1907). More recently, an alternative classification has been proposed, which reflects the chemical and genetic relationships of the component polypeptides of gluten (Tatham et al., 1985; Payne, 1987). The groups are: (1) sulfur-poor prolamins (ω-gliadins), (2) sulfur-rich prolamins (α-, β-, γ-gliadins; LMW glutenin subunits), and (3) HMW prolamins (HMW glutenin subunits). The broad correspondence between the two classification schemes relates to the identification of proteins present as monomers (single polypeptide chains) associated by hydrogen bonding and hydrophobic interactions as gliadins, and proteins present as polymers (multi-subunit aggregates) linked by interchain disulfide bonds as glutenins (Payne et al., 1985; Schofield, 1986; Payne, 1987). In this context, gliadins are lower MW proteins ($< 10^5$), which can adopt globular conformations with intramolecular disulfide bonds, α helical segments, and/or β turns, and act as the viscous, extensible, plasticizing component of gluten (Schofield, 1986; Payne, 1987; Blanshard, 1988; Slade et al., 1988). Glutenins are higher-MW ($> 10^5$) with broader MW distribution, highly elastic, multi-subunit proteins cross-linked by inter-subunit disulfide bonds (Schofield, 1986). Glutenins act as the elastic component (Lasztity, 1986; du Cros, 1987; Edwards et al., 1987; Ewart, 1987; Payne, 1987; Blanshard, 1988) of the viscoelastic network that forms in a flour–water dough via cross-linking by thiol–disulfide interchange reactions catalyzed by rheologically active thiols (Kasarda et al., 1978; Pomeranz, 1978; Bloksma, 1978; Schofield, 1986; Ewart, 1988). In this function, glutenins are similar to elastin, in that these are the only two protein systems known to exhibit high segmental mobility at high MWs and to form hydrated, three-dimensional, open-structured, rubberlike elastic networks via disulfide bonding (Schofield, 1986; Edwards et al., 1987; Ablett et al., 1988; Slade et al., 1988).

Behavior of Gluten as an Amorphous Thermosetting Polymer

Slade (1984) reported DSC results of Tg measurements from a study of commercial vital wheat gluten, approached as an amorphous, water- and lipid-compatible polymer. As is shown in Figure 5-39, isolated wheat gluten at low moisture (i.e., 6 w% "as is" water content) manifested a Tg of 66°C (curve *a*). Once it is heated through this glass transition, gluten has sufficient mobility, due to thermal and water plasticization, to form a thermoset network via disulfide cross-linking. This thermally irreversible thermosetting reaction was suggested (Levine and Slade, 1987) to be analogous to chemical "curing" of epoxy resin and "vulcanization" of rubber. As is described in the section

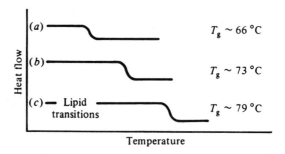

Figure 5-39. Perkin-Elmer DSC-2C heat flow curves of isolated native wheat gluten (*a*) as received, with 6 w% moisture; (*b*) 1:1 (w:w) comelt of gluten:triacetin; (*c*) 1:1 (w:w) comelt of gluten:B12K (lipids blend). (*From Levine and Slade, 1987, with permission.*)

Gelation Mechanism and Viscoelastic Properties of Thermosetting Amorphous Polymers, such classical thermosetting reactions of synthetic amorphous polymers are also possible only at $T > Tg$ (Prime, 1981; Sperling, 1986). In contrast to its response to plasticization by water alone, gluten remained thermoplastic (i.e., its glass-to-rubber transition was reversible) in 1:1 comelts with nonaqueous plasticizers such as triacetin (Fig. 5-39, curve *b*) and B12K (curve *c*), a blend of lipids commonly used as a bread shortening system. However, the higher Tg values of 73°C in curve *b* and 79°C in curve *c* demonstrated the more dramatic plasticizing effect on gluten of water at only 6 w%, compared to 50 w% triacetin or B12K. In the context of effects on bread baking performance, Slade (1984) deduced that the interaction of aqueous gluten with lipids represents another example of relative "antiplasticization" of a polymer by a higher-MW plasticizer (analogous to the effect described earlier of different sugar–water cosolvents on the gelatinization of starch), relative to plasticization by water alone. Slade's conclusion (1984) that the thermomechanical and rheological properties of wheat gluten can be explained based on the recognition of gluten proteins as a water- and lipid-plasticizable, highly amorphous polymer system, was verified in part by Tg measurements from a subsequent DSC study by Hoseney and co-workers (1986a). These investigators presented a graph of Tg versus weight percent water (shown in Fig. 5-15), which is a smooth curve from greater than 160°C for the glassy polymer at 1 w% moisture or less to 15°C for the rubbery polymer at 16 w% moisture. As has been mentioned, these results exemplified the typical extent of plasticization ($\simeq 10$°C/w%) by water, which was previously illustrated for several other water-compatible, amorphous food polymers. However, Hoseney and co-workers (1986a) did not observe the thermosetting behavior of aqueous gluten reported by Slade (1984), perhaps due to differences in DSC measurement techniques (Slade et al., 1988).

It has been suggested (Slade et al., 1988) that the enhanced ability of amorphous gluten in hard wheat flour–water doughs heated to $T > Tg$ to thermoset (i.e., undergo an irreversible structural transformation from a viscous liquid polymer with transient cross-links to an elastic, permanently cross-linked gel with a higher network Tg) represents a critical aspect of the baking/volume-determining mechanism for white pan bread and is partly responsible (along with starch gelatinization) for the mechanical and rheological differences between dough and baked bread. The dramatic increases in viscosity and elastic modulus of starch-free gluten–water doughs heated to temperatures between about 50°C and 100°C, leading to gelation via network formation by protein–protein aggregation at $T \gtrsim 80°C$ (Schofield et al., 1984; Davies, 1986; Ablett et al., 1988; Blanshard, 1988; Dreese et al., 1988a; Weegels et al., 1988), are strongly suggestive of a molecular mechanism involving a classical thermosetting process via covalent cross-linking in an amorphous polymer system. It is well known that the effect of shortening (mixed lipids) on flour–water bread doughs produces breads of increased baked loaf volume (BLV), by allowing greater ovenspring during baking (Moore and Hoseney, 1986). This effect has been interpreted (Slade et al., 1988) as arising from lipid plasticization of gluten to extend its period of thermoplasticity early in the baking process, prior to thermosetting of the water-plasticized gluten network at the end of baking, in addition to the better-known role of lipid to retard pasting of starch. It has also been suggested (Slade et al., 1988) that the characteristically rubbery/leathery texture produced by microwave reheating of bread and other baked products results in large part from a post-cure heat treatment of the thermoset gluten polymer, which effects a further increase in cross-linking density with a concomitant further elevation of network Tg. In a related vein, these authors have described (Slade et al., 1988) a new hypothesis (forthcoming section on structure–functional properties of gluten in wheat flour) for the role of gluten in the "spreading"/baking mechanism of sugar-snap cookies. This hypothesis differentiates, based on the functional properties of the flour, between certain glutens that thermoset during baking (as in bread) of inferior cookies made from poor cookie quality hard wheat flours and other glutens that remain thermoplastic until structural collapse occurs during baking of superior cookies made from excellent quality soft wheat flours.

In the context of the description in the earlier section on the gelation mechanism and viscoelastic properties of thermosetting amorphous polymers, the parallels seem obvious (Slade et al., 1988) among (1) the established behavioral characteristics of thermosetting synthetic amorphous polymers, (2) the equally well-documented behavior of thermosetting partially crystalline and paracrystalline keratin biopolymer systems (Whewell, 1977), and (3) the empirically known structure–function relationships of hard wheat glutens in the baking of excellent breads and inferior sugar-snap cookies (Bloksma, 1978; Kasarda et al., 1978; Pomeranz, 1978). Moreover, the general response of fully baked products to subsequent microwave heating is consistent with a post-cure reaction stage in a continuous, thermoset gluten network. Such a partially thermoset network, incompletely cured during baking, with incident $T_g = ultTg$ (as a result of removal from the oven rather than intervening vitrification), could have its curing process extended or even completed during microwaving, where superheated steam could provide internal temperatures greater than during oven baking. Consequently, the network Tg would increase to a higher $ultTg$ or even to $Tg\infty$. The resulting texture of a product after cooling to room temperature would be characterized by increased rubberiness/leatheriness, due to increased macroscopic viscosity, owing to the change in ΔT between the network Tg and room temperature (Slade et al., 1988). The maximum elevation of Tg effected by curing, from the molecular Tgo to network $Tg\infty$, varies significantly depending on the system; for example, from less than 30°C for an unidentified Du Pont elastomer (Maurer, 1981) to more than 170°C for a bisphenol-A epoxy (Prime, 1981). Considering the approximate value of Tgo for developed gluten in unbaked dough (i.e., gluten Tg' of about -7.5°C [Levine and Slade, 1988e]) and the comparative textural attributes at room temperature of the dough, the fresh oven-baked loaf, and the post-baking microwaved product, these authors have speculated (Slade et al., 1988) on the values of $ultTg$ after baking or after microwaving. For ΔT about 33°C (room $T - Tgo$), dough firmness is characteristically low. Fresh-baked and cooled bread is clearly elastic (diagnostic of a critical extent of network formation), but also quite deformable, so that ΔT must exceed zero. Thus baked $ultTg$ is between -7.5°C and 25°C. The leathery texture of microwave-reheated bread is diagnostic of $\Delta T \ll 20$°C. In the worst case, microwaved $ultTg$ may reach 25°C or above! As has been observed for the aging of commercial synthetic polymer thermosets, moisture plays a critical role in the aging of the baked product. Thus, $ultTg$ and product modulus could be elevated even further by moisture redistribution or drying (as an additional consequence of microwave heating) or depressed by humidification (Slade et al., 1988).

Such apparent analogies imply a relatively high potential for displacement of the distribution of free thiol groups and intramolecular disulfide bonds toward intermolecular disulfide cross-linking and network thermosetting during baking in glutens from hard wheat flours, without underestimating the contribution of leached-amylose networks (especially from chlorinated flours, as emphasized by Blanshard [1986]) or pentosan networks (Slade et al., 1988). Blanshard (1986) has described the progressive firming of cakes during cooling after baking in terms of the Tg of amylose and the plasticizing effect of moisture and short amylopectin branches. Because the cake crumb structure depends on leached amylose networks (so that replacement of wheat starch by waxy corn starch results in cake collapse [Blanshard, 1986], as described in Table 5-1), the operative Tg is the network Tg. In glutens from excellent cookie quality soft wheat flours used in optimized cookie formulations, the converse argument, suggested by the results to be reviewed in the section Structure–Functional Properties of Gluten in Wheat Flour, implies a lower potential for such a displacement of the distribution toward disulfide cross-linking during baking. These exploratory conclusions have been offered (Slade et al., 1988) to stress the importance of the network Tg, rather than the molecular Tg, for the rheological behavior of baked products, in the hope that they would stimulate future research.

Structure–Property Relationships of Stretched Gluten Films (Slade et al., 1988)

A nearly universal architecture occurs in partially crystalline and paracrystalline multicomponent systems in which biopolymers perform a structural role in nature. When the primary functional role of biological macromolecules is to provide structure, the prevailing architectural format consists of a relatively highly ordered or partially crystalline strutwork of higher polymers embedded in a relatively less ordered or amorphous matrix of lower polymers, equivalent to steel reinforcements embedded in concrete. The most familiar example of this architecture in the plant kingdom is the cellulosic cell wall, which has been reviewed in detail by Muhlethaler (1961). At the molecular level, the cellulose molecule is the historical prototype of the fringed micelle model for a semicrystalline polymer. A single cellulose molecule extends through multiple amorphous regions and microcrystalline domains, with typical microcrystallite dimensions of 73 Å in width and 460 Å in length (Sharples, 1966). About 2,000 cellulose molecules are laterally aligned into a microfibril with a diameter of 100 Å to 250 Å (Muhlethaler, 1961). The cellulose chain direction and crystallite orientation are parallel to the long direction of the microfibrils, which are aggregated further with parallel alignment into macrofibrils of about 0.5 μm diameter. The space-filling material among microfibrils and surrounding macrofibrils is a relatively amorphous matrix composed of hemicelluloses and pectic substances. Although the microcrystallites are well below the resolution of the light microscope, their orientation within the microfibrils and the orientation of the microfibrils within the macrofibrils allow ready detection of the highly anisotropic macrofibrils, whose dimensions are within the limit of resolution of polarization light microscopy. Placement of a red gypsum comparison crystal in the light path between crossed Nicol polarizing prisms at an angle of 45° permits observation of the positive birefringence of the cellulose fibrils; that is, the largest index of refraction lies parallel to the long axis of the macrofibrils and also to the chain direction in the microcrystallites. Lamellae in the cell wall, resulting from the apposition of parallel fibrils, are intensely positively birefringent and appear blue when the cell wall is oriented parallel to the gypsum crystal (the largest index of refraction along the long axis of a fibril is oriented parallel to the largest index of refraction of the gypsum crystal), or yellow when the cell wall is oriented perpendicularly. The largely amorphous middle lamella appears rose-colored like the isotropic background. Thus, polarization light microscopy provides an enchanting demonstration of the universal architecture of structural biopolymer systems.

Refinements of this universal architecture are required to furnish structural systems with suitable mechanical properties to accommodate variations in environment and secondary function or end use (Dickerson and Geis, 1969). Again, the cellulosic cell wall illustrates three commonly encountered properties of structural systems: orientation upon extension or drawing, fibrillation upon drawing, and differential swelling upon moisture uptake (Muhlethaler, 1961). Stretching of the cell wall during cell growth results in reorientation of the semicrystalline cellulose fibrils within the amorphous matrix of an apposed layer, with a concomitant increase in birefringence of the secondary cell wall. Elongation of the cell wall also results in a behavior called fibrillation, which involves lateral separation of the amorphous matrix between parallel fibrils within a layer, so that linear tears appear along the direction of extension, giving rise to striations about 0.4 μm wide. Increased total water content of the cell wall by humidification or soaking results in differential swelling of the amorphous matrix and the semicrystalline fibrils, exemplified by a 20% increase in diameter with only a 0.5% increase in length during swelling of a previously dried flax fiber. Thus, all three of these properties directly depend on the architectural features of orientable, partially crystalline struts embedded in an orienting, amorphous matrix.

Examples from nature of the universal architecture and mechanical refinements of structural biopolymer systems in the animal kingdom emphasize the importance of proteins, rather than polysaccharides. Microcrystalline fibrous proteins act as structural elements embedded in matrices composed of amorphous nonfibrous proteins. Each of the three principal classes of fibrous structural proteins is associated with a characteristic amino acid composition, consequent primary sequence repeats and sequences enriched in particular residues, and secondary conformation of individual polypeptide chains. Unstretched keratins depend on the α helix; silks (and stretched keratins) on the β-pleated sheet; and collagens on polyproline II helices (Dickerson and Geis, 1969). Crystallite dimensions and chain flexibility afforded by these molecular conformations, relative amounts of amorphous and crystalline regions in fibrillar structures and in the surrounding matrix, and dynamics of intermolecular cross-links all allow suitable variation in mechanical properties within the architectural theme. Silks produced by insects and spiders are typically composed of partially crystalline structural elements of the fibrous protein fibroin, cemented together and surrounded by an amorphous matrix of the viscous, sticky protein sericin (Otterburn, 1977). Silk threads are characteristically strong and flexible, but with relatively low extensibility (Dickerson and Geis, 1969). Wool and hair, as examples of keratins, demonstrate the same fundamental architecture as part of a complex hierarchy of structural levels. The recognized "two-phase structure" of keratins consists of birefringent crystallites of fibrous proteins embedded in a relatively amorphous protein matrix (Bradbury, 1973). Protofibrils (coiled-coil superhelices of α helices) aggregate with parallel alignment into microfibrils about 70 Å in diameter, and microfibrils are packed by parallel alignment in a small amount of matrix into macrofibrils with diameters up to 0.6 μm, surrounded by a larger amount of amorphous matrix (Swift, 1977; Bradbury, 1973; Fraser et al., 1972). The partially crystalline fibrous proteins are rich in α helices with relatively low sulfur contents, while the matrix proteins are believed to have a more compact "pseudo-globular" conformation with relatively high sulfur content and intramolecular disulfide bonds (Fraser et al., 1972). Mechanical properties directly related to the keratin architecture have been reviewed by Fraser and colleagues (1972). Differential swelling with a high value of volume expansion by the matrix is observed upon humidification of keratin fibers, such as porcupine quill with an 11% increase in microfibril volume and a 53% increase in matrix volume. The swelling of animal fibers is also highly anisotropic, with typical increases in diameter of 16%, compared to increases in length of only 1%. Preferential plasticization of the predominantly amorphous matrix upon hydration results in dramatic reduction of the modulus and enhancement of the extensibility of keratinous tissues at moisture contents above 10%. The effect of keratin architecture on the most important chemical reactions of keratin fibers, thiol–disulfide exchange reactions, is also attributed to their two-phase structure (Asquith and Leon, 1977). The equilibrium constant for exchange reactions of accessible thiols and disulfides of proteins in dilute solution is near unity, so kinetics and cooperativity of steric factors govern thiol–disulfide exchange and thiol–disulfide chain reactions in the concentrated environment of the keratin architecture (Friedman, 1973; Ziegler, 1977; Whewell, 1977). The final ratio of free thiols and new mixed disulfides and the occurrence of thermosetting gelation by a thiol–disulfide chain reaction at protein concentrations above 5% (Friedman, 1973) depend on the following factors:

1. steric factors related to the different tertiary structures of fibril proteins and matrix proteins
2. the kinetics of exposure of thiols and disulfides due to conformational changes allowed by increased moisture and temperature or the presence of redox agents, changing the instantaneous balance of disulfide content and catalytic thiols
3. the cooperativity of crystalline lattice energy or new hydrogen bonding patterns produced during the exchange reactions (Ziegler, 1977; Whewell, 1977).

Wool fibers are prized for their flexibility, extensibility, and elasticity, but they have relatively low strength (Dickerson and Geis, 1969). Connective tissue, an example of the collagens, also exhibits a hierarchy of structure based on the universal architecture. Detailed understanding of the collagen matrix and interactions between collagen fibrils and matrix components is lacking, but further research is considered critical for controlling the cooked texture of poorer quality cuts of meat (Pearson, 1987). In the formation of rat tail tendon, tropocollagen molecules (coiled-coil superhelices of polyproline II helices) spontaneously aggregate with parallel, staggered alignment into microfibrils, which are further aligned into macrofibrils with diameters from 0.1 μm to 0.5 μm (Escoubes and Pineri, 1980). Collagen fibers are noted for their great strength, but they have limited extensibility and flexibility (Dickerson and Geis, 1969). Finally, it is instructive to distinguish the roles of actin and tropomyosin in the thin filament complex of muscle fibers, although muscle fibers are not primarily structural systems. Pure actin is capable of forming long filamentous structures, consisting of beadlike globular subunits arranged along a two-start helix (Finch, 1975). In contrast, tropomyosin is a coiled-coil superhelix of α helices, and fibrous paracrystalline tactoids are formed by isolated tropomyosin in the presence of divalent cations (Ohtsuki et al., 1986). In the thin filament complex, the fibrous tropomyosin interacts stoichiometrically with the beadlike chains of compact actin subunits (Ohtsuki et al., 1986), providing a structural model that is intriguingly similar to the unconfirmed suggestion that the predominantly amorphous matrix of keratin shows some degree of arrangement of the "pseudo-globular" matrix proteins (Fraser et al., 1972).

Examples of the same universal architecture can be found in the realm of synthetic polymers, where the architecture becomes technologically important for the end use of the polymer system. Figure 5-40 shows photomicrographs of stretched films of a typical composition for a synthetic chewing gum base. The films were prepared by uniaxial extension of the chewing gum, after plasticization by chewing. The base composition contains an elastomer, polyisobutylene (PIB), which provides the oriented structural elements in the stretched film, and a resin, Staybelite, which provides the viscous orienting matrix. Figure 5-40 shows two orientations of the oriented sample, with respect to the crossed polarizing prisms of the microscope. In Figure 5-40B, the film is placed so that the direction of stretching is parallel to one of the polarizers, and the sample appears isotropic, except for the crystalline talc particles. The birefringence of the talc particles is unaffected by the placement of the sample with respect to the polarizers, because the dimensions (10μ to 40μ in diameter) of the particles are well above the limit of resolution of the microscope. Likewise, the absence of birefringence for the amorphous Staybelite matrix is unaffected by the placement of the sample. In contrast, observation of the birefringence of the oriented PIB macrofibrils, with a diameter of about 0.5μ, depends dramatically on the placement of the stretched

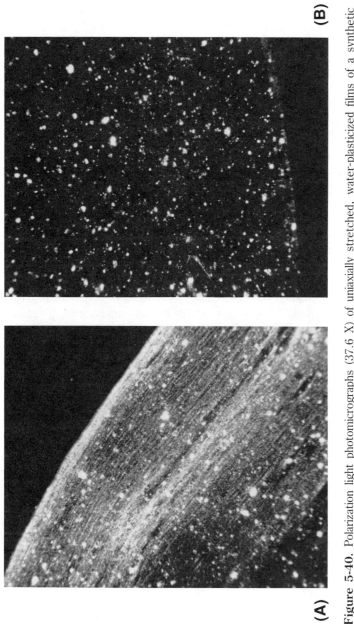

(A)

(B)

Figure 5-40. Polarization light photomicrographs (37.6 X) of uniaxially stretched, water-plasticized films of a synthetic chewing gum base containing an elastomer (polyisobutylene)/resin (Staybelite) blend: (A) stretching direction of film 45° to the crossed polarizing prisms; (B) stretching direction of film parallel to one of the polarizers. (*From Slade et al., 1988, with permission.*)

film, because the PIB microcrystallites have dimensions on the order of 100 Å .
The position of maximum birefringence of the microcrystalline PIB macrofibrils
occurs when the direction of stretching of the chewing gum film is 45° to
the crossed polarizing prisms, as illustrated in Figure 5-40A. Stretched, plasticized films of bubble gum base made with styrene butadiene rubber (photomicrographs not shown) exhibit exactly the same behavior. In both cases, use of a gypsum comparison crystal demonstrates that the elastomer
macrofibrils have negative birefringence, as expected for typical synthetic
polymer fibers (Sharples, 1966), because the chain direction in the microcrystalline regions of the fiber is perpendicular to the long axis of the fiber. These
chain-folded microcrystallites should be contrasted with the fringed micelle
structure previously described for cellulose macrofibrils, which exhibit positive birefringence. In addition to enhancement of birefringence by orientation
of partially crystalline macrofibrils due to elongation, the chewing gum films
share another mechanical property, fibrillation, which has been described for
growing cellulosic cell walls. The tears, which appear parallel to the direction of stretching in Figure 5-40A, result from lateral separation within the
amorphous matrix between PIB macrofibrils. Film strength is maximum in
the direction of extension, due to reinforcement by microcrystallization of the
macrofibrils (nucleation by orientation). Film strength is minimum perpendicular to the direction of extension due to the low modulus of the plasticized
amorphous matrix. The technological solution to the problem of anisotropy of
strength for films produced by uniaxial extension is reflected in the process for
production of polyethylene terephthalate plastic soda bottles by blow-molding
(E. E. Lewis, personal communication). Biaxial or spherical extension during
blow-molding allows isotropic orientation of microcrystalline macrofibrils, with
a concomitant isotropy of film strength.

Perhaps the most interesting example of this universal architecture in protein systems can be found in the area of food polymers. Ironically, gluten
proteins are not known to play a structural role in nature, where they perform
an important functional role as stable (in the glassy state) storage proteins in
the mature wheat endosperm to serve as an amino acid pool for the germi-
nating seed. However, the monumental uses of wheat gluten for technological
applications in the baking industry depend critically on the ability of gluten protein systems to attain the architecture that has been described. Figure 5-41
illustrates the now-familiar universal architecture in a uniaxially stretched film
of vital wheat gluten (IGP SG-80), plasticized with 5% aqueous ethanol, and
observed with polarization light microscopy. The photomicrographs of gluten
in Figure 5-41 show oriented macrofibrils with a diameter of about 0.5μ as
structural elements embedded in an orienting matrix which has sufficient viscosity to maintain the elongated ellipsoid shape of the trapped air bubbles,
and they emphasize the remarkable similarity of gluten proteins as an elastomer–resin system to the functionally optimized synthetic system of chewing
gum base shown in Figure 5-40. To facilitate comparison of the stretched

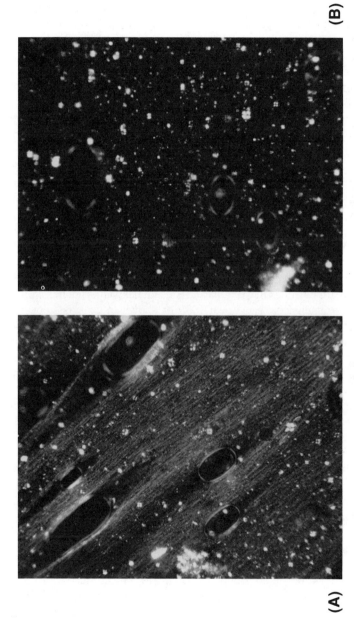

(A)

(B)

Figure 5-41. Polarization light photomicrographs (37.6 X) of uniaxially stretched films of vital wheat gluten (IGP SG-80) plasticized by 5% aqueous ethanol: (A) stretching direction of film 45° to the crossed polarizing prisms; (B) stretching direction of film parallel to one of the polarizers. (*From Slade et al., 1988, with permission.*)

gluten film to the synthetic elastomer–resin system, Figure 5-41B shows the gluten film placed so that the direction of stretching is parallel to one of the crossed polarizing prisms, and the sample appears isotropic, except for the partially crystalline native wheat starch granules. The birefringence of the starch granules is unaffected by the placement of the sample with respect to the polarizers, because the dimensions (4μ to 30μ in diameter) of the granules are well above the limit of resolution of the microscope. The native starch granules show a characteristic Maltese cross at any orientation of the sample to the crossed Nicol prisms, diagnostic of the spherical symmetry of the supramolecular structure of the granule. However, observation with a gypsum comparison plate shows that the starch granule is not a spherulite. The Maltese cross of a synthetic polymer spherulite exhibits negative birefringence, because the direction of the polymer chain in the microcrystallites is tangential to the surface of the spherulite (Sharples, 1966). The Maltese cross of the starch granule exhibits positive birefringence (quadrants with radial alignment of microcrystalline amylopectin parallel to the large index of refraction of the gypsum crystal appear blue) like cellulose macrofibrils, which also fail to show spherulitic growth (Sharples, 1966). The absence of birefringence for the viscous matrix of the stretched gluten film is unaffected by the placement of the sample, indicating that the matrix is essentially amorphous, with a lack of order on a dimensional scale near 0.5μ or above. In contrast, observation of the birefringence of the oriented macrofibrils is highly dependent on the placement of the stretched film, and the position of maximum birefringence of the microcrystalline macrofibrils occurs when the direction of stretching of the gluten film is 45° to the crossed polarizing prisms, as illustrated in Figure 5-41A. Use of a gypsum comparison crystal demonstrates that the gluten macrofibrils have positive birefringence, which is typical for biopolymer fibers such as gelatin, starch, and cellulose; all of these fibers have a fringed micelle structure with the chain direction in the microcrystalline regions of the fiber parallel to the long axis of the fiber. The stretched gluten film also shows that fibrillation has occurred as a result of uniaxial extension. The tears, parallel to the direction of stretching in Figure 5-41A, are especially evident near air bubbles, which interrupt the continuity of the perpendicularly weakened matrix. Fibrillation can be minimized during uniaxial extension of gluten films by the use of chemical agents that enhance the integrity of the macrofibrils. Cations of metal salts, such as Li^+, Na^+, and K^+, are particularly effective, because the affinity of these metal cations for peptide amides is greater than for water, esters, or amino groups (Kollman, 1978). Interaction of the metal cations with accessible peptide backbone amides has a stiffening effect on polypeptide chains, which is used to advantage in the processing of nylons and in the development of wheat flour dough formulations containing salt. Another method to minimize fibrillation of gluten involves biaxial extension of folded, sheeted doughs or spherical extension of air cells during fermentation of yeasted doughs to provide isotropic orientation of the microcrystalline gluten macrofibrils, with concomitant enhancement of film strength and gas retention. The analogy between cell walls of plant tissues and "gas cell walls" (actually air–matrix interfaces) in wheat flour doughs becomes apparent when proofed doughs are observed by polarization light microscopy with a gypsum comparison crystal and crossed polarizing prisms. The spherically oriented gluten macrofibrils embedded in the matrix surrounding the expanding gas bubble give an appearance identical to the cellulosic cell wall; that is, blue on the peripheral tangents parallel to the large index of refraction of the gypsum crystal, and yellow on the perpendicular peripheral tangents.

Birefringence of stretched gluten films has been observed previously, dating from 1938, as reviewed by Frazier and Muller (1967). Use of starch-free (less than 1%) gluten, stretched to 10 times its original length after plasticization by water, confirmed that the birefringence was due to gluten and not to starch. The loss of birefringence was retarded when aqueous potassium iodate solution was used to plasticize the film, but accelerated by thiol catalysts. The timeframes for the loss of birefringence were identical to those for the effects of these plasticizers on rheological relaxation of wheat flour doughs (Frazier and Muller, 1967). Birefringence has also been observed in solutions of isolated α-gliadins, which aggregate to form liquid–crystalline tactoids at protein concentrations above 5%, pH above 5, and ionic strength greater than 5 mM (Bernardin, 1975). This behavior of isolated, concentrated solutions of the relatively globular α-gliadin subunits (Bernardin, 1975) is reminiscent of the behavior of isolated actin (Ohtsuki et al., 1986), which can also form liquid crystalline tactoids, in contrast to the fibrous paracrystalline tactoids observed for the fibrillar subunits of isolated tropomyosin (Ohtsuki et al., 1986). The principal difference between these earlier reports of birefringence in gluten protein systems and the results shown in Figure 5-41 was the demonstration in Figure 5-41 of orientable microcrystalline macrofibrils with diameters near 0.5μ embedded in a viscous, orienting amorphous matrix, illustrating the universal architecture of structural polymer systems. Detailed experiments with stretched samples of commercial glutens, hand-washed glutens from varietal hard wheat flours, isolated gliadins, and isolated glutenins, observed by polarization light microscopy, have allowed identification of the composition of the macrofibrils as glutenin and the composition of the matrix as gliadins in gluten films. The macrofibrils are composed of HMW glutenin, which is the elastomer component of the elastomer–resin architecture of stretched gluten films. The elastomeric behavior of the relatively linear conformation of HMW glutenin subunits is believed to result from the β spiral conformation of the central portion of the polypeptide chain (Tatham et al., 1985; Schofield, 1986). In the relaxed state, the β-spiral is about 4 to 5 times shorter than a polyproline II helical conformation or a β-pleated sheet and about two times shorter than an α-helical conformation for an equivalent number of residues (Tatham et al., 1985; Krim and Bandekar, 1986; Dickerson and Geis, 1969), so that changes in length upon extension and subsequent relaxation are most dramatic for a β spiral conformation. The intermolecular disulfide cross-links in HMW glutenin are known to be located within the α helix enriched domains at the N- and C-termini (Tatham et al., 1985), and these domains are also likely to be involved in the microcrystalline regions of the glutenin macrofibrils. The matrix is composed of the relatively globular subunits of gliadins, which are the resin component of the elastomer–resin architecture of stretched gluten films. The ability of proteases, LMW thiols, surfactant lipids, hydrogen bond denaturants, and effective aqueous plasticizers to reduce the viscosity of the gliadin matrix, to the extent that glutenin macrofibrils cannot remain oriented, can be directly observed by polarization microscopy of stretched gluten films. The reduced effectiveness of aqueous plasticizers can be observed after heat treatment of stretched gluten films, which causes thermosetting gelation of glutenin and hardening of the gliadin matrix. The same considerations of kinetics, steric factors, and cooperativity, described above for the thiol–disulfide exchange and chain reactions in keratinous systems, are operative in the thermosetting reactions during processing of wheat flour doughs. Thus, investigation of this universal architecture of developed gluten protein systems provides valuable insights for the functional role of gluten in doughs and baked products (Slade et al., 1988).

Gluten as a Polymeric Cryostabilizer: Storage Stabilization of Frozen Bread Dough

Both the physiological role of collagen as the major structural protein of animal connective tissue and the functionality of gelatin, the technologically useful form of collagen, as a food ingredient and polymeric cryostabilizer reflect the same underlying structure–property relationships of this viscoelastic, partially crystalline polymer system at moisture contents above 35 w% (Slade et al., 1988; Slade and Levine, 1987b; Levine and Slade, 1988e). In contrast, the analogy between the physiological and technological roles of gluten is not so readily apparent. The physiological role of gluten as a plant storage protein depends mainly on the limited solubility of gluten at high concentrations. However, recognition of the constraints to diffusion processes afforded by the glassy solid state of gluten in the dry seed provides a clue to the technological utility of wheat gluten as a viscoelastic polymer system (Slade et al., 1988), unique among grain storage proteins.

Aspects of the functionality of endogenous hard wheat gluten in ordinary white pan bread making are well known, as reviewed by Bloksma (1978) and Pomeranz (1978). It is established practice in bread baking tests to measure BLV as a major criterion of functionality, which has been highly linearly correlated with flour protein content within individual hard wheat varieties.

However, the linear regression of BLV versus protein content differs among varieties, reflecting differences with respect to gluten in protein quality arising from differences in the composition of the LMW, extensible gliadin and HMW, elastic glutenin fractions. The effects on BLV of chemical additives or chemical modifications, selected for their specific action on various functional groups of proteins, have shown that the roles of (1) numerous weak hydrogen bonds and dispersion forces, which contribute to highly cooperative protein–protein interactions; (2) reactive, coexisting disulfides and thiol groups, which contribute to dynamic disulfide exchange; and (3) interchain disulfide cross-links, which contribute to supramolecular structure, are considered critical to dough processing and bread quality (Bloksma, 1978). Significant differences in gluten composition between bread wheat varieties have been shown in the composition of the gliadin fractions, and older reconstitution baking studies had suggested that these extensible gliadin proteins control the BLV potential of a wheat flour (Pomeranz, 1978). However, recent work in wheat genetics (Payne, 1987) has led to a qualified conclusion, which is now generally accepted. Breeding experiments with breadmaking wheat varieties to analyze the storage protein genes have shown that the dramatic biochemical complexity of their endosperm proteins and differences in gluten composition result from clusters of structural genes at only nine loci on six chromosomes (Payne and Holt, 1984). All of the genes exhibit sufficiently extensive allelic variation to allow identification of varieties by electrophoretic separation of the endosperm proteins. In particular, varietal protein quality for breadmaking is attributed to allelic variation at six of the nine loci (Payne and Holt, 1984). It has been concluded that, of the two gluten protein fractions, the HMW glutenin subunits are especially important in determining the BLV potential of a varietal wheat flour (Khan and Bushuk, 1979; Bushuk, 1984; Tatham et al., 1984; Schofield, 1986; Blanshard, 1988).

Commercial vital wheat gluten is often used by baking technologists as a supplement to increase the BLV in normal production of white pan bread formulated with an ordinary bread flour. All of the process steps involved in normal bread production occur at room temperature or above, and the thermomechanical behavior of doughs during mixing, proofing, and the early stage of baking depends on the softness and deformation afforded by the very large ΔT above the predominant Tg' and ΔW above Wg' of the relatively high-moisture system. Behavior in the later stage of baking and decreased softness of the baked loaf reflect contributions of thermoset network formation to a smaller ΔT (above the network Tg) and decreased water content to a smaller ΔW (Slade et al., 1988). Still, the fresh-baked loaf is quite soft, and diffusion-limited processes such as starch retrogradation proceed freely. In contrast, process steps and intermediate storage of doughs during the manufacture of frozen dough half-products involve freezer temperatures close to, and potentially below, the predominant Tg' of the dough formulation, where doughs exist in firm, rubbery or hard, solid states.

A recent application of gluten supplementation in the production of frozen bread dough involved a patented half-product, which goes from freezer to oven to table in one hour (Lindstrom and Slade, 1983; Larson et al., 1983, 1984). This yeast-leavened product contradicts the traditional dogma against proofing before frozen storage (Hoseney, 1986b); it is proofed before freezing, freezer-stored, and then baked directly from the frozen state. It has been demonstrated (Slade et al., 1988) that the technological basis of this novel product is the use of exogenous vital wheat gluten as a polymeric cryostabilizer to ensure at least three months' freezer storage stability at $-18°C$. Table 5-2 shows the protein compositions and low-temperature DSC results for a series of commercial glutens evaluated for potential use in such a frozen bread dough product. Total protein content (taken as gluten) was determined by Kjeldahl nitrogen analysis, and gliadin content was determined by spectrophotometric analysis of 70% ethanol extracts (Pomeranz, 1978), compared to a Sigma gliadin standard solution. Measurements of Tg' and Wg' were made on 1:1 (w:w) gluten:water mixtures. The results in Table 5-2 demonstrated the considerable compositional variability, even lot-to-lot, among commercial glutens, in terms of protein content and gliadin:glutenin ratio. Values of the latter, with

Table 5-2. Protein Compositions and Low-Temperature DSC Results for Commercial Vital Wheat Glutens

Gluten Sample	Percent Protein[a] (as is basis)	Gliadin/ Glutenin Ratio[b]	Tg'[c] (°C)	Wg' (g UFW/g)
IGP SG-80 (lot #323001)	76.6	2.53	− 8.5	0.38
IGP SG-80 (lot #234116)	79.6	0.80	− 8.5	0.34
IGP SG-80 (lot #264004)	78.8	0.87	− 7.5	0.11
IGP SG-80 (lot 11/81)	81.3	0.66	− 8.5	0.35
IGP SG-80 (lot 12/81)	71.7	0.73	− 7.5	0.07
IGP Extra Strong Prot.	74.5	0.82	− 7.5	0.31
IGP Whetpro-80	82.9	0.75	− 8	0.37
Henkel Pro-80	74.9	0.60	− 6.5	0.20
Henkel Pro-Vim	75.0		− 5	
Henkel E-35	33.6	1.12	−10	0.41
JR Short Gluten	45.3	1.29	− 7	0.20
Medimpex	78.3	0.92	− 7.5	0.38
Sigma Wheat Gluten	80.0		− 6.5	0.39

Source: From Slade et al. (1988), with permission.

[a] Protein content as w% of total sample.

[b] Ratio by weight of protein content soluble in 70% aqueous ethanol to protein content insoluble in 70% ethanol.

[c] Tg' of hydrated gluten sample (gluten/water, 100/100), prepared by hand-mixing gluten (as is basis) with equal weight of water.

an average of 1.01 within a broad range of 0.60–2.53, contrasted with a constant value of 1.13 reported for glutens from various hard wheat varieties with widely different breadmaking qualities (Pomeranz, 1978). The DSC results for Tg' and Wg' of aqueous glutens reflected the compositional variability of these commercial samples, but most of the values fell in the ranges of $-6.5°C$ to $-8.5°C$ and 0.3–0.4 g UFW/g typical of other high polymeric protein and carbohydrate cryostabilizers (Levine and Slade, 1988e). These physicochemical properties have been shown to underlie, in large part, the excellent functioning of vital wheat gluten (both endogenous and exogenous) as a stabilizer for frozen bread dough (Slade et al., 1988).

The data in Table 5-2 showed the generic, potential superiority of hydrated glutens as polymeric cryostabilizers, in terms of the narrow range of high values of Tg', despite the inferred functional variability of the samples, in terms of the obvious broad range of protein subunit composition. Essentially no correlation was observed between Tg' and protein content ($r = 0.34$) nor gliadin:glutenin ratio ($r = -0.27$). In order to refine the discriminative capability of DSC analysis to evaluate exogenous commercial glutens and endogenous glutens of varietal flours for their cryostabilizing potential, a low-temperature DSC assay was developed (Slade et al., 1988). The assay measured the Tg' of a hand-mixed sample of gluten:sucrose:water (100:10:100 parts by weight, prepared by addition of sucrose in solution to gluten), to examine the extent of depression by sucrose of Tg' in the ternary glass, compared to the Tg' of hydrated gluten in the simple binary glass, as reported in Table 5-2. Such an assay, using sucrose to provide a ternary glass with a lower value of $\overline{M}w$, has proved useful in the characterization of high polymers such as carbohydrate gums with values of Tg' near 0°C (Levine and Slade, 1988e). Tg' for the freeze-concentrated, entangled-polymer glass occurred as a superimposed shoulder on the low-temperature side of the ice melt. In the simplest case, when all of the added sucrose contributed to the ternary glass, depression of $\overline{M}w$ and Tg' was dramatic and quantitative, allowing deconvolution of Tg' from the Tm of ice, which was only colligatively depressed by more dilute sucrose upon melting. In the most complex cases for multicomponent systems, when only part of the added sucrose contributed to the major ternary glass, deconvolution was still possible, but the depression of Tg' was only qualitative. Table 5-3 illustrated the effect of sucrose in ternary glasses of aqueous gluten samples. It is important to remember that all of the test compositions in Table 5-3, as well as those in Table 5-2, contained excess water, which froze readily. Variation of the test system composition could have led to constraints in solubility and solute compatibility for different solute ratios and concentrations. The simplest case was exemplified by Henkel Pro-Vim. The ternary glass exhibited a single value of Tg' that was depressed compared to that for the binary glass of Table 5-2. For the same assay test system composition, gluten isolated from a high-protein varietal flour exhibited the same single value of

Tg'. In contrast, a more complex case was exemplified by the same sample of isolated gluten when a higher concentration of sucrose was used in the assay; here, *two* values of Tg' were observed. The expected further depression of Tg' for the ternary glass was observed as a smaller transition, but the major transition occurred near Tg' of sucrose itself ($-32°C$). For this same higher-sucrose composition in the test system, additional complexity was observed when another commercial gluten, equivalent to the IGP series described in Table 5-2, was assayed. Again, two values of Tg' were observed, but the minor transition occurred near the Tg' for sucrose itself.

Based on earlier studies (Slade and Levine, 1988a) of complex aqueous model systems, including mixtures of various proteins/sugars, polysaccha-rides/sugars, and many different multicomponent food ingredients and prod-ucts, multiple Tg' values have been attributed to the existence of two distinct aqueous glasses (Levine and Slade, 1988e; Slade et al., 1988). In a maxi-mally frozen matrix, the presence of multiple aqueous glasses, having differ-ent composition and heterogeneous spatial distribution (on a size scale of $\gtrsim$ 100 Å [Wunderlich, 1981]), is a result of partial disproportionation (i.e., par-titioning) of nonhomologous, molecularly incompatible solutes during freeze-concentration. By convention, multiple values of Tg' are listed in order of decreasing intensity of the transition (i.e., magnitude of the step change in heat flow), such that the first Tg' represents the major glass, and the second Tg', the minor glass. Experience has shown that the Tg' value of the predom-inant glassy phase largely determines the overall freezer-storage stability of a multicomponent food material (Levine and Slade, 1988c). In frozen dough, as in other complex, water-compatible food systems composed of mixtures of large and small proteins and carbohydrates (e.g., many dairy ingredients and products, typically dominated by large proteins and small sugars), the high-er-temperature Tg' of a doublet corresponds to a glass with higher protein and polysaccharide concentration, while the glass with the lower Tg' has a higher concentration of soluble sugars and amino acids (Slade et al., 1988). Saccharide or peptide oligomers of intermediate linear DP are critical to the specific partitioning behavior of lower oligomers and monomers. In assay test systems devised by these authors and in real food systems (Levine and Slade, 1988d), these intermediate oligomers can be used to increase the compatibil-ity of solutes in mixtures with an otherwise bimodal MW distribution. Then the tendency to form two separate ternary glasses is diminished or prevent-ed, and a single quaternary glass with an intermediate Tg' is observed. This increase in the underlying structural homogeneity at the molecular level is perceived macroscopically as an improvement in textural uniformity (smooth-ness) (Levine and Slade, 1988e).

In this context, the assay behavior of quaternary systems of gluten/10-DE maltodextrin/sucrose/water (shown in Table 5-3) was particularly informative. The Tg' of the simple binary glass of this 10-DE maltodextrin was $-9.5°C$, similar to the Tg' of a typical IGP gluten; that is, $-8.5°C$ to $-7.5°C$. However, addition of an equal weight of sucrose to IGP gluten led to a Tg' doublet, representing both a more mobile gluten-enriched glass ($Tg' = -16.5°C$) and partition of a separate sucrose glass ($Tg' = -33°C$). In contrast, addition of an equal weight of sucrose to 10-DE maltodextrin led only to a more mobile carbohydrate glass, without partitioning of a separate sucrose glass, owing to the greater compatibility of maltodextrin and sucrose. Compared to the effect of adding an equal weight of sucrose alone, the effect of adding an equal weight of maltodextrin and partial removal of sucrose on the Tg' ($-16.5°C$) of the gluten-enriched glass was not detectable; however, partition of maltodextrin into the separate sucrose-enriched glass was detected as a slight elevation of its Tg' ($-29.5°C$). To interpret the effect of increasing maltodextrin concentration on gluten glass behavior in the absence of sucrose, it was necessary to examine the glass properties of the endogenous water-soluble components of the gluten, whose partition behavior was modulated by maltodextrin as a compatible solute. Values of Tg' for 20 w% aqueous solutions of freeze-dried water extracts from commercial gluten and varietal flours ranged from $-37.5°C$ for the lower Tg' of a doublet to $-22°C$ for the higher Tg' or single Tg' value (Slade et al., 1988). The major constituents of the water solubles are albumins, peptides, and non-starch carbohydrates. Typical values of Tg' for 20% solutions of isolated water solubles are shown in Table 5-3. Addition of about 27 w% maltodextrin solution to gluten in the absence of sucrose, with a test system composition of gluten:maltodextrin:water (100:33:89), showed a reversal of the magnitudes of the doublet glasses. The major glass contained the endogenous water solubles, with an elevated Tg' ($-17.5°C$) due to the presence of maltodextrin, and the minor glass was highly enriched in gluten proteins ($Tg' = -8°C$). Addition of increased concentration of maltodextrin solution (50 w%) to gluten, with simultaneous increase in maltodextrin:gluten ratio (100:100:100), enhanced partition of maltodextrin into the gluten phase, so that Tg' of the minor glass was depressed by increased presence of maltodextrin, while Tg' of the major glass was depressed by decreased presence of maltodextrin. Finally, addition of about the same concentration of maltodextrin solution (53 w%), with much greater ratio of maltodextrin to gluten (100:300:267), allowed total compatibility of the solutes, so a single glass was observed with Tg' of $-9°C$.

A second generation of patented frozen dough products (Benjamin et al., 1985) relied on the cryostabilization provided by endogenous, rather than exogenous, gluten. Various types of breads, rolls, and pastries were produced from frozen doughs formulated with special high-protein ($\gtrsim 16\%$) varietal hard wheat flours. The low-temperature DSC assay previously described for the discriminative characterization of gluten quality was used to evaluate such flours for their cryostabilizing potential, on the basis of a flour:sucrose:water (100:10:100 by weight) hand-mixed "dough," as a model of a stressful condition for a lean bread dough formula (100:6:60 by weight) (Sultan, 1976). Results of this assay for thirty-four selected varietal flour samples, compared

Table 5-3. Low-Temperature DSC of Commercial and Hand-Washed Wheat Glutens — Tg' Results for Gluten/Sucrose/Water Assay Test System Samples

Gluten Sample	Assay Composition (parts by weight)				$Tg'(°C)^a$
	Gluten	Maltodextrin	Sucrose	Water	
Henkel Pro-Vim	100	0	10	100	− 9
Gluten-5409 (1979)[b]	100	0	10	100	− 9
Gluten-5409 (1979)[b]	100	0	100	133	−32.5 & −14
IGP SG-80 (lot 5/81)	100	0	100	133	−16.5 & −33
	0	100	100	133	−18
IGP SG-80 (lot 5/81)	100	100	50	133	−16.5 & −29.5
IGP SG-80 (lot 5/81)	100	33	0	89	−17.5 & − 8
IGP SG-80 (lot 5/81)	100	100	0	100	−19.5 & − 9
IGP SG-80 (lot 5/81)	100	300	0	267	− 9
10 DE maltodextrin[c]					− 9.5
Solubles-5409 (1981)[c]					−22.5
Solubles-IGP SG-80 (lot 5/81)[c]					−29.5 & −37.5

Source: From Slade et al. (1988), with permission.

[a] *Tg'* of assay test system samples (gluten/sucrose/water or gluten/10 DE maltodextrin/sucrose/water), with compositions as noted for individual samples. In each case, gluten (as is basis) was hand-mixed with the indicated total weight of sucrose syrup or carbohydrate syrup.

[b] Hand-washed gluten from 1979 crop year of 5409 varietal flour.

[c] *T_g'* of 20% aqueous solution.

to nine samples containing commercial blended bread-wheat flours (Pillsbury 4X Patent and High Gluten) and three flour samples from sprouted grains (malted barley, Brigand, and Bounty), are shown in Table 5-4. The results demonstrated the significant year-to-year and even farmer-to-farmer variability in both *Tg'* and protein content of many of these genetically constant flours. Among the flours analyzed for both protein content and gliadin:glutenin ratio, the former ranged from 11.5% to 19.8% and the latter from 0.96 to 4.55.

Most of the model "doughs" shown in Table 5-4 exhibited a pair of *Tg'* values. As deduced from the results of Table 5-3, the higher-temperature *Tg'* of the doublet correlated, to a first approximation, with the gluten contribution by a flour, and corresponded to a ternary gluten:sucrose:water glass dominated by the protein (referred to as the "gluten" phase). The lower *Tg'* corresponded to the glass dominated by the amount of added sucrose that did not partition into the ternary gluten/sucrose/water glass plus the endogenous small carbohydrates and other lower-MW, water-soluble solids, referred to as the "aqueous" phase. Generally, those samples in Table 5-4 with higher percent protein manifested the *Tg'* of the "gluten" phase as the only detectable transition or the major *Tg'* of a doublet, whereas samples with lower percent protein showed the *Tg'* of the "aqueous" phase as the more intense or only detectable transition. The results in Table 5-4 for individual varietal flours of different crop year and protein content, and relative results among the series of sixteen different hard wheat varieties, demonstrated an apparent positive correlation between total protein content and the value of the Tg' of the "gluten" phase, but with obvious exceptions, relating to the mutual effects on partition behavior of the endogenous water solubles and the added sucrose in the assay test system. Thus, it was deduced that the primary influence of flour composition on Tg' was due to the relative ratio of gluten content to content and type of endogenous water-soluble components (Slade et al., 1988). A secondary influence of differences in protein composition among individual flour samples, in terms of gliadin:glutenin ratio, was predicted to have a profound effect on the specific mechanical properties (such as absolute magnitude of the modulus, as reflected in Fig. 5-18, and ratio of G''/G') of the dough, while the relative mechanical relaxation times would depend on the ΔT that governs kinetics above Tg' (Slade et al., 1988).

For the patented frozen bread dough products discussed in this section, the major criterion used to assess their cryostability during extended storage was baking performance, in terms of BLV after baking from the frozen state. Freezer-storage instability was typically manifested by steadily decreasing BLVs with increasing storage times, which was especially pronounced during the first month under normal storage conditions at $-18°C$. This loss in BLV was generally caused by the combined effects of (1) loss of ovenspring, and (2) loss of dough volume (Slade et al., 1988).

Ovenspring is defined as the increase in volume during baking (measured in cc as the difference between BLV and volume of a 1 lb frozen dough blank), and its loss as a result of frozen storage was due primarily to the deteriorated condition of the gluten film around the air cells in the dough, which had been stressed by proofing before freezing. As was alluded to in the previous section, this use of the term "gluten film" is understood in the context of the polymer science of elastomer–resin systems (Ferry, 1980), in which elastomeric fibrils or networks are embedded in an extensible, film-forming resin matrix. When such elastomer–resin systems are foamed, biaxial extension due to expansion of nucleated air cells results in formation of a continuous cell structure in the film-forming matrix, reinforced by the embedded elastomer superstructure. The matrix/air cell interface in a closed-cell foam structure is equivalent to a continuous solid film. A reticulated, or open-cell, foam structure would be discontinuous at the air cell interface. Although a discrete "film" or "wall" does not exist (as has been pointed out by Hoseney [1986b]), in the form of a separable "balloon," the foam structure based on the functional roles of reinforcing elastomer and film-forming resin is critical to the understanding of structure–property relationships of doughs and baked products (Slade et al., 1988). (For example, the transformation during baking of ordinary yeast-leavened bread dough or chemically leavened cake batter can be described as a change from a closed-cell foam structure to an opened-cell foam or "sponge"

Table 5-4. Low-Temperature DSC of High-Protein Varietal Hard Wheat Flours and Bread Formula Additives—Tg' Results for Flour/Sucrose/Water (100/10/100) Assay Test System Samples

Varietal Flour (crop year—source)	Percent Protein (as is flour basis)	Gliadin/ Glutenin Ratio	Tg' (°C)
PLV (1979)	18.1		−16
PLV (1980)	14.5	2.65	−19
PLV (1981)	15.1		−10.5 & −21.5
3466 (1979)	19.2		−15.5
5471 (1979)	17.8		−11.5 & −18.5
5471 (1980)	18.8	1.42	−15.5
Eagle 717 (1980—Kansas)	13.4		−19
Eagle (1981)	12.8	4.55	−20.5
Glenlea (1980)	11.5	1.08	−20
Colorado 723 (1980)	12.4		−21
823 (1980—North Dakota)	15.3	2.95	−20.5 & −10
Waldron (1980—Dickinson)	17.2		−17.5
Eureka (1980—Dickinson)	16.8		−20.5 & −10.5
Olaf (1980—Dickinson)	16.4		−20 & −10
Olaf (1980—Wanner)	15.3	1.10	−18 & −11.5
Olaf (1981—Minot)	14.1		−13.5 & −24
Olaf (1981—Fargo)	12.9		−21 & −12.5
Len (1980—Wanner)	16.1		−11
Len (1980—Stang)	13.6	0.96	− 9.5 & −23
Coteau (1980—Stang)	17.2		−10.5

structure [Davies, 1986].) Two facets of the deterioration of the matrix/air cell interface were observed. Immediate loss of ovenspring after only 24 hours of frozen storage related to simple "freeze-thaw instability." Progressive loss of ovenspring during long-term frozen storage related to "freezer-storage (in)stability," which is the subject of this discussion.

Loss of dough volume resulted from two related processes, elastic shrinkage and viscous flow. Both of these phenomena are microscopic collapse processes, as described in Table 5-1, which are allowed to proceed whenever $Tf > Tg'$ of the pre-proofed, porous matrix of the dough blank during frozen storage. The rate of such collapse was governed by the ΔT between Tf and Tg', and increased dramatically with increasing ΔT above Tg' of the rubbery amorphous phase(s), as dictated by WLF kinetics. For example, in the case of a hypothetical dough exhibiting a major Tg' for its "gluten" phase and a minor Tg' for its "aqueous" phase, dough collapse would be: (1) essentially prohibited if $Tf < Tg'$ "aqueous" $< Tg'$ "gluten"; (2) permitted if Tg' "aqueous" $< Tf < Tg'$ "gluten"; and (3) most rapid if Tg' "aqueous" $< Tg'$ "gluten" $< Tf$.

Table 5-4. (*continued*)

Coteau (1980—Steffan)	15.0	1.30	−12 & −19
Coteau (1980—RK/S1)	17.8		−10 & −18
Coteau (1980—RK/S3)	16.7		−10 & −19
Coteau (1980—RR/E1)	17.0		−10 & −18.5
Coteau (1980—RR/E2)	18.4		−10 & −18
Coteau (1981—Williston)	17.3		−11.5 & −19.5
Pr2360 (1981—Williston)	16.6		−14 & −23
Pr2360 (1981—Fargo)	13.0		−12 & −20.5
Benito (1981—Fargo)	15.2		−12.5 & −21
Tracey (1981—Fargo)	11.2		−21
5409 (1979)	20.5		−10
5409 (1980)	17.9	2.40	−12
5409 (1981)	19.8	0.97	−10 & −21
5409 (1982)	?		−23 & −12
5409 (1980):0.5 DE Maltodextrin (10:1)	16.3		−21 & −13
Pillsbury 4X (1979)	12.9		−21.5
Pillsbury 4X (1980)	14.4		−20.5 & −14.5
Pillsbury High Gluten (PHG) (1980)	13.9		−18
PHG (1981)	15.8	2.09	−12 & −22
PHG (1980):Henkel Pro-80 (3:1)	29		−11
PHG (1980):Henkel Pro-80 (9:1)	20		−12.5
PHG (1980):0.5 DE Maltodextrin (9:1)	12.5		−20
PHG (1980):Pro-80:0.5 DE (9:1:1)	18.2		−20 & −14
PHG (1980):Whetsorb (10:1)	12.6		−10.5
Malted barley	?		−23.5
Brigand (1978—England)	10.0		−23.5 & −29
Bounty (1979—England)	10.0		−22.5 & −29

Source: From Slade et al. (1988), with permission.

Baking test results consistently revealed that cryostabilization against both loss of dough volume and loss of ovenspring could be achieved through formulation with specific polymeric cryostabilizers (e.g., exogenous low-DE maltodextrin and endogenous gluten in a high-protein varietal flour) having inherent and distinctive compatibilities and partition behavior with respect to the "aqueous" or "gluten" amorphous phases in frozen dough (Slade et al., 1988).

In one preliminary storage/baking study, "control" doughs made with Pillsbury High Gluten (1980 crop) flour were stored for up to 8 weeks at −30°C alongside experimental doughs formulated with added 10-DE maltodextrin (equal by weight to sucrose). Differential scanning calorimetry measurements of four-component model systems (Table 5-3) showed that the maltodextrin significantly elevated Tg' by partitioning mainly to the "aqueous" phase. Thus, this phase would have existed as a more stable, glassy solid, because Tf was well below its Tg', in the experimental dough, but as a less stable, rubbery fluid in the control. Consequently, results of baking tests done at 1, 3, 5, and

8 weeks demonstrated much improved storage stability for the experimental dough, in terms of more stable ovenspring and BLV. Between weeks 1 and 8, control loaves decreased by an average of 44% in ovenspring and 17% in BLV, while the experimentals decreased only 15% in ovenspring and 5% in BLV (Slade et al., 1988).

In another preliminary baking study, experimental doughs made from four high-protein varietal flours (1979 crop of 5409, PLV, 3466, and 5471, shown in Table 5-4) showed freezer-storage stability superior to that of a Pillsbury HG (1980) control dough, as reflected by BLVs that were 12–43% greater after 4 weeks at $-18°C$. Greater cryostabilization due to higher amounts of endogenous gluten produced a storage situation in which $Tf < Tg'$ of the "gluten" phase for the varietal flours, but not for Pillsbury High Gluten. Furthermore, results of this study suggested a predictive correlation between increasing value of this Tg' and increasing BLV across this group of varietal flours, and also showed that the extent of shrinkage of the frozen dough blanks after 4 weeks at $-18°C$ decreased with increasing "gluten" Tg' (Slade et al., 1988). The 5409 flour was especially noteworthy for its single, unusually high Tg' value of $-10°C$ and its 43% greater BLV.

Results of a major follow-up baking study (Slade et al., 1988) confirmed both the critical relevance of Tg' to the cryostability of frozen bread doughs and the molecular origin of Tg' in the flour test system, as determined by the DSC assay of high-protein varietal wheat flours. Nine different varietal flours (mostly 1980 crop), shown in Table 5-4 with their measured values of gliadin/glutenin ratio (0.96–4.55) and percent protein (11.5–18.8%), were evaluated versus a PHG (1981) control. One set of doughs was produced from these ten flours "as received" (i.e., with their different endogenous gluten contents), and another set from all the flours uniformly supplemented to 22% total protein content with a single source of vital gluten. The BLVs were measured after 3 weeks' storage at $-18°C$. Figure 5-42 shows a plot of log BLV as a function of the relative ratio of total water-soluble solids/gluten (determined analytically) for all twenty flour samples. Comparative values of residual BLV at a given storage time point represent relative rates of shrinkage for the frozen dough samples. Expressed as log values, these rates are expected to exhibit a curvilinear relationship with any composition parameter that reflects an underlying variation in Tg', and therefore in ΔT. The purpose of these storage studies was to establish a diagnostic tool to guide formulation of frozen products with improved storage stability, rather than to define the particular values of coefficients for the WLF equation to describe this curvilinear relationship. In this context, the plot in Figure 5-42 revealed the importance of the solubles/gluten ratio as a parameter to characterize flours and predict their functional behavior with regard to the storage stabilization of frozen doughs. The linear regression with $r = -0.84$ was used qualitatively to demonstrate the predictive correlation between increasing

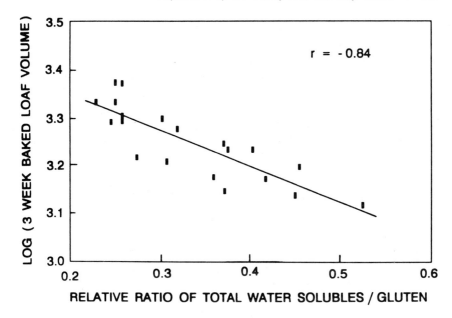

Figure 5-42. Variation of log (baked loaf volume), following three weeks of storage at −18°C, with the relative ratio of total water solubles/gluten in the flour, for frozen 1 lb bread dough blanks made with a series of gluten-supplemented and unsupplemented high-protein varietal wheat flours. (*From Slade et al., 1988, with permission.*)

BLV (i.e., increasing storage stability) and decreasing water solubles/gluten ratio (i.e., increasing cryostabilization) across this diverse set of flours. This correlation was superior to the corresponding one between log BLV and flour protein content alone. The analysis verified that the content of LMW soluble solids (mainly small endogenous carbohydrates) of a varietal flour, in addition to gluten content, had a significant impact on the cryostability and subsequent baking performance of frozen doughs.

Another plot (not shown), of log BLV versus Tg' in the flour assay test system for the ten unsupplemented flours, revealed that it is appropriate to interpret frozen dough storage stability as a mechanical relaxation process (Slade et al., 1988). In the cases of Tg' doublets, a single value of Tg' was calculated as a linear combination of the "gluten" and "aqueous phase" values, weighted by the relative amounts of the two phases. Such an interpretation of mechanical relaxation above Tg' has also been applied successfully to the storage stabilization of other frozen foods such as ice cream (Levine and Slade, 1988e). The plot showed a fair predictive relationship ($r = 0.78$) between log BLV and flour Tg' (as a reflection of ΔT relative to Tf) for this series of varietal flours. The combined results of this analysis and that of Figure 5-42 supported the premise that Tg' should increase with decreasing solubles/gluten ratio of a flour (i.e., with increasing $\overline{M}w$ of the system of water-compatible solids), since BLV increased with increasing Tg'. The loss of BLV after frozen storage was determined to be due in large part to the consequences of related mechanical relaxation processes in the rubbery frozen dough. The resulting rate of structural deterioration during storage (via collapse processes facilitated by water plasticization of the amorphous polymer matrix) was governed by WLF kinetics, which are applicable generically to frozen aqueous rubbery systems (Levine and Slade, 1988e).

The plot shown in Figure 5-43, of Tg' ("linear combination") in the flour model system versus the relative ratio of total water-soluble solids/gluten for the nine varietal flours evaluated in this baking study, confirmed the predicted relationship between these two parameters, and thus the molecular origin of flour Tg'. Tg' was highly linearly correlated ($r = -0.95$) with water solubles/gluten ratio of a flour across this series of varietal flours. The effect of flour solubles/gluten ratio on the thermomechanical behavior of frozen dough was also illustrated by DMA measurements of the temperature of maximum value of the loss modulus. The plot of $Tmax$ loss modulus versus solubles/gluten ratio also shown in Figure 5-43 revealed a linear correlation, as good as the one from DSC measurements of Tg', between increasing value of $Tmax$ loss modulus and decreasing solubles/gluten ratio. These DMA results demonstrated the correspondence between $Tmax$ loss modulus and Tg', and thus verified that the thermomechanical stability of frozen doughs increased with decreasing flour solubles/gluten ratio. Since log BLV and Tg' were both demonstrated to be highly correlated with, and dependent upon, flour solubles/gluten ratio, Tg' was thus identified as a key physicochemical property related to breadbaking performance. The predictive relationship between flour Tg' and log BLV, like the one between log BLV and solubles/gluten ratio, was much better than the corresponding correlation between flour protein content and baking performance after frozen storage. Thus, the major research finding revealed by this baking study was the potential utility of the DSC Tg' assay as a flour specification, more predictive than protein content, for high-protein varietal hard wheat flours with good functionality and storage stability in frozen bread dough applications (Slade et al., 1988).

Structure–Functional Properties of Gluten in Wheat Flour—Baking Mechanism of Sugar-Snap Cookies

The historical discrepancy in the selection of a reproducible quality parameter for characterizing hard wheat flours versus soft wheat flours provides a clue to the contrasting mechanisms of their functional behavior during baking. As was

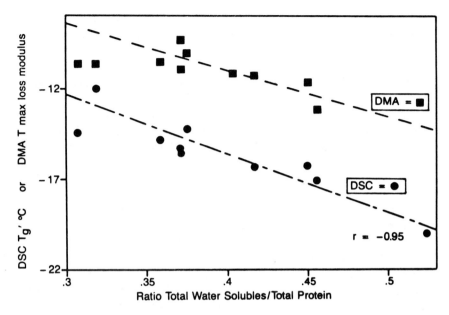

Figure 5-43. Variation of *Tg'* (of the flour:sucrose:water 100:10:100 model system), measured by DSC, with the relative ratio of total water solubles/endogenous gluten in the flour, in comparison to the variation of the loss modulus (of the flour:sucrose:water 100:10:100 model system), measured by DMA, with the relative ratio of total water solubles/endogenous gluten in the flour for the same series of high-protein varietal wheat flours. (*From Levine and Slade, 1988e, with permission.*)

mentioned in the previous section, it is established practice in the breadbaking test to characterize the functional quality of hard wheat flours (Pomeranz, 1978) by a volume parameter, BLV, highly correlated with varietal flour protein content and biochemistry. The corresponding tradition in the sugar-snap cookie baking test for soft wheat flours (Finney et al., 1950; Pomeranz, 1978; Yamazaki and Lord, 1978; Abboud et al., 1985; Doescher, Hoseney, and Milliken, 1987; Doescher, Hoseney, Milliken, and Rubenthaler, 1987) relies on a one-dimensional parameter, diameter (cookie "spread"), rather than overall cookie volume. Cookie spread potential has been accepted as the most meaningful quality criterion for soft wheat varieties, because cookie height is not adequately reproducible, even for test bakes that exactly replicate cookie diameter (Finney et al., 1950). The authors of this chapter have offered an alternative to this traditional approach, the conventional wisdom behind it, and especially the mechanistic interpretations it has led to, because they believe that this very lack of reproducibility in cookie height is key to understanding the functional contribution of good-quality soft wheat flours to cookie baking (Slade et al., 1988). These authors hoped that a new hypothesis and related speculations to explain flour functionality in sugar-snap cookies would provide others in the fields of cereal science and baking technology with food for thought.

This new hypothesis, for the underlying mechanism of baking performance and the functional roles of poor and excellent cookie-quality flours in the standard AACC cookie spread test (Finney et al., 1950), has been postulated (Slade et al., 1988) as a result of an analysis of the kinetics of three-dimensional expansion using data from time-lapse records of the baking process (Yamazaki and Lord, 1978). The hypothesis is based on a theoretical description of collapse phenomena, as discussed in the section Cryostabilization Technology—Collapse Processes and as illustrated in Table 5-1. Previously, conventional interpretations of baking data have been hampered by definitions of cookie "set" and "set time" as the conditions of no further increase in cookie diameter (Abboud et al., 1985; Doescher, Hoseney, and Milliken, 1987; Doescher, Hoseney, Milliken, and Rubenthaler, 1987). The analysis described below clearly shows that the cookie is *not* set when it reaches maximum diameter. Poor flour doughs exhibit controlled elastic expansion to maximum diameter, but this is followed by controlled elastic shrinkage (i.e., volume contraction, which results in minimum cracked surface) *after* the "set time." Excellent flour doughs allow viscous expansion accompanied by creep to maximum diameter, but this is followed by dramatic collapse (which results in maximum cracked surface) *after* the "set time." It should be noted that collapse refers here to irreversible, macroscopic structural collapse ("cave-in"), in contrast to microscopic collapse processes involving elastic shrinkage and viscous flow, which can be reversible (Levine and Slade, 1986, 1988e). The behavior of poor flour doughs is diagnostic of elastic recovery in a rubbery thermoset polymer system (Ferry, 1980); that is, expansion by composite film/network formation (which produces a self-supporting elastic network) and shrinkage above the thermoset network Tg. In contrast, the behavior of excellent flour doughs is diagnostic of structural collapse in a rubbery, predominantly thermoplastic polymer system; that is, expansion by film formation (with no significant functional network formation) and collapse above the molecular Tg. Spontaneous collapse during baking can occur at any temperature more than about 20°C above the instantaneous effective value of molecular Tg (Levine and Slade, 1986). Aspects of the mechanism of cookie baking postulated by the authors of this chapter are fully consistent with an earlier description by Manley (1983), which apparently has been overlooked by those who define a "set time" during cookie baking. Manley stated that

> The structure [of rich short doughs] relies on a sugary matrix which, of course, does not set appreciably as the temperature rises. Thus during baking, the dough becomes softer, some flow occurs, and a very large expansion is observed

followed by considerable collapse. As the dough piece loses moisture during baking, some contraction occurs and therefore some loss of product lift or spring is inevitable. In most cases, this is small compared with collapse of the internal structure due to ruptures of the gas bubbles, but it is worth noting that some contraction will continue even till the biscuit is completely charred if heating continues. The collapse is responsible for the cracked surface of biscuits.

Gluten, which is capable of undergoing its glass-to-rubber transition at temperatures and moisture contents relevant to cookie baking (Slade, 1984; Hoseney et al., 1986), has been proposed as the major functional flour component involved in the mechanism of sugar-snap cookie spreading (Doescher, Hoseney, and Milliken, 1987). This suggestion is especially reasonable in light of the fact that the other major flour component, native granular starch, does not gelatinize (i.e., the baking temperature does not exceed the effective T_g of starch [Slade and Levine, 1988b]) during successful baking of a standard high-sugar, low-moisture cookie dough (Kulp and Olewnik, 1984). Gluten comprises the highly amorphous, water-plasticizable polymer system that is strongly implicated as either the predominantly viscous thermoplastic or predominantly elastic thermoset material which controls cookie spread during baking. Damaged starch and pentosans are other network-forming flour polymers, which are, like glutenin, potentially detrimental to cookie quality (Gaines et al., 1988; Donelson, 1988). The presence of gliadin, while not essential for sugar-snap cookie baking performance (Donelson 1988), is potentially beneficial as a film-former. Compared to the disappearance of gliadins during the transformation from bread dough to high-quality white pan bread, the disappearance of gliadins during cookie baking is minimal (Pomeranz et al., 1987). Still, the small decrease in highly hydrophobic gliadins appeared to be more pronounced during the baking of cookies from good-quality cookie flours than from poor-quality cookie flours (Pomeranz et al., 1987). This hypothesis for the different functional roles of glutens in excellent and poor cookie-quality flours (Slade et al., 1988) is being tested by baking experiments performed with time-lapse photography (Saunders and Slade, 1989).

The AACC Micro Cookie Baking Procedure III (Finney et al., 1950), introduced by Finney's group in 1950, has become the industry standard test for evaluating soft wheat, with respect to functionality in cookie baking. Traditionally, only the final product has been evaluated, and performance has been evaluated based on measurement of cookie width only, or on the ratio of cookie width to height (Finney et al., 1950). Because better reproducibility was observed in cookie width (also called diameter or spread) than in height (also called thickness), the use of diameter measurements, rather than the diameter/thickness ratio, was recommended (Finney et al., 1950). The method has been improved by the addition of time-lapse photography (Yamazaki and Lord, 1978; Abboud et al., 1985; Doescher, Hoseney, Milliken, and Rubenthaler, 1987) to monitor the kinetics of changes in visual appearance and geometry during the baking process. In spite of the detailed information available from time-lapse photographic documentation of the baking process, the traditional treatment of data obtained from time-lapse profiles, based on measurement of cookie width only (Yamazaki and Lord, 1978; Abboud et al., 1985; Doescher, Hoseney, Milliken, and Rubenthaler, 1987), has persisted to the present time. Yamazaki and Lord (1978) described a mechanism for cookie spread based on evaluation of the increase in diameter in their time-lapse pictures:

> Competition between sugar and flour for the limited quantity of available water in the dough [i.e., 86/60/36, dry flour/sugar/water] and its resulting partition . . . low-hydration flour permits more sugar syrup to form in the dough, making it slacker at elevated temperature. This enables the leavening to expand the dough to a greater extent before setting takes place.

Analysis of the kinetics of one- and three-dimensional expansion during baking, that is, changes in diameter, height, and volume, led to a new hypothesis for the functional roles of cookie flours in the sugar-snap cookie spread test (Slade et al., 1988).

The time-lapse pictures published (and kindly provided to the authors of this chapter) by Yamazaki and Lord (1978), of cookies baked from poor and excellent cookie-quality flours, which were used as a source of data to examine the effect of flour selection on the kinetics of expansion during baking, are reproduced here in Figure 5-44 at about 40% of the actual size (Finney et al., 1950). Qualitative differences between the baking performance of a poor versus an excellent cookie-quality flour are immediately obvious. Both types of cookies expand in both width and height due to the production of leavening gases from the sodium/ammonium bicarbonate leavening system, and their expansion behavior is essentially identical for the first 2 minutes of baking. The cookie made from poor flour is capable of supporting its own weight throughout the entire baking cycle; structural collapse is not observed. In contrast, already by the third minute of baking, the cookie made from excellent flour is no longer able to support its own weight; continued expansion is accompanied by progressive structural collapse and lateral creep, and the most dramatic phase of collapse occurs in the last minute of the 10-minute baking cycle.

Values of cookie height and width were measured, and values of volume (as a cylinder) and width/height ratio were calculated, from each time point of the time-lapse baking profiles shown in Figure 5-44. The data are shown graphically in Figure 5-45, so that quantitative differences between the baking performance of a poor versus an excellent cookie-quality flour can be easily seen. A plot of cookie height versus baking time (Fig. 5-45A) confirms that

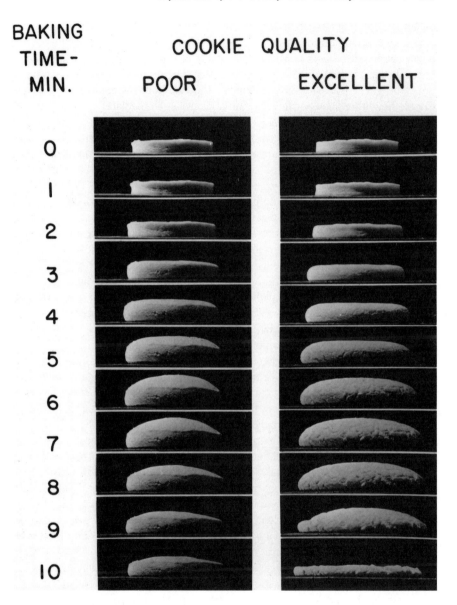

Figure 5-44. Time-lapse photographs during baking of sugar-snap cookies made with an excellent and a poor cookie quality flour. (*From Yamazaki and Lord, 1978, with permission.*)

initial baking behavior is similar for both poor and excellent flour doughs: no vertical expansion up to 2 minutes, then a significant increase in cookie height begins at minute 2 and continues to minute 6 (but with about 25% less net vertical expansion for the excellent flour than for the poor flour). After 6 minutes, the vertical expansion behavior for the two flours is distinctly different. Cookie height for the poor flour begins to decrease immediately, but at a lower rate than the preceding expansion, so that final cookie height is greater than the initial dough height. In contrast, the net height of the excellent flour cookie is maintained until about minute 8, when dramatic collapse predominates the expansion process, so that the final cookie height is smaller than the initial dough height. Cookie height varies continuously throughout the entire baking cycle for both flour types; at no point could the cookie be described as "set."

Figure 5-45*B* shows the conventional treatment of baking data, reported as the increase in cookie diameter versus baking time. As for vertical expansion, both flour types behave identically up to 2 minutes; unlike vertical expansion, lateral expansion begins immediately at time zero. The diameter of the cookie made from poor flour slowly increases to a maximum at about 5 minutes (when the height is still increasing rapidly). This maximum net diameter is maintained for about 3 minutes and then shrinkage predominates the expansion process, so that a slight decrease below the maximum cookie diameter is observed. Like the cookie height, the diameter of the cookie made from poor flour varies continuously (but over a smaller range) throughout the baking cycle, and no "set time" is observed. The diameter profile for the cookie made from excellent flour explains why baking data have been misinterpreted in the past. The diameter profile for the excellent flour cookie "appears" to consist of two linear phases: the initial phase of rapid lateral expansion up to minute 8, followed by a final phase of zero slope. The intersection of the two linear phases has been defined as the "set time" (Abboud et al., 1985; Doescher, Hoseney, and Milliken, 1987; Doescher, Hoseney, Milliken, and Rubenthaler, 1987). However, examination of the simultaneous vertical expansion which shows dramatic collapse, and direct observation of the photographs to see the consequent progressive surface cracking, clearly negate this interpretation. The apparent "set" of the excellent flour cookie diameter results from the opposition and final predominance of structural collapse (which is vertically anisotropic, due to gravity) over expansion. Because the cookie made from poor flour exhibits controlled shrinkage (i.e., elastic recovery, which is macroscopically reversible), rather than collapse (which is macroscopically irreversible, for reasons described in detail elsewhere [Levine and Slade, 1986, 1988c, 1988e]), no abrupt change in lateral expansion is observed.

Figure 5-45*C* demonstrates the necessity of examining cookie geometry throughout the entire baking cycle, rather than only for the final products. *Final* cookie width/height ratio, or diameter/thickness (D/T) ratio, called the "spread factor," has also been used traditionally as a diagnostic for cookie flour quality, but its use has been discouraged because of its irreproducibility (Finney et al., 1950). Lack of reproducibility for excellent flours is caused (quite predictably) by the uncontrolled collapse that occurs *after* the so-called "set time," compared to the controlled elastic shrinkage for poor flours. The plot of cookie width/height versus baking time in Figure 5-45C shows most readily that vertical expansion exceeds lateral expansion during the first half of the baking cycle, up to the point of limiting vertical expansion for both flour types. Overexpansion leads to elastic recovery for the poor flour, but to collapse for the excellent flour. The preferential vertical expansion is perpendicular to the direction of sheeting of the once-rolled dough and could reflect lower cohesiveness between small local lamina than within lamina, and/or fibrillation of partially developed gluten. Protein fibril formation upon flour wetting and protein hydration has been observed microscopically by Bernardin and Kasarda (1973). These investigators suggested that these viscoelastic protein (i.e., gluten) fibrils "are the structural elements which form the cohesive matrix of dough when worked mechanically." Such gluten development is undesirable in a cookie dough, as it leads to product toughness (Yamazaki and Lord, 1978). The microscopic observations of laminated puff pastry doughs recently reported by Telloke (1987) may also be relevant to the discussion of vertical expansion, in that they may point to parallels in the mechanism for poor-flour cookie doughs and excellent-flour puff pastry doughs that produce good pastry "lift." Puff pastry doughs require high-quality wheat varietal flours with strong glutens that exhibit optimum elastic properties. As in poor-flour cookie doughs that show controlled elastic expansion and then shrinkage during baking, excellent puff pastry doughs have sufficient "toughness" to prevent amalgamation of dough layers during lamination and loss of gas-holding capacity during baking, but at the expense of an acceptable degree of pastry shrinkage. Conversely, a weak-flour dough containing gluten of lower elasticity, which would show viscous expansion followed by collapse during baking, would constitute a poor puff pastry dough. The dramatic collapse seen in Figure 5-45A, and the very large value of D/T seen in Figure 5-45C, suggest the behavior of an excellent cookie flour with more leavening (60/30/3% sugar/shortening/bicarbonate, flour basis) than the standard diagnostic formulation (60/30/1.75%) (Finney et al., 1950).

The plot of cookie volume versus baking time in Figure 5-45D reaffirms several observations from previous plots. Cookie geometry is not set at the point where cookie diameter no longer increases. Controlled elastic expansion and shrinkage of the poor flour dough are emphasized and contrasted with the longer-lasting and greater expansion, followed by collapse, of the excellent flour dough. The graphical comparison of the quantitative time course for each baking parameter within each flour type, illustrated by the plots in Figure 5-45, reaffirms that no time point *during* the 10-minute baking cycle could be described aptly as a "set time." Results for the poor cookie emphasize that

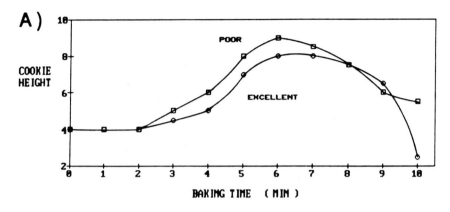

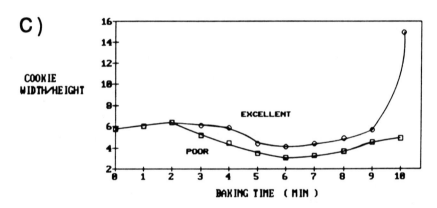

Figure 5-45. Data analysis of cookie dimensions measured from time-lapse photographs in Fig. 5-44: plots of (A) cookie height (mm) versus baking time (min); (B) cookie diameter (mm) versus baking time (min); (C) cookie width:height ratio versus baking time (min); and (D) cookie volume (mm³) versus baking time (min). (*From Slade et al., 1988, with permission.*)

the elastic volume expansion and contraction of the poor flour dough, although determined mostly by the vertical expansion behavior, reflect the uniformity of time courses for lateral and vertical dimensional changes. In contrast, results for the excellent cookie emphasize that the time courses of lateral and vertical dimensional changes are not uniform when a collapse process intervenes.

Doescher, Hoseney, and Milliken (1987) proposed a mechanism for cookie dough "setting." These authors suggested that when gluten in flour, hydrated by sugar solution in dough, "undergoes a glass transition, it expands to form a continuous matrix, and viscosity increases, causing cookie doughs to stop spreading." Abboud and colleagues (1985) had previously shown that both

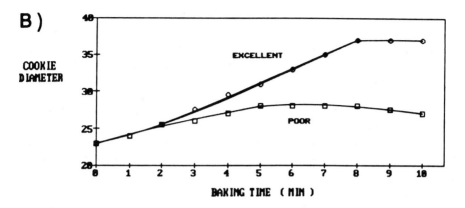

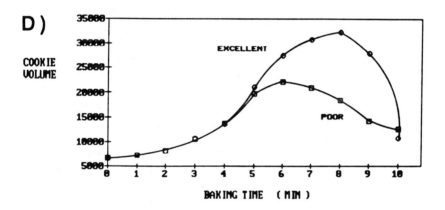

soft and hard flour cookie doughs undergo a decrease in viscosity as their temperature is raised, yet they proposed that a temperature-mediated change in viscosity controls the "set time" of cookie doughs, and believed that hard flour doughs set at a lower temperature than soft flour doughs, because hard flours have a "lower temperature change in viscosity," which stops cookie spread earlier in the baking cycle. Doescher, Hoseney, and Milliken (1987) reported that the Tg of water-plasticized gluten in hard wheat flour is lower than the corresponding Tg in soft wheat flour, thus accounting for the lower temperature change in viscosity and smaller cookie diameter of hard flour doughs. Paradoxically, the description by Doescher, Hoseney, and Milliken (1987) of a viscosity *increase* above Tg could be appropriate only for the performance of *poor* cookie-quality flour doughs, in which gluten exhibits the classic behavior of a rubbery thermosetting polymer. Actually, the analysis by the authors of this chapter demonstrated that gluten in an excellent cookie-quality flour dough behaves as an amorphous thermoplastic, rather than thermosetting, polymer system (Slade et al., 1988). Furthermore, the mechanism proposed by Doescher, Hoseney, and Milliken (1987) is opposed by their data for the effects of different sugar ingredients on the buildup of viscosity during baking (Doescher, Hoseney, and Milliken, 1987), compared to effects of the same sugars on cookie spread (Doescher, Hoseney, Milliken, and Rubenthaler, 1987). Conversely, their data strongly support the hypothesis (Slade et al., 1988) that gluten hydration (i.e., plasticization, which would be most constrained by sucrose–water, less by glucose–water, and least by fructose–water) allows dough expansion by two-dimensional extensible film, rather than three-dimensional elastic network, formation, followed by collapse, due to *decreasing* dough viscosity, as the dough temperature rises well above the molecular Tg of rubbery gluten.

The key to the functional difference between poor and excellent cookie-quality flours lies in the structure–activity relationship (SAR) of elastic composite film/network formation versus collapsible film formation (network absent or inelastic). In the former case, at baking temperatures above its effective molecular Tg, plasticized gluten appears to manifest predominantly the behavior of a rubbery thermoset, while in the latter, the contrasting behavior of a thermoplastic polymer. The authors of this chapter have speculated (Slade et al., 1988) that the underlying reason for this fundamental difference in behavior between glutens from excellent and poor flours may involve the relative potential for permanent intermolecular disulfide network formation during baking. Excellent cookie flours are understood to be limited to those derived from wheat varieties that manifest "softness" both genotypically and phenotypically. On the other hand, poor cookie flours derive from wheat varieties that are either genotypically and phenotypically "hard" or genotypically "soft" but phenotypically "hard." As a consequence, secondary sources of network formation by damaged starch and pentosans would exaggerate the contribution of gluten networks. The chemical, structural, and/or genetic bases for such distinctions in functional performance among glutens are unknown at present, but these authors believe that future study is warranted.

CONCLUSIONS

This chapter has emphasized a growing awareness, especially since 1980, of the value of a polymer science approach to the study of food materials, products, and processes among a small but increasing number of food scientists. In this respect, food science has followed the compelling lead of the synthetic polymers field. The emerging research discipline of food polymer science emphasizes the fundamental and generic similarities between synthetic polymers and food molecules, and provides a new theoretical and experimental framework for the study of food systems that are under kinetic control. On a theoretical basis of established structure–property relationships from the field of synthetic polymer science, this innovative discipline has developed to unify structural aspects of foods (conceptualized as kinetically metastable, completely amorphous or partially crystalline polymer systems) with functional aspects dependent upon mobility and conceptualized in terms of "water dynamics" and "glass dynamics," in order to explain and predict the functional properties of food materials during processing and product storage. Key elements of this theoretical approach to investigations of food systems whose behavior is governed by dynamics rather than by energetics include recognition of (1) the importance of the glass-to-rubber transition and the characteristic temperature Tg at which it occurs as physicochemical parameters that can control processing, product properties, quality, and stability in food polymer systems; (2) the central role of water as a ubiquitous plasticizer of natural and fabricated amorphous food ingredients and products; (3) the effect of water as a plasticizer on Tg and the resulting non-Arrhenius diffusion-limited behavior of amorphous polymeric, oligomeric, and monomeric food materials; and (4) the significance of nonequilibrium glassy and rubbery states (as opposed to equilibrium thermodynamic phases) in all "real world" food products and processes, and their effects on time-dependent properties related to quality and storage stability. This chapter has described how the recognition of these key elements of the food polymer science approach and their relevance to the behavior of doughs and baked products has increased markedly during this decade. This chapter has illustrated the perspective afforded by using this conceptual framework and demonstrated the technological utility of this new approach to understand and explain complex behavior, design processes, and predict product quality and storage stability, based on fundamental structure–property relationships. In future years, these authors expect to see much progress reported from this emerging, cross-disciplinary research area, especially with regard to the determination (both in the laboratory and in the plant), interpretation, and understanding of the rheological properties of doughs and the textural properties of baked products.

Bloksma and Bushuk (1988), in their recent review of dough rheology, concluded that "the expansion of our knowledge [in this field] is more quantitative than qualitative; it is more a matter of facts and precision than of understanding." The authors of this chapter suggest that recognition of the influences of the glassy and rubbery states on the thermal, mechanical, and structural properties of doughs and baked products offers a major opportunity to expand our qualitative understanding.

ACKNOWLEDGEMENT

We thank Nabisco Brands for approval to publish this chapter. We thank Susan R. Saunders for her contributions to the study of the baking mechanism of sugar-snap cookies, which will be the subject of future publications.

REFERENCES

Abboud, A. M., Hoseney, R. C., and Rubenthaler, G. L. 1985. Factors affecting cookie flour quality. *Cereal Chem.* **62**:130–133.

Ablett, S., Attenburrow, G. E., and Lillford, P. J. 1986. Significance of water in the baking process. In *Chemistry and Physics of Baking*, ed. J. M. V. Blanshard, P. J. Frazier, and T. Galliard, pp. 30–41, London: Royal Society of Chemistry.

Ablett, S., Barnes, D. J., Davies, A. P., Ingman, S. J., and Patient, D. W. 1988. C_{13} and pulse NMR spectroscopy of wheat proteins. *J. Cereal Sci.* **7**:11–20.

Alfonso, G. C., and Russell, T. P. 1986. Kinetics of crystallization in semicrystalline/amorphous polymer mixtures. *Macromolecules* **19**:1143–1152.

Apicella, A., and Hopfenberg, H. B. 1982. Water-swelling behavior of EVA copolymer. *J. Appl. Polym. Sci.* **27**:1139–1148.

Asquith, R. S., and Leon, N. H. 1977. Chemical reactions of keratin fibers. In *Chemistry of Natural Protein Fibers*, ed. R. S. Asquith, pp. 193–265, New York: Plenum Press.

Atkins, A. G. 1987. Basic principles of mechanical failure in biological systems. In *Food Structure and Behaviour*, ed. J. M. V. Blanshard and P. Lillford, pp. 149–176, London: Academic Press.

Atwell, W. A., Hood, L. F., Lineback, D. R., Varriano-Marston, E., and Zobel, H. F. 1988. Terminology and methodology associated with basic starch phenomena. *Cereal Foods World* **33**:306–311.

Bair, H. E. 1985. Curing behavior of epoxy resin above and below Tg. *Polym. Prep.* **26**:10.

Baro, M. D., Clavaguera, N., Bordas, S., Clavaguera-Mora, M. T., and Casa-Vazquez, J. 1977. Evaluation of crystallization kinetics by DTA. *J. Thermal Anal.* **11**:271–276.

Batres, L. V., and White, P. J. 1986. Interaction of amylopectin with monoglycerides in model systems. *J. Amer. Oil Chem. Soc.* **63**:1537–1540.

Batzer, H., and Kreibich, U. T. 1981. Influence of water on thermal transitions in natural polymers and synthetic polyamides. *Polym. Bull.* **5**:585–590.

Benjamin, E. J., Ke, C. I., Cochran, S. A., Hynson, R., Veach, S. K., and Simon, S. J. I. 1985. Frozen Bread Dough Product. European patent 145,367.

Berens, A., and Hopfenberg, H. B. 1980. Induction and measurement of glassy state relaxations by vapor sorption methods. *Stud. Phys. Theoret. Chem.* **10**:77–94.

Bernardin, J. E. 1975. Rheology of concentrated gliadin solutions. *Cereal Chem.* **52**:136r–145r.

Bernardin, J. E., and Kasarda, D. D. 1973. Hydrated protein fibrils from wheat endosperm. *Cereal Chem.* **50**:529–536.

Biliaderis, C. G. 1983. DSC in food research—review. *Food Chem.* **10**:239–265.

Biliaderis, C. G., and Galloway, G. 1989. Crystallization habit of amylose-V complexes: structure–property relationships. *Carbohydr. Polym.* **10**: in press.

Biliaderis, C. G., Maurice, T. J., and Vose, J. R. 1980. Starch gelatinization phenomena studied by DSC. *J. Food Sci.* **45**:1669–1680.

Biliaderis, C. G., Page, C. M., and Maurice, T. J. 1986a. Non-equilibrium melting of amylose-V complexes. *Carbohydr. Polym.* **6**:269–288.

Biliaderis, C. G., Page, C. M., and Maurice, T. J. 1986b. Multiple melting transitions of starch/monoglyceride systems. *Food Chem.* **22**:279–295.

Biliaderis, C. G., Page, C. M., Slade, L., and Sirett, R. R. 1985. Thermal behavior of amylose–lipid complexes. *Carbohydr. Polym.* **5**:367–389.

Biliaderis, C. G., Page, C. M., Maurice, T. J., and Juliano, B. O. 1986. Thermal characterization of rice starches: A polymeric approach to phase transitions of granular starch. *J. Agric. Food Chem.* **34**:6–14.

Billmeyer, F. W. 1984. *Textbook of Polymer Science*, 3rd ed., New York: Wiley–Interscience.

Biros, J., Madan, R. L., and Pouchly, J. 1979. Heat capacity of water-swollen polymers above and below 0°C. *Collection Czech. Chem. Commun.* **44**:3566–3573.

Bizot, H., Buleon, A., Mouhoud-Riou, N., and Multon, J. L. 1985. Water vapor sorption hysteresis on potato starch. In *Properties of Water in Foods*, ed. D. Simatos and J. L. Multon, pp. 83–93, Dordrecht, Holland: Martinus Nijhoff.

Blanshard, J. M. V. 1979. Physicochemical aspects of starch gelatinization. In *Polysaccharides in Food*, ed. J. M. V. Blanshard and J. R. Mitchell, pp. 139–152, London: Butterworths.

Blanshard, J. M. V. 1986. Significance of the structure and function of the starch granule in baked products. In *Chemistry and Physics of Baking*, ed. J. M. V. Blanshard, P. J. Frazier, and T. Galliard, pp. 1–13, London: Royal Society of Chemistry.

Blanshard, J. M. V. 1987. Starch granule structure and function: physicochemical approach. In *Starch: Properties and Potential*, ed. T. Galliard, pp. 16–54, Chichester, U. K.: Wiley.

Blanshard, J. M. V. 1988. Elements of cereal product structure. In *Food Structure—Its Creation and Evaluation*, ed. J. R. Mitchell and J. M. V. Blanshard, pp. 313–330, London: Butterworths.

Blanshard, J. M. V., and Franks, F. 1987. Ice crystallization and its control in frozen food systems. In *Food Structure and Behaviour*, ed. J. M. V. Blanshard and P. Lillford, pp. 51–65, London: Academic Press.

Bloksma, A. H. 1978. Rheology and chemistry of dough. In *Wheat Chemistry and Technology*, 2nd ed., ed. Y. Pomeranz, pp. 523–584, St. Paul, MN: American Association of Cereal Chemists.

Bloksma, A. H. 1986. Rheological aspects of structural changes during baking. In *Chemistry and Physics of Baking*, ed. J. M. V. Blanshard, P. J. Frazier, and T. Galliard, pp. 170–178, London: Royal Society of Chemistry.

Bloksma, A. H., and Bushuk, W. 1988. Rheology and chemistry of dough. In *Wheat Science and Technology*, 3rd ed., Vol. 2, Chap. 4, pp. 131–217, ed. Y. Pomeranz, St. Paul, MN: American Association of Cereal Chemists.

Blum, F. D., and Nagara, B. 1986. Solvent mobility in gels of atactic polystyrene. *Polym. Prep.* **27**(1):211–212.

Bone, S., and Pethig, R. 1982. Dielectric studies of the binding of water to lysozyme. *J. Mol. Biol.* **157**:571–575.

Bone, S., and Pethig, R. 1985. Dielectric studies of protein hydration and hydration-induced flexibility. *J. Mol. Biol.* **181**:323–326.

Bonner, D. C., and Prausnitz, J. M. 1974. Thermodynamic properties of some concentrated polymer solutions. *J. Polym. Sci. Polym. Phys. Ed.* **12**:51–73.

Borchard, W., Bremer, W., and Keese, A. 1980. State diagram of the water–gelatin system. *Colloid Polym. Sci.* **258**:516–526.

Boyer, R. F., Baer, E., and Hiltner, A. 1985. Concerning gelation effects in atactic polystyrene solutions. *Macromolecules* **18**:427–434.

Bradbury, J. H. 1973. Structure and chemistry of keratin fibers. In *Advances in Protein Chemistry*, ed. C. B. Anfinsen, J. T. Edsall, and F. M. Richards, Vol. 27, pp. 111–211, New York: Academic Press.

Braudo, E. E., Belavtseva, E. M., Titova, E. F., Plashchina, I. G., Krylov, V. L., Tolstoguzov, V. B., Schierbaum, F. R., and Richter, M. 1979. Struktur und Eigenschaften von Maltodextrin-Hydrogelen. *Staerke* **31**:188–194.

Braudo, E. E., Plashchina, I. G., and Tolstoguzov, V. B. 1984. Structural characterization of thermoreversible anionic polysaccharide gels by their elastoviscous properties. *Carbohydr. Polym.* **4**:23–48.

Bryan, W. P. 1987a. Thermodynamics of water–biopolymer interactions: irreversible sorption by two uniform sorbent phases. *Biopolymers* **26**:387–401.

Bryan, W. P. 1987b. Thermodynamic models for water–protein sorption hysteresis. *Biopolymers*, in press.

Brydson, J. A. 1972. The glass transition, melting point, and structure. In *Polymer Science*, ed. A. D. Jenkins, pp. 194–249, Amsterdam, The Netherlands: North Holland.

Buchanan, D. R., and Walters, J. P. 1977. Glass transition temperatures of polyamide textile fibers. *Text. Res. J.* **47**:398–406.

Buleon, A., Duprat, F., Booy, F. P., and Chanzy, H. 1984. Single crystals of amylose with a low degree of polymerization. *Carbohydr. Polym.* **4**:161–173.

Buleon, A., Bizot, H., Delage, M. M., and Pontoire, B. 1987. Comparison of X-ray diffraction and sorption properties of hydrolyzed starches. *Carbohydr. Polym.* **7**:461–482.

Bulpin, P. V., Cutler, A. N., and Dea, I. C. M. 1984. Thermally-reversible gels from low-DE maltodextrins. In *Gums and Stabilizers for the Food Industry*, Vol. 2, ed. G. O. Phillips, D. J. Wedlock, and P. A. Williams, pp. 475–484, Oxford, U.K.: Pergamon.

Burghoff, H. G., and Pusch, W. 1980. Thermodynamic state of water in cellulose acetate membranes. *Polym. Eng. Sci.* **20**:305–309.

Burros, B. C., Young, L. A., and Carroad, P. A. 1987. Kinetics of corn meal gelatinization at high temperature and low moisture. *J. Food Sci.* **52**:1372–1380.

Bushuk, W. 1984. *Flour Proteins: Structure and Functionality in Dough and Bread*, abstract 115, American Association of Cereal Chemists 69th Annual Meeting, Minneapolis, MN.

Cakebread, S. H. 1969. Factors affecting the shelf life of high boilings. *Manufact. Confect.* **49**:41–44.

Cantor, C. R., and Schimmel, P. R. 1980. *Biophysical Chemistry: Part I—The Conformation of Biological Macromolecules*, San Francisco, CA: W. H. Freeman.

Chan, R. K., Pathmanathan, K., and Johari, G. P. 1986. Dielectric relaxations in the liquid and glassy states of glucose and its water mixtures. *J. Phys. Chem.* **90**:6358–6362.

Chungcharoen, A., and Lund, D. B. 1987. Influence of solutes and water on rice starch gelatinization. *Cereal Chem.* **64**:240–243.

Cole, B. A., Levine, H., McGuire, M. T., Nelson, K. J., and Slade, L. 1983. Soft, Frozen Dessert Formulation. U.S. patent 4,374,154.

Cole, B. A., Levine, H., McGuire, M. T., Nelson, K. J., and Slade, L. 1984. Soft, Frozen Dessert Formulation. U.S. patent 4,452,824.

Cowie, J. M. G. 1973. *Polymers: Chemistry and Physics of Modern Materials*, New York: Intertext.

D'Arcy, R. L., and Watt, I. C. 1981. Water vapor sorption isotherms on macromolecular substrates. In *Water Activity: Influences on Food Quality*, ed. L. B. Rockland and G. F. Stewart, pp. 111–142, New York: Academic Press.

Davies, A. P. 1986. Protein functionality in bakery products. In *Chemistry and Physics of Baking*, ed. J. M. V. Blanshard, P. J. Frazier, and T. Galliard, pp. 89–104, London: Royal Society of Chemistry.

Derbyshire, W. 1982. Dynamics of water in heterogeneous systems with emphasis on subzero temperatures. In *Water: A Comprehensive Treatise*, Vol. 7, ed. F. Franks, pp. 339–430, New York: Plenum Press.

Dickerson, R. E., and Geis, I. 1969. *The Structure and Action of Proteins*, New York: W. A. Benjamin.

Doescher, L. C., Hoseney, R. C., and Milliken, G. A. 1987. Mechanism for cookie dough setting. *Cereal Chem.* 64:158–163.

Doescher, L. C., Hoseney, R. C., Milliken, G. A., and Rubenthaler, G. L. 1987. Effect of sugars and flours on cookie spread evaluated by time-lapse photography. *Cereal Chem.* 64:163–167.

Domszy, R. C., Alamo, R., Edwards, C. O., and Mandelkern, L. 1986. Thermoreversible gelation and crystallization of homopolymers and copolymers. *Macromolecules* 19:310–325.

Donelson, J. R. 1988. The contribution of high-protein fractions from cake and cookie flours to baking performance. *Cereal Chem.* 65:389–391.

Donovan, J. W. 1979. Phase transitions of the starch–water system. *Biopolymers* 18:263–275.

Donovan, J. W. 1985. DSC in food research. In *Proceedings 14th NATAS Conference*, pp. 328–333, San Francisco: NATAS.

Donovan, J. W., Lorenz, K., and Kulp, K. 1983. DSC of heat–moisture treated wheat and potato starches. *Cereal Chem.* 60:381–387.

Downton, G. E., Flores-Luna, J. L., and King, C. J. 1982. Mechanism of stickiness in hygroscopic, amorphous powders. *Ind. Eng. Chem. Fundam.* 21:447–451.

Dreese, P. C., Faubion, J. M., and Hoseney, R. C. 1988a. Effect of different heating and washing procedures on dynamic rheological properties of wheat gluten. *Cereal Foods World* 33:225–228.

Dreese, P. C., Faubion, J. M., and Hoseney, R. C. 1988b. Dynamic rheological properties of flour, gluten, and gluten–starch doughs. I. *Cereal Chem.* 65:348–353.

duCros, D. L. 1987. Glutenin proteins and gluten strength in durum wheat. *J. Cereal Sci.* 5:3–12.

Durning, C. J., and Tabor, M. 1986. Mutual diffusion in concentrated polymer solutions under a small driving force. *Macromolecules* 19:2220–2232.

Edwards, S. F., Lillford, P. J., and Blanshard, J. M. V. 1987. Gels and networks in practice and theory. In *Food Structure and Behaviour*, ed. J. M. V. Blanshard and P. Lillford, pp. 1–12, London: Academic Press.

Eisenberg, A. 1984. The glassy state and the glass transition. In *Physical Properties of Polymers*, ed. J. E. Mark, A. Eisenberg, W. W. Graessley, L. Mandelkern, and J. L. Koenig, pp. 55–95, Washington, D.C.: American Chemical Society.

Eliasson, A. C. 1985. Retrogradation of starch as measured by DSC. In *New Approaches to Research on Cereal Carbohydrates*, ed. R. D. Hill and L. Munck, pp. 93–98, Amsterdam, The Netherlands: Elsevier.

Eliasson, A. C. 1986. On the effects of surface active agents on the gelatinization of starch—A calorimetric investigation. *Carbohydr. Polym.* **6**:463–476.

Ellis, H. S., and Ring, S. G. 1985. Study of some factors influencing amylose gelation. *Carbohydr. Polym.* **5**:201–213.

Ellis, T. S. 1988. Moisture-induced plasticization of amorphous polyamides and their blends. *J. Appl. Polym. Sci.* **36**:451–466.

Ellis, T. S., Jin, X., and Karasz, F. E. 1984. Water-induced plasticization behavior of semi-crystalline polyamides. *Polym. Prep.* **25**(2):197–198.

Escoubes, M., and Pineri, M. 1980. Comparison of weight and energy changes in the absorption of water by collagen and keratin. In *Water in Polymers*, ed. S. P. Rowland, ACS Symposium Series 127, pp. 235–252, Washington, D.C.: American Chemical Society.

Evans, I. D. 1986. Investigation of starch/surfactant interactions using viscosimetry and DSC. *Staerke* **38**:227–235.

Ewart, J. A. D. 1987. Calculated molecular weight distribution for glutenin. *J. Sci. Food Agric.* **38**:277–289.

Ewart, J. A. D. 1988. Thiols in flour and breadmaking quality. *Food Chem.* **28**:207–218.

Fearn, T., and Russell, P. L. 1982. Kinetic study of bread staling by DSC—effect of loaf specific volume. *J. Sci. Food Agric.* **33**:537–548.

Ferry, J. D. 1948. Mechanical properties of substances of high molecular weight. *J. Amer. Chem. Soc.* **70**:2244–2249.

Ferry, J. D. 1980. *Viscoelastic Properties of Polymers*, 3rd ed., New York: Wiley.

Finch, J. T. 1975. Electron microscopy of proteins. In *The Proteins*, 3rd ed., Vol. 1, ed. H. Neurath, R. H. Hill, and C. L. Boeder, pp. 413–497, New York: Academic Press.

Finney, J. L., and Poole, P. L. 1984. Protein hydration and enzyme activity: role of hydration-induced conformation and dynamic changes in activity of lysozyme. *Comments Mol. Cell. Biophys.* **2**:129–151.

Finney, K. F., Morris, V. H., and Yamazaki, W. T. 1950. Micro versus macro cookie baking procedures for evaluating cookie quality of wheat varieties. *Cereal Chem.* **27**:30–49.

Flink, J. M. 1983. Structure and structure transitions in dried carbohydrate materials. In *Physical Properties of Foods*, ed. M. Peleg and E. B. Bagley, pp. 473–521, Westport, Conn.: AVI.

Flory, P. J. 1953. *Principles of Polymer Chemistry*, Ithaca, NY: Cornell University Press.

Flory, P. J. 1974. Introductory lecture—Gels and gelling processes. *Faraday Disc. Chem. Soc.* **57**:7–18.

Flory, P. J., and Weaver, E. S. 1960. Phase transitions in collagen and gelatin systems. *J. Amer. Chem. Soc.* **82**:4518.

Forsyth, M., and MacFarlane, D. R. 1986. Recrystallization revisited. *Cryo-Letters* **7**:367–378.

Franks, F. 1982. The properties of aqueous solutions at subzero temperatures. In *Water: A Comprehensive Treatise*, ed. F. Franks, Vol. 7, pp. 215–338, New York: Plenum Press.

Franks, F. 1983a. Solute–water interactions: Do polyhydroxy compounds alter the properties of water? *Cryobiology* **20**:335.

Franks, F. 1983b. Bound water: Fact and fiction. *Cryo-Letters* **4**:73–74.

Franks, F. 1985a. *Biophysics and Biochemistry at Low Temperatures*, Cambridge: Cambridge University Press.

Franks, F. 1985b. Complex aqueous systems at subzero temperatures. In *Properties of Water in Foods*, ed. D. Simatos and J. L. Multon, pp. 497–509, Dordrecht, Holland: Martinus Nijhoff.

Franks, F. 1985c. Water Activity and Biochemistry—Specific Ionic and Molecular Effects. Presented at Discussion Conference on "Water Activity: A Credible Measure of Technological Performance and Physiological Stability?", Faraday Division, Royal Society of Chemistry, July 1–3, Cambridge, UK.

Franks, F. 1986. Unfrozen water: Yes; Unfreezable water: Hardly; Bound water: Certainly not. *Cryo-Letters* **7**:207.

Franks, F. 1987. Nucleation: A maligned and misunderstood concept. *Cryo-Letters* **8**:53–55.

Franks, F., Reid, D. S., and Suggett, A. 1973. Conformation and hydration of sugars and related compounds in dilute aqueous solution. *J. Solution Chem.* **2**:99–118.

Franks, F., Asquith, M. H., Hammond, C. C., Skaer, H. B., and Echlin, P. 1977. Polymeric cryoprotectants in the preservation of biological ultrastructure, I. *J. Microsc.* **110**:223–238.

Fraser, R. D. B., MacRae, T. P., and Rogers, G. E. 1972. *Keratins—Their Composition, Structure and Biosynthesis*, Springfield, IL: Charles C Thomas.

Frazier, P. J., and Muller, H. G. 1967. Birefringence of stretched gluten. *Cereal Chem.* **44**:558–559.

French, D. 1984. Organization of starch granules. In *Starch: Chemistry and Technology*, 2nd ed., ed. R. L. Whistler, J. N. BeMiller, and E. F. Paschall, pp. 183–247, Orlando, FL: Academic Press.

Friedman, M. 1973. *Amino Acids, Peptides and Proteins*, New York: Pergamon Press.

Fuzek, J. F. 1980. Glass transition temperature of wet fibers: Its measurement and significance. In *Water in Polymers*, ed. S. P. Rowland, ACS Symposium Series 127, pp. 515–530, Washington, D.C.: American Chemical Society.

Gaeta, S., Apicella, A., and Hopfenberg, H. B. 1982. Kinetics and equilibria associated with the absorption and desorption of water and LiCl in an ethylene–vinyl alcohol copolymer. *J. Membr. Sci.* **12**:195–205.

Gaines, C. S., Donelson, J. R., and Finney, P. L. 1988. Effects of damaged starch, chlorine gas, flour particle size, and dough holding time and temperature on cookie dough handling properties and cookie size. *Cereal Chem.* **65**:384–389.

Ghiasi, K., Hoseney, R. C., and Varriano-Marston, E. 1983. Effects of flour components and dough ingredients on starch gelatinization. *Cereal Chem.* **60**:58–61.

Ghiasi, K., Hoseney, R. C., Zeleznak, K., and Rogers, D. E. 1984. Effect of waxy barley starch and reheating on firmness of bread crumb. *Cereal Chem.* **61**:281–285.

Gidley, M. J. 1987. Factors affecting crystalline type of native starches and model materials. *Carbohydr. Res.* **161**:301–304.

Gidley, M. J., and Bulpin, P. V. 1987. Crystallization of maltaoses as models of the crystalline forms of starch. *Carbohydr. Res.* **161**:291–300.

Gidley, M. J., Bulpin, P. V., and Kay, S. 1986. Effect of chain length on amylose retrogradation. In *Gums and Stabilizers for the Food Industry* 3, ed. G. O. Phillips, D. J. Wedlock, and P. A. Williams, pp. 167–176, London: Elsevier.

Graessley, W. W. 1984. Viscoelasticity and flow in polymer melts and concentrated solutions. In *Physical Properties of Polymers*, ed. J. E. Mark, A. Eisenberg, W. W. Graessley, L. Mandelkern, and J. L. Koenig, pp. 97–153, Washington, D.C.: American Chemical Society.

Guegov, Y. 1981. Phase transitions of water in products of plant origin at low temperatures. *Adv. Food Res.* **27**:297–360.

Guilbot, A., and Godon, B. 1984. Le Pain Rassis. *Cah. Nut. Diet.* **19**:171–181.

Hill, A. S., Skerritt, J. H., and Watson, A. R. 1988. *Simple and Rapid Test Method for Gluten in Starches, Bakery Raw Materials and Processed Foods*, abstract 66, American Association of Cereal Chemists 73nd Annual Meeting, San Diego, CA.

Hiltner, A., and Baer, E. 1986. Reversible gelation of macromolecular systems. *Polym. Prep.* **27**(1):207.

Hizukuri, S. 1986. Polymodal distribution of chain lengths of amylopectins and its significance. *Carbohydr. Res.* **147**:342–347.

Hoeve, C. A. J. 1980. The structure of water in polymers. In *Water in Polymers*, ed. S. P. Rowland, ACS Symposium Series 127, pp. 135–146, Washington, D.C.: American Chemical Society.

Hoeve, C. A. J., and Hoeve, M. B. J. A. 1978. The glass point of elastin as a function of diluent concentration. *Organ. Coat. Plast. Chem.* **39**:441–443.

Hopfenberg, H. B., Apicella, A., and Saleeby, D. E. 1981. Factors affecting water sorption in and solute release from EVA copolymers. *J. Membr. Sci.* **8**:273–282.

Hoseney, R. C. 1984. DSC of starch. *J. Food Qual.* **6**:169–182.

Hoseney, R. C. 1986a. Component interaction during heating and storage of baked products. In *Chemistry and Physics of Baking*, ed. J. M. V. Blanshard, P. J. Frazier, and T. Galliard, pp. 216–226, London: Royal Society of Chemistry.

Hoseney, R. C. 1986b. *Principles of Cereal Science and Technology*, p. 238, St. Paul, MN: American Association of Cereal Chemists.

Hoseney, R. C., Zeleznak, K., and Lai, C. S. 1986. Wheat gluten: A glassy polymer. *Cereal Chem.* **63**:285–286.

Huang, W. J., Frick, T. S., Landry, M. R., Lee, J. A., Lodge, T. P., and Tirrell, M. 1987. Tracer diffusion measurement in polymer solutions near Tg by forced Rayleigh scattering. *AIChE J.* **33**:573–582.

Imberty, A., and Perez, S. 1988. A revisit to the three-dimensional structure of B-type starch. *Biopolymers* **27**:1205–1221.

Jane, J. L., and Robyt, J. 1984. Structure studies of amylose-V complexes and retrograded amylose. *Carbohydr. Res.* **132**:105–118.

Jankowski, T., and Rha, C. K. 1986a. Retrogradation of starch in cooked wheat. *Staerke* **38**:6–9.

Jankowski, T., and Rha, C. K. 1986b. DSC study of wheat grain cooking process. *Staerke* **38**:45–48.

Jin, X., Ellis, T. S., and Karasz, F. E. 1984. The effect of crystallinity and crosslinking on the depression of the glass transition temperature in nylon 6 by water. *J. Polym. Sci. Polym. Phys. Ed.* **22**:1701–1717.

Jolley, J. E. 1970. The microstructure of photographic gelatin binders. *Photogr. Sci. Eng.* **14**:169–177.

Juliano, B. O. 1982. Properties of rice starch in relation to varietal differences in processing characteristics of rice grain. *J. Jap. Soc. Starch Sci.* **29**:305–317.

Kainuma, K., and French, D. 1972. Naegeli amylodextrin and its relationship to starch granule structure. II. Role of water in crystallization of B-starch. *Biopolymers* **11**:2241–2250.

Kakivaya, S. R., and Hoeve, C. A. J. 1975. The glass transition of elastin. *Proc. Nat. Acad. Sci. (USA)* **72**:3505–3507.

Karel, M. 1985. Effects of water activity and water content on mobility of food components, and their effects on phase transitions in food systems. In *Properties of Water in Foods*, ed. D. Simatos and J. L. Multon, pp. 153–169, Dordrecht, Holland: Martinus Nijhoff.

Karel, M. 1986. Control of lipid oxidation in dried foods. In *Concentration and Drying of Foods*, ed. D. MacCarthy, pp. 37–51, London: Elsevier.

Karel, M., and Flink, J. M. 1983. Some recent developments in food dehydration research. In *Advances in Drying*, Vol. 2, ed. A. S. Mujumdar, pp. 103–153, Washington, D.C.: Hemisphere.

Karel, M., and Langer, R. 1988. Controlled release of food additives. In *Flavor Encapsulation*, ed. S. J. Risch and G. A. Reineccius, ACS Symposium Series 370, pp. 177–191, Washington, D.C.: American Chemical Society.

Kasarda, D. D., Bernardin, J. E., and Gaffield, W. 1968. Gliadin and glutenin fractions of wheat gluten. *Biochemistry* **7**:3950–3957.

Kasarda, D. D., Nimmo, C. C., and Kohler, G. O. 1978. Proteins and amino acid composition of wheat fractions. In *Wheat Chemistry and Technology*, 2nd ed., ed. Y. Pomeranz, pp. 227–299, St. Paul, MN: American Association of Cereal Chemists.

Keinath, S. E., and Boyer, R. F. 1981. Thermomechanical analysis of Tg and T > Tg transitions in polystyrene. *J. Appl. Polym. Sci.* **26**:2077–2085.

Kelley, S. S., Rials, T. G., and Glasser, W. G. 1987. Relaxation behavior of the amorphous components of wood. *J. Materials Sci.* **22**:617–624.

Khan, K., and Bushuk, W. 1979. Structure of wheat gluten in relation to functionality in breadmaking. In *Functionality and Protein Structure*, ed. A. Pour-El, ACS Symposium Series 92, pp. 191–206, Washington, D.C.: American Chemical Society.

Kollman, P. 1978. Affinity of metal cations for peptide amides. *Chem. Phys. Lett.* **55**:555–559.

Kowblansky, M. 1985. Calorimetric investigation of inclusion complexes of amylose with long-chain aliphatic compounds. *Macromolecules* **18**:1776–1779.

Krimm, S., and Bandekar, J. 1986. Vibrational spectroscopy and conformation of peptides, polypeptides, and proteins. In *Advances in Protein Chemistry*, Vol. 38, pp. 181–364, ed. C. B. Anfinsen, J. T. Edsall, and F. M. Richards, New York: Academic Press.

Krueger, B. R., Knutson, C. A., Inglett, G. E., and Walker, C. E. 1987. DSC study on effect of annealing on gelatinization behavior of corn starch. *J. Food Sci.* **52**:715–718.

Krusi, H., and Neukom, H. 1984. Untersuchungen uber die Retrogradation der Starke in Konzentrierten Weizenstarkegelen. *Starke* **36**:300–305.

Kuge, T., and Kitamura, S. 1985. Annealing of starch granules—Warm water treatment and heat–moisture treatment. *J. Jap. Soc. Starch Sci.* **32**:65–83.

Kulp, K., and Ponte, J. G. 1981. Staling of white pan bread: Fundamental causes. *CRC Crit. Rev. Food Sci. Nutr.* **15**:1–48.

Kulp, K., and Olewnik, M. 1984. *Starch Functionality in Cookie Systems*, abstract 83, American Association of Cereal Chemists 69th Annual Meeting, Minneapolis, MN.

Kuprianoff, J. 1958. Fundamental aspects of the dehydration of foodstuffs. In *Conference on Fundamental Aspects of the Dehydration of Foodstuffs*, pp. 14–23. Society of the Chemical Industry, Aberdeen, Scotland, March 25–27.

Kwak, Y. T., and Winter, W. T. 1988. Low temperature exotherm in starch/water systems: A retrogradation marker. *J. Appl. Polym. Sci.* **35**:2091–2098.

Labuza, T. P. 1985. Water binding of humectants. In *Properties of Water in Foods*, ed. D. Simatos and J. L. Multon, pp. 421–445, Dordrecht, Holland: Martinus Nijhoff.

l'Anson, K. J., Miles, M. J., Morris, V. J., Ring, S. G., and Nave, C. 1988. Study of amylose gelation using synchrotron X-ray source. *Carbohydr. Polym.* **8**:45–53.

Larson, R. W., Lou, W. C., DeVito, V. C., and Neidinger, K. A. 1983. Frozen Bread Dough Product and Process. U.S. patent 4,406,911.

Larson, R. W., Lou, W. C., DeVito, V. C., and Neidinger, K. A. 1984. Frozen Bread Dough Product and Process. U.S. patent 4,450,177.

Lasztity, R. 1986. Recent results in the investigation of the structure of the gluten complex. *Die Nahrung* **30**:235–244.

Lelievre, J. 1976. Theory of gelatinization in a starch–water–solute system. *Polymer* **17**:854–858.

Lenchin, J. M., Trubiano, P. C., and Hoffman, S. 1985. Converted Starches for Use as a Fat- or Oil-Replacement in Foodstuffs. U.S. patent 4,510,166.

Leubner, I. H. 1987. Crystal nucleation under diffusion-controlled conditions. *J. Phys. Chem.* **91**:6069–6073.

Levine, H., and Slade, L. 1986. A polymer physicochemical approach to the study of commercial starch hydrolysis products. *Carbohydr. Polym.* **6**:213–244.

Levine, H., and Slade, L. 1987. Water as a plasticizer: Physicochemical aspects of low-moisture polymeric systems. In *Water Science Reviews*, Vol. 3, ed. F. Franks, pp. 79–185, Cambridge, U.K.: Cambridge University Press.

Levine, H., and Slade, L. 1988a. Interpreting the behavior of low-moisture foods. In *Water and Food Quality*, ed. T. M. Hardman, pp. 71–134, London: Elsevier.

Levine, H., and Slade, L. 1988b. Thermomechanical properties of small carbohydrate-water glasses and "rubbers": Kinetically-metastable systems at subzero temperatures. *J. Chem. Soc. Faraday Trans.* I**84**:2619–2633.

Levine, H., and Slade, L. 1988c. Principles of cryostabilization technology from structure–property relationships of water-soluble food carbohydrates—a review. *Cryo-Letters* **9**:21–63.

Levine, H., and Slade, L. 1988d. A food polymer science approach to the practice of cryostabilization technology. *Comments Agric. Food Chem.* **1**:315–396.

Levine, H., and Slade, L. 1988e. Cryostabilization technology: Thermoanalytical evaluation of food ingredients and systems. In *Thermal Analysis of Foods*, ed. C. Y. Ma and V. R. Harwalkar, London: Elsevier, in press.

Levine, H., and Slade, L. 1988f. Collapse phenomena—A unifying concept for interpreting the behavior of low-moisture foods. In *Food Structure—Its Creation and Evaluation*, ed. J. R. Mitchell and J. M. V. Blanshard, pp. 149–180, London: Butterworths.

Lillford, P. J. 1988. The polymer/water relationship—Its importance for food structure. In *Food Structure—Its Creation and Evaluation*, ed. J. R. Mitchell and J. M. V. Blanshard, pp. 75–92, London: Butterworths.

Lindstrom, T., and Slade, L. 1983. A Frozen Dough for Bakery Products. U.S. patent 4,374,151.

Lioutas, T. S., Baianu, I. C., and Steinberg, M. P. 1986. O-17 and deuterium NMR studies of lysozyme hydration. *Arch. Biochem. Biophys.* **247**:68–75.

Lookhart, G. L., Menkovska, M., and Pomeranz, Y. 1987. Changes in the Gliadin Fraction(s) During Breadmaking: Isolation and Characterization by HPLC and PAGE, abstract 24, American Association of Cereal Chemists, 72nd Annual Meeting, Nashville, TN.

Longton, J., and LeGrys, G. A. 1981. DSC studies on crystallinity of aging wheat starch gels. *Staerke* **33**:410–414.

Lund, D. B. 1983. Applications of DSC in foods. In *Physical Properties of Foods*, ed. M. Peleg and E. B. Bagley, pp. 125–143, Westport, Conn.: AVI.

Lund, D. B. 1984. Influence of time, temperature, moisture, ingredients, and processing conditions on starch gelatinization. *CRC Crit. Revs. Food Sci. Nutr.* **20**:249–273.

Luyet, B. 1939. Ice recrystallization in frozen aqueous solutions of sugars and polyols. *J. Phys. Chem.* **43**:881–885.

Luyet, B. 1960. On various phase transitions occurring in aqueous solutions at low temperatures. *Ann. NY Acad. Sci.* **85**:549–569.

Luyet, B., and Rasmussen, D. H. 1968. Study by DTA of the temperatures of instability of rapidly cooled solutions of glycerol, ethylene glycol, sucrose, and glucose. *Biodynamica* **10**:167–192.

Ma, C. -Y., and Harwalkar, V. R. 1988. *Thermal Analysis of Foods*, London: Elsevier.

MacKenzie, A. P. 1975. Collapse during freeze drying—Qualitative and quantitative aspects. In *Freeze Drying and Advanced Food Technology*, ed. S. A. Goldlith, L. Rey, and W. W. Rothmayr, pp. 277-307, New York: Academic Press.

MacKenzie, A. P. 1977. Non-equilibrium freezing behavior of aqueous systems. *Phil. Trans. Roy. Soc. Lond.* **B278**:167–189.

MacKenzie, A. P., and Rasmussen, D. H. 1972. Interactions in the water–PVP system at low temperatures. In *Water Structure at the Water–Polymer Interface*, ed. H. H. G. Jellinek, pp. 146–171, New York: Plenum Press.

Magnuson, K. M. 1985. Uses and functionality of vital wheat gluten. *Cereal Foods World* **30**:179–181.

Mandelkern, L. 1986. Thermoreversible gelation and crystallization from solution. *Polym. Prep.* **27**(1):206.

Manley, D. J. R. 1983. Technology of Biscuits, Crackers and Cookies, p. 311, Chichester, U.K.: Ellis Horwood Ltd.

Marsh, R. D. L., and Blanshard, J. M. V. 1988. The application of polymer crystal growth theory to the kinetics of formation of the B-amylose polymorph in a 50% wheat starch gel. *Carbohydr. Polym.* **9**:301–317.

Marshall, A. S., and Petrie, S. E. B. 1980. Thermal transitions in gelatin and aqueous gelatin solutions. *J. Photogr. Sci.* **28**:128–134.

Mashimo, S., Kuwabara, S., Yagihara, S., and Higasi, K. 1987. Dielectric relaxation time and structure of bound water in biological materials. *J. Phys. Chem.* **91**:6337–6338.

Matsukura, U., Matsunaga, A., and Kainuma, K. 1983. Contribution of amylose to starch retrogradation. *J. Jap. Soc. Starch Sci.* **30**:106–113.

Matsuoka, S., Williams, G., Johnson, G. E., Anderson, E. W., and Furukawa, T. 1985. Phenomenological relationship between dielectric relaxation and thermodynamic recovery processes near the glass transition. *Macromolecules* **18**:2652–2663.

Maurer, J. J. 1981. Elastomers. In *Thermal Characterization of Polymeric Materials*, ed. E. A. Turi, pp. 571–708, Orlando, FL: Academic Press.

Maurice, T. J., Slade, L., Sirett, R. R., and Page, C. M. 1985. Polysaccharide–water interactions—Thermal behavior of rice starch. In *Properties of Water in Foods*, ed. D. Simatos and J. L. Multon, pp. 211–227, Dordrecht, Holland: Martinus Nijhoff.

Mayer, E. 1988. Hyperquenching of water and dilute aqueous solutions into their glassy states: An approach to cryofixation. *Cryo-Letters* **9**:66–77.

Menkovska, M., Pomeranz, Y., Lookhart, G. L., and Shogren, M. D. 1987. *Gliadin Fractions in Breadmaking: Differences in Flours Varying in Breadmaking Potential*, abstract 141, American Association of Cereal Chemists 72nd Annual Meeting, Nashville, TN.

Mestres, C., Colonna, P., and Buleon, A. 1988. Gelation and crystallization of maize starch after pasting, drum-drying, or extrusion cooking. *J. Cereal Sci.* **7**:123–134.

Miles, M. J., Morris, V. J., Orford, P. D., and Ring, S. G. 1985a. Roles of amylose and amylopectin in gelation and retrogradation of starch. *Carbohydr. Res.* **135**:271–281.

Miles, M. J., Morris, V. J., Orford, P. D., and Ring, S. G. 1985b. Gelation of amylose. *Carbohydr. Res.* **135**:257–269.

Mitchell, J. R. 1980. The rheology of gels. *J. Text. Stud.* **11**:315–337.

Moore, W. R., and Hoseney, R. C. 1986. Effects of flour lipids on expansion rate and volume of bread baked in a resistance oven. *Cereal Chem.* **63**:172–174.

Morozov, V. N., and Gevorkian, S. G. 1985. Low-temperature glass transition in proteins. *Biopolymers* **24**:1785–1799.

Moy, P., and Karasz, F. E. 1980. The interactions of water with epoxy resins. In *Water in Polymers*, ed. S. P. Rowland, ACS Symposium Series 127, pp. 505–513, Washington, D.C.: American Chemical Society.

Muhlethaler, K. 1961. Cellulose structure in the plant cell wall. In *The Cell—Biochemistry, Physiology, Morphology*, Vol. 2, *Cells and Their Component Parts*, ed. J. Brachet and A. E. Mirsky, pp. 85–134, New York: Academic Press.

Nakazawa, F., Noguchi, S., Takahashi, J., and Takada, M. 1984. Thermal equilibrium state of starch–water mixture studied by DSC. *Agric. Biol. Chem.* **48**:2647–2653.

Nakazawa, F., Noguchi, S., Takahashi, J., and Takada, M. 1985. Retrogradation of gelatinized potato starch studied by DSC. *Agric. Biol. Chem.* **49**:953–957.

Neogi, P. 1983. Anomalous diffusion of vapors through solid polymers. *Amer. Inst. Chem. Eng. J.* **29**:829–839.

Ng, P. K. W., and Bushuk, W. 1987. *Relationship Between High Molecular Weight Subunits of Glutenin and Breadmaking Quality*, abstract 142, American Association of Cereal Chemists, 72nd Annual Meeting, Nashville, TN.

Ohtsuki, I., Maruyama, K., and Ebashi, S. 1986. Regulatory and cytoskeletal proteins of vertebrate skeletal muscle. In *Advances in Protein Chemistry*, Vol. 38, pp. 1–67, ed. C. B. Anfinsen, J. T. Edsall, and F. M. Richards, New York: Academic Press.

Osborne, T. B. 1907. *The Proteins of the Wheat Kernel*, Publ. No. 84, Washington, D.C.: Carnegie Institute.

Otterburn, M. S. 1977. Chemistry and reactivity of silk. In *Chemistry of Natural Protein Fibers*, ed. R. S. Asquith, pp. 53–80, New York: Plenum Press.

Pace, R. J., and Datyner, A. 1981. Temperature dependence of Langmuir sorption capacity factor in glassy polymers. *J. Polym. Sci. Polym. Phys. Ed.* **19**:1657–1658.

Paton, D. 1987. DSC of oat starch pastes. *Cereal Chem.* **64**:394–399.

Payne, P. I. 1987. Genetics of wheat storage proteins and the effect of allelic variation on bread-making quality. *Ann. Rev. Plant Physiol.* **38**:141–153.

Payne, P. I., and Holt, L. M. 1984. *Genetic Analysis of Storage Protein Genes of Wheat*, abstract 36, American Association of Cereal Chemists 69th Annual Meeting, Minneapolis, MN.

Payne, P. I., Holt, L. M., Jarvis, M. G., and Jackson, E. A. 1985. Fractionation of endosperm proteins of bread wheat. *Cereal Chem.* **62**:319–326.

Pearson, A. M. 1987. Summary of "Collagen as a Food" symposium. In *Advances in Meat Research*, Vol. 4, ed. A. M. Pearson, T. R. Dutson and A. J. Bailey, pp. 377–383, Westport, Conn.: AVI.

Petrie, S. E. B. 1975. The problem of thermodynamic equilibrium in glassy polymers. In *Polymeric Materials: Relationships Between Structure and Mechanical Behavior*, ed. E. Baer and S. V. Radcliffe, pp. 55–118, Metals Park, OH: American Society of Metals.

Phillips, A. J., Yarwood, R. J., and Collett, J. H. 1986. Thermal analysis of freeze-dried products. *Anal. Proceed.* **23**:394–395.

Pikal, M. J. 1985. Use of laboratory data in freeze-drying process design. *J. Parent. Sci. Technol.* **39**:115–138.

Pikal, M. J., Shah, S., Senior, D., and Lang, J. E. 1983. Physical chemistry of freeze-drying. *J. Pharmaceut. Sci.* **72**:635–650.

Pomeranz, Y. 1978. Composition and functionality of wheat-flour components. In *Wheat Chemistry and Technology*, 2nd ed., ed. Y. Pomeranz, pp. 585–674, St. Paul, MN: American Association of Cereal Chemists.

Pomeranz, Y., Lookhart, G. L., Rubenthaler, G. L., and Alberts, L. A. 1987. *Changes in Gliadin Proteins During Cookie Making*, abstract 140, American Association of Cereal Chemists, 72nd Annual Meeting, Nashville, TN.

Poole, P. L., and Finney, J. L. 1983a. Sequential hydration of a dry globular protein. *Biopolymers* **22**:255–260.

Poole, P. L., and Finney, J. L. 1983b. Hydration-induced conformational and flexibility changes in lysozyme at low water content. *Int. J. Biol. Macromol.* **5**:308–310.

Pouchly, J., Biros, J., and Benes, S. 1979. Heat capacities of water-swollen hydrophilic polymers above and below 0°C. *Makromol. Chem.* **180**:745–760.

Prime, R. B. 1981. Thermosets. In *Thermal Characterization of Polymeric Materials*, ed. E. A. Turi, pp. 435–569, Orlando, FL: Academic Press.

Quinquenet, S., Grabielle-Madelmont, C., Ollivon, M., and Serpelloni, M. 1988. Influence of water on pure sorbitol polymorphism. *J. Chem. Soc., Faraday Trans.* I**84**:2609–2618.

Rasmussen, D., and Luyet, B. 1969. Complementary study of some non-equilibrium phase transitions in frozen solutions of glycerol, ethylene glycol, glucose, and sucrose. *Biodynamica* **10**:319–331.

Reid, D. S. 1985. Correlation of the phase behavior of DMSO/NaCl/water and glycerol/NaCl/water as determined by DSC with their observed behavior on a cryomicroscope. *Cryo-Letters* **6**:181–188.

Reid, D. S., and Charoenrein, S. 1985. DSC studies of starch–water interaction in the gelatinization process. In *Proceedings 14th NATAS Conference*, San Francisco: NATAS, pp. 335–340.

Reuther, F., Damaschun, G., Gernat, C., Schierbaum, F., Kettlitz, B., Radosta, S., and Nothnagel, A. 1984. Molecular gelation mechanism of maltodextrins investigated by wide-angle X-ray scattering. *Colloid Polym. Sci.* **262**:643–647.

Reutner, P., Luft, B., and Borchard, W. 1985. Compound formation and glassy solidification in the system gelatin–water. *Colloid Polym. Sci.* **263**:519–529.

Richardson, M. J. 1978. Quantitative DSC. In *Developments in Polymer Characterization-1*, ed. J. V. Dawkins, pp. 205–244, London: Applied Science.

Richter, M., Schierbaum, F., Augustat, S., and Knoch, K. D. 1976a. Method of Producing Starch Hydrolysis Products for Use as Food Additives. U.S. patent 3,962,465.

Richter, M., Schierbaum, F., Augustat, S., and Knoch, K. D. 1976b. Method of Producing Starch Hydrolysis Products for Use as Food Additives. U.S. patent 3,986,890.

Ring, S. G. 1985a. Observations on crystallization of amylopectin from aqueous solution. *Int. J. Biol. Macromol.* **7**:253–254.

Ring, S. G. 1985b. Some studies on starch gelation. *Staerke* **37**:80–83.

Ring, S. G., and Orford, P. D. 1986. Recent observations on retrogradation of amylopectin. In *Gums and Stabilizers for the Food Industry* 3, ed. G. O. Phillips, D. J. Wedlock, and P. A. Williams, pp. 159–165, London: Elsevier.

Ring, S. G., Colonna, P., l'Anson, K. J., Kalichevsky, M. T., Miles, M. J., Morris, V. J., and Orford, P. D. 1987. The gelation and crystallization of amylopectin. *Carbohydr. Res.* **162**:277–293.

Roussis, P. P. 1981. Diffusion of water vapor in cellulose acetate, I. *Polymer* **22**:768–773.

Rowland, S. P. 1980. *Water in Polymers*, ACS Symposium Series 127, Washington, D.C.: American Chemical Society.

Russell, P. L. 1983. Kinetic study of bread staling by DSC. *Staerke* **35**:277–281.

Russell, P. L. 1987a. Gelatinization of starches of different amylose/amylopectin content—A DSC study. *J. Cereal Sci.* **6**:133–145.

Russell, P. L. 1987b. Aging of gels from starches of different amylose/amylopectin content studied by DSC. *J. Cereal Sci.* **6**:147–158.

Salmen, N. L., and Back, E. L. 1977. Influence of water on Tg of cellulose. *Tappi* **60**:137–140.

Saunders, S. R., and Slade, L. 1989. A new approach to time-lapse photography analysis for investigation of the mechanism of cookie baking. Part I. Methods development. *J. Cereal Sci.*, in press.

Scandola, M., Ceccorulli, G., and Pizzoli, M. 1981. Water clusters in elastin. *Int. J. Biol. Macromol.* **3**:147–149.

Schaefer, J., Garbow, J. R., Stejskal, E. O., and Lefelar, J. A. 1987. Plasticization of poly(butyral-co-vinyl alcohol). *Macromolecules* **20**:1271–1278.

Schenz, T. W. 1987. Glasses in Aqueous Systems. Paper presented at 24th Society of Cryobiology Meeting, Edmonton, Alberta, June 22–26.

Schenz, T. W., Rosolen, M. A., Levine, H., and Slade, L. 1984. DMA of frozen aqueous solutions. In *Proceedings of the 13th NATAS Conference*, ed. A. R. McGhie, pp. 57–62, Philadelphia, PA: NATAS.
Schofield, J. D. 1986. Flour proteins: Structure and functionality in baked products. In *Chemistry and Physics of Baking*, ed. J. M. V. Blanshard, P. J. Frazier, and T. Galliard, pp. 14–29, London: Royal Society of Chemistry.
Schofield, J. D., Bottomley, R. C., LeGrys, G. A., Timms, M. F., and Booth, M. R. 1984. Effects of heat on wheat gluten. In *Gluten Proteins*, ed. A. Graveland and J. H. E. Moonen, pp. 81–90, Wageningen, Holland: TNO.
Sears, J. K., and Darby, J. R. 1982. *The Technology of Plasticizers*, New York: Wiley–Interscience.
Shalaby, S. W. 1981. Thermoplastic polymers. In *Thermal Characterization of Polymeric Materials*, ed. E. A. Turi, pp. 235–364, Orlando, FL: Academic Press.
Shannon, J. C., and Garwood, D. L. 1984. Genetics and physiology of starch development. In *Starch: Chemistry and Technology*, ed. R. L. Whistler, J. N. BeMiller, and E. F. Paschall, 2nd ed., pp. 25–86, Orlando, FL: Academic Press.
Sharples, A. 1966. *Introduction to Polymer Crystallization*, London: Edward Arnold.
Shiotsubo, T., and Takahashi, K. 1986. Changes in enthalpy and heat capacity associated with gelatinization of potato starch. *Carbohydr. Res.* **158**:1–6.
Sichina, W. J. 1988. Predicting mechanical performance and lifetime of polymeric materials. *Amer. Lab.* **20**(1):42–52.
Siljestrom, M., Bjorck, I., Eliasson, A. C., Lonner, C., Nyman, M., and Asp, N. G. 1988. Effects on polysaccharides during baking and storage of bread. *Cereal Chem.* **65**:1–8.
Simatos, D., and Karel, M. 1988. Characterization of the condition of water in foods—Physicochemical aspects. In *Food Preservation by Moisture Control*, ed. C. C. Seow, pp. 1–41, Amsterdam, The Netherlands: Elsevier.
Skerritt, J. H., Temperley, L. G., and Lew, P. Y. 1988. *Probing Cereal Grain Protein Structure using Monoclonal Antibodies (MCA)*, abstract 22, American Association of Cereal Chemists, 73rd Annual Meeting, San Diego, CA.
Slade, L. 1984. *Starch Properties in Processed Foods: Staling of Starch-Based Products*, abstract 112, American Association of Cereal Chemists, 69th Annual Meeting, Minneapolis, MN.
Slade, L., and Levine, H. 1984a. *Thermal Analysis of Starch and Gelatin*, abstract 152, American Chemical Society, Northeast Regional Meeting 14, June 12, Fairfield, CT.
Slade, L., and Levine, H. 1984b. Thermal analysis of starch and gelatin. In *Proceedings of the 13th NATAS Conference*, p. 64, ed. A. R. McGhie, Philadelphia, PA: NATAS.
Slade, L., and Levine, H. 1987a. Recent advances in starch retrogradation. In *Industrial Polysaccharides—The Impact of Biotechnology and Advanced Methodologies*, ed. S. S. Stivala, V. Crescenzi, and I. C. M. Dea, pp. 387–430, New York: Gordon and Breach.
Slade, L., and Levine, H. 1987b. Polymer–chemical properties of gelatin in foods. In *Advances in Meat Research*, Vol. 4, *Collagen as a Food*, ed. A. M. Pearson, T. R. Dutson, and A. Bailey, pp. 251–266, Westport, CT: AVI
Slade, L., and Levine, H. 1988a. Non-equilibrium behavior of small carbohydrate–water systems. *Pure Appl. Chem.* **60**:1841–1864.

Slade, L., and Levine, H. 1988b. Non-equilibrium melting of native granular starch. I. Temperature location of the glass transition associated with gelatinization of A-type cereal starches. *Carbohydr. Polym.* **8**:183–208.

Slade, L., and Levine, H. 1988c. A food polymer science approach to aspects of starch gelatinization and retrogradation. In *Frontiers in Carbohydrate Research: Food Applications*, ed. J. N. BeMiller, London: Elsevier, in press.

Slade, L., and Levine, H. 1988d. Thermal analysis of starch. In *Scientific Conference Proceedings*, pp. 169–244. Washington, D.C.: Corn Refiners Association.

Slade, L., and Levine, H. 1988e. Structural stability of intermediate moisture foods — A new understanding? In *Food Structure—Its Creation and Evaluation*, ed. J. R. Mitchell and J. M. V. Blanshard, pp. 115–147, London: Butterworths.

Slade, L., Oltzik, R., Altomare, R. E., and Medcalf, D. G. 1987. Accelerated Staling of Starch-Based Products. U.S. patent 4,657,770.

Slade, L., Levine, H., and Finley, J. W. 1988. Protein–water interactions: Water as a plasticizer of gluten and other protein polymers. In *Protein Quality and the Effects of Processing*, ed. D. Phillips and J. W. Finley, pp. 9–124, New York: Marcel Dekker.

Snyder, E. M. 1984. Industrial microscopy of starches. In *Starch: Chemistry and Technology*, 2nd ed., ed. R. L. Whistler, J. N. BeMiller, and E. F. Paschall, pp. 661–689, Orlando, FL: Academic Press.

Soesanto, T., and Williams, M. C. 1981. Volumetric interpretation of viscosity for concentrated and dilute sugar solutions. *J. Phys. Chem.* **85**:3338–3341.

Sperling, L. H. 1986. *Introduction to Physical Polymer Science*, New York: Wiley–Interscience.

Starkweather, H. W. 1980. Water in nylon. In *Water in Polymers*, ed. S. P. Rowland, ACS Symposium Series 127, pp. 433–440, Washington, D.C.: American Chemical Society.

Starkweather, H. W., and Barkley, J. R. 1981. Effect of water on secondary dielectric relaxations in nylon 66. *J. Polym. Sci. Polym. Phys. Ed.* **19**:1211–1220.

Suggett, A. 1976. Molecular motion and interactions in aqueous carbohydrate solutions. III. A combined NMR and dielectric relaxation strategy. *J. Solution Chem.* **5**:33–46.

Suggett, A., and Clark, A. H. 1976. Molecular motion and interactions in aqueous carbohydrate solutions. I. Dielectric Relaxation Studies. *J. Solution Chem.* **5**:1–15.

Suggett, A., Ablett, S., and Lillford, P. J. 1976. Molecular motion and interactions in aqueous carbohydrate solutions. II. NMR studies. *J. Solution Chem.* **5**:17–31.

Sultan, W. J. 1976. *Practical Baking*, 3rd ed., p. 75, Westport, CT: AVI.

Swift, J. A. 1977. Histology of keratin fibers. In *Chemistry of Natural Protein Fibers*, ed. R. S. Asquith, pp. 81–146, New York: Plenum Press.

Tait, M. J., Suggett, A., Franks, F., Ablett, S., and Quickenden, P. A. 1972. Hydration of monosaccharides: A study by dielectric and nuclear magnetic relaxation. *J. Solution Chem.* **1**:131–151.

Tatham, A. S., Shewry, P. R., Field, J. M., Kasarda, D. D., Forde, J., Forde, B. G., Fry, R. P., and Miflin, B. J. 1984. *Contribution of High MW Gluten Polypeptides to Gluten Structure*, abstract 34, American Association of Cereal Chemists 69th Annual Meeting, Minneapolis, MN.

Tatham, A. S., Miflin, B. J., and Shewry, P. R. 1985. Beta-turn conformation in wheat gluten proteins and gluten elasticity. *Cereal Chem.* **62**:405–412.

Telloke, G. 1987. Puff pastry under the microscope. *Chorleywood Digest* **64**:22.

Thom, F., and Matthes, G. 1986. Ice formation in binary aqueous solutions of ethylene glycol. *Cryo-Letters* **7**:311–326.

To, E. C., and Flink, J. M. 1978. "Collapse," a structural transition in freeze-dried carbohydrates, I–III. *J. Food Technol.* **13**:551–594.

Tomka, I. 1986. Thermodynamic theory of polar polymer solutions. *Polym. Prepr.* **27**(2):129.

Tomka, I., Bohonek, J., Spuhler, A., and Ribeaud, M. 1975. Structure and formation of the gelatin gel. *J. Photogr. Sci.* **23**:97–103.

Tsourouflis, S., Flink, J. M., and Karel, M. 1976. Loss of structure in freeze-dried carbohydrate solutions. *J. Sci. Food Agric.* **27**:509–519.

Turi, E. A. 1981. *Thermal Characterization of Polymeric Materials*, Orlando, FL: Academic Press.

van den Berg, C. 1981. Vapor sorption equilibria and other water–starch interactions: A physicochemical approach. Doctoral Thesis, Agricultural University, Wageningen, Holland.

van den Berg, C. 1986. Water activity. In *Concentration and Drying of Foods*, ed. D. MacCarthy, pp. 11–36, London: Elsevier.

van den Berg, C., and Bruin, S. 1981. Water activity and its estimation in food systems: Theoretical aspects. In *Water Activity: Influences on Food Quality*, ed. L. B. Rockland and G. F. Stewart, pp. 1–61, New York: Academic Press.

Walton, A. G. 1969. Nucleation in liquids and solutions. In *Nucleation*, ed. A. C. Zettlemoyer, p. 225, New York: Marcel Dekker.

Watt, I. C. 1980. Adsorption-desorption hysteresis in polymers. *J. Macromol. Sci.-Chem.* **A14**:245–255.

Weegels, P. L., Verhoek, A., and Hamer, R. J. 1988. *A Comprehensive Study of the Effect of Heat on Wheat Gluten*, abstract 133, American Association of Cereal Chemists, 73rd Annual Meeting, San Diego, CA.

Welsh, E. J., Bailey, J., Chandarana, R., and Norris, W. E. 1982. Physical characterization of interchain association in starch systems. *Prog. Food Nutr. Sci.* **6**:45–53.

Wesson, J. A., Takezoe, H., Yu, H., and Chen, S. P. 1982. Dye diffusion in swollen gels by forced Rayleigh scattering. *J. Appl. Phys.* **53**:6513–6519.

Whewell, C. S. 1977. The chemistry of wool finishing. In *Chemistry of Natural Protein Fibers*, ed. R. S. Asquith, pp. 333–370, New York: Plenum Press.

Whistler, R. L., and Daniel, J. R. 1984. Molecular structure of starch. In *Starch: Chemistry and Technology*, ed. R. L. Whistler, J. N. BeMiller, and E. F. Paschall, 2nd ed., pp. 153–182, Orlando, FL: Academic Press.

White, G. W., and Cakebread, S. H. 1966. The glassy state in certain sugar-containing food products. *J. Food Technol.* **1**:73–82.

Williams, M. L., Landel, R. F., and Ferry, J. D. 1955. Temperature dependence of relaxation mechanisms in amorphous polymers and other glass-forming liquids. *J. Amer. Chem. Soc.* **77**:3701–3706.

Wolf, W., Spiess, W. E. L., and Jung, G. 1985. Standardization of isotherm measurements. In *Properties of Water in Foods*, ed. D. Simatos and J. L. Multon, pp. 661–679, Dordrecht, Holland: Martinus Nijhoff.

Wright, D. J. 1984. Thermoanalytical methods in food research. *Crit. Rev. Appl. Chem.* **5**:1–36.

Wunderlich, B. 1973. *Macromolecular Physics*, Vol. 1, *Crystal Structure, Morphology, Defects*, New York: Academic Press.

Wunderlich, B. 1976. *Macromolecular Physics*, Vol. 2, *Crystal Nucleation, Growth, Annealing*, New York: Academic Press.

Wunderlich, B. 1980. *Macromolecular Physics*, Vol. 3, *Crystal Melting*, New York: Academic Press.

Wunderlich, B. 1981. The basis of thermal analysis. In *Thermal Characterization of Polymeric Materials*, ed. E. A. Turi, pp. 91–234, Orlando, FL: Academic Press.

Wynne-Jones, S., and Blanshard, J. M. V. 1986. Hydration studies of wheat starch, amylopectin, amylose gels and bread by proton magnetic resonance. *Carbohydr. Polym.* **6**:289–306.

Yamamoto, M., Harada, S., Sano, T., Yasunaga, T., and Tatsumoto, N. 1984. Kinetic studies of complex formation in amylose–SDS–iodine ternary system. *Biopolymers* **23**:2083–2096.

Yamazaki, W. T., and Lord, D. D. 1978. Soft wheat products. In *Wheat Chemistry and Technology*, 2nd ed., ed. Y. Pomeranz, pp. 743–776, St. Paul, MN: American Association of Cereal Chemists.

Yannas, I. V. 1972. Collagen and gelatin in the solid state. *J. Macromol. Sci.-Revs. Macromol. Chem.* **C7**:49.

Yost, D. A., and Hoseney, R. C. 1986. Annealing and glass transition of starch. *Staerke* **38**:289–292.

Zeleznak, K., and Hoseney, R. C. 1986. Role of water in retrogradation of wheat starch gels and bread crumb. *Cereal Chem.* **63**:407–411.

Zeleznak, K. J., and Hoseney, R. C. 1987a. The glass transition in starch. *Cereal Chem.* **64**:121–124.

Zeleznak, K. J., and Hoseney, R. C. 1987b. Characterization of starch from bread aged at different temperatures. *Staerke* **39**:231–233.

Ziegler, K. 1977. Crosslinking and self-crosslinking in keratin fibers. In *Chemistry of Natural Protein Fibers*, ed. R. S. Asquith, pp. 267–300, New York: Plenum Press.

Zobel, H. F. 1984. Gelatinization of starch and mechanical properties of starch pastes. In *Starch: Chemistry and Technology*, ed. R. L. Whistler, J. N. BeMiller, and E. F. Paschall, 2nd ed., pp. 285–309, Orlando, FL: Academic Press.

Zobel, H. F. 1988. Starch crystal transformations and their industrial importance. *Starke* **40**:1–7.

Zobel, H. F., Young, S. N., and Rocca, L. A. 1988. Starch gelatinization: An X-ray diffraction study. *Cereal Chem.* **65**:443–446.

Chapter 6

Basic Principles of Food Texture Measurement

Malcolm C. Bourne

The three main acceptability factors for evaluating foods are appearance, flavor, and texture. If any of these three factors fails to reach expectations, the food will not be consumed, or if it is consumed, it will provoke negative consumer response. Food technologists have good reason to devote considerable time to measuring and controlling the acceptability factors for foods.

The cereal foods industry has a long history of studying the rheology of work in process, especially for dough. The field of dough rheology is well developed. Measuring the texture of the finished cereal products that are marketed has not developed to the same extent, however. This is an appropriate time for the cereal industry to pay more attention to measuring and specifying the textural properties of finished products.

Texture is primarily a sensory attribute. Brean (1980) stated, "Texture is a sensory attribute, perceived by the senses of touch, sight, and hearing. Thus the only direct method of measuring texture is by means of one of the senses."

Another pioneer in the field of texture technology stated, "By definition, texture is a sensory property. Thus it is only the human being that can perceive, describe and quantify texture. Furthermore, it is generally recognized that texture, just like flavor, is a multiparameter attribute." (Szczesniak, 1987)

It is preferable to talk about textural properties (plural) rather than texture because there are a number of different textural properties. Texture is not a single-dimensional attribute; it is a multidimensional attribute. The textural properties of foods have the following characteristics:

1. They are a *group* of physical properties.
2. They derive from the structure of the food.
3. They belong under the mechanical or rheological subheading of physical properties.

4. They are sensed by the feeling of touch.
5. Objective measurement is usually by means of functions of mass, distance, and time. For example, force has the dimensions MLT^{-2}.

Most food scientists approach a texture problem by thinking in terms of which instruments should be used for their work. This is an inefficient and often unsatisfactory route to solving problems in texture measurement. A more effective way is a process consisting of four steps (Bourne, 1982):

1. Consider all of the principles that can be used in texture measurement.
2. Select the test principle that is best suited for the product of interest.
3. Now select an instrument that uses this test principle.
4. Finally, refine the test procedure to give the highest correlation with sensory evaluation of texture.

PRINCIPLES OF TEXTURE MEASUREMENT

Table 6-1 lists principles of texture measurement. Most texture measurements are based on force measurements, but other principles such as distance, time, energy, and miscellaneous tests may be used.

The *puncture* principle involves measuring the force required to push a probe into a food. This principle is widely used for fruits, vegetables, and gels. The author of this chapter believes that the technique has the potential for wider use on baked goods.

The *extrusion* principle involves applying force to a food until it flows through an outlet that is usually in the form of one or more slots or holes. This test principle is seldom used on finished baked goods, although it has been used for frostings and fillings.

Cutting shear involves cutting across a piece of food. This test principle is often mistakenly called a shear test. It is often used for meat and meat products.

Gentle compression is the application of a small nondestructive force to measure deformability. There are two ways of performing this test: (1) measure the force required to achieve a standard compression (the American Association of Cereal Chemists Bread Firmness Test uses this principle), and (2) measure the distance the product deforms under a standard force.

The *crushing* principle involves subjecting the food to a high degree of compression until it breaks up. This destructive test is the basis of the texture profile analysis procedure (TPA), which will be described more fully below.

The *tensile* principle involves the force required to break a food in tension. Although it has been occasionally used for evaluating bread and dough, it is infrequently used with other foods.

Table 6-1. Principles of Food Texture Measurement

Principle	Examples
1. Force (dimensions MLT^{-2})	
a) puncture	Magness-Taylor Effe-Gi, Stevens
b) extrusion	Shear press, tenderometer
c) cutting-shear	Warner-Bratzler shear
d) gentle compression	Bread firmness
e) crushing (heavy compression)	Texture profile analysis
f) tensile	
g) bending–snapping	Structograph
h) torque	Farinograph, mixograph
2. Distance (dimensions l, l^2, or l^3)	
a) length (l)	Penetrometers
b) area (l^2)	Grawemeyer and Adams consistometers
c) volume (l^3)	Bread volume, juice volume
3. Time (dimension t)	British Baking Industries Research Association biscuit tester
4. Multiple (can measure force, distance, slope, work, etc.)	Instron, Lloyd, Ottawa texturometer
5. Miscellaneous	Optical density, crushing sounds

The *bending–snapping* principle involves measuring the force required to bend and snap a food that is usually in the shape of a bar, cylinder, or sheet. This principle is used in the Brabender Struct-O-Graph™. Bruns and Bourne (1975) showed in snapping tests on cookies and other snappy foods that the snapping force is directly proportional to the width and the square of the thickness, and inversely proportional to the length of the test specimen.

The *torque* principle involves measuring the torsional force needed to rotate or twist one part of a material around an axis with respect to other parts. This principle is used widely in the study of dough rheology but has little application to solid foods. The rotational viscometers that are used to study liquids and semiliquids use this principle.

The *distance* principle is the second most frequently used principle for measuring food texture. This principle involves the measurement of distance, which may be a length (l), an area (l^2), or a volume (l^3). The measurement of the volume of a standard loaf of bread is well known in the cereal industry.

The *time* principle involves the measurement of time, mainly by eflux viscometers, and has little application to baked goods. The British Baking

Industries Research Association Biscuit Texture Meter measures the time for a small circular saw to cut through a stack of cookies or crackers (Wade, 1968).

Multiple measuring instruments record the entire force–time history of a texture test, which allows a number of texture parameters to be measured, including force, rate of change of force (slope), distance, and work (area). The Instron was the first universal testing machine to be adapted to use with foods (Bourne et al., 1966). Other very satisfactory universal testing machines are now available, such as the Lloyd, Ottawa, and Zwick, which cost less than an equivalent Instron.

A wide range of test principles have been described above. Every one of these principles has been successful with some foods, and no principle has been successful with all foods. It is important to identify as early as possible the correct test principle that should be used for each particular application. Considerable time can be wasted if an incorrect principle is used.

Texture profile analysis is a special kind of test that should be set apart from the single-point tests described above. The TPA procedure was developed by a group at the General Foods Corporation Technical Center (Friedman et al., 1963; Szczesniak et al., 1963). Most researchers now use a universal testing machine to perform TPA (Bourne, 1968), and many use a digitizer interfaced to a computer or a direct computer readout of the data (Bourne et al., 1978).

This test consists of compressing a bite-size piece of food two times in a reciprocating motion that imitates the action of the jaw and extracting from the resulting force–time curve a number of textural parameters that correlate well with sensory evaluation of those parameters. Figure 6-1 shows a generalized texture profile analysis curve obtained in the Instron Universal Testing Machine. Since this test simulates mastication, the degree of compression should be high; 80–90% compression levels are usually used in this highly destructive test.

SELECTION OF A SUITABLE TEST PRINCIPLE

In view of the large number of test principles and instruments that can be used for measuring the texture of food, a cereal chemist can easily become bewildered when faced with the problem of developing a suitable procedure for measuring the textural properties of baked goods. The procedure outlined below is designed to guide one through the necessary steps to select the correct test principle for each particular application in the shortest possible time (Bourne, 1982).

Step 1: Determine the nature of the product.   The kind of material (crisp, aerated, homogeneous, plastic, brittle, heterogeneous) affects the type of test principle that should be used. For example, the extrusion test is unsuited

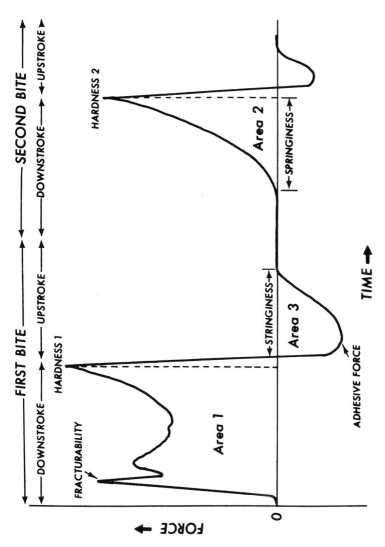

Figure 6-1. A generalized texture profile analysis curve showing the parameters that are measured. The ratio: area 2/area 1 gives the cohesiveness. *(From Bourne, 1978; copyright © 1978 by the Institute of Food Technologists.)*

for baked goods, because baked goods do not flow under force. Viscosity measuring instruments cannot be used on baked goods, but they can be used on doughs and precooked cereals made into baby food.

Step 2: Define the purpose of the test.   Will the test be used for quality control, for product development, for setting official legal standards, or for basic research? The answer to this question is an integral part of the selection process.

Step 3: Determine the degree of accuracy required.   A large sample size or a greater number of replicate tests gives a higher degree of accuracy, but this approach requires more product, a higher force range, and more time to perform the test. A compromise has to be made between the cost and time of the test and the degree of accuracy required.

There is a large inherent variability from unit to unit in the same sample lot for most foods. This variation is inherent in the food and is not a defect of the instrument if the instrument is correctly operated. Because of this high variability, it is recommended that as many replicate tests as possible be performed. A high degree of precision is a secondary consideration in most texture work, because there is little point to attempting to measure some textural parameter to a high degree of precision when results for replicate samples may vary by 20%, 50%, or more.

Step 4: Decide if the test is to be destructive or nondestructive. Destructive tests ruin the structure of the sample, rendering it unsuitable for repeating the tests or for using the product for other purposes. Nondestructive tests leave the food in a condition close to its original state, so the test can be repeated on the same item. Both types of tests are used in the food industry. Because mastication is a highly destructive process, it seems logical that destructive tests should be the predominant type of test to be used on foods.

Step 5: Figure the costs.   The cost of a test includes the initial cost of the instrument and its maintenance and operating costs, and the labor costs for the operator of the instrument. A simple instrument can be operated by unskilled or semiskilled personnel, whereas sophisticated instruments need to be operated by a person with higher qualifications. Although the initial cost is higher, automated readout and calculations from the data are often more economical in the long run because of labor savings and the reduced risk of errors.

Step 6: Consider time.   The amount of time that can be spent on the test should be taken into account. Routine quality control tests require an instrument that gives results rapidly, while some basic research requires sophisticated tests and the amount of time required is not of great consequence.

Step 7: Decide on a location.   Where will the instrument be used? An instrument may need to withstand heat, dust, vibration, and other hazards on the plant floor. Some instruments cannot withstand such an environment. All instruments can be used inside a clean, quiet laboratory.

Selection of Test Procedures

After the seven steps enumerated above are considered, the field should be narrowed to about two or three promising test principles. The unsuitable test principles are given no further consideration.

In selecting the most promising two or three test principles, it often helps to observe what kind of test principle people use in their sensory evaluation of textural quality. For example, if people judge textural quality by gently squeezing the food in their hand (as in measuring the firmness of bread), the deformation test principle is an obvious candidate for further work; if people use a snapping test, then the bending-snapping test principle should be given consideration for further study.

Each test principle that is selected as having the highest promise then needs to be tested in practice over the full range of texture variability of the food, from excellent quality to poor quality. A minimum of three levels of textural quality are needed. Each test principle is used over the full quality range. The data points should be plotted to make a simple scatter diagram, and the results should be examined for trend lines. An examination of the diagrams should quickly identify which of the test principles are best for a particular application.

Figure 6-2 shows examples of test principles that are good, marginal, and poor. All of the good tests (left-hand column) have two essential features: (1) a steep slope, and (2) a close fit of the data points to the line. The line may have a positive or a negative slope and may be rectilinear or curved; the curve may be concave or convex, but the test is satisfactory as long as the two essential features are present. The curvilinear relationships can be rendered rectilinear by suitable mathematical manipulation; for example, by taking logarithms on one or both axes. Taking logarithms is recommended before correlation coefficient is calculated, because the correlation coefficient will be deceptively low when it is calculated for a curve.

The marginal tests (center column) are less satisfactory. Seeking an improved test is recommended before a marginal test is used. However, using these tests may be necessary if no better test can be found after a reasonable search. The problems here are either a shallow slope or a wide scatter of the data points.

The poor tests (right-hand column) are not worth using. Using no test at all may be better, because at least no false sense of security will result from relying on such poor tests. These tests are characterized by (1) a very wide scatter of the data points (top curve), (2) a shallow slope and a moderate or high degree of scatter (center curve), or (3) a change in slope of the curve (bottom curve).

When it is completed, this exercise should have identified the best of the two or three promising principles that were examined. This principle is then used for further work.

338 / *M. C. Bourne*

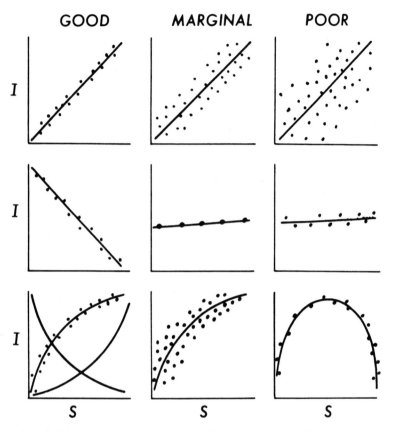

Figure 6-2. Schematic correlations between instrumental tests (I) and sensory tests (S) on foods. *(From Bourne, 1982; copyright © 1982 by Academic Press.)*

The results of tests used to test a principle are shown in Figure 6-3, which was replotted from the data of Rao and co-workers (1986). This figure shows how the consistency of chapati dough is affected by the amount of added water. Three instruments were used to measure the consistency: the farinograph, the General Foods Texturometer and the research water absorption meter (RWAM). Each instrument gives a rectilinear plot when log (instrument reading) is plotted against percent (added water). Each instrument gives an excellent fit of the data points to a straight line. However, the RWAM gives a slope that is more than twice as steep as those for the other two instruments (0.0565 for RWAM versus 0.0200 for the farinograph and 0.0235 for the GF Texturometer). Therefore, the RWAM will give the most sensitive resolution for the samples.

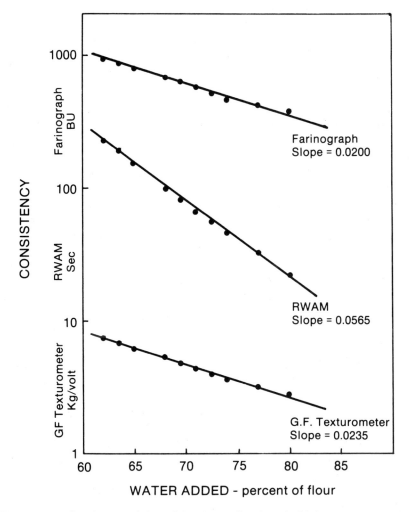

Figure 6-3. Consistency of chapati dough as a function of added water measured by three instruments. *(After Rao et al., 1986.)*

The procedure described above should identify the test principle most suited for a particular application. Sometimes, none of the established procedures give satisfactory results. In these cases researchers should develop new test procedures or apparatus suitable for their own purposes including new ones devised by researchers. If none of these test procedures work, looking at texture profile analysis may be worthwhile to identify which texture property or group of properties is most important.

Optimizing the Test Procedure

The final step of developing a test procedure is to refine the test conditions. Operating parameters such as sample size, temperature, number of replicates, and speed of the test should be checked to see which conditions give the highest correlation with the texture.

At this point one can begin examining the instruments that use a particular test principle to determine which instrument is most suited for each particular application. Notice that the instrument is selected late in the process. If an instrument that uses an inappropriate test principle is selected in the beginning, considerable time will be expended without obtaining satisfactory results.

A useful guide of correlation for quality purposes was given by Kramer (1951). When the simple coefficient of correlation between the instrument test and the sensory score is ± 0.9 to ± 1.0, the instrument can be used with confidence as a predictor of the sensory texture score. When the correlation coefficient lies between ± 0.8 and ± 0.9, the test can be used as a predictor but with less confidence. Attempting to refine the test conditions further is worthwhile to bring the correlation coefficient above 0.9. When the correlation coefficient lies between ± 0.7 and ± 0.8, the test is of marginal use as a predictor, and when it is less than about ± 0.7, the test is of little use for predictive purposes.

Most food technologists are familiar with statistical procedures. Statistical significance is not nearly as important in the selection of an instrument as predictive reliability, which is measured by the correlation coefficient. Unless the correlation coefficient is high, the test is not a good predictor regardless of how good the statistical significance. If the sample size is large enough, a statistically significant relationship between an instrumental test and a sensory test may be found even when the correlation coefficient is low. A correlation coefficient of 0.5 may be statistically highly significant, for example, but it is useless for predictive purposes. To be satisfactory, an objective test must show a high correlation with sensory evaluation, because the ultimate calibration of any test must be against the human senses.

REFERENCES

Bourne, M. C. 1967. Deformation testing of foods. I. A precise technique for performing the deformation test. *J. Food Sci.* **32**:601–605.

Bourne, M. C. 1968. Texture profile of ripening pears. *J. Food Sci.* **33**:223–226.

Bourne, M. C. 1978. Texture profile analysis. *Food Technol.* **32**(7):77, 78, 80.

Bourne, M. C. 1982. *Food Texture and Viscosity: Concept and Measurement.* 325 pp. New York: Academic Press.

Bourne, M. C., Kenny, J. F., and Barnard, J. 1978. Computer-assisted readout of data from texture profile analysis curves. *J. Text. Stud.* **9**:481–494.

Bourne, M. C., Moyer, J. C., and Hand, D. B. 1966. Measurement of food texture by a universal testing machine. *Food Technol.* **20**:522–526.

Brennan, J. G. 1980. Food texture measurement. In *Developments in Food Analysis Techniques*, Vol. 2, ed. R. D. King, pp. 1–78. Essex, England: Applied Science Publishers.

Bruns, A. J., and Bourne, M. C. 1975. Effect of sample dimensions on the snapping force of crisp foods. Experimental verification of a mathematical model. *J. Text. Stud.* **6**:445–458.

Friedman, H. H., Whitney, J. E., and Szczesniak, A. S. 1963. The Texturometer—A new instrument for objective texture measurement. *J. Food Sci.* **28**:390–396.

Kramer, A. 1951. Objective testing of vegetable quality. *Food Technol.* **5**:265–269.

Rao, P. H., Leelavathi, K., and Shurpalekar, S. R. 1986. Objective measurements of the consistency of chapati dough using a research water absorption meter. *J. Text. Stud.* **17**:401–420.

Szczesniak, A. S. 1987. Correlating sensory with instrumental texture measurements—An overview of recent developments. *J. Text. Stud.* **18**:1–15.

Szczesniak, A. S., Brandt, M. A., and Friedman, H. H. 1963. Development of standard rating scales for mechanical parameters of texture and correlation between the objective and sensory methods of texture evaluation. *J. Food Sci.* **28**:397–403.

Wade, P. 1968. A texture meter for the measurement of biscuit hardness. In *Rheology and Texture of Foodstuffs*, Society of Chemical Industry Monograph No. 27, pp. 225–234, London.

Chapter 7

Application of Rheology
in the Bread Industry

Ronald Spies

While it has long been recognized that dough rheology plays an important role in almost every step of the breadmaking process, the application of fundamental rheological principles to the breadmaking industry is still in its infancy. A major reason for this delay is a lack of fundamental knowledge about basic dough rheology.

One reason for this lack of knowledge is the difficulty in using a bread dough for basic rheological studies. Not only is bread dough a multiphase system, it also contains a number of ingredients, many of which affect its rheological properties. An additional complication is that one of those ingredients, yeast, is a live biological system that is constantly altering the rheology of the dough as well as its shape. These changes make sample presentation difficult and complicate the use of fundamental rheological equations that rely on the assumption that these parameters are constant.

Faced with a need to use fundamental knowledge not yet in existence, the baking industry has been forced to resort to using empirical measurements in quality control and research situations. In practical terms, this means that a separate test and instrument must be developed to measure how a dough will react during each step in a breadmaking process. With the increasing diversity and scale of the industry, this approach is proving less and less satisfactory.

In a sponge-and-dough process employed by most of the bread industry, approximately 7–8 hours elapse from the start of a sponge mix until a loaf of bread from that sponge is wrapped. Most of the rheological properties of a dough will have been determined by the time the mixing step is completed. The ingredients used in the sponge and the dough, the fermentation of the sponge, and the work input during dough mixing will all contribute to the rheology of a dough. Approximately 3 hours will elapse from the time when a dough is mixed to the time problems that may be related to the rheology of a product become apparent at the wrapper. With the throughput of some larger bakeries, this would mean that by the time a problem such as excessive keyholing manifests itself during cooling, over 36,000 additional loaves that also have that problem would already be in the system. Clearly, the time delay between mixing and wrapping precludes the use of final product quality as input to a feedback control loop for the mixing operation. It also points to the need for a better understanding of how fundamental aspects of the rheology of a dough at each processing step will affect its performance in later stages of the breadmaking process.

Another area in which rheology plays an important role in breadmaking is that of bread staling. A major change during the aging of bread is an increased resistance to compression (firming). Fundamental studies of this rheological change may prove beneficial in obtaining a better understanding of the staling process and how to control it.

INGREDIENTS

A large portion of the bread made in the United States is produced by using a sponge and dough process (Figure 7-1). The rheological properties of a piece of dough, which govern how it will respond to machining steps in the process, will be determined almost entirely during the early steps of the process. Many of these rheological properties are a result of ingredient interactions that occur during fermentation and dough mixing. Nearly all of the ingredients affect the rheology of a dough to a certain extent, but most of the properties are derived from the flour, water, yeast, and air. Several minor ingredients, such as dough conditioners, salt, fats, enzymes, and emulsifiers, affect the rheology to a lesser degree.

Considerable effort has been spent studying how flour components affect the rheology of a flour–water dough. Most studies have centered on the role of the gluten protein fractions glutenin and gliadin. The gliadins are a group of smaller proteins (molecular weights of about 40,000) which contribute primarily to the viscous behavior of a viscoelastic dough. Glutenin proteins have a larger average molecular weight (about 3 million) and are responsible for the elastic component of a dough's behavior.

It is an oversimplification, however, to assume that *only* gluten proteins affect the rheology of a dough. It is only after these proteins interact with other components in the flour and dough that the viscoelastic nature of a dough emerges. Of these interactions, the flour–water interaction is the most important. Upon contact with water, gluten proteins form fibrils spontaneously (Bernardin and Kasarda, 1973). These fibrils then become aligned during the mixing process to form the gluten matrix.

The level of water in a dough also has a strong influence on rheological properties. Many ingredients in a dough compete with flour for the available water. Likewise, there is competition among various flour components for water. If there is insufficient water to meet the hydration needs of all of the dough ingredients, the gluten does not become fully hydrated and the elastic nature of the dough does not become fully developed. Conversely, an excessive level of free water in the dough results in the domination of the viscous component of a dough, with a decreased resistance to extension, increased extensibility, and the development of a sticky dough.

Air incorporation is a major function of the mixing step. The oxygen present in the air plays an important role in determining the rheology of a dough. Doughs mixed in the presence of oxygen are more elastic and offer more

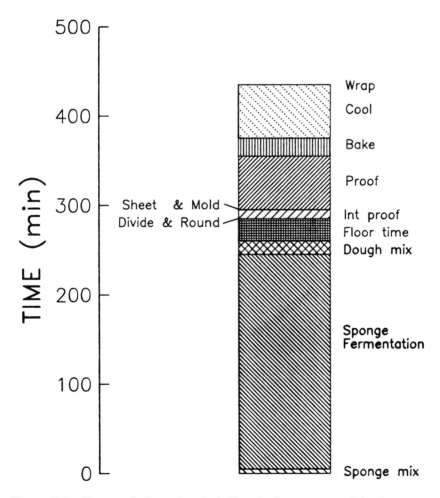

Figure 7-1. Time required to make a loaf of bread using a sponge and dough process.

resistance to extension than those mixed in the absence of oxygen (Smith and Andrews, 1952). Oxygen is also involved in the breakdown of a dough during overmixing, as doughs mixed in the absence of oxygen do not exhibit the decrease in elasticity normally found when a dough is mixed past its point of minimum mobility in air (Baker and Mize, 1937). It is interesting to note that doughs mixed in the absence of oxygen still develop the look of an overmixed dough, a wet sheen that is normally associated with a release of water by the dough, even though the excessively viscous nature of an overmixed dough does not appear.

Fermentation by yeast cells also produces certain desirable rheological changes in bread dough. A major product of yeast fermentation is carbon dioxide. As carbon dioxide dissolves in the dough water, its pH decreases, resulting in a slight change in the configuration of the gluten proteins and therefore an alteration in their rheological properties (Tanaka et al., 1967). The change in pH also affects the function of various enzyme systems that act on gluten proteins. Finally, the evolution of carbon dioxide, and the resulting decrease in pH, may also have an effect on the reaction rate of certain dough oxidation agents. This effect may be part of the reason that the action of potassium bromate is not seen until during fermentation, as it reacts faster at lower pH values.

Other rheological changes are brought about by yeast fermentation. As fermentation progresses, the ratio of elastic to viscous components of the dough changes, with the elastic component becoming more dominant (Hoseney et al., 1979). It is important to note that while yeast is necessary for this change to occur, the products of yeast fermentation may not be solely responsible for this change. Hoseney and co-workers (1979) showed that addition of yeast fermentation products to a dough system that does not contain yeast does not produce the rheological changes seen in a fermenting dough. The mechanism by which yeast alters the rheology of a fermenting dough is still unclear.

Several other minor ingredients also affect the rheology of a dough. These effects occur primarily through the action of the minor ingredients on, or interaction with, the major constituents of the dough. Salt changes water interactions between components, and it alters the configuration of gluten proteins because of its competition for water. The combination of these actions results in an increase in the mixing time for the dough. To overcome this effect, addition of salt to the dough is normally delayed until late in the mixing step.

Oxidizing agents are commonly added to bread dough to increase the ratio of elastic to viscous components (Dempster et al., 1956). While reaction rates and action patterns vary with the type of oxidizer used, the end result is the same as far as the end product is concerned. Conversely, it is sometimes necessary to add reducing agents such as cysteine to reduce the predominance of the elastic component (to mellow a bucky dough). It is also possible to achieve the same type of reaction through the addition of proteolytic enzymes, although they have no effect on mixing properties of a dough.

Another class of compounds frequently used in the breadmaking industry is the dough strengthening or dough conditioning surfactant. The effects of these compounds on the fundamental rheology of a dough have not been widely studied; however, their ability to increase what is often called dough strength would indicate that they have an effect on the elastic component of a dough. Furthermore, their ability to improve the tolerance of a dough to machining may indicate an action involving the stress relaxation properties of the dough.

MIXING

The mixing step is the last processing step in which the rheological properties of a dough can be significantly altered by the baker. The major functions of mixing are to blend ingredients into a quasi-homogeneous mixture, to develop the gluten matrix, and to incorporate air. It is only after a dough has been developed to its optimum point that the full breadmaking potential of that dough can be realized.

Several steps in the mixing process allow a dough to reach this optimum state. The first step involves hydration of flour particles. The high shear rate in a dough mixer helps to speed this hydration by removing the outer layers of the flour particles as they become hydrated, exposing a new surface for further hydration. As gluten becomes hydrated, it forms fibrils that are aligned into a matrix by the repeated shearing action in the mixer. During this process, a dough exhibits an increasing resistance to extension. At some point during mixing, resistance to extension no longer increases, and the dough starts to break down. This point is called the point of minimum mobility or optimum mixing time.

Some of the molecular forces involved in the development of a gluten matrix are of a secondary or noncovalent bonding type. These bonding forces include hydrogen bonding and hydrophobic interactions. Evidence indicating that more than one bonding force is involved is found in the fact that dough development is partially reversible (Tipples and Kilborn, 1975). The unmixing phenomenon occurs if mixing is stopped and the dough is allowed to relax. When the mixer is started again, the resistance of the dough to extension is not at the same level as when the mixer was stopped. Some additional mixing is required before the previous level is reached, indicating that the process is partially reversible, and that covalent bonds are not the only type of bonds present.

If mixing is carried past the optimum development time, a dough becomes wet and sticky. This overmixed dough is commonly referred to as "broken down." This phenomena has been related to shear thinning, a process in which continued mixing action lines up the gluten fibrils and provides less resistance to extension. This explanation is probably not valid, as breakdown does not occur if a dough is mixed under a nitrogen atmosphere instead of one containing oxygen (Baker and Mize, 1937), or if the water-soluble fraction of flour is removed (Schroeder and Hoseney, 1978); these factors are likely to be involved in a shear thinning process. A more likely theory involves oxidation of some part of the water-soluble fraction and free radical formation.

The most critical factors in the mixing stage that affect the rheology of a dough are flour quality (an elusive term to be sure), the amount of water added, and the magnitude of work done on the developing dough. Several instruments are used in the bread industry to measure these factors. The most commonly used are the farinograph and the mixograph. Both instruments are torque-measuring devices, but differ slightly in their mixing action. The farinograph provides a gentler kneading type of mixing, while the mixograph uses a harsher pin mixing method. Both instruments provide empirical information about the mixing properties of a flour by recording the resistance of the dough to the mixing blades during prolonged mixing. A schematic representation of a farinograph curve and measurements usually taken from it are shown in Figure 7-2. A schematic of a mixograph curve and its measurements are shown in Figure 7-3.

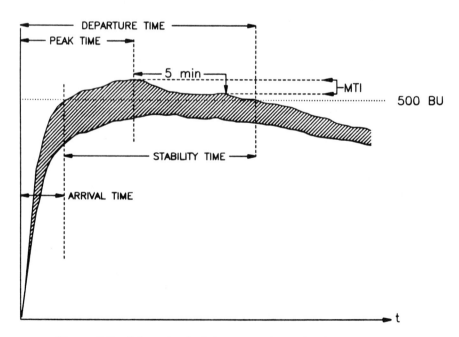

Figure 7-2. Schematic of a farinogram and associated measurements.

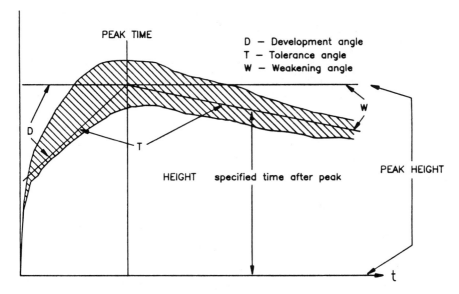

Figure 7-3. Schematic of a mixogram and associated measurements.

The most common pieces of information taken from these curves are the development time of a flour, its tolerance to overmixing, and optimum water absorption. The time for a dough to reach the point of minimum mobility (optimum mix) is called the mixing, or peak, time. With a farinograph, the tolerance of a flour to overmixing is included in several measurements: the stability, mixing tolerance index, and departure time (D'Appolonia and Kunerth, 1984). For a mixograph, the tolerance of a flour can be seen in the weakening angle, the area under the curve, the height of the curve at a specified time after peak, and the angle between the ascending and descending portions of the curve (Kunerth and D'Appolonia, 1985).

An additional piece of information usually obtained with these instruments is the amount of water needed to produce a dough of "optimum" absorption. For both instruments the absorption percentage obtained should be used only as a general indicator, as its derivation is subjective. With a faringraph the absorption percentage is obtained by measuring the amount of water required to produce a dough of an arbitrarily chosen consistency. The strength of the correlation between this value and the optimum absorption in a bakery needs to be determined on an individual basis, as this method makes the assumption that water level is the only factor that changes the consistency of a dough at its optimum development point. This is not the case if there are large differences in protein levels or starch damage among the flours tested.

The situation for a mixograph is more complicated, as no predetermined "optimum" consistency exists that doughs can be compared to. This means that some other method must be used to determine optimum absorption. In most cases the determination is accomplished by evaluating curve width, height, and shape, coupled with a lot of experience. Even so, complications can arise, because with this instrument, the amount of added water has a significant effect on the measurable mixing parameters. For example, absorption has been shown to be strongly correlated with peak time, area, and peak height (Kunerth and D'Appolonia, 1985). One method that might be used to overcome this problem is that employed at the U.S. Department of Agriculture/Agricultural Research Service Northern Wheat Quality Lab (Kunerth and D'Appolonia, 1985). All mixograms are run at a constant absorption. Baking absorption is then obtained from an experimentally derived chart that is based on peak height. True mixing time is also calculated by correcting the peak time by a factor that is also based on peak height.

While several numbers are usually taken from curves of both instruments, it should be realized that they do not represent fundamental rheological values. Rather, they give information about how a flour reacts during mixing. With experience, it is possible to take this information and predict mixing time and optimum absorption when the tested flour is used in a full bread formula and mixed in a larger mixer. In a bakery, where uniformity of the flour supply is very important, it is more common to use these instruments to verify that mixing parameters of a particular shipment of flour are not significantly different from those of previous shipments.

In most cases instruments such as the mixograph and farinograph are used only to measure parameters of simple flour–water doughs. It is assumed that other dough ingredients either have a minimal effect on dough rheology or, at the very least, have a constant influence. With the increasing production of variety breads, this rationale must be questioned. The inclusion of fiber in the formula, particularly from non-wheat sources, and the consequent addition of gluten to compensate for the presence of this fiber, provides a much greater opportunity for interactions to occur between the flour and other dough ingredients. If these interactions affect dough rheology, then the levels of these interactions may vary between flour shipments. It may, therefore, be beneficial to test these interactions in a farinograph or mixograph.

An extension of the above-described methods of gathering information about the mixing characteristics of a flour, which is becoming more popular with the increased use of computers, is recording curves on a computer through substitution of a transducer–computer combination for the pen recording system. This practice allows for a more rapid, reproducible, and accurate calculation of mixing parameters from the curves (Rubenthaler and King, 1986).

A slightly different approach is the use of a wattmeter to measure the work done by the mixing motor. This technique can also be easily computerized and may be used on a farinograph, mixograph, or full-scale mixer on a bakery floor (Kilborn and Preston, 1985). Provided that the motor is not oversized for the mixer, the output from this meter closely resembles that from the torque-measuring devices ordinarily found on both instruments. This procedure has been advocated as a means of monitoring ingredient addition (time of salt addition), scaling errors (incorrect levels or omissions), as well as dough development time.

During the past 25 years there have been many attempts to correlate values obtained from these torque-measuring instruments with various parameters of the final baked product. The fact that they have not been particularly successful points again to the limited application of their results, and to the need for a better understanding of how fundamental rheology is involved in the mixing and baking process.

As the study of fundamental rheology is becoming more popular in bread-making research, the use of instruments utilizing dynamic rheological measurements is increasing. These instruments can be used to follow the changes in fundamental rheological properties during the mixing process, and to study the effects of various ingredients on those properties (Abdelrahman and Spies, 1986).

PROCESSING

Once a dough is mixed, its final rheological characteristics are set. This is not to say that the dough does not change any more during further processing; in fact, its fundamental mechanical properties change substantially. Rather, nothing further can be added to the dough to alter those characteristics. Any adjustments that are needed to correct processing problems must be made in the settings of the processing equipment or in the time spent during various stages.

The rheological properties of a dough play important roles during several processing steps after mixing. A valid generalization is that any time work is done on a dough, it needs some time to recover from that action before the next processing step is encountered. During these periods, the relaxation properties of a dough have a strong influence on the rheological state of the product as it enters the next processing step.

As a dough piece is divided and rounded, the balance of its viscoelastic properties is critical. A dough that is too viscous flows too much during this step and does not maintain the desired final shape. If the elastic component is too dominant (often termed a bucky dough), it is difficult to round into the desired shape and the final product also does not have the desired shape.

Another rheological property, stickiness, is also important at the dividing and rounding stage. A dough that is too sticky does not leave the rounder properly and produces an excessive amount of "doubles." If, on the other hand, a dough is not sticky enough, it is not formed properly by the rounder and the resulting product does not have the desired crumb structure.

A correct balance of viscoelastic properties is also important during the sheeting and moulding steps. If a dough is too extensible (its viscous component too dominant), the result is a piece of dough that is too long for the moulding step. If the elastic component is excessive (bucky), the dough springs back too far after sheeting and produces a piece that is too short to fill the pan completely. An appropriate level of stickiness is also needed to help hold the folded layers of dough together and prevent large holes in the final baked product.

As a dough piece proofs prior to baking, its rheological properties help determine the final shape of the loaf of bread. The desired loaf shape results when the dough piece expands to fill the pan, not when it flows. This behavior implies that elastic behavior should be dominant during this phase. If the viscous nature of the dough is too strong, it flows to fill the corners of the pan, resulting in a loaf with sharp edges and a flat top. On the other hand, if the dough is excessively elastic, it does not expand sufficiently during proofing or baking. In this case, the final loaf will have a low volume and a shape that resembles the initial moulded piece rather than the loaf pan.

Compared to the volumes of literature available on the empirically measured rheology of dough during mixing, relatively little work has been reported on the rheological properties of dough after mixing. A major reason for this lack of information is the difficulty in working with a fermenting dough. The constantly changing dimensions and physical properties of the system makes measurements much more difficult.

Measurements of the rheological properties of a dough after mixing are usually made using instruments that measure stress–strain relationships, thus providing information about elasticity. The most common instruments used for this type of measurement are the Brabender Extensigraph and the Chopin Alveograph. Because of the complexity of these instruments and the time involved in operating them, they have not been widely used for quality control applications, but rather in research situations.

The extensigraph gives information about resistance to stretching and extensibility of a dough. It does so by measuring the force required to pull a hook through a rod-shaped piece of dough. An example of an extensigraph curve and its measurements are shown in Figure 7-4. The resistance to extension is related to the elastic properties of a viscoelastic dough, and the extensibility is related to the viscous component. The most commonly used values obtained from this instrument are the area under the curve, and the ratio of stretching resistance to extensibility. The ratio is a good indication of the balance of elastic and viscous components of the dough, and the area is related to the absolute levels of these components. It is also possible to run the same type of test with an Instron Universal Testing Machine, if the proper testing heads are used (Frazer, Fitchett, and Russell Eggit, 1985).

The alveograph provides similar information by measuring the pressure required to blow a bubble in a sheeted piece of dough. An alveograph curve and the measurements taken from it are shown in Figure 7-5. An often cited advantage of the alveograph over the extensigraph is its mode of expansion. In contrast to the constant rate of extension in only one direction with the extensigraph test, the alveograph expands the dough in two directions and the rate of expansion varies as the bubble grows. It is felt that this action more closely resembles the action on a piece of dough during fermentation and the early stages of baking (Launay and Bure, 1977).

There is still some controversy over what type of rheological data is contained in some of the numbers taken from an alveograph curve. Several workers have found the values for H (maximum height) and P (overpressure) to be related solely to test geometry and relaxation time, and not to a rheological property of the dough (Hlynka and Barth; 1955, Bloksma 1957). It is a

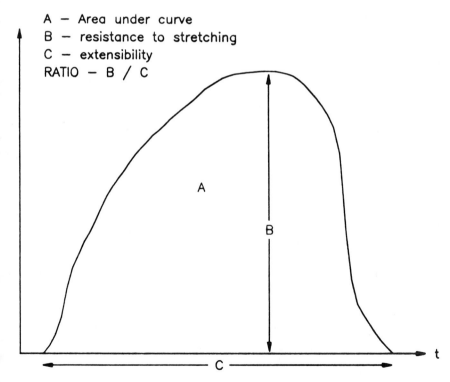

A — Area under curve
B — resistance to stretching
C — extensibility
RATIO — B / C

Figure 7-4. Example of an extensigram and associated measurements.

generally accepted definition for P, however, that P is an indicator of the resistance to deformation. Using that definition, information about the elastic component of a dough is contained in the maximum height of the curve (P and H), and the length of the curve (L) is related to the viscous portion of a dough. It follows that the P/L ratio is another indication of the balance of elastic to viscous components in a dough.

A drawback of both instruments is the relative complexity of the testing procedures. Both methods require that the dough be formed into a reproducible shape, with a reproducible amount of work. This is not easy to do. To help overcome this problem, both methods average results of tests run on multiple dough pieces. Even so, a skilled operator is still required to obtain results with a satisfactory level of reproducibility.

It should be noted that, while both instruments have established methods, it may be beneficial to modify these methods to more closely fit a specific application. Modifications may be in the form of different dough formulas, mixing methods, resting periods, levels of yeast, or fermentation periods.

Although both instruments are primarily used empirically, it is theoretically possible to obtain true rheological values from both. With the extensigraph, this requires knowledge of the cross-sectional area of the portion of dough directly under the hook at all times during the test. This area is extremely

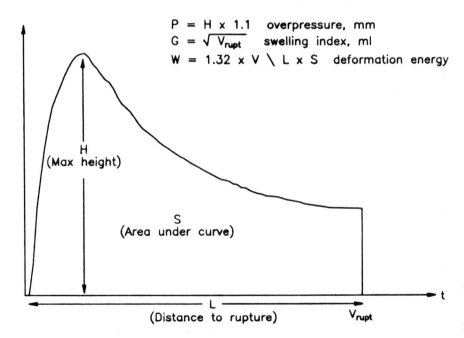

Figure 7-5. Example of an alveogram and associated measurements.

difficult to measure, as it changes continually during the stretching action, in a manner that is not calculable or predictable. For the alveograph, the calculation of theoretical values requires measurements of the bubble membrane thickness during the test. This also is not an easy measurement, because of irregularities in the bubble's surface during expansion (Rasper and Danihelkova, 1986).

As with the the torque-measuring devices, there has been little success in developing strong correlations between the measurements from these instruments and the final breadbaking performance of a flour. There are several possible reasons for this lack of correlation. One reason, alluded to earlier, may be the use of methods that are not based on the actual baking process. Another may be the fact that both methods rely on measurements taken when a dough is stressed beyond its elastic limit. A dough does not experience this level of stress after mixing, and it may not be important in determining the final baking performance. Another set of complications are the heat-induced changes in the rheological properties that occur during baking. It is likely that flours with similar rheological properties respond in a different manner to oven heating, either by setting at different temperatures or by losing their elastic character at different rates. These differences point to the fact that little work has been done on the measurement of true rheological parameters, and how they relate to final baking performance.

The spread test is a relatively easy empirical test that provides information regarding rheological changes during fermentation. A simplified version of a creep test, the spread test involves measuring the width and height of a dough cylinder during fermentation (Hoseney et al., 1979). The advantages of the spread test are that it is very simple to run and requires minimal instrumentation; the disadvantages are that it is lengthy and it does not provide fundamental rheological information.

A simple method of measuring the effects of dough rheology on processing parameters that may be beneficial in a plant environment is the monitoring (by a transducer) of the force developed when a dough piece passes through a sheeter (Kilborn and Preston, 1985). It is possible to measure both the length of a sheeted dough piece and the work done on that piece. With this information, the relative extensibility (buckiness) of a dough can be determined. This information could be used in an automatic feedback system to control the gap of the sheeting rolls.

BAKING

The rheology of a dough changes dramatically during baking, as it is converted into a loaf of bread. The nature of these changes and the times or temperatures at which they occur appear to play a large role in determining the final volume of a loaf of bread (Moore and Hoseney, 1986, Junge and Hoseney, 1981). The inability, so far, to successfully correlate any empirical rheological measurements during the mixing or processing of dough to the final loaf volume may be an indication that these changes are not related to the prior rheological history of the dough.

The ratio of the viscous to the elastic components drops significantly during heating, particularly as the dough is heated from 55°C to 75°C (Dreese et al., 1988). Most of the drop in the ratio appears to be due to the large increase in the elastic nature of the dough. As gluten is responsible for the elastic nature of the dough, changes in the rheology of gluten would appear to be responsible for the final setting of the loaf.

The exact reasons for this change in gluten elasticity are still unclear. It is unlikely that gluten denaturation is the cause, as was previously thought, because recent attempts to measure changes in gluten during heating have revealed no evidence of a denaturation process (Arntfield and Murry, 1981, Eliasson and Hegg, 1980). Other possible explanations center around chemical changes that occur in gluten during heating, and interactions between gluten and various lipid fractions in the flour and dough (Schofield et al., 1983).

It is interesting to note that starch gelatinization is a prerequisite for these changes, whatever they are. Dreese and co-workers (1988) showed that the extent of the increase in the elastic component of the dough was directly related to the amount of starch that was present. In fact, if most of the starch was removed from the system, there was no increase in the elasticity of the gluten.

BREAD TEXTURE

The final area of rheological importance in breadbaking is final product texture. Many components are part of the texture of a slice of bread, such as firmness, stickiness, and springiness, but firmness is the one most often measured. The major reason for measuring firmness is the strong correlation between crumb firmness and consumer perception of the freshness of white pan bread (Axford et al., 1968). This is shown in Figure 7-6.

Many instruments can be used to measure the firmness of bread crumb, but most use some type of plunger to deform a slice of bread and then measure either the compression distance resulting from an application of a fixed force (softness), or the force needed to compress the sample a specific distance (firmness) (Bice and Geddes, 1949).

Several relatively inexpensive instruments do a satisfactory job of measuring either bread firmness or softness. These include the Baker Compressimeter, the Bloom Gelometer, the Precision Penetrometer, and the Voland-Stevens Texture Analyzer (Baker et al., 1987, Kamel and Rasper, 1986).

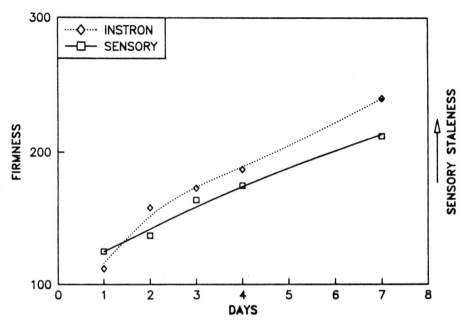

Figure 7-6. Changes in bread texture during storage as measured by a sensory panel and Instron firmness (unpublished data).

The most versatile (and most expensive) of the commonly used instruments is Instron's Universal Testing Machine (Baker and Ponte, 1987). Through the use of various testing heads and configurations on this instrument, it not only is possible to measure most components of a final product's texture, but it is also possible to measure many fundamental rheological properties of a dough. An example of a curve that could be obtained from any of these instruments is shown in Figure 7-7.

While the American Association of Cereal Chemists (AACC) has published standard methods for measuring the firmness of a slice of bread (AACC 74-09, 1986; AACC 74-10, 1961), many labs use slight variations of their own design. Several factors, independent of the type of instrument used, must be taken into account when a method is being developed that will be accurate and reproducible. A slice of bread presents a nonhomogeneous surface to the testing instrument. The best possible test would, therefore, use the largest possible platen surface on the compressing probe. Also of great importance is the need to avoid the areas close to the crust, as they will have a disproportionately large effect on the compressibility test. Some methods avoid this problem by removing the crust immediately before compression testing, but this practice can be extremely time consuming.

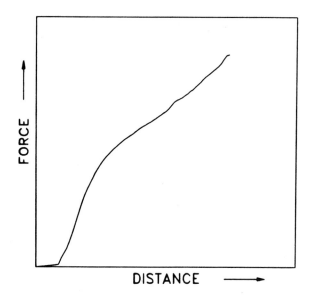

Figure 7-7. Compression curve of a slice of bread taken with an Instron Universal Testing Machine.

Another factor that must be taken into account is the variation in compressibility that is found at different locations in a loaf of bread (Hibberd and Parker, 1985). The crumb is firmer in the center of the loaf than it is towards the ends. This variation is usually accounted for by consistently testing slices from the same location in a loaf.

The amount of compression used in the test is another variable. When a slice of bread is compressed, there is a point at which the elastic limit of the crumb is exceeded and the slope of the compression curve changes (Figure 7-7). It is more desirable, in terms of reproducibility, to use conditions that result in readings being taken *after* that change in slope (Baker and Ponte, 1987).

While firmness is the component most often measured in assessing bread texture, several other components also contribute to overall texture. A method often used to measure other components of food texture is repeated compressions of the same sample (Bourne, 1978). If the force is measured throughout the compression–release–compression cycle, it is possible to evaluate adhesiveness as the negative force caused by the sample sticking to the platen during release, or cohesiveness as the ratio of the second compression force to the first compression force of the bread crumb.

Another texture factor, which is being looked at more frequently as microwaving becomes more popular, is the rubberiness, or elasticity, of the crumb as it is torn. Study of this factor requires replacement of the normal compression head with a type of head that shears the slice as it is compressed (Dahle and Sambucci, 1987).

APPLICATIONS OF FUNDAMENTAL RHEOLOGY

As more fundamental knowledge is gathered about the rheology of dough throughout the breadmaking process, and how the properties at one stage will affect either the final product quality or processing problems in later stages, it will become easier to use this knowledge in a quality control program. It should be possible to establish a more complete set of specifications for the incoming ingredients. Such a set of specifications would allow better control of processing tolerances and final product quality.

Process development would benefit from the knowledge of how each piece of equipment affects the important rheological properties of dough. This knowledge would help in the design of new processes and in troubleshooting at existing plants. It would also allow a greater use of feedback loops for process control.

Professionals involved in product development would also benefit from an increase in the knowledge of fundamental rheology of dough. As more becomes known about how each ingredient affects the rheology of a dough, it should be easier to build a desired set of rheological properties into a product, either for processing ease or for final product properties.

As more knowledge is developed about the fundamental rheology of the breadmaking process, it should be possible to more easily merge the "art"of breadmaking with the "science" of breadmaking.

REFERENCES

Abdelrahman, A., and Spies, R. 1986. Dynamic rheological studies of dough systems. In *Fundamentals of Dough Rheology*, ed. H. Faridi and J. Faubion, pp. 87–103. St. Paul, MN: American Association of Cereal Chemists.

American Association of Cereal Chemists. 1961. Staleness of bread compression test with Baker compressimeter. Method 74-10. In *Approved Methods of the AACC*. St. Paul, MN: American Association of Cereal Chemists.

American Association of Cereal Chemists. 1986. Bread firming by Universal Testing Machine. Method 74-09. In *Approved Methods of the AACC*. St. Paul, MN: American Association of Cereal Chemists.

Arntfield, S. D., and Murry, E. D. 1981. The influence of processing parameters on food protein functionality. I. Differential scanning calorimetry as an indicator of protein denaturation. *Inst. Canadian Sci. Tech.* **14**:436–451.

Axford, D., Colwell, K., Conford, S., and Elton, G. 1968. Effect of loaf specific volume on the rate and extent of staling in bread. *J. Sci. Food Agric.* **19**:95.

Baker, A. E., Dibben, R. A., and Ponte, J. G., Jr. 1987. Comparison of bread firmness measurements by four instruments. *Cereal Foods World* **32**:486–489.

Baker, A. E., and Ponte, J. G., Jr. 1987. Measurement of bread firmness with the Universal Testing Machine. *Cereal Foods World* **32**:491–493.

Baker, J. C., and Mize, M. D. 1937. Mixing dough in a vacuum and in the presence of various gases. *Cereal Chem.* **14**:721–734.

Bice, C., and Geddes, W. 1949. Studies of bread staling. IV. Evaluation of methods for measurement of changes which occur during bread staling. *Cereal Chem.* **26**:440–465.

Bernardin, J. E., and Kasarda, D. D. 1973. Hydrated protein fibrils from wheat endosperm. *Cereal Chem.* **50**:529–536.

Bloksma, A. H. 1957. A calculation of the shape of the alveograms of some rheological model substances. *Cereal Chem.* **34**:126–136.

Bourne, M. C. 1978. Texture profile analysis. *Food Technol.* **32**:62–66.

Dahle, L. K, and Sambucci, N. 1987. Application of devised universal testing machine procedures for measuring texture of bread and jam-filled cookies. *Cereal Foods World* **32**:466–470.

D'Appolonia, B. L., and Kunerth, W. H. 1984. *The Farinograph Handbook.* St. Paul, MN: American Association of Cereal Chemists.

Dempster, C. J., Cunningham, D. K., Fisher, M. H., Hlynka, I., and Anderson, J. A. 1956. Comparative study of the improving action of bromate and iodate by baking data, rheological measurements, and chemical analysis. *Cereal Chem.* **32**:221–239.

Dreese, P. C., Faubion, J. F., and Hoseney, R. C. 1988. Dynamic rheological properties of flour, gluten, and gluten–starch doughs. I. Temperature-dependent changes during heating. *Cereal Chem.* **65**:348–353.

Eliasson, A. C., and Hegg, P. O. 1980. Thermal stability of wheat gluten. *Cereal Chem.* **57**:436–437.

Frazier, P. J., Fitchett, C. S., and Russell Eggit, P. W. 1985. Laboratory measurement of dough development. In *Rheology of Wheat Products*, ed. H. Faridi, pp. 151–176. St. Paul, MN: American Association of Cereal Chemists.

Hibberd, G., and Parker, N. 1985. Measurements of the compression properties of bread crumb. *J. Text. Stud.* **16**:97–107.

Hoseney, R. C., Hsu, K. H., and Junge, R. C. 1979. A simple spread test to measure the rheological properties of fermenting dough. *Cereal Chem.* **56**:141–143.

Hlynka, I., and Barth, F. W. 1955. Chopin alveograph studies. II. Structural relaxation in dough. *Cereal Chem.* **32**:472–480.

Junge, R. C., and Hoseney, R. C. 1981. A mechanism by which shortening and certain surfactants improve loaf volume in bread. *Cereal Chem.* **58**:408–412.

Kamel, B., and Rasper, V. F. 1986. Comparison of Precision Penetrometer and Baker Compressimeter in testing bread crumb firmness. *Cereal Foods World* **31**:269–274.

Kilborn, R. H., and Preston, K. R. 1985. Grain Research Laboratory instrumentation for studying the breadmaking process. In *Rheology of Wheat Products*, ed. H. Faridi, pp. 133–150. St. Paul, MN: American Association of Cereal Chemists.

Kunerth, W. H., and D'Appolonia, B. L. 1985. Use of the mixograph and farinograph in wheat quality evaluation. In *Rheology of Wheat Products*, ed. H. Faridi, pp. 27–50. St. Paul, MN: American Association of Cereal Chemists.

Launay, B., and Bure, J. 1977. Use of the Chopin alveograph as a rheological tool. II. Dough properties in biaxial extension. *Cereal Chem.* **54**:1152–1158.

Moore, W. R., and Hoseney, R. C. 1986. The effects of flour lipids on the expansion rate and volume of bread baked in a resistance oven. *Cereal Chem.* **63**:172–174.

Rasper, V. F., and Danihelkova, H. 1986. Alveography in fundamental dough rheology. In *Fundamentals of Dough Rheology*, ed. H. Faridi, and J. Faubion, pp. 169–180. St. Paul, MN: American Association of Cereal Chemists.

Rubenthaler, G. L., and King., G. E. 1986. Computer characterization of mixograms and their relationship to baking performance. In *Fundamentals of Dough Rheology*, ed. H. Faridi, and J. Faubion, pp. 131–168. St. Paul, MN: American Association of Cereal Chemists.

Schofield, J. D., Bottomley, R. C., Timms, M. F., and Booth, M. R. 1983. The effect of heat on wheat gluten and the involvement of sulphydryl–disulfide interchange reactions. *J. Cereal Sci.* **1**:241–253.

Schroeder, L. F., and Hoseney, R. C. 1978. Mixograph studies. II. Effect of activated double bond compounds on dough mixing properties. *Cereal Chem.* **55**:348–359.

Shuey, W. C. 1975. Practical instruments for rheological measurements on wheat products. *Cereal Chem.* **52**:42r–81r.

Smith, D. E., and Andrews, J. S. 1952. Effect of oxidizing agents upon dough extensigrams. *Cereal Chem.* **29**:1–17.

Tanaka, K., Furukawa, K., and Matsumota, H. 1967. The effect of acid and salt on the farinogram and extensigram of dough. *Cereal Chem.* **44**:675–680.

Tipples, K. H. and Kilborn, R. H. 1975. "Unmixing"—The disorientation of developed breaddoughs by slow speed mixing. *Cereal Chem.* **52**:248–256.

Chapter 8

Application of Rheology in the Cookie and Cracker Industry

Hamed Faridi

Cookie and cracker manufacturing lends itself to extensive mechanization, and it is now entering on an era of increasing automation. Although mechanization has made it possible to reduce labor costs and to eliminate many tedious and repetitive jobs, it has also created a great demand for better understanding of factors that affect the cookie/cracker manufacturing process (Manley, 1983).

Crackers, and to a lesser degree cookies, are made from viscoelastic doughs that undergo large changes in their rheological properties during processing. Although time-dependent changes in dough properties make it difficult to make useful measurements during processing, determining the rheological properties of a dough yields valuable information concerning the quality of the raw materials, the machining properties of the dough, and possibly the textural characteristics of the finished product. In this chapter, factors affecting cookie and/or cracker dough rheology are discussed and major instruments or techniques available for measuring short dough rheology are briefly described.

COOKIES

In general, cookies are products made from soft and weak wheat flours. They are characterized by a formula high in sugar and shortening and low in water. Cookie doughs are cohesive but to a large degree lack the extensibility and elasticity characteristic of bread doughs. Relatively high quantities of fat and sugar in the dough allow dough plasticity and cohesiveness without the formation of a gluten network (Hoseney, 1986). In addition, and again depending on the formulation, cookie dough tends to become larger and wider as it bakes rather than to shrink as does cracker dough. This increase in size, or "spread," is the greatest single problem in process control (Manley, 1983).

The most tender eating cookies can be obtained from a given short dough only if the amount of mixing is kept to a minimum. Minimal gluten development is achieved by carrying out the mixing process in two or even three stages. The mixing technique for cookie production was classified by Kramer (1953) into three categories: (1) the multistage or creaming method in which several steps are used, (2) the simplified single stage, and (3) the continuous. Obtaining a dough of correct consistency at the end of mixing is important, and the mixing step can therefore be considered critical. Compared to the "art" of cookie baking, process control engineering is a relatively new technology in which the mixing process is neither well defined nor understood. The bulk of the published research refers to the mixing of flour–water doughs or formulas more representative of bread than of either cookies or crackers. Measuring (quantifying) rheological changes during the mixing process is crucial for process automation. What constitutes an optimally mixed cookie or cracker dough has yet to be determined.

Perhaps the best way to classify cookies is by the way the dough is placed on the baking band. Such a classification allows us to divide cookies into three major types.

Rotary Molded Cookies

For rotary molded cookies, the dough is forced into molds on a rotating roll. As the roll completes a half turn, the dough is extracted from the cavity and falls on the baking oven band. The consistency of the dough must be such that it feeds uniformly and readily fills all of the crevices of the roll cavity under the pressure existing in the feeding hopper. At the same time, it must be possible to extract the dough pieces cleanly from the cavity without their undergoing distortion or forming "tails." Rotary molded dough must also be sufficiently cohesive to hold together during baking. Dough spread and rise should usually be minimized. Doughs formulated to meet these requirements are usually fairly high in sugar and shortening but low in moisture. The development of gluten is definitely to be avoided. The typical dough is crumbly, lumpy, and stiff, with virtually no elasticity. Much of the cohesiveness of this type of dough comes from the plastic shortening used (Hoseney, 1986).

Wire-Cut Cookies

In the production of wire-cut cookies, a relatively soft dough is extruded through an orifice and cut to size, usually by a reciprocating wire. The dough must be cohesive enough to hold together on the band, yet short enough to separate cleanly when cut by the wire. The rate of extrusion is related to the dough consistency. Here, dough spread during baking is desirable but must be closely controlled. The bulk of the published research on the chemistry or technology of the cookie system has been done using the wire-cut system or its model. Another type of extruded dough is similar to wire-cut except that the extrusion of dough is continuous without wires and the orifices are usually designed to produce strips rather than round shapes. Fig bars are made by extruding a fig paste within a tube of dough of similar consistency.

Cutting Machine Cookies

The process and formulation of cutting machine cookies produces the familiar Christmas cookie. A dough with slightly less fat and sugar but more water than rotary mold dough is sent through multiple sheeting rolls and made into a continuous sheet. The sheeting and cutting operations for the short doughs are essentially the same as for cracker doughs. The dough shows very little elasticity, so shrinkage before cutting is not a problem. Cutting can be similar to the production of crackers in that docking and imprinting of a name or simple pattern is performed before the outline is cut. One major rheological problem with this method is that due to the thickness of the dough sheet, the amount of scrap resulting from the process is excessive. Since the scrap dough is both more dense and more worked (developed) than fresh dough, its incorporation into a new dough can be critical. Due to this difficulty, relatively few cookies are made by this method (Matz and Matz, 1978).

CRACKERS

In general, crackers contain little or no sugar but moderate levels of fat. The dough generally contains low levels of water. There are two major types of crackers, saltine and snack. Crackers, particularly saltines, are deceptively simple food systems. In fact, the process required to produce acceptable saltines is both lengthy and complex—so complex that the significant changes occurring in the physical properties of the dough are just now being understood in detail. A brief description of the process follows.

Saltine crackers are made by a sponge dough process that requires about 24 hours, much of it for fermentation. Since cracker sponges are mixed just long enough to wet the flour, gluten development occurs only to a limited extent at this stage. During the next 19 hours of sponge fermentation, the consistency of the sponge changes drastically and the sponge becomes less elastic. Results of extensigraph and farinograph studies conducted on cracker sponges by

Doescher and Hoseney (1985) and Faridi (1975) showed that resistance to extension and extensibility, as well as cohesiveness (farinograph band thickness), decreased with fermentation time (Figure 8-1). After fermentation, the sponge is mixed with other dough ingredients and the dough-up flour. The dough is allowed to relax for 4 to 6 hours.

After the relaxation/resting period is over, the dough is taken to the hopper of the sheeter. Sheeting has several functions, the most obvious of which is to compact and gauge the mass of dough into a sheet of even thickness and at the full width of the band. The dough sheet must have no significant holes, and the edges should be smooth, not ragged. Often the sheeter also enables the incorporation of dough returned from the cutter, known as cutter scrap, with fresh dough (Manley, 1983).

The second, and often overlooked, function of sheeting has major consequences for the rheological properties of the dough. Dough enters the sheeter as a dry, poorly cohesive mass and exits as a continuous cohesive sheet. Clearly, gluten development occurs during sheeting. Sheeters under the dough hoppers almost always have three rolls. Front and back discharge three-roll sheeters are made. The back discharge system is preferred for all extensible doughs, but front discharge is required when the dough is weak and short and needs to be well supported as the sheet leaves the sheeter.

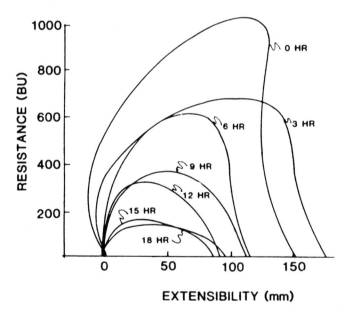

Figure 8-1. Effect of fermentation time on the extensigram properties of cracker sponges. Each sponge was adjusted to pH 7.0 with soda. BU = Brabender unit. *(From Doescher and Hoseney, 1985, with permission.)*

It should be noted that the greater the height of dough in the hopper, the more the pressure at the sheeter and therefore the more extrusion of dough through the machine (i.e., more dough passes through for a given number of revolutions of the rolls). Thus, for process control purposes, it is advisable to try to maintain a constant level of dough in the hopper. The best way to do this is via a presheeter that meters to the three-roll sheeter. However, it is quite common to drop whole batches of 1,500 pounds of dough into the hopper at intervals. In this case one must expect a change in delivery rate as the hopper empties (Manley, 1983).

Little, if any, research has been published on the effects on cracker quality of different amounts of work done on the dough in sheeters. This lack is unfortunate, since the physical properties of the dough are profoundly changed by the sheeting process. It is clear, however, that the surface of the sheet as it emerges from the sheeter is of great importance to both the baked cracker surface appearance, and often to the degree of lift obtained in the oven. It seems that a rough, rippled, or holey sheet surface cannot be satisfactorily "repaired" during subsequent gauging.

Cracker doughs are laminated after exiting the sheeting rolls. The purpose of laminating the sheeted dough (into eight layers) is fourfold. First, it constitutes a method of repairing a dough sheet with holes or tears. Second, by turning the folded dough through 90°, stresses in the dough are made more uniform in two directions. Third, the repetitive rolling and folding of the dough accomplishes a significant amount of work on the gluten. This process makes the dough more suitable for baking a delicate and layered structure characteristic of saltine crackers. Fourth, by introducing another material, like fat, between layers of dough, a characteristic flaky structure is produced after baking.

Multiple pairs of heavy steel rolls gradually reduce the dough sheet thickness to that desired for cutting. Typically, there are two or three pairs, although only one pair may be used for short doughs and more than three may be used where very gentle reductions in dough thickness are necessary. As a rule of thumb, the reduction in thickness should be about 2:1 for each pass through a roll pair, although ratios of up to 4:1 are used. Obviously, the greater the ratio, the more work and stress is put into the dough and the more its physical properties change.

Dough emerging from a gauge roll is always slightly thicker than the gap it has come through. This greater thickness results because of the elastic properties of the dough ("spring") and also because some extrusion, as compared with rolling, occurs through the nip. Tension on the dough should be both minimal and as constant as possible, to ensure the maintenance of a full sheet which is accomplished by allowing a slight loop in the dough on the feed and discharge sides of the gauge rolls. If the dough is pulled away from the rolls, tension is produced in the dough. Such tensions are maximal at the edges of the dough sheet (Manley, 1983).

Once it has passed the final gauge roll, the dough is allowed to relax and shrink before going to the cutter, by means of grossly overfeeding it onto an intermediate web to form ripples. The intermediate web then feeds the cutting web, whose speed is such that the ripples are just taken up and that a smooth sheet is presented to the cutter.

After sheeting, the dough is cut and docked. Cutting is usually done with a rotary cutter, which cuts individual crackers to size but leaves them together in a continuous sheet. Docking is done with docker pins that have blunt ends; the purpose of docking is to pin the dough together so that it will not separate into layers.

EFFECT OF MAJOR FORMULA INGREDIENTS ON COOKIE OR CRACKER DOUGH RHEOLOGY

Flour

Many of the physical properties of flour or its components affect the rheological properties of cookie and cracker doughs and consequently finished product quality. The principal properties and components are discussed below.

Flour Moisture

The moisture content of a flour is defined as the percent weight loss of a sample after it is heated evenly at 130°C for one hour. Variations in dough consistency are commonly blamed on the formula's flour and are corrected for by addition or reduction of the dough water level (Manley, 1983). Since the flour moisture changes over a typical range stated in each manufacturer's flour specification, this practice in itself may profoundly affect the consistency of any particular dough. For example, a typical cookie or cracker batch formula may contain 600–800 pounds of soft wheat flour. Seven hundred pounds of flour at 13% moisture has 609 pounds of dry solid content. If the flour moisture content increases by only 0.5%, an extra 3.5 pounds of flour dry solids is added to the formula, and this amount of additional solids is known to affect the rheological properties of the resulting doughs. Even so, in practice, ranges greater than 0.5% moisture are common in the industry. This is particularly true if a bakery receives flours from various suppliers.

Steele (1977) studied the role of flour moisture content on cookie dough consistency, and he demonstrated that a marked increase in dough firmness resulted from a small decrease in flour moisture content. The change in flour moisture content and the consequent change in dough consistency resulted in a substantial change in dough piece weight (density). His studies also indicated that added water (recipe) and water already present in flour are not equivalent. Therefore, compensating for flour moisture content by changing the formula water level does not yield doughs of equal consistency. Similar observations were reported by Doescher and Hoseney (1985).

Flour Protein Content

Although mixing and gluten development are not major issues in cookie and cracker production, the quality and quantity of flour protein are important factors in cookie and cracker dough rheology. By blending the wheat as well as air classification and stream selection of the resulting flour, the miller may achieve almost any desired flour protein level between 8% and 13%. Baking technologists, however, have long realized that protein quantity is not sufficient for predicting the baking performance of a flour. Experience has repeatedly shown that, although two flours may have identical ash and protein contents, they can perform very differently under comparable baking conditions. Understandably, baking technologists and cereal chemists have been looking for an alternative to such chemical tests. The most promising alternative has been the objective measurement of the dough's physical properties. During the past 100 years, therefore, sensitive instruments have been evolving that are designed to produce records of the mixing, absorption, fermentation, and oxidation characteristics of flours.

Several instruments are on the market today for measuring the protein "strength" of flour in somewhat different manners. The more widely used physical test methods and instruments may be divided into three broad categories, as follows.

Recording Dough Mixers. Mixers such as the farinograph and mixograph record the torque that results from mixing a dough at constant speed or, what is essentially the same thing, the resistance to mixing. Provided that both scales of the record are linear, the area under the curve is proportional to the energy required for mixing (Bloskma and Bushuk, 1988).

The recorded curves yield information about changes in rheological properties during mixing. These curves typically consist of a rising part, reflecting an increase in resistance with mixing time, and a more or less identifiable peak followed by a decrease in resistance. As could be expected from the end uses and the gross characteristics of the doughs, these curves can be quite different for hard (bread) and soft (cookie) wheat flours. Figure 8-2A shows representative farinograph records, or farinograms, of a hard and a soft flour. The height of the middle of the band, indicating consistency, is expressed in arbitrary Brabender units (BU). The mixograph uses a different, more severe, mixing action to produce a curve with the same three general features as for a farinograph.

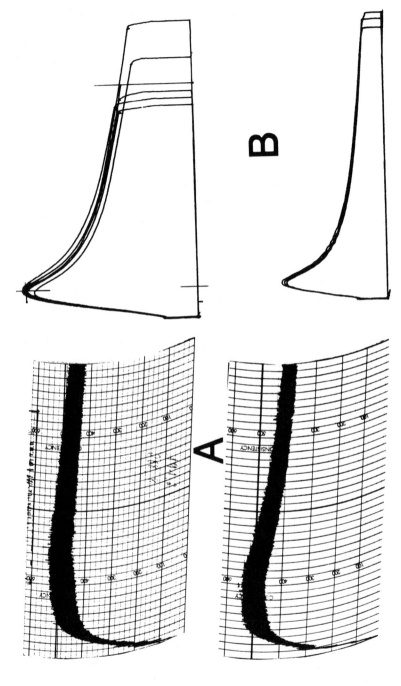

Figure 8-2. Farinograms (*A*) and alveograms (*B*) of typical strong (bread) and medium (cookie/cracker) flours. (*Courtesy of Brabender Instruments and Tripette et Renaud S.A., respectively.*)

Load-Extension Instruments. Instruments are available to assess the physical properties of nonfermenting dough when it has been mixed. The most common are the extensigraph and alveograph, in which a test piece of dough is stretched or expanded until it ruptures and the force versus extension curve is recorded. From these records, resistance to deformation and extensibility can be read. Whereas in the Brabender extensigraph, the extension of the test piece is in only one direction, or uniaxial, in the alveograph expansion is in multiple directions, or biaxial, in the surface of a bubble.

The surface area under the alveogram or extensigram is proportional to the energy that is required to rupture the test piece. The true reason for its importance is that flour strength is associated with both a large resistance and a large extensibility. Therefore, determining the surface area under the curve is a convenient method for characterizing flour strength.

Typical alveograph records, or alveograms, for soft and hard flours are shown in Figure 8-2B. The remarkable sharp maxima in the curves do not reflect changes in dough properties. These maxima can be explained by the geometries of the test pieces and their response to biaxial deformation. The usual interpretation of an alveogram is similar to that of an extensigram. The maximum height of the curve is a measure of resistance, and its length a measure of extensibility. Since in this method the doughs are made with the constant addition of water regardless of flour type, the resistance value is strongly affected by the absorption of water by the flour. Instead of the surface area under the curve itself, this area is multiplied by a constant factor; the product is called the W value.

Tests with the instruments described above provide information about flour strength and consistency. In recording dough mixers a high water absorption, a long dough development time, and a small degree of softening or mixing tolerance index, and in load-extension tests a large resistance and a large extensibility, are indicative of a strong flour. Flours lacking most of these properties are called weak and are more suitable for cookie production (Bloskma and Bushuk, 1988).

Viscometers. This broad category contains many types of viscometers, many of which may be used in cookie/cracker plants for measuring the viscosities of incoming flours as slurries or suspensions. The most popular instruments are the Brabender amylograph and the Brookfield viscometer. The former heats the test slurry through a programmed increase to 90°C, while the latter operates at constant temperature. High amylogram values indicate a lack of amylolytic activity or greater pasting properties of the sample flour or starch. The viscosity of an acidulated slurry of flour can be measured by a Brookfield instrument. In a dilute lactic acid solution, flour gluten swells considerably. This swelling increases the viscosity of the flour–water suspension, in direct relation to the swelling properties and the quantity of gluten present in the test flour.

Flour Starch Damage

During the milling process, as the endosperm is fractured and then crushed to flour, some of the starch granules are physically damaged. The amount of damage varies with the severity of the grinding operation and the physical hardness of the wheat kernel. Damaged starch has a profound effect on the water absorption capacity of the flour when a dough is made, because the capacity of damaged starch to absorb water is many times higher than that of the intact granular starch (Farrand, 1964). In the production of most types of cookies and crackers, since the final product must be almost dry (2–4% moisture), the amount of water used to make the dough should be at a minimum; therefore, flours of low water absorption and hence low starch damage are more suitable. High starch damage is a negative factor in soft wheat flour used as cookie flour.

Flour Ash Content

Bran contains a higher mineral content than endosperm and germ. Ash content, therefore, is a reflection of the bran content of the white flour. While it is known that flours with higher ash contents and darker color cause difficulty in dough handling and processing, neither ash nor color is responsible for adverse effects. Pieces of bran may physically damage the gluten network. In addition, the presence of non-starchy networks forming polysaccharides such as pentosans affects the rheological properties of wheat flour doughs.

Sugars

The quantity and quality of sweeteners have significant effects on the rheology of a cookie dough, as well as on the texture, appearance, and flavor of the finished product. Machining properties and the response of the dough piece to oven conditions are also closely related to the type and quantity of sweetener employed (Matz and Matz, 1978). Sucrose acts as a hardening agent by crystallizing as the cookie cools, which makes the product crisp (Hoseney, 1986). The properties and types of sugars that affect cookie and cracker dough rheology include the following.

Particle Size

The finer the sugar, the faster the rate of solubilization in water. On the average, only about half the formula sugar is actually dissolved during mixing.

The lack of available water causes the remaining sugar to stay in solid form until the dough is heated in the oven, at which time the sugar dissolves. When sugar dissolves, it increases the volume of solution in the system. Each gram of sugar, when dissolved in 1 ml of water, produces 1.6 ml of total solution. The increase in total solution volume has a pronounced effect on the cookie dough, tending to make it sticky. Thus, if a manufacturer attempts to produce cookies using a syrup rather than crystalline sucrose, sticky doughs are produced that do not machine well. Large amounts of sugar in the formula tend to make the dough sticky and hinder release from dies and wires (Hoseney, 1986).

Crystalline State

The proportion of formula sugar present in the crystalline state affects its solubility in water and therefore affects dough consistency. Generally, the greater the extent of crystallinity, the lower is the solubility in water.

The amount of sucrose that can be replaced by high-fructose corn syrup (HFCS) in a cookie formula is often limited by the fact that the amount of water in the syrup exceeds the total water in the formula. Curley and Hoseney (1984) reported that replacement of sucrose with HFCS in a sugar-snap cookie formula affected dough rheology (stickiness), surface cracking, and the characteristic snap associated with this type of cookie.

The two Instron curves shown in Figure 8-3A indicate the compression and tension cycles of cookie doughs made with 0% and 100% dissolved sucrose. The dough with 0% dissolved sucrose was firm and manageable. During measurement on the Instron, the plunger made a quick and clean release from the dough. The dough with 100% dissolved sucrose was very sticky, making it unmanageable for a production situation. During the tension cycle on the Instron, the plunger released slowly, carrying part of the dough with it (Curley and Hoseney, 1984). When 50% of the granular sucrose was replaced with HFCS, the resultant dough was as sticky and unmanageable as the dough made with 100% dissolved sucrose. In addition, the dough was very soft. Instron measurement of that dough gave a curve (solid line, Figure 8-3B) similar to that obtained for the dough with 100% dissolved sucrose (Figure 8-3A). When water content in the HFCS dough system was reduced from 22.7% to 17.7% (based on the flour weight), the dough lost most of its soft and sticky characteristics. The dough was still slightly tacky but not tacky enough to affect its manageability. Olewnik and Kulp (1984) reported that changes in dough consistency due to sugar content were dependent on the type of cookie dough evaluated. The consistency of wire-cut doughs remained fairly constant, between 30% and 45% sugar (on flour basis). Sugar content variations in the consistency of "deposit" doughs were more pronounced. Doughs with high levels of added sugar showed sharp increases in both consistency

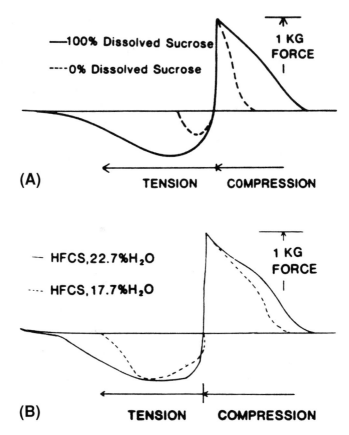

Figure 8-3. *(A)* Instron tension curves for dough stickiness. - - - = 0% dissolved sucrose _____ = 100% dissolved sucrose. *(B)* Effect of water level (percent based on flour weight) on stickiness of cookie doughs containing high-fructose corn syrup. *(From Curley and Hoseney, 1984, with permission.)*

and cohesive properties (Figure 8-4). This finding is in agreement with the findings of Steele (1977) and Miller (1985), who reported that increased sugar levels cause dough firming.

Fats

Fats are the third largest components after flour and sugar in most cookie formulas. During dough mixing, the aqueous phase and fat compete for the surfaces of the flour particles. When fat coats the flour prior to hydration, gluten network formation is inhibited. If the fat level is high, the lubricating

Sucrose Variations
Wire Cut

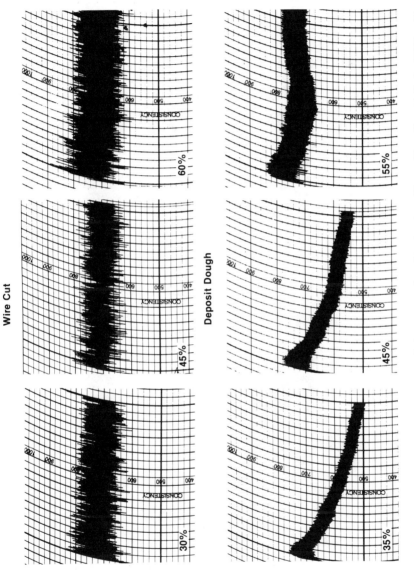

Figure 8-4. Effect of sucrose levels on farinograms of cookie doughs. *(From Olewnik and Kulp, 1984, with permission.)*

function in the dough is so pronounced that little water is required to achieve a desired consistency (Figure 8-5). The rheological properties of a dough in large part reflect those of the solid phase of the shortening present. Miller (1985) reported that increasing formula fat levels has a softening effect and lowers the consistency of short doughs. Similar results were reported by Olewnik and Kulp (1984).

It is important to know the melting characteristics (solid fat index) of the fats used in a cookie formula. Fat melting temperature determines the form of the fat at a given dough temperature when dough pieces are formed. The rheological properties of the plastic shortenings used in cookies are extremely temperature sensitive. It is not surprising, therefore, that uniform

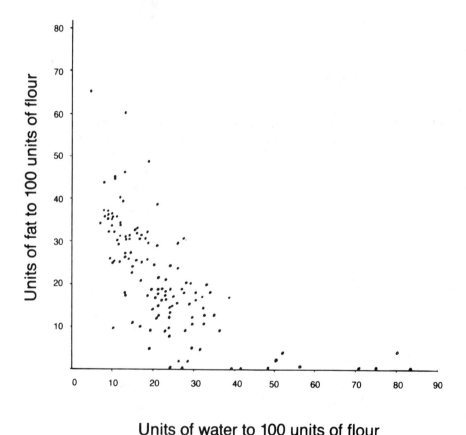

Figure 8-5. Relationship of fat and water levels in cookie doughs. *(From Manley, 1983, with permission.)*

temperatures during various steps of processing are essential for making uniform cookies. The temperature of the dough affects its physical properties as well as cookie spread and surface appearance. To obtain the best results, dough temperature should be below the upper limit of the plastic range of the shortening—preferably between 70°F and 80°F, even lower if the shortening content is high.

OBJECTIVE MEASUREMENT OF COOKIE AND CRACKER DOUGHS

Variability in measuring dough consistency is considered to be a major source of problems in controlling dough piece weight and cookie/cracker dimensions. Subjective assessment of dough texture is the routine method of measuring dough consistency in bakeries. Two subjective tests are common in many bakeries. In one, the mixer operator takes a sample of dough in his/her palm and then, either compressing it with his/her fingers or inserting his/her thumb into the dough ball, the operator judges the dough to be too soft, acceptable, or too tight. The second test involves pulling the dough ball apart as in a test for extensibility related to gluten development. The first test is more common.

To date, no method is generally accepted for objective measurement of the mechanical properties of cookie and cracker doughs. However, some of the methods used for measuring the rheological properties of short doughs are described below.

Remixing

Various types of recording mixers may be used to measure the rheological properties of cookie and cracker doughs after they have been mixed in a commercial mixer. One major drawback of this approach is that remixing subjects the dough to a high-shear work input that changes the physical character of the dough to the extent that its behavior may no longer be indicative of its performance during cookie and cracker production. Olewnik and Kulp (1984) used the Brabender Farinograph to study the effect of mixing time and ingredient variation on the rheological properties of cookie doughs; the dough consistency (stiffness) and band width (cohesion) of the farinograms were used as appropriate test values. All doughs were made in a Hobart mixer equipped with a 12-quart bowl and cake paddle. After mixing was completed, 300 g doughs were transferred into the farinograph bowl for measuring consistency.

Compression–Extrusion Tests

Compression–extrusion tests consist of applying force to a food in a test cell of defined geometry until the food flows through an outlet that may be in the form of one or more slots or holes. As an example of such tests, the capillary rheometer (Figure 8-6) has been employed to measure the viscosity of doughs by an extrusion process. The pressure required to obtain flow at a constant rate is measured by a compression load cell and recorded. By the use of a variety of plunger speeds and capillary sizes, a wide range of viscosities and shear rates may be measured. While viscosities from about 5 to 5×10^6 poise may be measured, corrections for dough compressibility, entrance effects, capillary length to diameter ratios, and other parameters may be necessary

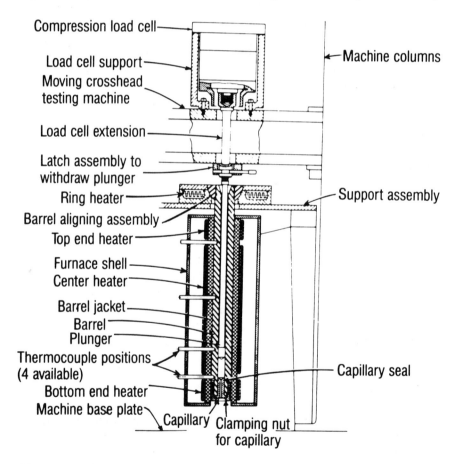

Figure 8-6. Cross section of an Instron capillary rheometer. *(From Johnson et al., 1975, with permission.)*

to obtain meaningful results. Another method for measuring the rheological properties of cookie and cracker doughs is back-extrusion. In back-extrusion, two physical movements are involved: a cylindrical plunger is forced down into a dough and the dough flows upward through a concentric annular space (Figure 8-7). The back-extrusion technique is simple but does require equipment such as the Instron. Quality control is an important area of application in which a large number of samples could be quickly tested at low unit costs, the method could be adapted for automated rheological testing (Osorio and Steffe, 1987).

In addition, tests exist that employ only compression to evaluate cookie doughs. The Instron instrument may be used in many ways to measure the mechanical properties of cookie and cracker doughs. The simplest method is that of attaching a compression cell load. Abboud and co-workers (1985) studied the rheological properties of cookie doughs with a compression cell of 2-kg maximum load. The tester was adjusted to give a full-scale deflection of 250 g. The head speed was set at 5 cm/min. The cookie dough was placed on the tester, and the force required to press the probe 3 cm into the dough was taken as the compression force. The stainless steel probe was covered with a 0.8 cm plastic tube attached to increase the surface area (Figure 8-8).

Miller (1985) developed a method to measure the consistency of short doughs by using a Stevens-Leatherhead Food Research Association (LFRA)

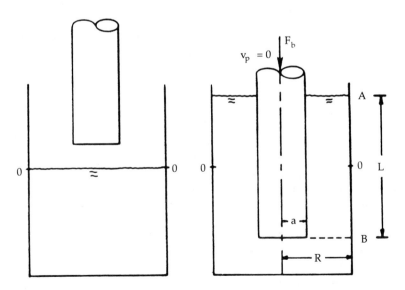

Figure 8-7. Schematic diagram of a simple cell for back-extrusion. Position of the plunger is shown before and after completion of testing. *(From Osorio and Steffe, 1987, with permission.)*

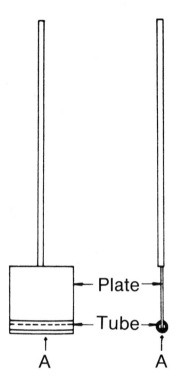

Figure 8-8. A 16 cm long stainless steel probe with a 3.2 × 3.2 × 0.1 cm plate at the end. A 0.6 cm plastic tube (A) is attached to increase surface area of contact in compression text. *(From Abboud et al., 1985, with permission.)*

Texture Analyzer. The method of preparing the dough samples included a simple means of removing randomly distributed pockets of air while minimizing handling of the dough. A 105-g sample of dough is weighed into a sample cup and compressed with the spiked lid to remove all air pockets (Figure 8-9). The surface of the dough is then flattened by the smooth surfaced lid. This will also make a dough with a predetermined specific volume. The dough consistency is obtained as the average of six compression measurements using two cups of dough (three each).

    Miller (1985) reported that the above method is operator dependent (Figure 8-10). The source of this dependency was dough sample preparation. Using a wide range of cookie flours and water levels, during a 3-year period, this investigator demonstrated that Stevens-LFRA dough consistency measurements were reasonably ($r = 0.85$) correlated with subjective assessment by determining the feel of the dough.

Figure 8-9. Sample preparation equipment, showing a compressed dough after flattening of the sample surface. (*From Miller, 1985, with permission.*)

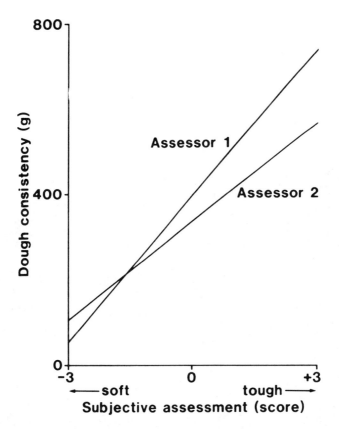

Figure 8-10. The relationships between dough consistency measurements and subjective scores recorded by two assessors. (*From Miller, 1985, with permission.*)

Fundamental Tests

Obtaining fundamentally sound rheological values for cookie and cracker doughs has been a challenge to dough rheologists. The first problem is that cookie and cracker doughs are difficult systems to work with, as they exhibit both viscous and to a lesser extent elastic properties. However, fundamental rheological values such as viscosity and elasticity can be measured by using some of the instruments that have been described. Published material on the fundamental properties of cookie and cracker doughs is scarce. Bagley and Christianson (1986) studied the response of commercial chemically leavened doughs to uniaxial compression using the Instron equipped with a 50 kg load cell. Yasukawa and co-workers (1986) and Mizukoshi (1986) studied the rheological properties of cake batters during expansion and heat setting.

On-Line Tests

Cookie and cracker doughs are viscoelastic materials, and they exhibit major changes in rheological properties during processing. Time-dependent changes in these dough properties make it difficult to obtain useful off-line measurements.

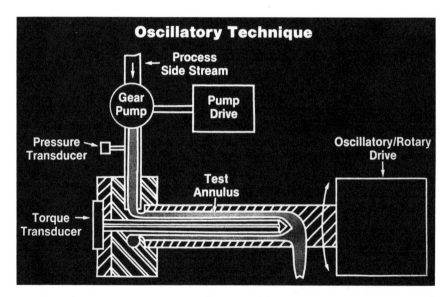

Figure 8-11. Schematics of an on-line rheometer. *(Courtesy of Rheometrics Inc.)*

Convincing evidence exists that bakeries of the future may no longer be able to afford off-line and/or laboratory instrumentation. On-line sensors will assess differences in the mechanical properties of the doughs, and an associated control system will make the necessary adjustments for the next batch. One of the instruments available for on-line dough rheology testing, the on-line rheometer, is manufactured by Rheometrics, Inc. (Piscataway, NJ). The principle behind the on-line rheometer is simple. A small side stream of dough is diverted from the manufacturing line, supplying dough to the gear pump of the rheometer (Figure 8-11). The pump feeds the dough at a controlled flow rate into the sample chamber, where the dough fills the gap between two concentric cylinders. A DC torque motor oscillates the outer cylinder, deforming the material; the material, in turn, exerts a torque on the stationary inner cylinder. The torque is measured, and the built-in microprocessor then calculates the viscosity and the viscous and elastic moduli. The test results can be printed and plotted in graphic format. Light-emitting diode (LED) displays on the console also display the real-time results. Steffe and Morgan (1987) have identified several problems with this on-line rheometer, such as the following: high pressure required to move material through the unit, internal dead spots causing buildup of material, and cleaning problems due to difficult disassembly. The manufacturer apparently is working on the design of the instrument to overcome these difficulties.

REFERENCES

Abboud, A. M., Hoseney, R. C., and Rubenthaler, G. L. 1985. Factors affecting cookie flour quality. *Cereal Chem.* **62**:130–133.

Bagley, E. B., and Christianson, D. D. 1986. Response of commercial chemically leavened doughs to uni-axial compression. In *Fundamentals of Dough Rheology*, ed. H. Faridi and J. M. Faubion, pp. 27–36. St. Paul, MN: American Association of Cereal Chemists.

Bloskma, A. H., and Bushuk, W. 1988. Rheology and chemistry of dough. In *Wheat Chemistry and Technology*, 3rd ed., vol. 2, ed. Y. Pomeranz, pp. 131–218. St. Paul, MN: American Association of Cereal Chemists.

Bourne, M. C. 1982. *Food Texture and Viscosity*. New York: Academic Press.

Curley, L. P., and Hoseney, R. C. 1984. Effects of corn sweeteners on cookie quality. *Cereal Chem.* **61**:274–278.

Doescher, L. C., and Hoseney, R. C. 1985. Saltine crackers: Changes in cracker sponge rheology and modification of a cracker sponge procedure. *Cereal Chem.* **62**:158-162.

Faridi, H. A. 1975. Physical and Chemical Changes During Saltine Cracker Lengthy Fermentation. Ph.D. diss., Kansas State University, Manhattan, KS.

Farrand, E. A. 1964. Flour properties in relation to modern bread processes in the United Kingdom, with special reference to alpha-amylase and starch damage. *Cereal Chem.* **41**:98–111.

Hoseney, R. C. 1986. *Principles of Cereal Science and Technology*. St. Paul, MN: American Association of Cereal Chemists.

Johnson, J. F., Martin, J. R., and Porter, R. S. 1975. *Determination of Viscosity of Food Systems in Theory. Determination and Control of Physical Properties of Food Materials*, ed. C. Rha, Boston, MA: Reidel.

Kramer, B. R. 1953. Cookie doughs: New approaches to the mixing operation. *Proc. Am. Soc. Bakery Eng.*, p. 259.

Manley, D. J. 1983. *Technology of Biscuits, Crackers and Cookies*. Chichester, U.K.: Ellis Horwood.

Matz, S. A., and Matz, T. D. 1978. *Cookie and Cracker Technology*. Westport, CT: AVI.

Miller, A. R. 1985. The use of a penetrometer to measure the consistency of short doughs. In *Rheology of Wheat Products*, ed. H. Faridi, pp. 117–132. St. Paul, MN: American Association of Cereal Chemists.

Mizukoshi, M. 1986. Rheological studies of cake baking. In *Fundamentals of Dough Rheology*, ed. H. Faridi and J. M. Faubion, pp. 73–86. St. Paul, MN: American Association of Cereal Chemists.

Olewnik, M. C., and Kulp, K. 1984. The effect of mixing time and ingredient variation on farinograms of cookie doughs. *Cereal Chem.* **61**:532–537.

Osorio, F. A., and Steffe, J. F. 1987. Back extrusion of power law fluids. *J. Tex. Stud.* **18**:43–63.

Steele, I. W. 1977. The search for consistency in biscuit doughs. *Baking Ind.* **9**:21, 23–24, 33.

Steffe, J. F., and Morgan, R. G. 1987. On-line measurement of dynamic rheological properties during food extrusion. *J. Food Proc. Eng.* **10**:21–26.

Yasukawa, T., Mizukoshi, M., and Aigami, J. 1986. Dynamic viscoelastic properties of cake batter during expansion and heat setting. In *Fundamentals of Dough Rheology*, ed. H. Faridi and J. M. Faubion, pp. 63–72. St. Paul, MN: American Association of Cereal Chemists.

Chapter 9

Application of Rheology in the Pasta Industry

David H. Hahn

The use of rheological methods in the pasta industry is very much in its infancy. Few, if any, pasta manufacturers utilize quality assurance (QA) tests based on the rheological properties of the raw materials. Consequently, the manufacture of pasta remains more an art than a science. Large, highly auto-mated production lines still rely on the expertise of the extruder operator to determine the proper water:semolina ratios. In contrast, cereal chemists have used the rheological properties of durum wheat for many years to eval-uate the potential of durum wheat varieties for pasta production. Most of the rheological methods used have been adapted from those used in the bak-ing industry. These methods use the mixograph, farinograph, extensigraph, alveograph, amylograph, and sodium dodecyl sulfate (SDS) sedimentation test. Many researchers have shown that, as is the case for bread wheats, gluten strength is an important factor in making high-quality pasta products from durum wheat. For this reason, there is good justification for the use of these methodologies. They have not proved to be entirely satisfactory for predicting the pasta-making potential of flour or semolina, however. The best test of the potential of a wheat flour or semolina for pasta production remains to actually make pasta.

Objective rheological methods are not commonly used by the pasta industry as QA procedures to evaluate the quality of the finished pasta. Subjective evaluation of the cooking performance of finished product is done periodically by most pasta manufacturers. The use of objective techniques to evaluate the cooking quality of spaghetti has been used in the literature.

These tests appear to have promise as QA procedures for the pasta indus-try, ultimately replacing or supplementing subjective cooking tests. These tests have not been commonly used by the pasta industry, however. The main stumbling block for their widespread use is the cost of instrumentation.

The ultimate test of the quality of a flour or semolina for pasta production is the eating quality of the finished pasta. Rheological methods are useful only in their ability to predict or measure this quality. To a great extent, the structure of fresh, dry, and cooked pasta determines the rheological and the physical properties of a pasta product. Because of this, an understanding of pasta manufacture and structure is essential in order to obtain rheological measurements of semolina and pasta quality. Therefore, this chapter first discusses pasta production and structure, and then reviews the status of rheological methods for evaluating pasta and the application of these methods in the pasta industry. Areas where new rheological techniques are needed by the pasta industry are then discussed.

PASTA PRODUCTION

Durum semolina, durum flour, and hard wheat flour are used to produce pasta products. High-quality pasta is made with 100% durum wheat milled into semolina. This pasta has a bright yellow color and the best eating quality. While additional ingredients (eggs, spinach powder, and tomato powder) can be added to color and flavor specialty pastas, this discussion deals only with pasta made with semolina and water.

Large, highly automated production lines are used to produce pasta commercially. Production rates range from a low of approximately 110 lb/hr to a high of approximately 7,000–8,000 lb/hr (Baroni, 1988). The basic equipment includes a continuous pasta press (mixer and extruder), shaker/spreader, predrier, final drier, and storage unit. Developments over the last few years have greatly improved production efficiencies and product quality. Most prominent among these developments are the use of Teflon®-lined dies, the increased use of high-temperature drying, the development of microwave drying, and the development of more precise automatic controls.

Mixing

Warm water and semolina are introduced into a twin shaft mixing chamber at a ratio to bring the moisture level to approximately 28–31%. Water to semolina ratio (i.e., optimum moisture) is determined by the extruder operator based on his experience. Operators use appearance in the mixer, appearance of the freshly extruded pasta and the load on the extruder motor to determine water level.

The two shafts of the mixing chamber rotate in opposite directions, creating a degree of back mixing. This action allows for thorough mixing and also limits the amount of clumping or balling that can occur. Water and semolina are blended together in the mixing chamber. Mixing time depends primarily on the particle size. Optimum particle size for semolina is between 488 and 142 μ (Manser, 1984). Mixing only wets the semolina particles and does not significantly change the microstructure of the mixture (Resmini and Pagoni, 1983). With the moisture available, the energy supplied by the mixer is insufficient to develop gluten interactions.

Extrusion

Wet, mixed semolina is fed from the mixing chamber into the screw of the extruder. The screw kneads the semolina/water mixture into a stiff dough and forces the material through the die. Friction generated in the extruder barrel and die warms the dough, often making it necessary to use external barrel cooling. For best results, the pasta dough should be extruded at 115°. At temperatures greater than 140°F, the gluten proteins begin to denature and detrimental effects on final pasta quality are produced. The mixing chamber and/or screw are maintained under vacuum to remove air trapped in the dough. If the air is not removed, small air bubbles in the finished product result in a chalky appearance and reduced mechanical strength.

Pasta is extruded in a variety of sizes and shapes (Table 9-1). Short goods are cut by a rotary knife that scrapes the surface of the die. The flow rate of dough through the die is important, since any fluctuation in flow rate across the die causes variations in product size. The same is true for long goods. These variations in product size increase the amount of waste and the cost of production.

Table 9-1. Some Examples of the
Variety of Pasta Shapes

Long Goods	Short Goods	
Spaghetti	Ditali	Mostaccioli
Spaghettini	Ditalini	Shells
Vermicelli	Elbow macaroni	Rigatoni
Capellini	Rings	Rotini
Fettucini	Elbow spaghetti	Vermicelli
Linguini	Tubettini	Bow ties
Lasagna	Ziti	Butterflies
Perciatelli	Pastini	Egg noodles
Noodles	Manicotti	Wheels
Fusilli	Penne	Orzo
Mafalde	Radiatose	Quadratini

Drying

Drying is the most critical as well as the most difficult step in pasta production. The objective of drying, to reduce the moisture of the extruded pasta from 31% to 10–12%, is deceptively simple. Even though many types and designs of driers are used, the problems of selecting proper temperature/humidity profiles are the same. Pasta spoils due to microbiological growth if it is dried too slowly. If pasta is dried too quickly, moisture gradients are created that crack, or check, the pasta. Checking can occur during the drying cycle, but it can also occur up to several weeks after the pasta leaves the drier. This means the product can check after it is packaged and sold. Mok and Dick (1987) found that the difference between environmental relative humidity (RH) and pasta RH was the driving force in this type of checking. It therefore is critical to monitor and control the drying and storage conditions of pasta very closely.

Drying is usually done in stages. Long goods, the most difficult type of pasta to dry, are dried in four stages, normally requiring a total of 20 to 24 hours (Table 9-2). Short goods can be dried much more quickly and are normally dried in three stages taking 4 to 6 hours (Table 9-3).

Long Goods Drying

After extrusion, long goods are separated and placed on rods by the spreader. Immediately after it goes through the spreader, the pasta goes into a predrier (150°F and 65% RH). The predrier rapidly reduces moisture in the pasta from 31% to approximately 25%, thereby avoiding fermentation, which could damage the pasta. At this moisture level, pasta is still flexible and checking does not develop. Predrying also sets the shape of the pasta, preventing deformation during the remaining stages of drying.

Table 9-2. Traditional Drying of Long Goods

Process Stage	Temperature (°F)	Relative Humidity (%)	Duration (hr)	Moisture (%)
Extrusion	120	—	—	28–31
Pre-drier	150	65	0.75–1	25
Final Drier				
Stage 1	130	95	1.5–2	25
Stage 2	130	83	4–6	18
Stage 3	110	65	8–12	10–12
Collector	Ambient		1+	10–12

Table 9-3. Traditional Drying of Short Goods

Process Stage	Temperature (°F)	Relative Humidity (%)	Duration (hr)	Moisture (%)
Extrusion	120	—	—	28–31
Shaker	—	—	0.25	20–25
First drier	140–150	75	3.5	17–18
Final drier	140–150	68	3.5	10–12
Collector	Ambient		1 +	10–12

After the initial moisture has been removed, the remaining moisture in the product is removed in three stages. The first stage of the final drier "sweats," or equilibrates, the pasta for 1.5–2 hours at 130°and 95% RH. The second stage reduces the moisture of the pasta to approximately 18% over the course of 4–6 hours at 130°F and an RH of 83%. During the third stage, the remaining moisture is removed over a period of 8 to 12 hours at 110°and an RH of 70%. After drying, the pasta is removed to a storage unit, where it is cooled to room temperature prior to packaging.

Short Goods Drying

Short goods are conveyed on a shaker from the extruder to the first drying stage. Warm air is blown over the pasta in the shaker. The pasta is dried quickly to about 20–25% moisture in a manner similar to the method used in the predrier for long goods. This process dries the surface of the pieces, which prevents agglomeration and deformation of the pieces during the subsequent drying stages. After it leaves the shaker, the pasta enters a pair of driers, each at a temperature of 140–150°. The first drier equilibrates the pasta and dries it to 17–18% moisture in approximately 3.5 hours at 75% RH. The second drier completes drying the pasta to 10–12% moisture in approximately 3.5 hours at 68% RH. The pasta is moved from the drier into a storage unit to cool to room temperature prior to packaging.

High-Temperature Drying

Several methods for "high-temperature" drying of pasta products have been developed during the past 20 years. The term "high-temperature drying" in the pasta industry implies drying temperatures of 140 to 200°F (Buhler-Miag,

Inc., 1979; Pavan, 1980; Braibanti, 1980). High-temperature drying is done in stages similar to conventional drying for both long and short goods except that higher temperatures are used. High-temperature drying has the following advantages: (1) bacterial control during drying, (2) marked reduction in drying time, (3) improved cooking quality of the final product, and (4) improved color of the final product. Color and cooking quality are affected most dramatically when poorer quality raw materials (i.e., non-durum wheat flour or farina) are used for pasta production utilizing high-temperature drying.

Microwave Drying

Microwave driers for pasta have recently been developed for short goods and are currently being used by several companies. Technical problems prevent their use in the drying of long goods at this time. These problems may soon be overcome, however (Banasik, 1981). The main advantages of microwave drying are space savings, reduced drying time, and low microbial counts. Short goods can be dried with microwave driers in 1.5 hours, compared to 4–8 hours for high-temperature and conventional driers.

PASTA MICROSTRUCTURE

The surface of freshly extruded pasta is a continuous protein film. The inner portion of pasta is a compact structure of starch granules embedded in an amorphous protein matrix (Resmini and Pagoni, 1983). Starch granules and protein are aligned in layers parallel to the protein film on the surface of the pasta. After proper drying, there is essentially no change in this structure. Improper drying or elevated dough temperatures during extrusion disrupt the continuity of the protein film and the underlying matrix.

While pasta is being cooked, protein hydration and starch gelatinization move from the outside toward the center of the piece (Grzybowski and Donnelly, 1977). During the cooking of good-quality pasta, the protein absorbs water and swells more rapidly than the starch. This process results in a continuous fibrillar network of denatured protein surrounding gelatinized starch granules (Resmini and Pagoni, 1983). In poor-quality pasta, the proteins aggregate in discrete masses rather than in a continuous matrix. Hydration of the protein fraction of pasta before the beginning of starch gelatinization appears to be important to produce a firm, good-quality cooked pasta. Pasta produced with lower-protein semolina cooks faster than pasta produced with a higher-protein semolina. The rate of protein hydration appears to be the rate-limiting factor in the cooking time of good-quality pasta.

MEASUREMENT OF RHEOLOGICAL PROPERTIES

Predictive Tests for Raw Materials

It becomes apparent in reviewing the literature dealing with pasta rheology that protein quantity and quality are important in the production of high-quality pasta products. Protein quantity, however, does not always correlate with good pasta-making characteristics. Reconstitution and chromatographic studies indicate that gluten quality is a major factor in determining good cooking quality (Dick, 1985; Wasik and Bushuk, 1975). More specifically, the glutenin protein fraction appears to be the most important. High glutenin:gliadin ratios correlate well to good cooking quality (Walsh and Gilles, 1971; Wasik and Bushuk, 1975). Strong gluten flour or semolina does not necessarily guarantee good cooking quality, however. Gluten of medium strength can produce pasta of very high quality (Matsuo and Irvine, 1970). Finally, hard wheat flours with very strong gluten produce poor-quality pasta. Clearly, gluten quality for pasta manufacture and breadmaking are not exactly the same.

The same chemical and rheological tests that have been used in the baking industry to determine wheat flour quality for breadmaking have been applied to durum flour/semolina. There is some basis for this practice due to the importance of gluten quality in producing good-quality pasta. As has been mentioned, however, gluten quality that is desirable for pasta production is not necessarily the same as for bread production. Understandably, the application of these tests, usually with some modifications, has had mixed results for predicting pasta-making quality. The mixograph, farinograph, SDS sedimentation, and alveograph tests have been the most popular (Dick, 1985; Cubadda, 1988; Dick and Youngs, 1988). The use of these tests for predicting pasta-making quality is reviewed below.

Mixograph

The mixograph is a high-speed recording mixer that can measure the rate of dough development, dough resistance to mixing, and dough tolerance to extended mixing. Its primary use has been to evaluate the quality of hard wheats for bread baking. North Dakota State University (NDSU) currently uses a mixograph procedure to evaluate the relative gluten strengths of experimental durum wheat lines (Dick, 1985). The NDSU method uses 10 g of semolina on an as is moisture basis and 5.8 ml of water. Semolina and water are mixed in the 10 g mixing bowl on a spring setting of 8 at approximately 26°C. The mixogram is compared to a set of standard curves (Fig. 9-1), and gluten strength is rated from 1 (weak) to 8 (strong). This rating system cor-

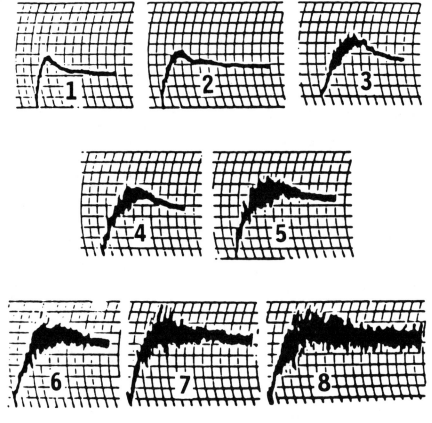

Figure 9-1. Reference mixograms for durum wheat. Higher number indicates stronger curve type. (*From Dick, 1985, with permission.*)

relates with the quality of the pasta produced from tested flours and semolinas (Dick and Quick, 1983; Joppa et al., 1984). It should be noted that the procedure evaluates gluten quality at a water absorption level that is much higher than that used for pasta production. Therefore, only relative information on pasta-making potential can be obtained from this test. The procedure gives no indication of the optimum water absorption required for pasta manufacture.

Farinograph

The farinograph curve gives an indication of water absorption at a fixed dough consistency. Peak time, stability, and mixing tolerance are important parameters that can be obtained from the farinogram. Dexter and Matsuo (1979) found that these parameters could be used to predict the potential for pasta production of durum wheat varieties. Farinograms run at 52–65% absorption give a good indication of gluten strength and could be related to the cooking quality of the resulting pasta (Dexter and Matsuo, 1979). Grzybowski and Donnelly (1979) developed an eight-point scale, similar to that discussed for the mixograph, using farinograms run at 52–65% absorption. Peak time, stability, and mixing tolerance are strongly influenced by protein content and may not always be correlated with pasta-making quality. Bandwidth appears to be a better indicator of protein quality for pasta manufacture, as it is less affected by protein content (Dexter and Matsuo, 1980). Farinograms have been obtained near pasta-making absorptions (30%) and have been used to predict extrusion properties (Matsuo et al., 1982). Their results indicate that a modified farinograph may be useful for determining optimum processing conditions for a semolina. More research is needed, however, before this type of test can be used to estimate water absorption or predict processing conditions.

Viscoelastograph and Aleurograph

The viscoelastographic and aleurographic tests have been proposed as methods to determine the quality of semolina in Europe (Cubadda, 1988). The viscoelastograph measures changes in the thickness of a cooked pasta disk during and after the application of pressure. Resistance to flattening has been claimed to be related to good-quality pasta. The aleurograph measures the cohesiveness of a cooked pasta disk exposed to water pressure. The presumption has been that the higher the cohesiveness, the better the pasta quality. These tests, however, have not been shown to have a high correlation with subjective cooking quality tests (Cubadda, 1988).

Gluten Measurement

Gluten proteins have been characterized by several methods in attempts to more precisely predict the pasta-making quality of durum wheat. Durum wheat proteins have been characterized by sedimentation tests, electrophoresis, and chromatography. Wet and dry gluten contents have been used as measures of quality (Dick, 1985). The SDS sedimentation test (Dexter et al., 1980; Quick and Donnelly, 1980; Dick and Quick, 1983) appears to be a useful indicator of gluten quality. The SDS sedimentation test is especially useful when only small samples are available. The microsedimentation test developed by Dick and Quick (1983) requires only 1 g of sample.

The alveograph and extensigraph have been used to evaluate the strength of gluten (Matsuo and Irvine, 1970; Matsuo, 1978). The alveograph measures the force required to break a bubble of wet gluten, and the extensigraph measures the force required to break a gluten strand. The breaking force required for the gluten is correlated to its strength. Damidaux and co-workers (1980) measured the elastic recovery of a cooked gluten disk. Extracted gluten was cooked in a cell of known dimensions. After cooling, the gluten disk was deformed under a constant load and for a set time. The amount of recovery after the load was removed was related to gluten quality. Gluten from good-quality durum wheat showed a recovery value of greater than 1.6 mm.

Breaking Strength of Dry Pasta

Breaking strength gives a measure of the strength of the dry pasta. The strength of the dry pasta determines how well the product tolerates shipping and may indicate how well a product holds together upon cooking. The gluten strength/quality of the parent semolina may determine the dry strength of the pasta. Therefore, breaking strength is indirectly related to cooking quality. Processing conditions have a much greater effect on breaking strength than flour/semolina quality, however. For this reason, breaking strength has seldom been used in the literature to predict the cooking quality of pasta. Determining breaking stress may, however, be useful as a QA procedure for the pasta industry. Measuring breaking stress can be done more rapidly than a cook test, and it can be used as a near-line procedure. The breaking strength of dry spaghetti and noodles can be determined by using the Instron Universal Testing Instrument (Oh et al., 1985; Voisey and Wasik, 1978). A single strand of spaghetti is suspended across a span of known distance. A force is applied to the center of the span and the breaking stress calculated. Additional research is needed to determine the suitability of this procedure as a replacement or supplement to a cook test to determine finished product quality.

Cooking Quality

Rheological measurements can be useful tools for predicting the end use quality of durum flour and semolina for pasta. Final judgment of pasta quality is based on evaluation of cooked texture, however. This evaluation is best done by subjective texture panels. Subjective cook tests are commonly used in the industry to evaluate the final quality of pasta products. Many variables can influence cooking tests, including the type of water, use of salt, ratio of water to pasta, cooking temperature and time, and method of draining cooked pasta. Since no standard laboratory cook test exists, the subjective tests show a great deal of variability. Objective instrumental methods have been widely used in the literature, but industrial applications have been limited. The objective tests used have shown good correlation with evaluation by subjective methods and show promise for use as QA methods. Objective measurement of pasta cooking quality have relied on the measurement of one of three parameters: firmness/tenderness, compressibility, and stickiness. The absence of standard laboratory cook tests also affects the repeatability of objective measurements. Although water hardness does not affect cooked pasta firmness, it has a marked effect on surface stickiness (Matsuo, 1988). Therefore, Matsuo (1988) has recommended that prepared water of constant hardness and pH be used for standard cooking tests.

Firmness/Tenderness Measurement

Instrumental methods for measuring cooked spaghetti firmness or tenderness have been developed that correlate well with taste panel scores (Matsuo and Irvine, 1969; Walsh, 1971; Voisey et al., 1978; Oh et al., 1983). Walsh (1971), Matsuo and Irvine (1969), and Oh and co-workers (1983) all used similar blades to perform a single-blade cut test. Walsh (1971) and Walsh and Gilles (1971) described a procedure to measure cooked spaghetti firmness in which a single strand of cooked spaghetti was placed on a Plexiglas plate and sheared at 90° with a Plexiglas tooth. The spaghetti strand was sheared with a tooth speed of 0.018 cm/sec. The area under the shearing curve was taken as a measure of spaghetti firmness. Matsuo and Irvine (1969) also described a cut test to measure the firmness of a single strand of spaghetti. In their test a modified Buhler Bending Stress Tester was used to measure the rate of cutting and the force required to cut the strand of cooked spaghetti. The rate of shear was calculated as a measure of tenderness. Oh and colleagues (1983) described a cut test similar to the two tests discussed above, but using three strands of cooked spaghetti. The three strands of cooked spaghetti were cut with a Plexiglas blade (Fig. 9-2) similar to that used by Walsh (1971). The maximum cutting force per unit area was used as a measure of spaghetti firmness. The results for all of these tests correlated well with data obtained by subjective cooking tests.

The American Association of Cereal Chemists (AACC) is currently evaluating a standard method based on the single-blade cutting test of Walsh (1971) for measuring spaghetti firmness. The results of initial collaborative studies indicate that the method will work well (Jacobs, 1987). These tests have been applied only to spaghetti, however. As has been discussed, pasta is made in many shapes and sizes (Table 9-1). While the single-blade cut test will likely work well for most long goods, it may not work for short goods. An alternative may be the method of Voisey and colleagues (1978), who utilized a shear

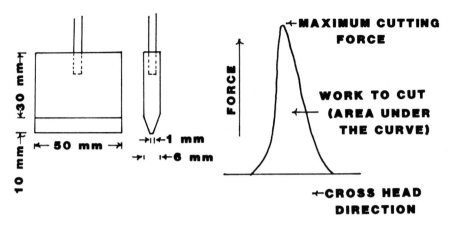

Figure 9-2. Blade for firmness measurement and Instron curve. (*From Oh et al., 1983, with permission.*)

cell with ten cutting blades to shear through multiple strands of spaghetti. The use of this type of multiblade shear cell may be applicable to both long and short goods. More research is needed to adapt rheological procedures for evaluating the quality of cooked spaghetti for use with other pasta shapes.

Compressibility

Oh and co-workers (1983) described a method to measure the compression and recovery of three cooked spaghetti strands by using a flat blade (Fig. 9-3). The strands were compressed to a maximum stress of 1.3 kg/cm². The force was removed at this point, and the compression slope (taken as the slope of the tangent at 50% of the maximum force/blade contact area), resistance to compression (initial thickness − A in Fig. 9-3), and recovery (C/B in Fig. 9-3) were calculated. Resistance to compression and recovery were found to be correlated with chewiness as measured by a trained texture panel. Chewiness was defined by the panel as the time (seconds) required to reduce a sample of spaghetti to pieces that were small enough to swallow.

Matsuo and Irvine (1971) compressed a strand of cooked spaghetti with a blunt blade at a constant force and for a constant time; the force was then released and the strand was allowed to reach its resting thickness (Fig. 9-4). Compressibility and recovery were measured. Compressibility was defined as the ratio of blade penetration to the original diameter of the cooked spaghetti (Y/X). Recovery was defined as the ratio of the distance the blade is forced back to the original penetration of the blade. Samples with low compressibility and high recovery were judged by the authors to be of good quality.

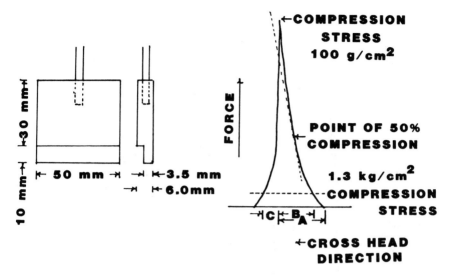

Figure 9-3. Blade for compression test and Instron curve. (*From Oh et al., 1983, with permission.*)

Samples with high compressibility and low recovery were judged to be of poor quality. These conclusions were drawn on very limited organoleptic data, however.

Stickiness

Stickiness is a third parameter that is important in evaluating the quality of cooked pasta. Procedures for the objective measurement of the stickiness of cooked pasta have been described in the literature (Voisey et al., 1978; Dexter et al., 1980; Dexter et al., 1983). All of the previously cited procedures placed cooked spaghetti strands in a single layer on an aluminum plate (100 mm × 100 mm), with a grooved surface (0.15 mm deep, spaced at 3.6 mm) and an aluminum retainer plate (6 mm thick, 165 g), with a 43 mm × 22 mm hole placed on top of the strands. A 40 mm × 19 mm aluminum plunger applied a compression force of 5200 N/m^2 at a plunger speed of 4 mm/min. Once the maximum compression force was reached, the plunger was lifted until the pasta was released. The maximum depression of the curve recorded during lifting of the plunger was defined as stickiness. This method could be used to measure the surface stickiness of cooked pasta up to the point of mastication. Both authors stress the importance of standardizing the test conditions for this procedure. Meaningful comparisons could be made only if the deformation rate, maximum compression, final compression force, elapsed time after cooking, and relaxation time were carefully controlled. This method did not correlate with the stickiness of the pasta in the mouth during chewing.

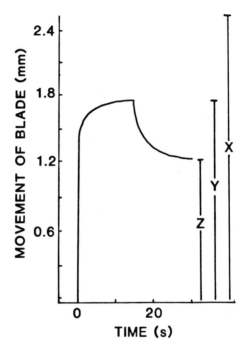

Figure 9-4. Values derived from compressibility curve: X = diameter of cooked spaghetti, Y = blade penetration, and Z = distance blade is forced back. (*From Matsuo and Irvine, 1971, with permission.*)

PROSPECTS FOR RHEOLOGY IN THE PASTA INDUSTRY

Testing of Raw Materials

Nearly all of the durum wheat used for pasta production in the United States is grown in a four-state region (North Dakota, South Dakota, Minnesota, and Montana). With this small production area the durum crop is vulnerable to the effects of annual climatic variations. In addition, the annual acreage planted depends on many factors, including prices during the previous year, competition with other cash crops, and demand for exports. For these reasons, the year to year availability of adequate amounts of good-quality durum wheat is uncertain. Therefore, pasta manufacturers would like to be able to evaluate durum wheats from other areas, both foreign and domestic, for their suitability for pasta production. There is a need for predictive rheological tests for evaluating semolina and flour that would be suitable for use as specifications and QA procedures.

The predictive tests discussed above, though very useful for the cereal chemist evaluating new durum wheat varieties, have not been widely used by the commercial pasta producer. A suitable QA test must be practical, fast, and relatively inexpensive to perform. All of these predictive tests measure gluten strength or related properties. An interesting possibility has been suggested by Dick and Youngs (1988), who reported that a combination of the SDS microsedimentation results, mixogram score, and protein content could predict the cooked firmness of spaghetti ($r^2 = 0.713$). Protein is commonly used in specifications by the pasta industry, and the microsedimentation and mixogram tests are fast and relatively inexpensive. For these reasons, the results of this combination of tests seem to have the best potential as specifications to ensure the suitability of semolina or flour for pasta production. However, additional research is required specifically to develop the database necessary to use these tests as QA procedures.

Other tests also need to be developed that can be used in this manner. More emphasis should be placed on measuring the rheological properties of durum wheat that are specifically related to pasta-making quality, rather than adapting methods used in the baking industry. New, innovative testing procedures based on the distinctive rheological properties needed for pasta production need to be developed. Future standards and specifications should be based more on such rheological tests.

In addition to QA testing procedures, methods that can accurately predict the optimum absorption of a lot of semolina are needed. This capability would allow better process control. Predictive tests to determine the correct water:semolina ratio would reduce the subjectivity involved in the production of pasta. None of the predictive tests currently used in the evaluation of semolina can be used to predict water absorption. More research is needed to develop predictive tests that can be used in this manner.

Testing of Finished Product

The subjective tests of cooking quality currently used by the pasta industry to evaluate the quality of pasta are highly variable. Operator-induced differences and the lack of standardized cooking procedures are the main reasons for this variability. The instrumental methods that have been developed should be useful as replacements for these highly variable tests. The instrumental test currently being evaluated by the AACC for measuring spaghetti firmness is repeatable and correlates well with subjective tests of cooking quality. It is simple, fast, and precise. The cost of the instrumentation required for the test prohibits its use as a QA procedure, however. The development of inexpensive instrumentation with the required precision is needed. In addition, objective cooking quality tests currently used, and described in the literature, work well for spaghetti but have not been applied to other pasta shapes; that is, they likely work well for most long goods, but may not work as well for short goods. These tests need to be adapted for evaluating pasta shapes other than long goods.

Determining the breaking stress of uncooked pasta is another rheological test that may have application as a QA procedure. This test may make it possible to detect potential quality problems resulting from improper processing conditions directly out of the drier. Breaking stress objectively measures pasta strength, which might be used to supplement a cooking or firmness test.

Rheological tests capable of monitoring pasta quality during processing would be useful to the pasta industry. At this time no suitable in-line tests are available. The rheological procedures discussed above have not yet been evaluated near-line. Knowing the production process and the current rheological tests available, it is possible, however, to speculate on the potential application of these tests as near-line QA procedures. The firmness tests for evaluating cooking quality appear to have the best potential of any of the procedures currently used as near-line tests for pasta evaluation. Evaluation of freshly extruded pasta before drying might identify potential quality problems before drying. This practice would save not only the time but also the expense of drying what could be an unacceptable product. This type of test may also be useful in optimizing the processing conditions for a given production run. However, considerable research is still needed to develop the tests and prove the effectiveness of near-line QA procedures such as this.

Process Control

Direct process control by using in-line rheological measurements does not appear to be practical in the near future. In current pasta extruders, a dough with measurable properties is not formed until it is actually in the extruder barrel. Since this occurs anywhere from 10 to 30 minutes after the water and semolina have entered the mixer, it is too far removed from ingredient addition to be useful for primary process control. Evaluating the dough at the extruder barrel may have use for monitoring the extrusion process. In addition, in-line testing procedures need to be developed for pasta extrusion. Motor load and die pressure are usually measured as monitoring tools on current pasta extruders, but they do not appear to relate directly to the optimum water absorption of a semolina. A more precise technique is needed to monitor the rheological behavior of a pasta dough in the extruder. Considerable research is needed to apply the known rheological properties of pasta dough to an in-line test.

The next generation in process control will likely be measurement of the moisture content of the incoming semolina and adjusting the water: semolina ratio to a preselected level. A similar option would be to measure the moisture in the mixer and adjust it to a preselected level. The technology for this type of extruder control is currently available, but it has not been used in an operating extruder. This type of process control, however, does not eliminate the subjective aspect of pasta extrusion without a test to estimate the optimum water absorption of a semolina.

Research Needs

Much progress has been made in understanding the rheology of pasta products at all stages of the production and cooking processes. Rheological methods have been developed that can measure the quality of raw materials and pasta relatively well, and these methods have proved useful in durum wheat breeding and research programs. Additional research is needed to adapt and standardize these methods for the QA needs of the pasta industry. An information database developed utilizing these methods in production situations is needed. Rheological tests that can monitor product quality during processing also need to be developed.

Mok and Dick (1987) appear to have a good understanding of the mechanism of delayed checking. However, additional research is needed to determine the effects of ingredients and processing conditions on the phenomenon of delayed checking. Processing, environmental, or packaging parameters that promote delayed checking need to be better defined so that checking can be prevented. Quality assurance procedures need to be developed to detect susceptible product before it leaves the warehouse.

Protein quality is not the only factor that determines the final quality of a pasta product. Very little research has been done to determine the effect of starch, the major constituent of semolina. What chemical and physical changes does starch undergo during processing, and how do these changes affect the quality and rheology of the finished pasta product? The effects of the minor constituents of semolina, such as lipids, soluble and insoluble pentosans, minerals, and enzymes, have not been thoroughly investigated. Many of these minor constituents have major effects on the rheology of bread dough. Do they have similar effects in the production of pasta?

A great deal of information on the effects of ingredients on pasta quality has been accumulated. Very little information is available on the effects of processing on pasta rheology and overall quality, however. Studies on the interactions of raw materials and processing need to be continued. Additional emphasis needs to be placed on non-wheat ingredients such as whole eggs, egg whites, vegetable powders, and spices used in specialty pastas.

Advances in processing technology can change the quality requirements of the raw materials and affect the quality of the final product. High-capacity extruders require semolina with a smaller particle size. High-temperature drying appears to improve the final quality of product made with raw materials of marginal quality. The physical and chemical effects of processing conditions on pasta products need to be investigated further. The reasons for the apparent quality improvement during high-temperature drying need to be determined. Investigation of new and innovative processing techniques must continue. The development of microwave and other accelerated drying processes should also continue. There is considerable interest in more convenient, quicker-cooking pasta products. Cooker extruders have recently been applied to produce these quick-cooking pasta products. The textural quality of this product needs to be improved. Additional research is needed to understand the chemical and physical changes that occur in quick-cooking products and their effects on product quality.

CONCLUSIONS

Consumer interest in pasta and pasta products has been rising steadily over the past few years. The perception of pasta as a "healthy" food has been growing as well. Pasta products are relatively low in cost and easy to prepare, and they are extremely versatile because of the wide variety of shapes and sizes. In addition to an increase in demand for pasta, the perception of quality in the United States has changed. Consumers now demand a firmer, or al dente, bite. High-quality durum semolina remains the best raw material for the production of high-quality pasta. Rheological methods to evaluate the quality of semolina and pasta products will play a significant role in developing better durum wheat varieties and standards to maintain high-quality raw materials and products.

REFERENCES

Banasik, O. J. 1981. Pasta processing. *Cereal Foods World* **26**(4):166–169.

Braibanti, E. 1980. New Developments in Pasta Drying Technology. *Macaroni J.* **61**: 48–50.

Baroni, D. 1988. Manufacture of pasta products. In *Durum Wheat: Chemistry and Technology*, ed. G. Fabriani and C. Lintas, pp. 191–216. St. Paul, MN.: American Association of Cereal Chemists.

Buhler-Miag, Inc. 1979. "High temperature" drying of pasta products. *Macaroni J.* **60**:30–32.

Cubadda, R. 1988. Evaluation of durum wheat, semolina, and pasta in Europe. In *Durum Wheat: Chemistry and Technology*, ed. G. Fabriani and C. Lintas, pp. 217–228. St. Paul, MN: American Association of Cereal Chemists.

Damidaux, R., Autran, J. C., and Feillet, P. 1980. Gliadin electrophoregrams and measurements of gluten viscoelasticity in durum wheats. *Cereal Foods World* **25**(12):754–756.

Dexter, J. E., Kilborn, R. H., Morgan, B. C., and Matsuo, R. R. 1983. Grain research laboratory compression tester: Instrumental measurement of cooked spaghetti stickiness. *Cereal Chem.* **60**(2):139–142.

Dexter, J. E. and Matsuo, R. R. 1979. Effect of water content on changes in semolina proteins during dough-mixing. *Cereal Chem.* **56**(1):15–19.

Dexter, J. E. and Matsuo, R. R. 1980. Relationship between durum wheat protein properties and pasta dough rheology and spaghetti cooking quality. *J. Agric. Food Chem.* **28**(5):899–902.

Dexter, J. E., Matsuo, R. R., Kosmolak, F. G., Leisle, D., and Marchylo, B. A. 1980. The suitability of the SDS-sedimentation test for assessing gluten strength in durum wheat. *Can. J. Plant Sci.* **60**:25–29.

Dexter, J. E., Matsuo, R. R., and Morgan. B. C. 1983. Spaghetti stickiness: Some factors influencing stickiness and relationship to other cooking quality characteristics. *J. Food Sci.* **48**:1545–1551.

Dick, J. W. 1985. Rheology of durum. In *Rheology of Wheat Products*, ed. by H. Faridi, St. Paul, MN: American Association of Cereal Chemists.

Dick, J. W. and Quick, J. S. 1983. A modified screening test for rapid estimation of gluten strength in early-generation durum wheat breeding lines. *Cereal Chem.* **60**(4):315–318.

Dick, J. W., and Youngs, V. L. 1988. Evaluation of durum wheat, semolina, and pasta in the United States. In *Durum Wheat: Chemistry and Technology*, ed. G. Fabriani and C. Lintas, pp. 237–248. St. Paul, MN: American Association of Cereal Chemists.

Grzybowski, R. A., and Donnelly, B. J. 1977. Starch gelatinization in cooked spaghetti. *J. Food Sci.* **42**(5):1304–1305.

Grzybowski, R. A., and Donnelly, B. J. 1979. Cooking properties of spaghetti: Factors affecting cooking quality. *J. Agric. Food Chem.* **27**(2):380–384.

Jacobs, J. 1987. Report on pasta quality collaborative study to Pasta Analysis Committee. Paper presented at AACC 72nd Annual Meeting, November 1–5, Nashville, TN.

Joppa, L. R., Josephides, C., and Youngs, V. L. 1984. Chromosomal location of genes affecting quality in durum wheat. *Proc. 6th Int. Wheat Genet. Symp.*, Kyoto, Japan. pp. 297–304.

Manser, J. 1984. Degree of fineness of milled durum products from the viewpoint of pasta manufacture. Paper presented at 11th Durum and Pasta Convention, April 5–6, at Detmold, Switzerland.

Matsuo, R. R. 1978. Note on a method for testing gluten strength. *Cereal Chem.* **55**(2):259–262.

Matsuo, R. R. 1988. Evaluation of durum wheat, semolina, and pasta in Canada. In *Durum Wheat: Chemistry and Technology*, ed. G. Fabriani and C. Lintas, pp. 249–261. St. Paul, MN.: American Association of Cereal Chemists.

Matsuo, R. R., Dexter, J. E., Kosmolak, F. G., and Leisle, D. 1982. Statistical evaluation of tests for assessing spaghetti-making quality of durum wheat. *Cereal Chem.* **59**(3):222–228.

Matsuo, R. R., and Irvine, G. N. 1969. Spaghetti tenderness and testing apparatus. *Cereal Chem.* **46**:1–6.

Matsuo, R. R., and Irvine, G. N. 1970. Effect of gluten on the cooking quality of spaghetti. *Cereal Chem.* **47**:173–180.

Matsuo, R. R., and Irvine, G. N. 1971. Note on an improved apparatus for testing spaghetti tenderness. *Cereal Chem.* **48**:554–558.

Mok, C., and Dick, J. W. 1987. Moisture sorption and cracking of spaghetti during storage. Paper presented at AACC 72nd Annual Meeting, November 1–5, Nashville, TN.

Oh, N. H., Seib, P. A., and Chung, D. S. 1985. Noodles. III. Effects of processing variables on quality characteristics of dry noodles. *Cereal Chem.* **62**(6):437–440.

Oh, N. H., Seib, P. A., Deyoe, C. W., and Ward, A. B. 1983. Noodles. I. Measuring the textural characteristics of cooked noodles. *Cereal Chem.* **60**(6):433–438.

Pavan, G. 1980. High temperature drying improves pasta quality. *Food Eng. Int'l.* **5**(2):37–39.

Quick, J. S., and Donnelly, B. J. 1980. A rapid test for estimating durum wheat gluten quality. *Crop Sci.* **20**:816–818.

Resmini, P., and Pagoni, M. A. 1983. Ultrastructure studies of pasta. A review. *Food Microstruc.* **2**:1–12.

Voisey, P. W., and Wasik, R. J. 1978. Measuring the strength of uncooked spaghetti by the bending test. *J. Inst. Can. Technol. Aliment.* **11**:34–37.

Voisey, P. W., Larmond, E., and Wasik, R. J. 1978. Measuring the texture of cooked spaghetti. 1. Sensory and instrumental evaluation of firmness. *J. Inst. Can. Sci. Technol. Aliment.* **11**(3):142–148.

Voisey, P. W., Wasik, R. J., and Loughheed, T. C. 1978. Measuring the texture of cooked spaghetti. 2. Exploratory work on instrumental assessment of stickiness and its relationship to microstructure. *J. Inst. Can. Sci. Technol. Aliment.* **11**(4): 180–188.

Walsh, D. E. 1971. Measuring spaghetti firmness. *Cereal Sci. Today* **16**(7):202–205.

Walsh, D. E. and Gilles, K. A. 1971. The influence of protein composition on spaghetti quality. *Cereal Chem.* **48**(5):544–559.

Wasik, R. J., and Bushuk, W. 1975. Relation between molecular-weight distribution of endosperm proteins and spaghetti-making quality of wheats. *Cereal Chem.* **52**(3):322–328.

Chapter 10

Application of Rheology in the Breakfast Cereal Industry

Jimbay Loh and Wesley Mannell

Ready-to-eat (RTE) cold breakfast cereal accounts for 75% of the total breakfast cereal consumption in the United States, nearing a value of $4.8 billion in 1986. These cereals are made mainly from corn, wheat, rice, and oats. A raw material blend is first plasticized with a combination of moisture and heat to gelatinize the native starch, and the gelatinized mass is then shaped and sized into the desired configuration. The product may be expanded and dried to develop structure and a crispy texture. Various processes such as flaking, puffing, extruding, shredding, toasting, frying, and coating are used to achieve specific product characteristics. Finally, the product may be fortified with vitamins and minerals. Based on differences in processing techniques, at least six basic types of RTE cereal are on the market: flaked, gun-puffed, oven-puffed, extruded, shredded, and granola-type products (Fast, 1987).

In comparison to the baking and other cereal industries, the use of rheological methods in the RTE cereal industry is extremely limited. In addition, little information is contained in the literature. Few manufacturers of RTE cereal routinely utilize rheological tests for evaluating or specifying raw material quality. The methods used are generally shared by other cereal industries and are not specifically designed for RTE cereal. These methods include use of the farinograph, mixograph, amylograph, falling number, and others. Rheometers capable of quantifying fundamental rheological properties may also be in use, although the importance of rheology in RTE cereal manufacturing has not been demonstrated by cereal scientists. The rheology of the material in process is commonly recognized as an important factor governing many processing steps, such as flaking, extruding, and shredding. Little or no work has been published on the rheological properties of specific materials in process, probably because of the great diversity of materials, processes, and products. In contrast, it is widely believed that the importance of the rheology of finished products is crucial to consumer acceptance and the success of the product. With the increasing use of fruit in RTE cereal, the texture of fruit pieces and fillings and its changes during storage are becoming of great interest to the industry. This subject will not be covered, however, to avoid drawing attention away from the main subject, cereal.

Because of the lack of industrial application of rheological methods to the characterization of raw and in-process materials and the overwhelming importance for the texture of finished RTE cereals, the following discussion emphasizes rheological methods used for finished RTE cereals. The discussion also includes a brief description of the manufacturing processes for selected RTE cereals and a microstructure model for RTE cereal.

RTE CEREAL PRODUCTION

Flaked cereals are traditionally made by using a batch process, from whole grains (e.g., wheat), chunks of grains (e.g., corn) or flour of single or multiple grains. To facilitate effective penetration of moisture, sugar, and flavors, whole grains are first bumped. Bumping is accomplished by mild steaming followed by gentle flattening of the kernel between rolls, just enough to break the bran coat and create fissures into the endosperm. The bumped grains are usually cooked in a pressurized steam cooker to an optimum moisture content and machinability, with or without added flavor materials such as sugar, salt, and malt. Cooked particles are then separated, sized, and dried slowly to a moisture content of 10–13% under controlled humidity. For floury material, a cooking extruder is necessary to cook and shape the grits in one continuous operation. After adequate tempering to allow moisture equilibration, the grits are flaked between rolls and sometimes are only bumped, depending on the desired thickness of the finished product. Toasting in an oven develops the structure, texture, color, and flavor of the finished product. Toasting also serves for final drying to lower the moisture content to 2–8%, which is essential to stabilize the porous structure and maintain shelf life.

Gun-puffed RTE cereals are made from whole grains of rice or wheat by heating the grains in a high-pressure vessel. Moisture and heat cause the starch to gelatinize. Upon rapid release of the pressure after cooking, rubbery starch masses of individual kernels expand, as steam flashes out, resulting in simultaneous expansion, dehydration, and cooling of the kernels. If flour is used as the originating material, a cooking extruder is used to mix, knead, and cook the dough, and then to shape and form the pellets. The extruded pellets are tempered and gun-puffed. Certain RTE cereals can be puffed by sudden release of pressure at the extruder die.

In making shredded RTE cereal, whole grain is first cooked in boiling water. Completely cooked grain is cooled and tempered to a certain uniform moisture distribution. Shredding is done by compacting soft/cooked grain between a smooth and a grooved roll. The packed mass in the groove is removed by a comb that is matched to the grooved roll, and the resulting shreds are laid down on a moving conveyor belt. Several layers of shreds are laid to form a web, which is then cut into bite-size pieces and baked. Continuous cooking extrusion using farinaceous material is also used to form shreds.

Granola-type RTE cereal is made from a blend of rolled oats, nuts, coconut, oil, malt, sugar, water, and other flavor materials. The mixture is first formed and made into a sheet and is then baked in a continuous oven. After baking and cooling, the sheet is broken into the desired size.

In general, the rheology of the material in process, as well as of the raw material, is important to the production of RTE cereals. For example, the cohesiveness, plasticity/formability, and surface stickiness of cooked kernels and grits often govern the efficiency of the flaking process. Proper heat and moisture treatments assure optimum bumping. In forming and shaping processes using an extruder with or without cooking, the rheology of the dough is often critical to the subsequent machinability and finished product quality. Optimum oven-puffing requires careful tempering to obtain a viscoelastic mass that is capable of rapid expansion and setting of the developed structure. This process bears some resemblance to bread baking, except that the entire expansion process takes as little as 30 seconds.

Rheological changes during the processing of RTE cereal are assessed based on the texture and structure of the finished product. Understanding the rheological aspects of the key processing steps used in RTE cereal production and the rheological properties of the materials in process can provide new and more effective means of process control. Development of meaningful rheological method for characterizing the raw materials and specific intermediate products appears to be the necessary first step towards reaching this goal.

STRUCTURE

The mechanical properties of breakfast cereals are affected by their structure at all levels, including macrostructure (as perceived by the naked eye), microstructure (as resolved by microscopes), and molecular structure. Tremendous differences in the density, size, and shape of commercial products are easily noticeable by simple visual observation. In spite of macroscopic differences, the microstructure of common breakfast cereals can be generally characterized as a solid foam analogous to the cellular structure of the expanded structural foams of thermoplastics. This generalization is based mainly on microscopic evidence and partially on mechanical properties that have been observed. As they relate to development of the cellular structure, the functions of the key processing steps employed in breakfast cereal production fall into the following categories:

1. Plasticizing (mixing, tempering, cooking, etc.)
2. Forming and shaping (flaking, extrusion, shredding, etc.)
3. Expansion (puffing, toasting, extrusion, etc.)
4. Stabilization (drying, cooling, etc.)

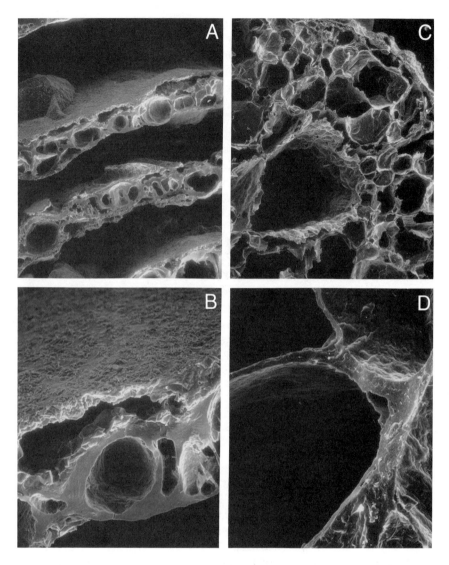

Figure 10-1. Microstructure of sugar-coated corn flakes: *(A)* and *(B)* at 8× and 39× magnification, respectively; and puffed cereals: *(C)* and *(D)* at 78× and 293× magnification, respectively; showing foamlike cellular structure.

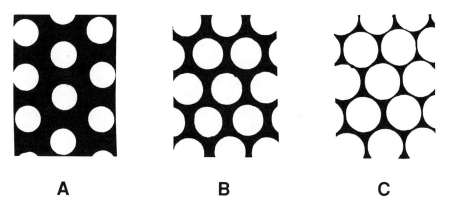

A   **B**   **C**

Figure 10-2. Idealized two-dimensional foam structure of increasing degree of puffing or decreasing density ratio from *(A)* to *(C)*.

The resulting structure is an interspersed network of polyhedral to spherical cells varying in both number and size. This structure is supported by a series of struts and walls (Figure 10-1). Above a critical density ratio (i.e., foam density divided by base materials density), the openness of the cells decreases as the density ratio increases (Figure 10-2). Structural inhomogeneity and anisotropy are common in commercial breakfast cereals and are related to the expansion process. The maximum cell dimension is usually in the direction of expansion.

The mechanical properties of expanded cereals depend on the cell structure and are governed mainly by the density ratio, and the modulus and strength (i.e., rupture strain under tension) of the base material. In general, the modulus of the foam is proportional to the modulus of the base material and the square of the density ratio (Hilyard and Young, 1982). The predominant base material of breakfast cereals is gelatinized starch in which proteins, lipids, fiber, and other minor components are either embedded in or complexed with the starch. Under compressive or shear loading, the cell walls often undergo unstable failure before gross fracture happens. Since the structural elements, that is, walls and struts, are made primarily of starch, this behavior suggests that the composition and physical state of the starch—as affected by processing, the presence of the other ingredients, moisture content, and the age of the product—are important to the mechanical properties of the product. The thermodynamic state and structure of the starch are characterized by analytical techniques such as differential scanning calorimetry (DSC), nuclear magnetic resonance (NMR) spectroscopy, X-ray diffraction, and other non-rheological techniques. Differences in starch composition and physical state (crystalline vs. amorphous) are well known to affect the rate of moisture penetration as the cereal comes in contact with milk and saliva under the conditions of normal consumption.

TEXTURE

Brittle fracture of breakfast cereal varies. Fracture patterns are determined mechanistically by the mechanical strength of the base material, the cellular structure, the size/shape of the product, and the external stress applied through mastication. Consequently, the perceived fracture pattern can be described variously as crispy, crunchy, brittle, crackly, crumbly, hard, rubbery, soggy, mushy, compactible, and so on. An excellent review of the mechanics of brittle fracture and crack propagation is that of Pugh (1967) Sensory perception of the texture of breakfast cereals is further complicated by the addition of milk and the human factor.

The most commonly used textural descriptor for cold breakfast cereals is crispness, as is evident by the frequent mention of crispness in commercial advertisements. Crispness is one of the most complex of the textural parameters, and it is regarded as almost synonymous with wholesomeness for all food types in general and for the textural acceptability of cold breakfast cereals in particular. Crispness has been related to the following: high brittleness and friability (Amerine et al., 1965); low cohesiveness (Szczesniak, 1963); moderate hardness (Le Magnen, 1964); sudden yield with a characteristic sound (Jowitt, 1974); low deformation to rupture and a high modulus of elasticity (Brennen et al., 1974); and a crushing noise with a high frequency, high volume, and long duration (Vickers, 1984). Audio parameters often complement mechanical parameters in predicting sensory crispness. Audio aspects of food crispness have been discussed by Kapur (1971) and Vickers (1981).

There are many other important textural parameters of RTE breakfast cereals, in addition to crispness. These parameters include hardness, fracturability, roughness, grittiness (especially in products with added fiber), dryness, cohesiveness, adhesiveness, gumminess, ease of swallowing, molar packing, and others. Generally speaking, the textural quality of a given RTE cereal is judged primarily by the consumer and manufacturer alike, based on initial crispness and the ability to retain that crispness after immersion in milk.

MECHANICAL MEASUREMENT

The mechanical testing methods currently used for breakfast cereals are far from satisfactory. Over the years, attempts have been made to quantify the crispness, or parameters relating to the crispness, of low-moisture, brittle foods such as RTE cereals (Anderson et al., 1973; Brennen et al., 1974; Burns and Bourne, 1976; and Shearman and Deghaidy, 1978). The application of bending, crushing (with a blade), and shearing forces have been preferred. None of these methods, however, have proved to be readily applicable for industrial quality control or product research uses for breakfast cereals.

Commercial production of breakfast cereal creates, often deliberately, large variations in the size, shape, and density of individual pieces within the same product. The purpose of these variations is to create a more "natural" appearance. In addition, the individual pieces used as test samples are usually too small to be handled conveniently or tested properly without slipping or rocking during testing. For this reason as well as cost, bulk sampling is generally preferred over sampling individual pieces for industrial applications. Therefore, the use of a good "averaging" test cell capable of testing bulk samples is normally used.

There are two major disadvantages of this type of test cell. The first disadvantage is the poor reproducibility of sample packing into the cell. The second is the multiple, complex, and often uncontrollable modes of force application. Such loadings usually involve a combination of compression, shear, cutting, and extrusion forces. Gromley (1984) described a procedure using a shear press for measuring the crispness of corn flakes. This approach is more suitable for puffed cereals because of the difficulty of packing the flakes reproducibly into a shear/compression cell without flakes falling through the slots at the bottom of the test cell. How fast the blade descends through the test sample usually is not critical for dry, brittle cereals. Crosshead speed becomes important if the sample is flexible and quick to relax (e.g., staled, soggy samples or milk-soaked samples). A crosshead speed of 3.35 mm/s was used to generate the data in Figure 10-3, which shows the test fixture and the resulting force versus distance curve for a puffed cereal sample. The initial portion of the curve reflects the resistance of the sample to compressive deformation, followed by sample fracturing, compaction, or usually both. The

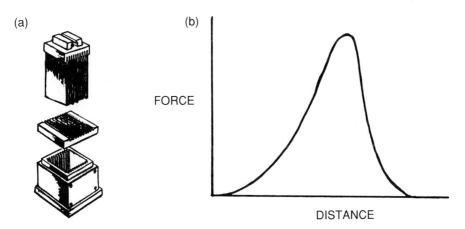

(a)   (b)

FORCE

DISTANCE

Figure 10-3. Test arrangement *(a)* using standard shear/compression cell (Food Technology Corporation, Maryland) and a typical force/deformation curve *(b)* of a puffed cereal sample.

sample is finally sheared and extruded. The shear/extrusion resistance is indicated by the peak force, which is used loosely as an index of the overall mechanical strength (not the crispness) of a dry cereal. Low peak force usually, but not necessarily, indicates a friable/crispy product. High peak force is related to ductile/noncrispy, and sometimes staled/soggy products. Since breakfast cereal is normally consumed after being mixed with cold milk, it is of primary importance that samples be tested after exposure to conditions that are similar to soaking in milk for a specific period of time. The experience of the authors of this chapter indicates that relative retention of the initial texture of a dry product after 3 minutes' soaking in milk allows better correlation with evaluation of the sensory crispness of the same product under realistic use conditions.

The alternate test for samples in the form of flakes, small granules, and so on, uses the General Foods (GF) Texturometer and a cup test arrangement. An aluminum drying dish (e.g., 5.08 cm in diameter and 2.22 cm in height) is filled with the sample without tapping the cup. After excess sample is removed and the cup is centered on the platform, the sample is then compressed and unloaded with a 1.27 cm (D) and 2.54 cm (L) cylindrical lucite plunger at a speed of 12 chews/min. The clearance (i.e., the space between the platform and plunger position at the end of the stroke) is critical to the proper measurement of peak force; this clearance must be set precisely at 2.0 mm. The inclusion of milk soaking in the cup method yields a similar benefit in

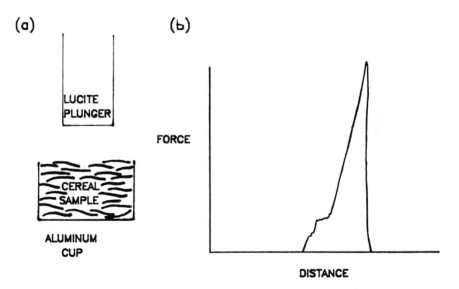

Figure 10-4. Test arrangement *(a)* using cup method and General Foods Texturometer and a typical force/deformation curve *(b)* of a commercial bran flakes sample.

terms of sensory correlation. Figure 10-4 shows the test setup and a typical force/deformation curve.

Percent texture retention may be calculated by taking the ratio of the peak force for milk-soaked sample over the peak force for dry sample, times 100. Plots of texture retention versus soaking time in milk (Figure 10-5) give a clear product comparison based on the resistance of the products to becoming soggy in the presence of milk.

For practical purposes, these three parameters—namely, peak force dry (i.e., peak force of the dry cereals), peak force wet (i.e., peak force for the same cereals immediately after 3 minutes' soaking in milk), and percent texture retention—are usually sufficient for differentiating commercial RTE cereal products and for providing guidance for product development.

Direct product comparison using either cup or shear press methods is limited to products of similar geometry (e.g., size and thickness of flakes, size and shape of puffed cereals, etc.) Sampling and sample packing techniques

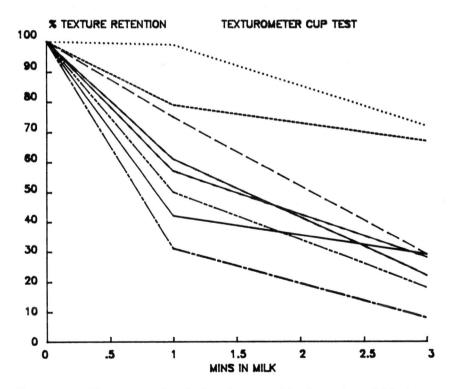

Figure 10-5. Texture retention of selected commercial and experimental flaked cereals in milk soaking. The various lines show drastic differences among samples varying in formulation and processing.

are important in dealing with breakfast cereals that show high variability within and between samples. In order to minimize effects due to sample geometry, oversized as well as undersized pieces are normally excluded from test samples. Breakfast cereals vary appreciably in specific density and bulk density. Whenever possible, the quantity of the sample to be used for testing is standardized by bulk volume rather than by weight to assure the same strain conditions for each sample during testing. This procedure is also better for sensory evaluation. In actual practice when the equal volume procedure is used, however, one often finds great variation due to the test operator. Therefore, the equal weight procedure is preferred. Results obtained may be expressed in force per unit volume of the product for better correlation with sensory data. This is accomplished by determining sample density separately and using that value to convert force per unit weight to force per unit volume.

A crispy product is characterized by a relatively low peak force (dry) and high texture retention after soaking in milk. The detected differences in peak force have been attributed to sample moisture content (Figure 10-6), sample

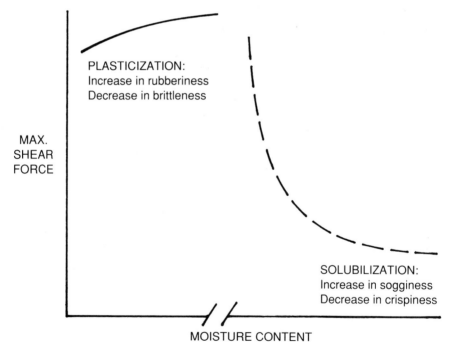

Figure 10-6. Generalized plot showing effect of sample moisture content on peak force. Left and right portion of the curve reflect dry cereals gaining or losing moisture via contact with air and milk, respectively.

age, and other product characteristics (e.g., formulation, process condition, coating, geometry, density, etc.). Note that peak forces for dry and milk-soaked samples are independent of one another. For example, samples A and B may have identical dry peak forces or crispness, but after soaking in milk, sample A may become mushy while sample B may remain crispy.

Testing large numbers of individual pieces of breakfast cereals to derive a useful average is often prohibitively costly. If product uniformity becomes a central issue, however, the Pabst Texture Tester, an automatic, self-cleaning instrument, can be used for testing individual bite-size pieces of breakfast cereal. The test principle for the Pabst Texture Tester is similar to that for the shear press, with both using a hydraulically driven shear–compression cell. Applications of the Pabst Texture Tester have been reported by Szczesniak and colleagues (1974). An example comparing results for two corn flake samples prepared by different processes is given in Figure 10-7.

All of the mechanical tests described so far are empirical and inadequate for quantifying subtle, qualitative differences in crispness. Due to the obvious difficulties in defining and monitoring sample geometry, exact mode, degree of deformation at any point during testing, and other factors, the usefulness of the force/deformation curve is limited. Only curves generated under conditions of equal sample volume and similar sample geometry should even be

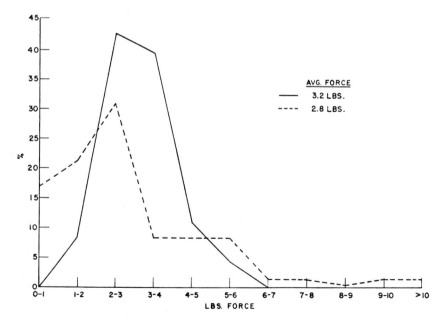

Figure 10-7. Shear force distribution curves for puffed corn flakes made by two processes. (From Szczesniak, et al., 1974.)

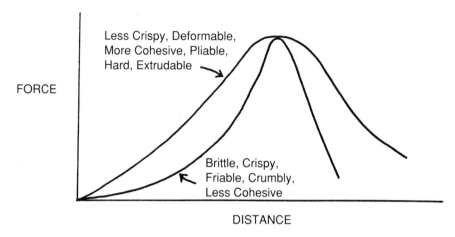

Figure 10-8. Qualitative textural difference among cereal samples of similar peak force values.

compared. In such comparisons, force/deformation curves are normally interpreted by considering the relationships between the mode of deformation and the pattern of sample deformation or failure at a corresponding portion of the curve. A hypothetical example is given in Figure 10-8 to illustrate the textural differences of two samples having identical shear press peak forces. A sample having a lower initial resistance is normally more friable and crispy than a sample having a higher initial resistance. This correlation can be confirmed by examining the particle sizes of the tested samples or by performing audio tests. Particle size analysis may be done subjectively, by using a sieving method or by employing a digital image analyzer with the capability of statistical particle sizing.

AUDIO MEASUREMENT

The difficulty normally encountered in the mechanical testing of small pieces of granular and flaky breakfast cereals prompted the development and use of audio methods for objective crispness measurement. A test setup, referred to as the General Foods Audio Crispometer (Figure 10-9), was developed for generating, recording, analyzing, and quantifying sounds produced by crushing a test sample. The instrument consists of a mechanized wooden crushing roller assembly housed in a soundproof box, a microphone attached to the roller, an instrumentation-grade audio tape recorder, and a statistical noise analyzer.

Sound is characterized by its frequency, intensity, and duration. Sound is normally captured by this instrument over a time duration in the form of discrete signals for measuring the intensity. Consequently, the results are expressed as statistical sound intensity in reference to the signal population. Samples can be compared based on histograms of sound intensity distribution. Figure 10-10 shows the effect of moisture on the crushing sound obtained through the GF Audio Crispometer. The L_{10} value in decibels, that is, the sound level corresponding to the tenth percentile of the measured signals, was found to be useful as a simple parameter for quick sample comparison. Reproducibility of the test is excellent. The intensity of the generated sounds correlates well with sensory panel judgments of product crispness. Since sound intensity is expressed on a logarithmic decibel scale, conversion of sound intensity in decibels to a linear loudness scale in some greatly improves sensory correlation. This conversion is simply done based on the following equation (ISO, 1959):

$$\text{Perceived loudness} = 2^{(\text{sound intensity in decibels} - 40)/10}$$

Sensory perception of crispness is influenced by the mechanical properties of the cereal and by the sound produced due to rupture of the cereal in the mouth by mastication. To more fully define crispness, it is desirable to objectively measure both of these properties. The test and instrumentation

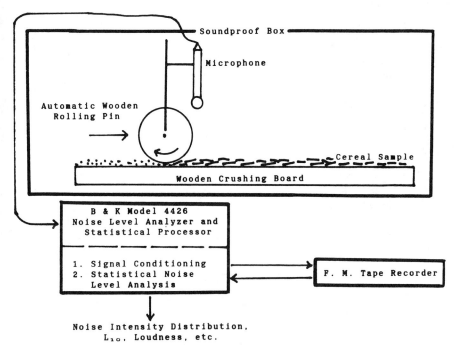

Figure 10-9. Schematic of the General Foods Audio Crispometer.

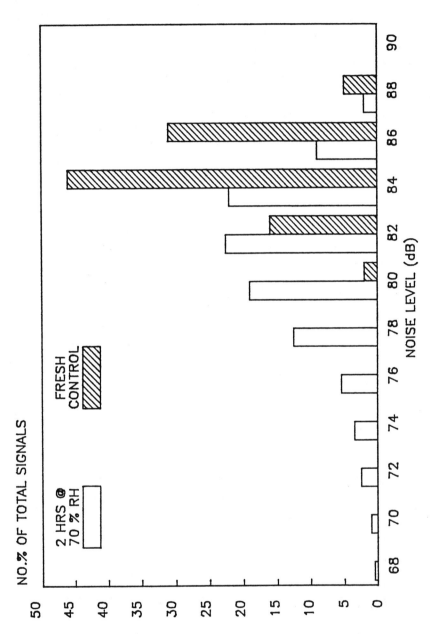

Figure 10-10. Sound intensity distribution of corn flakes.

described above are more practical, thus more suitable, for characterizing breakfast cereals than are those described independently by Drake (1963, 1965) Edmister and Vickers (1985). This greater suitability is due mainly to the reproducible crushing mechanism specially designed for small food pieces or particles.

CONCLUSION

Rheological testing methods, as currently employed with breakfast cereals in industry, have been described. Tests related to characterizing raw materials are basically similar to those used in other cereal industries, for example, the baking industry. Consequently, the tests are not tailored specifically for breakfast cereals. Improvements in test procedures, test conditions, and data interpretation are desirable. Both the mechanical and audio methods used are less than satisfactory and are empirical in nature. In addition, the use of these tests for in-plant quality control purposes needs to be encouraged.

Since advanced processing technologies are adopted rapidly by the breakfast cereal industry (evident by the replacement of old batch processes by continuous processes and manually controlled processes by automated turnkey processes), the need for a better understanding of raw materials and the effects of processing on finished product crispness as well as human perception of crispness becomes increasingly important.

Specifically, a fundamental study on major breakfast cereal types to define the interrelationships of sensory crispness, structure, and mechanical and audio responses as affected by materials and processes will contribute greatly to the future success of the breakfast cereal industry.

REFERENCES

Amerine, M. A., Pangborn, R. N., and Roessler, E. B. 1965. *Principles of Sensory Evaluation of Food.* New York: Academic Press.

Anderson, Y., Drake, B., Granquist, A., Johanason, B., Pangborn, R. N., and Akesson, C. 1973. Fracture force, hardness and brittleness in crisp bread, with a generalized regression analysis approach to instrumental sensory comparisons. *J. Tex. Stud.* 4:110–143.

Burns, A. J., and Bourne, M. C. 1976. Effect of sample dimensions on the snapping force of crisp foods. *J. Tex. Stud.* 6:445–458.

Brennen, J. G., Jowitt, R., and Williams, A. 1974. Sensory and Instrumental Measurement of "Brittleness" and "Crispness" in Biscuits. Paper presented at the 4th International Congress of Food Science and Technology, Madrid, Spain, Sept. 22–27.

Drake, B. K. 1963. Food crushing sounds: An introductory study. *J. Food Sci.* 28:233.

Drake, B. K. 1965. Food crushing sounds: Comparisons of objective and subjective data. *J. Food Sci.* **30**:556.

Edmister, J. A., and Vickers, Z. M. 1985. Instrumental acoustical measures of crispness in foods. *J. Text. Stud.* **16**:153–167.

Fast, R. B. 1987. Breakfast cereals: Processed grains for human consumption. *Cereal Foods World* **32**(3):241.

Gromley, R. 1984. Food texture and consumer. *Farm Food Res.* **15**(5):141.

Hilyard, N. C., and Young, J. 1982. In *Mechanics of Cellular Plastics*, ed. N. C. Hilyard. New York: Macmillan.

International Organization for Standardization. 1959. *Expression of Physical and Subjective Magnitudes of Sound or Noise*, ISO Recommendation: R/131. (Available from ANSI, New York.)

Jowitt, R. 1974. The terminology of food texture. *J. Tex. Stud.* **5**:351–358.

Kapur, K. K. 1971. Frequency spectrographic analysis of bone conducted chewing sounds in process with natural and artificial dentitions. *J. Tex. Stud.* **2**:50–61.

Le Magnen, J. 1964. Vocabulaire tecnique des caracters organoleptiques et de la degustation des produits alimentaires. *Ann. Nutr. Aliment.* **18**:81.

Pugh, S. F. 1967. The fracture of brittle materials. *Brit. J. Appl. Phys.* **18**:129–162.

Sherman, P., and Deghaidy, F. S. 1978. Force-deformation conditions associated with the evaluation of brittleness and crispness in selected foods. *J. Tex. Stud.* **9**:437–459.

Szczesniak, A. S. 1963. Classification of textural characteristics. *J. Food Sci.* **38**:385–389.

Szczesniak, A. S. Einstein, M., and Pabst, R. E. 1974. The texture tester, principles and selected applications. *J. Tex. Stud.* **5**:299–316.

Vickers, Z. M. 1981. Crackliness: Relationships of auditor judgements to tactile judgements and instrumental acoustical measurement. *J. Tex. Stud.* **15**:49–58.

Vickers, Z. M. 1984. Crispness and crunchiness: A difference in pitch. *J. Tex. Stud.* **15**:157–163.

Chapter 11

Influence of Extrusion Processing on In-Line Rheological Behavior, Structure, and Function of Wheat Starch

Bernhard van Lengerich

Extrusion cooking of foods has developed, in recent years, into an important process technique. This technique makes the manufacture possible not only of new products or products with new attributes, such as breakfast cereals, snacks, modified starches, and baby foods, but also of surrogates such as flat bread.

In the manufacture of extruded products, a powdery pre-mix of the raw materials is in most cases introduced into the heated screw barrel of an extruder. The mixture is transported and compressed in this channel by means of one or two rotating screws.

The specific thermal and mechanical energy introduced into the product causes the material being transported to be plasticized. When a food is directly extruded into a puffed product, the plasticized mass reaches a temperature of over 100°C before it leaves the extruder. The pressure on the mass at this point is greater than the vapor pressure resulting from the mass temperature. Sudden evaporation of water results in an expansion of the plasticized material after it exits the die opening and gives the extruded product its characteristic internal structure and external form.

Under the conditions of extrusion cooking, and depending on the raw materials and the process variables used, raw material transformations and energy conversion take place that can lead to products of a most diverse nature. The changes of greatest importance that take place during the substance conversion include gelatinization of starch, coagulation of proteins, and the formation of Maillard products and amylose–lipid complexes (Harper, 1981; Mercier et al., 1980).

The reactions in the extruder depend on a large number of variable machine and raw material control parameters. These parameters determine reactions that are rather complex. Consequently, differentiating between the influences of the individual variables on changes in the characteristics of the materials is possible only to a limited extent. Thus, the use of extrusion technology to modify the characteristics of starchy materials is based predominantly upon empirically acquired knowledge.

The scientific literature on extrusion has, hitherto, principally described the physical and chemical properties of the extruded products; these properties result from the use of specific extrusion conditions in various machines (Mercier and Felliet, 1975; Linko et al ., 1981; Faubion and Hoseney, 1982). This approach does not consider the reaction behavior of starch, under the influence of specific energy, as it is related to the operational variables. Yet, the reaction behavior of starch is of great significance in the analysis/optimization of extrusion technology and the modification of starch or starch-containing raw materials.

The empirical application of extrusion technology is related to the design features of the extruder. Therefore, the results gained with one extruder cannot be transferred to another at will.

The extrusion process was analyzed more systematically by Meuser and colleagues (1982). The proposed model differentiates between independent process parameters, system parameters, structure parameters, and target parameters in the extrusion process (Fig. 11-1).[*] The independent process parameters correspond to the operational variables for the extruder. These parameters govern the specific thermal and mechanical energy introduced into the extruded product. Both types of energy have different influences on structural changes of the product. Structural changes can be characterized at the molecular level and can be described as the formation and dissolution of hydrogen bonds, the disaggregation of valence bonds, and the formation of complexes between amylose and amylopectin. It is these changes in the starch structure which cause various changes in the functional properties of the extruded starch.

The importance of the proposed system analytical model lies in the fact that it uses the mechanical and thermal introduction of energy as the factors that affect changes of the internal structure of a particular product. The final functional product properties are solely and directly the result of structural changes and are not, as has previously been stated in the literature (Mercier, 1980; Kim, 1982; Seiler and Schuy, 1983), the result of machine parameter changes. The extrusion process therefore can be described independently from the machine used. As long as the time, temperature, and shear history of a product within an extruder are the same, the resulting structural and functional properties should not be different.

---

[*] All of the figures in this chapter are from van Lengerich, 1984.

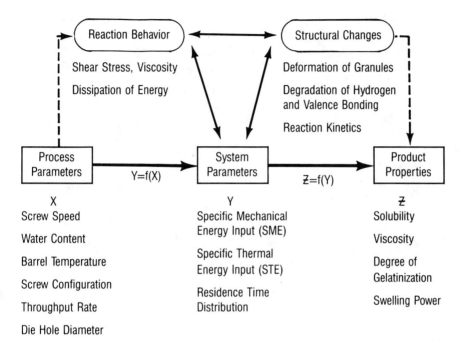

Figure 11-1. System analytical model for the extrusion of starches.

The first part of this chapter describes the development of a method to measure rheological behavior during the extrusion process. Use of this method to determine flow and viscosity curves of plasticized wheat starch under various extrusion conditions is then discussed. This type of data is particularly interesting, because physical properties such as puffing behavior are expected to depend to a large extent upon the rheological properties of a product after plasticization yet before puffing. On the other hand, the rheological properties of the plasticized mass may indicate time, temperature, and shear history effects in an extruder. They are, therefore, essential for gaining information related to the reaction kinetics in the system. Viscosity data generated in this manner allow the prediction of pressure buildup in production-scale extruders and are essential for reliable scale-up of lab-scale processes.

The second part of this chapter describes the influence of independent extrusion parameters (screw speed, throughput rate, barrel temperature, die hole diameter, screw configuration and product water content) on specific mechanical and thermal energy inputs and residence time distributions. The influences of these parameters on changes in structure and function of extruded wheat starch are also considered.

The extrusion tests were conducted using a Werner & Pfleiderer Twin Screw Extruder, Continua 37, with a $1/d$ ratio of 12. The rheological mea-

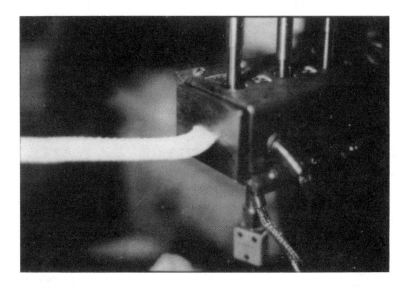

Figure 11-2. Slit die in operaton.

surements were conducted with a slit die with a 1/*h* ratio of 150 that was attached to the die plate of the extruder via a transition piece (Fig. 11-2). The dimensions of the slit die are shown in Figure 11-3.

WHAT HAPPENS INSIDE AN EXTRUDER?

The introduction of energy causes various changes in starch structure. First, water diffuses into the starch granules during the first few seconds of transportation in the screw chamber. The degree of fill of the screw chamber in this first stage is rather low; therefore, only a slight increase in temperature of the hydrated product takes place during its transportation to the shear or melting stage in the extruder. When the product enters the melting zone, a short section of reverse pitch screw elements generally lets the degree of fill increase. Starch granules are compressed, are deformed, and start to melt and lose their crystalline regions. At this point, pressure buildup causes leakage flow. Viscous dissipation of mechanical energy takes place, leading to breakdown of the starch molecules, particularly amylopectin. As the product passes through the shear zone, friction heat in combination with external heat contributes to the phenomenon generally known as gelatinization. Starch granules or parts of them swell. Hydrogen bonds between starch molecules disaggregate, and mobile water diffuses to the anhydrous glucose units to form loose bondings with available hydroxyl groups. Even though gelatinization is generally understood to be a process in which water is an essential reaction

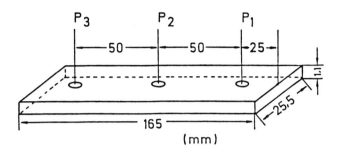

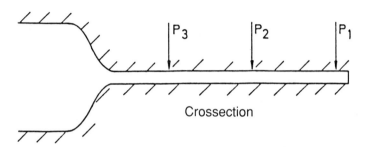

Figure 11-3. Schematic representation of the slit viscosimeter.

partner, extrusion of starch can result in similar phenomena even without any additional water (van Lengerich, 1984).

After it leaves the shear or melting zone, the starch is usually completely gelatinized. Starch granules are disrupted and do not show birefringence. Further confirmation of this fact is that the enthalpy required to complete gelatinization after extrusion (measured by differential scanning calorimetry [DSC]) is zero. Finally, the enzymatic hydrolysability of the starch indicates complete gelatinization. The product, which now is a plasticized mass, appears as distinctive coherent doughlike agglomerates and is transported into the die section. Here, the mass is compressed again and shaped by the geometry of the die before the pressure and temperature drop cause the material to expand more or less as it exits the extruder die (van Lengerich, 1984).

APPLIED RHEOLOGY IN EXTRUSION

Fundamental Relationships

The specific mechanical energy (SME) introduction can be calculated by using the equation derived from Meuser and colleagues (1982).

$$\text{SME} = \frac{M_d \, \omega}{\dot{m}} \text{ [Wh kg}^{-1}\text{]} \tag{11-1}$$

where

M_d = torque [Nm]
ω = angular velocity [s^{-1}]
$\dot{m}$ = throughput [kg h^{-1}]

The shear stress calculation results directly from the momentum balance and can be calculated by Equation 11-2:

$$\tau = \frac{\Delta P}{L} \cdot \frac{H}{2} \text{ [Pa]} \tag{11-2}$$

where

ΔP = pressure drop per distance [Pa]
L = distance between pressure measurements [mm]
H = slit height [mm]

The apparent shear rate was calculated by Equation 11-3.

$$\dot{\gamma}_{app} = \frac{3\dot{Q}}{2B \cdot h^2} \text{ [s}^{-1}\text{]} \tag{11-3}$$

where

$\dot{Q}$ = volumetric mass flow $\left[\dfrac{mm^3}{s}\right]$
B = width of slit [mm]
h = 1/2 slit height [mm]

This value was corrected by using Equation 11-4 according to the Rabinowitsch method (Rabinowitsch, 1929).

$$\dot{\gamma}_{corr} = \left(\frac{2n + 1}{3n}\right)\left(\frac{3\dot{Q}}{2B \cdot h^2}\right) \text{ [s}^{-1}\text{]} \tag{11-4}$$

where

n = flow index
$= (\partial \ln \tau)/(\partial \ln \dot{\gamma}_{app})$

The viscosity (η) was calculated by using the power law equation (Eq. 11-5).

$$\eta = m \cdot \dot{\gamma}_{\text{corr}}^{n-1} \quad [\text{Pa} \cdot \text{s}] \tag{11–5}$$

where m = consistency index. Chemical analysis of the extruded starch was conducted as described previously.

Determining Flow Curves with a Slit Rheometer Attached to a Corotating Twin-Screw Extruder

On-line viscosity measurements of high-viscosity fluids processed with corotating twin-screw extruders and transported through a slit viscosimeter with a high l/h ratio are generally difficult. This difficulty is because the intermeshing, corotating twin screws are designed as an axially open system that operates in a pressure-dependent manner. Thus, if the die pressure becomes excessive, the enormous back pressure buildup prevents further product flow and causes a high-pressure shutdown, or the product is not transported, fills the machine, and causes a high-torque shutdown. The slit viscosimeter employed (Haake design) was equipped with a circular entrance opening but had to be attached to the figure-eight–shaped screw barrel. For this reason, a transition piece was designed to provide a smooth hydrodynamic transition from the extruder into the die entrance.

Experience obtained with non-food polymers and single-screw extruders indicates that it is advisable to keep all extrusion parameters constant and to generate various shear rates simply by changing the screw speed. The resulting shear stresses, together with the shear rates, can then be used to calculate the rheological parameters of interest. The authors of this chapter performed tests following this procedure, and the screw speed was changed without altering any of the remaining extrusion parameters. The shear stress was plotted against the screw speed to obtain the resulting curves (Fig. 11-4). All curves showed a decrease in shear rate with increasing screw speed. An increase in screw speed is expected to result in a proportional increase of shear stress; however, for pseudoplastic behavior the increase in shear rate should be larger than the increase in shear stress.

These results indicated that the procedure used for polymer extrusion with single-screw extruders is not transferable to twin-screw extrusion. This nontransferability is because the screw part in the feeding zone of a single-screw extruder is always completely filled with product. Therefore, an increase in screw speed automatically leads to a proportional increase in

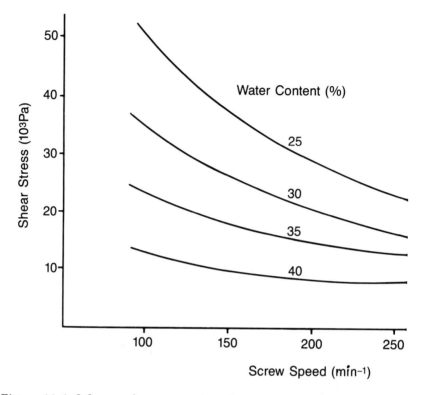

Figure 11-4. Influence of screw speed on shear stress at various water contents (shear rate: 360 s^{-1}).

throughput, which leads to an increase in the volumetric flow and a subsequent increase in the shear rate. Twin-screw extruders, however, are fed gravimetrically or volumetrically. The degree of fill in the feed zone is always smaller than 100%. The throughput becomes an independent variable and can, therefore, be changed independent of the screw speed. This means that with an increase in screw speed, no change in throughput (and subsequently shear rate) in the slit die occurs. The results shown in Figure 11-4 can then be interpreted as follows. Constant throughput leads to a constant shear rate, in this case 360 s^{-1}. Since apparent viscosity is determined as the ratio between shear stress and shear rate, the curves actually show the influence of screw speed on apparent viscosity at various water contents. Lower water content results in higher viscosities; an increase of the screw speed leads to a stronger shear thinning effect, the lower the water content. At high water contents (40%), shear thinning effects become very small and can be neglected at screw speeds above 200 rpm. This result is important insofar as it explains the tendency of starch molecules to disaggregate depending on water content. At low water contents, a high amount of mechanical energy can be introduced. This leads to a significantly higher level of molecular disruption than takes place at high water contents, where the amount of mechanical energy that can be introduced is limited.

The influence of various shear rates in the slit die on the resulting shear stresses of the starch can be investigated by setting the throughput rate at various levels and measuring the pressure drop along the slit die. However, changing the throughput rate would also result in a change in the degree of fill. To keep the degree of fill in the screw chamber constant for each individual flow curve, the screw speed was adjusted proportionally to a change in throughput rate. This was done by defining a specific throughput, which describes the ratio between the amount of starch gravimetrically fed into the screw chamber and the rpm of the screw. For example, the specific throughput rate was calculated to be 1.1 g/screw rotation. The actual throughput is, then, determined by the rpm; for example, 100 rpm would require 110 g/min, 300 rpm would require 330 g/min, and so on. This technique allows a change in shear rate, which leads to a proportional change in rpm, and yet allows the degree of fill to remain constant for all tests within one flow curve.

Effect of Extrusion Parameters on Rheological Properties

Effect of Water Content

Using the technique described above, tests were conducted with water contents in the range from 25% to 40%. The experimental results are shown in Table 11-1. The flow curves and the viscosity curves are shown in Figures 11-5 and 11-6.

The results suggest (as will be described later) that at low water levels, a significant structural breakdown of the starch granules takes place. The degree of pseudoplasticity, which is expressed in the small n values, is relatively high and results in the previously described strong shear thinning effect at low water levels. At the same time, consistency indices (m values) are extremely high. These values, together with the n values, indicate a significant change in the molecular structure of the starch.

With increasing moisture contents, the zero shear rate viscosity (consistency index) decreases drastically and the flow index increases (Table 11-1). The viscosity decrease explains the inability of the starch mass to resist the introduction of mechanical energy at high water levels. At 40% moisture, the flow index was determined to be 1.01. This index indicates a shift from pseudoplastic to Newtonian flow behavior.

As can be seen from the column in Table 11-1 that shows the expansion ratio of the extruded starches, expansion increases with lower water

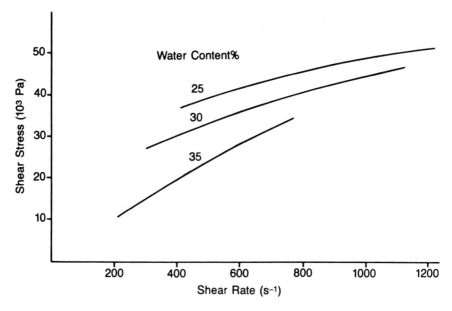

Figure 11-5. Flow curves at various water contents.

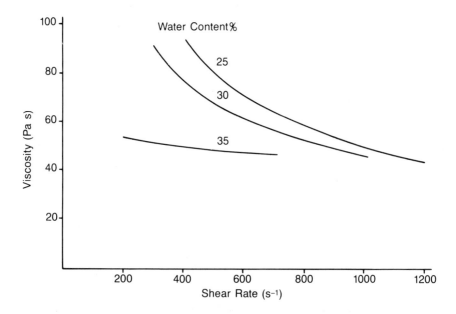

Figure 11-6. Viscosity curves at various water contents.

Table 11-1. Experimental Results from Test with Various Water Contents

Water Content (°C)	Screw Speed (min⁻¹)	$\Delta P/L$ (Pa cm⁻¹ × 10⁵)	Q (cm³s⁻¹)	τ (Pa)	$\dot\gamma_{corr}$ (s⁻¹)	Expansion Ratio $\left(\dfrac{\rho_{extrudate}}{\rho_{starch}}\right)$	Flow Curve Data		
							n	m	r^2
25	250	9.30	3.513	51,200	1,186.5	7.4			
	200	8.98	2.798	49,400	940.2	8.8			
	150	8.33	2.165	45,800	727.5	7.4	0.310	5,841	0.99
	100	7.58	1.552	41,700	521.5	5.0			
	75	6.58	1.185	36,200	398.2	2.7			
30	250	8.33	3.655	45,815	1,031.1	4.2			
	200	7.78	2.924	42,790	825.0	4.0			
	150	6.70	2.256	36,850	636.4	2.9	0.425	2,412	0.99
	100	5.88	1.566	32,340	441.8	2.3			
	75	4.23	1.441	23,270	207.0	2.2			
35	200	5.96	3.077	23,800	615.2	1.6			
	150	5.08	2.372	27,900	474.2	1.5	0.922	91.3	0.99
	100	3.68	1.688	20,200	337.5	1.4			
	75	2.67	1.282	14,700	256.3	1.0			
40	250	5.18	4.200	28,490	815.3	1.4			
	200	4.38	3.346	24,090	650.0	0.8			
	150	3.92	2.581	21,560	500.0	0.9	1.010	35	0.98
	100	2.40	1.706	13,200	331.2	1.5			
	75	1.58	1.334	8,690	259.0	1.7			

Notes: Degree of fill: 1.66 g/rpm, temperature: 170°C. $\Delta P/L=$ pressure loss along slit die; $Q=$ volumetric rate; $\tau=$ shear stress; $\dot\gamma_{corr}=$ corrected shear rate.

contents even if the shear rate remains relatively constant. Within one flow curve, however, expansion increases if viscosity decreases. Apparently, bubble growth is inhibited as the viscosity of the plasticized mass increases at a constant water level. Since high viscosities caused by low water contents also showed high expansion ratios, it may be valid to assume that expansion relates directly to the shear stress of the plasticized mass. Extremely high shear rates ($>1,000$ s⁻¹) cause a predominant molecular disruption of amylopectin molecules, which seems to inhibit a further increase of expansion. This phenomenon explains the decrease in the expansion after a shear rate maximum is passed at low water contents.

Effect of Barrel Temperature

An increase in barrel temperature increases the product temperature and decreases its viscosity. These changes decrease the torque at the screw shafts, which leads to a lower specific mechanical energy input and decreases molecular disaggregation of the extruded starch. Table 11-2 and Figures 11-7 and 11-8 show the effect of barrel temperature on viscosity. The data (Table 11-2) indicate that the lowest barrel temperature results in the highest n values. This correlation can be explained by the direct effect of temperature on viscosity. The low flow index reflects a high level of pseudoplasticity, which abates with increasing barrel temperature. The flow and viscosity curves (Figs. 11-7, 11-8) clearly show the shear thinning behavior of the starch, which decreases at higher barrel temperatures. Expansion reacts as expected. Increasing barrel temperatures (at similar shear rates) cause an increase in expansion. On the other hand, when barrel temperatures are kept low (160°C), higher shear stresses cause larger pressure drops and consequently greater expansion.

Effect of Degree of Fill

The influence of the degree of fill was investigated by adjusting the specific mass flow to three different levels. One major effect of differences in mass flow is, according to Equation 11-1, a proportional change in the introduction of specific mechanical energy. Regardless of whether it was achieved through an increase in mass flow at a constant screw speed, or through lowering the screw speed at a constant mass flow, a higher degree of fill always

Table 11-2. Experimental Results from Test with Various Barrel Temperatures

Barrel Temperature (°C)	Screw Speed (min⁻¹)	ΔP/L (Pa cm⁻¹ × 10⁵)	Q (cm³s⁻¹)	τ (Pa)	$\dot{\gamma}_{corr}$ (s⁻¹)	Expansion Ratio $\left(\dfrac{\rho_{extrudate}}{\rho_{starch}}\right)$	Flow Curve Data n	m	r²
160	250	7.18	2.574	39,490	1,063.5	8.2			
	200	6.85	2.083	37,680	860.7	7.4			
	150	6.40	1.593	35,200	658.2	6.1	0.229	8,017	0.99
	100	5.90	1.083	32,450	447.5	3.8			
	75	5.06	0.919	27,830	229.3	3.2			
170	250	6.28	2.471	34,540	760.6	7.4			
	200	5.81	2.104	31,960	647.7	6.8			
	150	5.38	1.537	29,590	484.2	5.0	0.364	3,073	0.99
	100	4.56	1.083	25,080	333.4	3.4			
	75	4.20	0.817	23,100	251.5	2.7			
200	250	4.12	2.590	22,660	444.4	4.2			
	150	2.50	1.593	13,750	273.3	4.2	1.046	2	0.94
	100	1.33	1.083	7,315	185.8	3.3			
	75	0.67	0.817	3,685	146.2	3.2			

Notes: Degree of fill: 1.11 g/rpm, water content: 25%. ΔP/L = pressure loss along slit die; Q = volumetric rate; τ = shear stress; $\dot{\gamma}_{corr}$ = corrected shear rate.

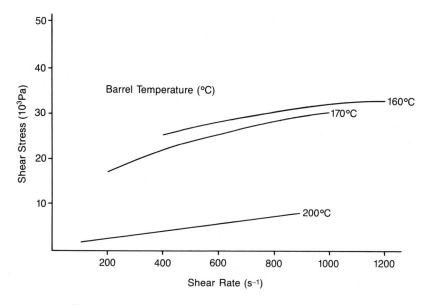

Figure 11-7. Flow curves at various barrel temperatures.

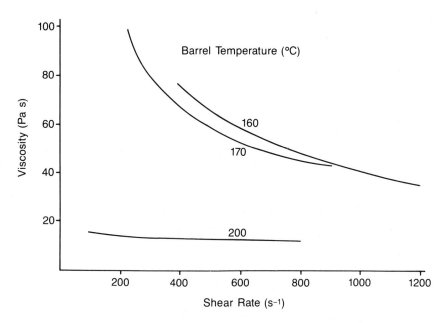

Figure 11-8. Viscosity curves at various barrel temperatures.

corresponded to a lower SME. When the degree of fill was increased by lowering the screw speed only, the shear rate in the slit die remained constant. In this case, a reduction of the screw speed caused a decrease in SME. This decrease led to reduced intensive granular disruption of the starch mass in the extruder, which in turn resulted in a higher viscosity in the slit die.

The data shown in Tables 11-3 and 11-4 and the curves in Figures 11-9 and 11-10 illustrate these results. Increasing the degree of fill causes only a very slight decrease in the flow indices but a clear increase in the consistency indices (Table 11-3). It can be shown that the molecular breakdown of the starch is more extensive at high screw speeds (low degree of fill). This breakdown causes lower viscosities in the capillary die (Table 11-4). A comparison of the SME values supports this conclusion. The data in Table 11-4 also show that residence time does not have a significant influence on the molecular breakdown of starch in the extruder. At 200 rpm, a relatively high amount of mechanical energy (150 Wh kg^{-1}) is introduced in a very short period of time, whereas at 75 rpm the SME is lowered (120 Wh kg^{-1}) but the residence time is 40 seconds, almost twice as long. The average molecular weights decreased with increasing SME values. It appears from the data that a high SME introduced in a relative short period of time affects molecular breakdown more than a low SME introduced over an extended (40 second) time period.

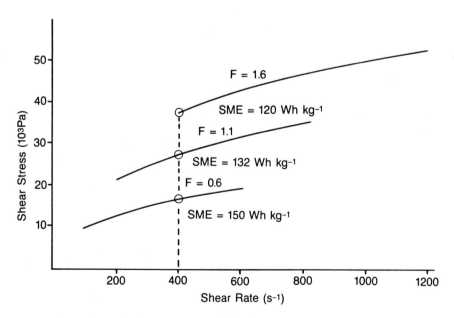

Figure 11-9. Flow curves at various degrees of fill (temp. 170°C).

Table 11-3. Experimental Results from Test with Various Degrees of Fill

Degree of Fill (g/rpm)	Screw Speed (min⁻¹)	$\Delta P/L$ (Pa cm⁻¹ ×10⁵)	Q (cm³s⁻¹)	τ (Pa)	$\dot{\gamma}_{corr}$ (s⁻¹)	Expansion Ratio $\left(\frac{\rho_{extrudate}}{\rho_{starch}}\right)$	Flow Curve Data n	m	r^2
0.66	250	3.10	1.593	17,050	475.3	3.7			
	200	2.96	1.328	16,280	396.2	2.3	0.384	1,639	0.96
	150	2.79	0.960	15,350	286.4	2.0			
	100	2.18	0.654	11,990	195.1	1.5			
1.11	250	6.28	2.471	34,540	760.6	7.3			
	200	5.81	2.104	31,960	647.7	6.8			
	150	5.38	1.573	29,590	484.2	5.0	0.364	3,073	0.99
	100	4.56	1.083	25,080	333.4	3.4			
	75	4.20	0.817	23,100	251.5	2.7			
1.66	250	9.30	3.513	51,200	1,180.5	7.4			
	200	8.98	2.798	49,499	940.2	8.8			
	150	8.33	2.165	45,800	727.5	7.4	0.310	5,841	0.99
	100	7.58	1.552	41,700	521.5	5.0			
	75	6.58	1.185	36,200	398.2	2.6			

Notes: Moisture content: 25%, Temperature: 170°C. $\Delta P/L$ = pressure loss along slit die; Q = volumetric rate; τ = shear stress; $\dot{\gamma}_{corr}$ = corrected shear rate.

Table 11-4. Effect of Screw Speed on Degree of Fill and Viscosity at Constant Shear Rate

Shear Rate (s⁻¹)	Screw Speed (min⁻¹)	Degree of Fill (g/rpm)	Residence Time (s)	SME (Wh kg⁻¹)	Viscosity (Pa·s)	Average Molecular Weight (10⁶ g mol⁻¹)
400	200	0.66	21	150	41	1.8
400	125	1.11	29	132	69	2.7
400	75	1.66	40	120	94	3.5

Influence of Extrusion Parameters on Shear Stress and the Apparent Viscosity of Starch

To summarize the results for all rheological tests, a polynomial regression analysis was conducted to show all observed data in the entire test range. Figure 11-11 shows ISO-curves for three different shear stresses. The figure shows, on one hand, what combinations of extrusion conditions lead to a specific shear stress and how the shear stress changes with changes in one or more extrusion conditions. At the same time, this figure shows how to

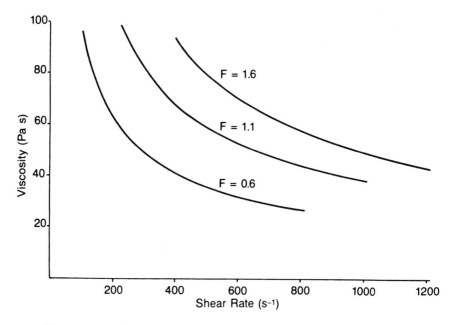

Figure 11-10. Viscosity curves for various degrees of fill (temp. 170°C).

compensate for a change in one extrusion parameter to maintain a constant shear stress; for example, at a water content of 28%, a barrel temperature of 150°C, and a degree of fill of 0.6, the shear stress is 21,000 Pa. If for some reason the water content decreases, the shear stress can be maintained by increasing the barrel temperature proportionally.

Figure 11-12 illustrates the influence of the tested extrusion parameters on apparent viscosity. The treatment of the starch in the screw chamber influences the apparent viscosity of the starch in the slit viscosimeter. The barrel temperature and the product water content have a direct influence in that higher barrel temperatures and water contents decrease the apparent viscosities over the entire test range. Increases in the degree of fill at the same shear rate can be accomplished only by decreasing the screw speed, which in turn decreases granular disruption and molecular disaggregation of the starch and leads to higher viscosity values.

Transformation of the Results into the System Analytical Model

To describe the ability of starch to resist mechanical stress, the SME values for all tests conducted with the same degree of fill were plotted against the

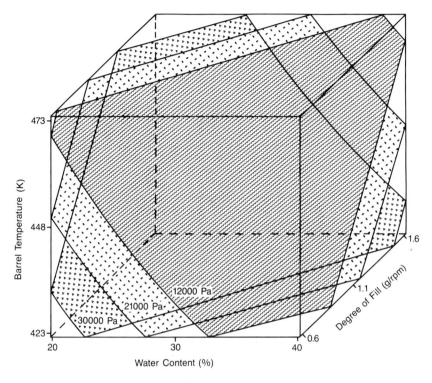

Figure 11-11. Influence of extrusion parameters on shear stress.

appropriate shear stresses (Fig. 11-13). The results show that a change in any extrusion parameter that leads to an increase in shear stress simultaneously causes a proportional change in the SME. This relationship implies that the shear stress can be used as a key parameter to characterize the ability to introduce mechanical energy into the starch. From a system analytical viewpoint using the model described above (Fig. 11-1), it can be concluded that water content and barrel temperature have the most significant influence on the mechanical energy dissipation capacity of the starch. These two parameters determine how much mechanical energy can be introduced, by increasing the screw speed, lowering the die cross-sectional area, or strengthening the screw configuration. Product water contents of more than 35% lead to an almost Newtonian flow behavior and limit viscous energy dissipation. By altering the barrel temperature, one can directly influence the product temperature and adjust the shear stress to a level at which the required amount of mechanical energy is introduced.

The third important influence on the energy dissipated into the starch results from the degree of fill. At a constant shear rate, the higher the specific mass flow (g/rpm), the higher the shear stress in the mass. With respect to

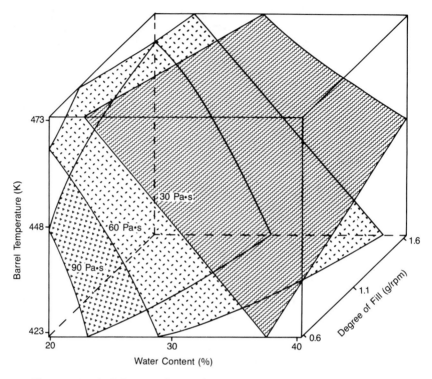

Figure 11-12. Influence of extrusion parameters on the apparent viscosity of the starch.

the reaction behavior of the plasticized starch mass in the extruder, it can be concluded that high levels of mechanical energy can be introduced only if the starch is able to maintain a high shear stress. This is the case when either the water content and/or the barrel temperature is low, or the degree of fill is high.

INFLUENCE OF EXTRUSION PARAMETERS ON THE INTRODUCTION OF SPECIFIC MECHANICAL AND THERMAL ENERGY

Experimental Plan

The results of rheological measurements have shown how extrusion parameters influence the flow and viscosity behavior of wheat starch within an extruder. To a certain extent, these results also provide an explanation of

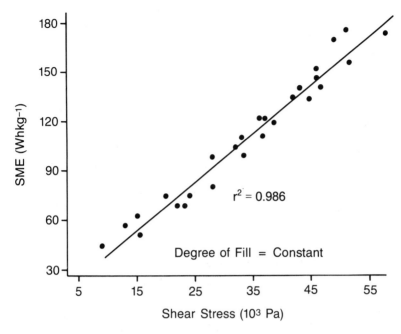

Figure 11-13. Correlation between SME and shear stress in the slit die.

the effects of the major extrusion parameters on changes in the introduction of energy during starch extrusion. The next important step to verify the system analytical model was to analytically characterize the relationships between extrusion parameters and specific energy introduction as well as their influence on the structural and functional properties of the extruded starch. All possible independent parameters for the extruder must be considered to establish these relationships. To accomplish this, a statistical fractionated factorial design was developed that defined the experiments to be conducted. The test range covered extrusion conditions from no or partial gelatinization to maximum molecular breakdown of the starch. A complete 3^n design was combined with a modified Box–Wilson design to achieve the desired analysis of linear, quadratic, and interactive effects on specific mechanical and thermal energy introduction. Mass flow, die hole diameter, barrel temperature, and screw speed were considered as quantifiable machine parameters. Product water content was included as a parameter related to the raw material.

The effects of screw geometry were studied qualitatively by conducting each extrusion test with two different screw configurations; these configurations produced different shear intensities. Screw A consisted primarily of transport elements that exert a slight shear effect and convey the product into the die area relatively rapidly. With this screw, the mass is plasticized only shortly before it enters the die, and product temperature is influenced to a great extent by changing the barrel temperature. In Screw C, 30% of the transport elements were replaced with kneading blocks and reverse pitch elements in the middle part of the screw to accomplish a faster and more severe plasticization. This configuration results in greater mechanical strain than with Screw A, as well as an increased production of frictional heat.

To determine residence time distribution, a screw configuration was chosen with a shear intensity that was intermediate to configurations A and C. All levels between the extrusion parameters were selected to be appropriately large in order to cover a wide experimental range. The set points of the variables used during the extrusion experiments, as well as the transformation equations employed to calculate the statistical levels of those variables, are given in Figure 11-14. The statistical experimental design used is shown in Figure 11-15.

Influence of Extrusion Parameters on Energy Input

The relationships between the extrusion parameters and energy input were characterized initially by a polynomial regression equation derived from the experimental data. Table 11-5 shows the regression coefficients calculated for screw configurations A and C that are valid for the test range studied. Only those values that were more than 95% statistically significant were used for the regression equation that describes the functional relationship. Since possible interactions were also considered in the calculation of the regression

Extrusion Parameter	Unit	Transformation Formula	Level		
			-1	0	+1
x_1 = Throughput Rate	$g\ min^{-1}$	$x_1 = \dfrac{m - 250}{100}$	150	250	350
x_2 = Screw Speed	min^{-1}	$x_2 = \dfrac{n - 180}{90}$	90	180	270
x_3 = Barrel Temperature	°C	$x_3 = \dfrac{Tg - 160}{80}$	80	160	240
x_4 = Die Hole Diameter	mm	$x_4 = \dfrac{D_r - 3.5}{1.5}$	2	3.5	5
x_5 = Water Content	%	$x_5 = \dfrac{Prod.-H_2O}{5}$	15	20	25

Figure 11-14. Transformation formulas for extrusion parameters.

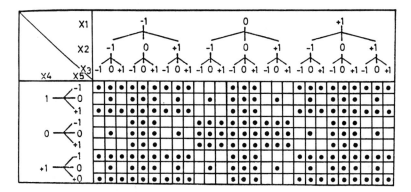

Figure 11-15. Combination of a complete with a fractional statistical experimental design.

equation, the resulting equation is composed of twenty different terms, each representing the product of a regression coefficient and an X value. The X values X_1–X_5 correspond to the levels of operating variables used and can assume only values of $+1$, 0, and -1 (see conversion equations, Fig. 11-14). The same is also true for the remaining quadratic and interactive effects, whose X values are composed exclusively of factors from X_1 to X_5. The constant in the equation, X_0, results when all of the operating parameters are adjusted to the middle level. Since the transformed values of the extrusion parameters then assume the value zero, all other terms of the equation drop out.

Influence of Extrusion Parameters on the SME

Graphic representation of the influence of n parameters on a dependent target dimension requires $n + 1$ dimensions. Therefore, the influence of two variables can be represented at the same time. If the values of the dependent variable are kept constant, however, the influence of three independent variables can be shown graphically. In this way, the influences of product water content, die diameter, and barrel temperature on the SME are shown in Figure 11-16.

The cube shown in Figure 11-16 represents the total test region studied. The figure assumes a constant screw speed and mass flow. The results show three extrusion parameters adjusted to three different levels. If the values of all possible SME values are calculated by using the regression equation, and all values with the same SME are combined, the resulting isosurfaces that are shown are formed. The shape of the isosurfaces in the test region indicates the influence of the extrusion parameters.

Table 11-5. Regression Coefficients to Characterize the Influence of Extrusion Parameters on Product Temperatures and the Specific Mechanical Energy Input

| Extrusion Parameter | Effect | Regression Coefficient | | | |
| | | Screw A | | Screw C | |
		SME (Wh kg^{-1})	T_P (°C)	SME (Wh kg^{-1})	T_P (°C)
	Constant Value	66.0***	119.0***	119.0***	134.8***
Throughput		−12.5***	0.7	−30.3***	0.2
Screw speed		16.7***	4.5***	33.5***	6.8***
Barrel temperature	Linear	−13.8***	17.1***	−24.1***	19.4***
Product water content		−20.3***	−16.9***	−29.6***	−14.5***
Die hole diameter		−15.8***	−10.3***	− 7.0***	− 5.4***
Throughput		2.1	−1.5	8.9***	3.4***
Screw speed		−1.1	−0.9	−2.5	−1.5
Barrel temperature	Quadratic	9.0***	−0.5	0.9	1.2
Product water content		−5.1**	2.7*	2.7	0.4
Die hole diameter		−0.7	−3.1**	3.6	−0.6
Throughput— screw speed		−6.2**	−0.5	−10.7***	1.6*
Throughput— barrel temperature		−3.8***	−1.5*	2.3	−0.8
Throughput— product water content		−0.2	−4.2**	8.4***	−1.7**
Throughput— die hole diameter		1.2	−0.4	−1.9	−0.3
Screw Speed— barrel temperature	Interactions	5.4***	−0.2	−0.5	−1.7**
Screw Speed— product water content		1.3	1.6	−2.2	−1.7**
Screw Speed— die hole diameter		−3.3**	−2.3**	−2.5	−0.3
Barrel temperature— product water content		−6.1***	5.1***	4.7**	2.3**
Barrel temperature— die hole diameter		2.1	0.8	3.2*	−0.2
Product water content— die hole diameter		3.3**	1.8**	−2.3	0.7
	r^2:	92.4%	95.5%	95.5%	97.1%

Source: From van Lengerich, 1984.
*** = $P > 99.9\%$
** = $P > 99\%$
* = $P > 95\%$

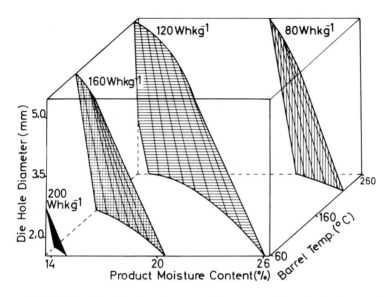

Figure 11-16. The effect of extrusion parameters on specific mechanical energy input.

It can be seen that SME increased with decreasing water content, reaching 200 Wh kg^{-1} at about 15%, when the barrel temperature and the die hole diameter were simultaneously adjusted to the lower level. Thus, the SME could decrease when the barrel temperature, the product water content, or the die hole diameter was reduced. The slightly bent shape of the surfaces indicates that the relationships were not linear, but were influenced by non-linear effects and interactions. In addition, Figure 11-16 shows that reaching a certain SME is not related to one specific combination of variables, but can be achieved by a nearly infinite combination of these three variables alone. If an SME outside this test region is required, other extrusion parameters, for example, the screw speed, the product mass flow, or the screw geometry, could be changed to accomplish this.

The test region shown in Figure 11-16 reflects behavior at intermediate screw speed and product mass flow values. By additional variation of these two extrusion parameters, the resulting SME can be raised or lowered. The SME values added each time for each combination of screw speed and product mass flow can be taken from the diagram isolines in Figure 11-17, which shows the effects of screw speed and product mass flow on SME. It follows from the figure that, starting at the middle level, a maximum increase in SME of 80 Wh kg^{-1} is possible. All of the SME values shown in Figure 11-16 increase by this value when the screw speed is raised to the upper level and the product mass flow is adjusted to the lower level. The opposite effect takes place at low screw speeds and high mass flows. The three-dimensional figure also shows the direct influence of these two extrusion parameters on SME. Thus, SME

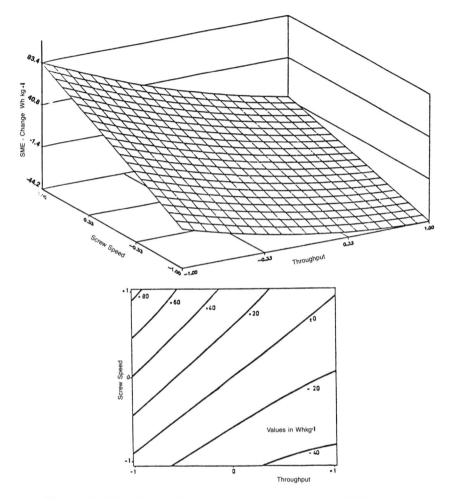

Figure 11-17. Influence of screw speed and throughput on SME change.

can be increased by raising the screw speed, by reducing the product mass flow, or by a combination of both.

Influence of Extrusion Parameters on Product Temperature

The influences of product water content, die hole diameter, and barrel temperature on product temperature are shown in Figure 11-18. The shape of the isosurfaces indicates that reductions in die hole diameter led to increases in product temperature. As expected, a change in the barrel temperature led in each case to a change in the product temperature. Comparison of this figure with Figure 11-16 makes it clear that, in the test region shown here, almost any combination of thermal and mechanical energy input can be obtained. In particular, at 140°C, a temperature that is of interest for the extrusion of starch, a wide range of SME values can be covered by suitable choices of extrusion parameters. An expanded representation of the influence of screw speed and product mass flow on product temperature is given in Figure 11-19.

While an increase in the screw speed always leads to an increase in the product temperature, Figure 11-19 shows that no significant change in the product temperature resulted from a change in the mass flow, although temperature changes were observed during the experiments. This observation can be explained by the fact that the temperature change resulted each time from the combined effects of the remaining operating variables.

The influence of mass flow on product temperature could, therefore, be derived from the quadratic and interactive effects that occurred among the mass flow and the remaining extrusion parameters.

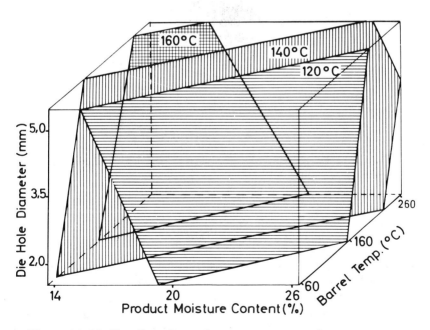

Figure 11-18. The effect of extrusion parameters on product temperature.

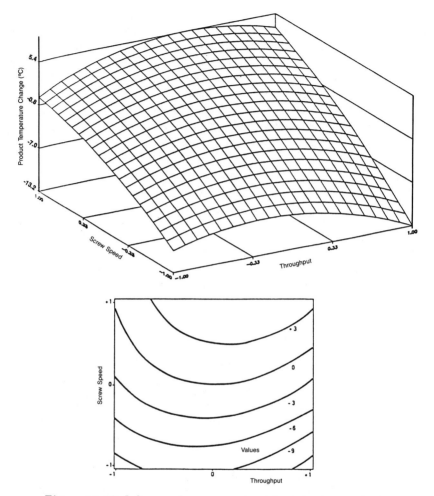

Figure 11-19. Influence of screw speed and throughput on product temperature change.

Screw Configuration

The geometry of the screws used cannot be described quantitatively. The level of mechanical and thermal energy input was shifted towards higher product temperatures and higher SME values with screw configuration C compared to screw configuration A, however. In addition to the experiments with these two screw configurations, extrusion experiments were done according to a fractionated 3^5-factorial experimental design with a third screw, which was made with conveying, mixing, kneading, and compression elements. The effects of different screw configurations on energy input can be determined by qualitatively comparing the regression coefficients of the linear main effects of the extrusion parameters. The constant terms of the equation were used to estimate the general intensity of the screws in mechanical and thermal energy introduction. The linear main effects for Screws A, B, and C are given in Table 11-6. As expected, the values of the constant terms increase with increasing shear intensity. A comparison of the regression coefficients of the effect of die diameter makes it clear that the effect on SME becomes smaller with increased shear intensity, which also results in less viscous dissipation and friction heat formation. This observation explains the decrease in the regression coefficients for product temperature. As the regression coefficients illustrate, the effect of water content becomes still stronger with increasing shear intensity of the screws. The screw configuration influences the effect of the barrel temperature. The product temperature can be influenced the most by the barrel temperature when Screw C is used. This result is due to the extended product residence time using this screw configuration.

Figure 11-20 shows the effect of a change in the screw configuration on the residence time distribution. The different arrangements of the screw elements increase the average residence time and thus contribute to a more intensive back-mixing of the mass. Since the SME also increases, a greater amount of friction heat has been introduced.

In summary, changes in the screw speed, product water content, or barrel temperature have a greater effect on the thermal and mechanical energy input into a substance when the screw is more shear intensive and provides a longer residence time of the mass in the barrel.

Table 11-6. Comparison of Regression Coefficients of Main Effects of Various Screw Configurations

Extrusion Parameter	Effect	SME $(Wh\ kg^{-1})$			T_P (°C)		
		A	B	C	A	B	C
	Constant	66	115	119	119	126	135
Throughput		−13	−24	−30	—	—	—
Screw speed		17	24	32	5	6	7
Barrel temperature	Linear	−14	−20	−24	17	15	19
Product water content		−20	−23	−30	−17	−17	−15
Die hole diameter		−16	−12	− 7	−10	−10	− 5

Note: $P > 99.9\%$

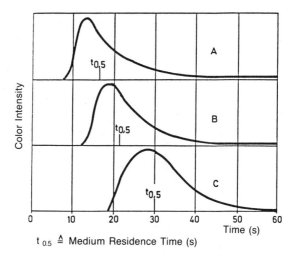

Figure 11-20. Influence of screw configuration on residence time distribution.

Influence of Energy Input on Structural Changes in Starch

The results that have been presented show that extrusion parameters affect the forms of energy input, and that the effects overlap. Functional relationships between extrusion parameters and thermal and mechanical energy input, as well as their influence on reaction behavior, have been characterized. The question to be answered now is how the specific mechanical and thermal energy input influences the structure of the starch.

The decomposition of both the starch granule and its paracrystalline structure was characterized by scanning electron microscopy and X-ray diffraction. Residual gelatinization enthalpy was determined indirectly by means of DSC studies, and viscous behavior was measured by gelatinization curves in a coaxial rheometer. In addition, the molecular weight distribution of the starch polymers was determined in an attempt to determine how the starch molecules had been changed by the disruption of their main valence bonds. Changes in amylose were determined by measuring iodine binding capacity.

Starch Granular Structure

During extrusion cooking, the starch granules are exposed to pressure and shear stress. These effects lead to plasticization of the starch mass and usually result in a complete disruption of the granule structure. However,

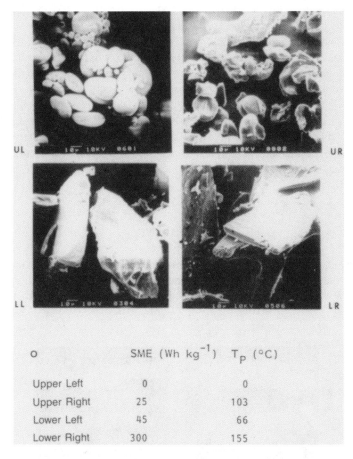

	SME (Wh kg^{-1})	T$_p$ ($^\circ$C)
Upper Left	0	0
Upper Right	25	103
Lower Left	45	66
Lower Right	300	155

Figure 11-21. Influence of energy input on the microscopic structure of extruded starch.

by selection of suitable extrusion parameters (those leading to extremely low energy input), starch granule structure was not completely lost. Several starches, extruded under different conditions, are shown in Figure 11-21. At relatively low thermal and mechanical energy inputs (see Fig. 11-24, position UR), the granules were only compressed, and somewhat deformed, and the original granular structure was largely maintained. With increased thermal energy input, a slight increase in granule size could be observed. Under extrusion conditions, this partial swelling process was combined with increasing deformation and partial destruction of the granule structure (van Lengerich, 1984). For SME values above about 50 Wh kg^{-1}, the granule structure was completed disrupted (Fig. 11-21, position UL). The input of this much mechanical energy led to the production of a homogeneous plasticized mass in which no intact starch granules could be detected microscopically.

Expansion of the extrudate after it left the die opening resulted in porous products with pore walls that were formed very differently depending on the energy input. The higher the energy input, the more the products expanded and the less the pore wall thickness of the extrudates. This result can also be seen from a comparison of the preparations shown in Figure 11-22. The material that had been extruded with a high energy input consisted mainly of disk-shaped, thin, bent pore walls; in the product extruded with a low energy input, only fragments of the hardened, slightly expanded, compact mass were detected.

Crystallinity

The changes in granular structure produced by strain in the extruder suggested that under extrusion conditions a disruption of the paracrystalline molecular organization should also occur. Mercier (1980) established that the

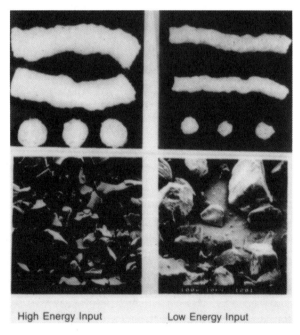

High Energy Input   Low Energy Input

Figure 11-22. Influence of energy input on the inner structure of extrudates.

crystallinity of potato and corn starch can be changed in specific ways by extrusion at different temperatures. Crystalline portions of corn starch clearly decreased at extrusion temperatures of 70°C or above. At 135°C a new structure was formed that was shown by the appearance of three new peaks in the X-ray diffraction spectrum. The largest new peak was found at an angle (h) of 9°54′. The spectrum that was formed could be identified as a V pattern.

Extrusion at 170°C led to a new structure with a primary peak at 9°03′. Higher extrusion temperatures led to a greater formation of this peak with a simultaneous decrease in the V structure. In the results published by Mercier (1980), no distinction was made between the thermal and the specific mechanical energy inputs.

X-ray diffraction spectra for several wheat starches extruded under conditions of various specific mechanical energy inputs are shown in Figure 11-23. The native wheat starch (#1) showed the characteristic spectrum of the A pattern. Introduction of thermal energy alone into the starch within an aqueous medium by pressure cooking caused a complete loss of the spectrum (#2). Even at low thermal and mechanical energy inputs in the extruder, the amount of crystallinity remaining in the granule was reduced dramatically (#3). This reduction indicated that even at a very low energy input, the paracrystalline structure of the starch had largely come apart and the starch had changed into an amorphous state. An increase in energy input led to the formation of a new crystalline structure with a primary peak at $h = 9°54′$ (#4). The formation of this crystalline structure could be detected only when a high thermal and mechanical energy were put into the starch simultaneously. Neither thermal energy input alone at 120°C nor high mechanical strain (#6) led to this spectrum. Comparison of extruded wheat starch with wheat starch extruded with the addition of glycerin monostearate showed that the newly formed crystalline region could be attributed to the formation of a complex between the amylose of starch and the fatty acid part of the added glycerol monostearate. This result also agrees with the values obtained by Mercier (1980), who described this structure as the E structure. A comparison of diagrams 4 and 6 in Figure 11-23 makes it clear that development of the E structure is not strictly a function of product temperature, but also occurs due to an increase in the SME when simultaneously the mass temperature is reduced.

Residual Gelatinization Enthalpy and Gelatinization Behavior

Since the energy put into the starch during extrusion caused a partial to complete disruption of both its granular structure and paracrystalline molecular organization, it could be assumed that some of the hydrogen bonds, through which the starch molecules are associated with one another, are also

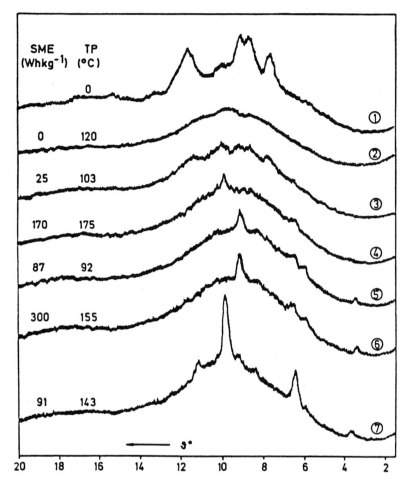

Figure 11-23. Influence of energy introduction on the crystalline structure of wheat starch extruded under various extrusion conditions characterized by SME and TP.

disrupted. The disruption of these bonds is an endothermal process, which can be evaluated thermodynamically by quantitative measurement of the gelatinization enthalpy by DSC. Thus, when the hydrogen bonds are disrupted by the addition of energy during extrusion, the disruption is detectable by a reduction in the energy still needed for complete gelatinization. To determine this energy, starches extruded at different energy inputs were analyzed by DSC.

The measurement results, some of which are given in Figure 11-24, show that native wheat starch gelatinizes over the temperature range between 58°C and 69°C and has an enthalpy value of 10 J g^{-1}. In addition to the

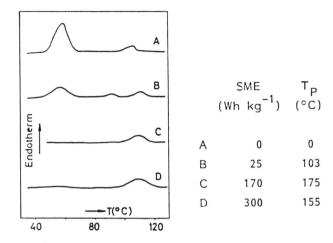

Figure 11-24. DSC curves for various extruded wheat starches.

gelatinization peak, there was a second peak in the thermograms at temperatures of 105–110°C. This peak has already been descrbed by Eberstein and co-workers (1980). The area of this second peak is a measure of the enthalpy of disaggregation of the amylose–lipid complex preexisting in native starch.

Extrusion with a low energy input caused a clear decrease in the gelatinization enthalpy. The peak for the decomposition of the amylose–lipid complex at first increased with increasing energy input before reaching a maximum and remaining constant with further increases in energy input. The influences of product temperature and SME on residual gelatinization enthalpy are shown in Figure 11-25. It can be seen that an increase in both forms of energy input first led to a large decrease in residual enthalpies. Residual enthalphy values could no longer be detected at an extrusion temperature of 90°C and an SME of 80 Wh kg^{-1}. Thus, the energy necessary for the partial gelatinization process is already added at a very early stage of the energy input during extrusion. It is also clear that the effects of a higher energy introduction during extrusion, as related to the residual gelatinization enthalpy, cannot be differentiated further by thermoanalytical methods.

The amylograph curves presented in Figure 11-26 gave further indications of the degree of disruption of secondary bonds due to extrusion. The starch extruded at low energy inputs exhibited a clear increase in paste viscosity in the temperature region of gelatinization. With increasing energy inputs in the extruder, the portion of nongelatinized starch decreased. This decrease led to lower viscosities at temperatures between 50°C and 75°C. The curves also showed that starches extruded with higher energy inputs (Fig. 11-26, No. 5 and 6) no longer developed a capacity for gel formation and had extremely

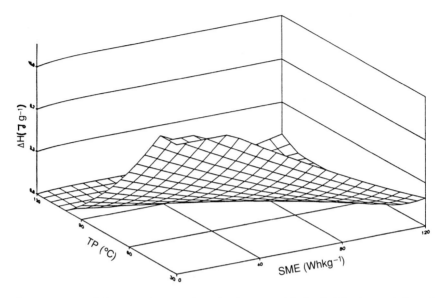

Figure 11-25. Influence of energy introduction on residual gelatinization enthalpy of extruded wheat starches.

low viscosities over the entire temperature range. The results noted above permitted the conclusion that secondary molecular structure was already disrupted to a great extent under extrusion conditions at relatively small energy inputs. The analytical identification of a further disruption of secondary bonds could be accomplished only indirectly by characterizing the hydration behavior of the extruded starches in an aqueous medium. To do this, the solubility of the extrudate in a formamide–sodium sulfate solution, the apparent viscosity of starch pastes, and the swelling behavior of ground extrudates were determined. As has been outlined in the system analytical model, these structure-describing characteristics were also formulated as functional properties and were described as target parameters, because they are of significant interest for technical applications. Thus, the influence of energy input on these functional properties will be considered separately.

Relative Molecular Weight Distribution

As model studies on the thermal energy of starch–water systems have shown (Henning and Lechert, 1974; Henning, 1977; Lechert, 1976; Lechert and Schwier, 1982; Nasa, 1979), a disruption of noncovalent bonds does not involve a change in the molecular size of the amylopectin and the amylose. With the addition of mechanical energy under extrusion conditions, however,

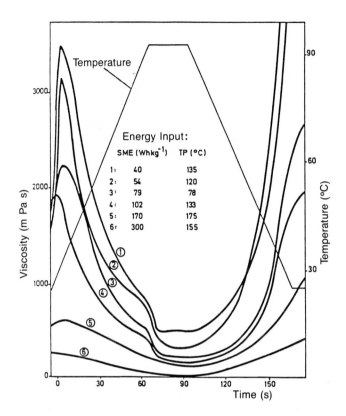

Figure 11-26. Influence of energy input on the gelatinization behavior of pastes of extruded wheat starches.

this type of structural change might be expected in starch. The extent of this change would depend on the energy input.

The reduction in molecular size caused by disruption of the primary valence bonds was studied by determining relative molecular weight distributions of extrudates (Klingler, 1978). The starches were first dispersed in aqueous dimethylsulfoxide (DMSO), and the molecules were separated according to their size by high-pressure gel permeation chromatography on porous glass. The column was calibrated with dextrans and synthetic amyloses of known molecular weights. This procedure made it possible to use the elution diagrams to determine the relative molecular weight distributions of the separated molecules. As can be seen from Figure 11-27, the chromatogram for native starch has two pronounced peaks (amylopectin and amylose). The amylopectin is characterized by the larger peak, the amylose by the smaller peak. No sharp boundary could be established between the two molecules, but only a continuous transition.

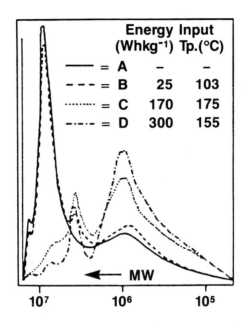

Figure 11-27. Influence of energy introduction on the relative molecular weight distribution of wheat starch extruded under various extrusion conditions characterized by SME and TP.

With extruded starch, a very small energy introduction during extrusion already resulted in a change in the measured molecular weight distribution. The amylopectin peak was smaller and the amylose peak was increased. In addition, a new peak appeared between the two native starch peaks, which indicated the formation of a new fraction of intermediate molecular size. If energy input during extrusion was increased, the result was a more severe decrease in the amylopectin peak, an enlargement of the intermediate peak, and an enlargement of the amylose peak. This result suggests that glycosidic bonds between glucose units in starch molecules are broken by the extrusion process. In particular, amylopectin molecules are affected by this molecular depolymerization. The larger fragments were intermediate in size between amylopectin and amylose (i.e., the middle peak on the chromatograms). However, the smaller molecule fragments were the same (or nearly the same) size as the amylose molecules and thus caused an increase in the amylose peak.

To further differentiate the influences of the forms of energy input on molecular weight distribution, the extent to which molecular depolymerization was influenced by changes in reaction behavior was assessed. For this study, extrudate samples were used that had been extruded with different barrel temperatures and different product water contents (Samples C and D in Fig. 11-27). It was shown earlier that higher barrel temperatures lead to a lower viscosity of the plasticized mass in the extruder because of an increase in the mass temperature. Analysis of the extrudates indicated that, in spite of higher mass temperatures (Sample C in Fig. 11-27), the starches extruded with less SME showed less molecular depolymerization. This result allowed the conclusion that under extrusion conditions, the amount of energy dissipated, expressed as SME, exerts a large influence on the depolymerization of the glycosidic bonds. Conversely, the mass temperature plays only a subordinate role in this process.

In Figure 11-28, the integral distribution of the molecular weights of the starches studied shows that the average molecular weights decreased proportionally to the increase in energy input and could be reduced by a power of ten (Fig. 11-28, Sample 6) compared to native starch.

By plotting average molecular weight versus SME, the course of molecular depolymerization could also be estimated (Fig. 11-29). For SME values up to about 200 Wh kg^{-1}, the decreases in average molecular weights can be described by the exponential function shown in the figure. The shape of the curve shows that in this region of energy input, molecular depolymerization due to a constant increase in SME always depends on the remaining concentrations of the average molecular weights. For SME values above 200 Wh kg^{-1}, the fit of the determined curve is not as good; however, the measured value

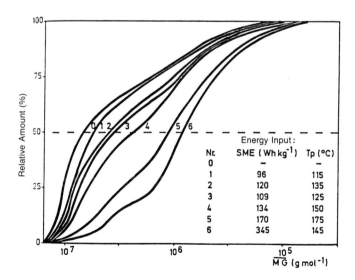

Figure 11-28. Integral distribution curves of relative molecular weights of extruded wheat starches.

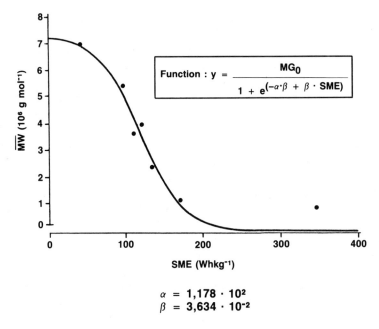

$$y = \frac{MG_0}{1 + e^{(-\alpha \cdot \beta + \beta \cdot SME)}}$$

Function : above

$\alpha = 1{,}178 \cdot 10^2$
$\beta = 3{,}634 \cdot 10^{-2}$

Figure 11-29. Influence of SME on the medium molecular weight of extruded wheat starch.

for an SME of 350 Wh kg^{-1} indicates that further molecular depolymerization should not be expected. The chromatograms in Figure 11-27 also indicate that with this amount of energy input the amylopectin molecules are depolymerized to the greatest extent. These results permitted the conclusion that after maximum depolymerization of the amylopectin molecules, at least under the conditions of extrusion cooking, significant molecular degradation no longer occurs.

To what extent the structure of the amylose molecules has been changed by energy input should be detectable by determining the iodine binding capacity of the extruded starches, as is discussed in the next section.

Iodine Binding Capacity

The iodine binding capacity of amylose is about 20%, while that for amylopectin is only about 1%. Thus, if the amylose molecules are depolymerized by extrusion, the iodine binding capacity of the extrudate will be reduced. Table 11-7 shows the iodine binding capacity of starches extruded at various levels of energy input up to 350 Wh kg^{-1}. This range includes the region that produced no significant reduction in the average molecular weights of the

Table 11-7. Influence of Energy Introduction
on Iodine Binding Capacity

Energy Input		Iodine Binding
SME (Wh kg⁻¹)	T_P (°C)	Capacity (%)
—	—	5.4
25	103	5.4
37	135	5.4
56	105	5.5
64	118	5.2
79	123	5.5
100	136	5.3
123	152	5.1
134	150	5.2
150	155	5.3
164	119	5.6
170	175	5.3
182	140	5.2
210	122	5.1
300	155	5.3
343	146	5.5

Source: From van Lengerich, 1984

extrudates. Thus, if depolymerization of the amylose molecules occurred, it would have been detected in these extrudates. Across all of the test region, no significant dependence of iodine binding capacity on energy input was indicated. Thus, it could be concluded that, if the amylose molecules were depolymerized under extrusion conditions, it was to a very limited extent only.

Influence of Structural Changes in Extruded Starches on Their Functional Properties

Extrudate Solubility in Formamide-Sodium Sulfate

From a systems analysis viewpoint, the structural characteristics of extruded starch described so far are those parameters that influence the functional properties of the final product. The choice of these parameters was based on technical application criteria. The solubility of the extrudates and their swelling and viscosity behavior are of special interest.

The quantifiable structural characteristics described thus far are the average molecular weight and residual gelatinization enthalpy. The first characterizes the degree of decomposition of covalent bonds. The second can be used only in a limited region of the energy input as a measurement of the dissolution of nonconvalent bonds, starch melting, and disruption of the crystalline regions of the granules. Since the disruption of covalent bonds is already known to lead to changes in cold water solubility and viscosity behavior, the values determined for these target dimensions were plotted against average molecular weight. Figure 11-30 indicates that the solubility of the extrudates increased with decreasing molecular weight and approached a final value of 70%.

Apparent Viscosity of the Hot Paste

The ability of starch to form a viscous paste in an aqueous medium upon the addition of heat depends on the formation of a three-dimensional network; in this network, the disruption of intermolecular and intramolecular hydrogen bonds leads to hydration of the starch molecules and to an orientation and association of the water molecules (Henning, 1977; Henning and Lechert, 1974; Lechert, 1976; Lechert and Schwier, 1982; Nasa, 1979). The newly formed intermolecular bonds between the starch and the water molecules cause a greater shear stress at a given shear rate and thus lead to a higher apparent viscosity. Since the formation of this network is related to the presence of intact covalent bonds within the starch molecules, a change in the

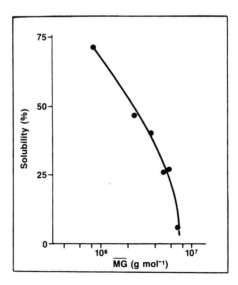

Figure 11-30. Influence of molecular degradation on the solubility of extruded wheat starch.

covalent bonds must also result in a change in the viscosities of hot stirred pastes. Therefore, the measurements obtained for the structural viscosities of hot pastes were compared to the average molecular weights (Fig. 11-31).

Molecular depolymerization caused by the introduction of mechanical energy at low levels led to a clear decrease in viscosity. This trend continued to a final value of 10 mPa·s. Thus, the apparent viscosity of the hot paste can also be used as an indicator of the disruption of covalent bonds.

Apparent Viscosity (Cold Paste) and Swelling Behavior

While the breakage of glycosidic bonds can be determined relatively well by analytical methods, no sufficiently accurate analytical method exists to date for the direct detection of the disruption of hydrogen bonds between starch molecules. Information can be obtained by indirect methods; for example, by measuring the residual gelatinization enthalpy through DSC. Within the region of energy input where partial loss of residual gelatinization enthalpy occurs, it should be possible to derive a relationship between the enthalpy values of the extrudates and their swelling behavior. Since gelatinization enthalpy can be reduced by thermal energy input alone or by mechanical energy input alone, measurements of sediment volume, which were used as a measure of

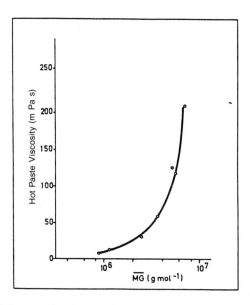

Figure 11-31. Influence of molecular degradation on the apparent hot paste viscosity of extruded wheat starch.

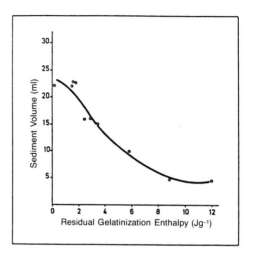

Figure 11-32. Influence of residual gelatinization enthalpy on the sediment volume of extruded wheat starch.

the hydration capacity of starch, were compared with measured enthalpy values. The results (Fig. 11-32) show the expected increases in sediment volume with increasing disruption of hydrogen bonds, as expressed by residual gelatinization enthalpy.

Within the region of energy input represented here, these results confirm that under extrusion conditions, with increasing disruption of hydrogen bonds,

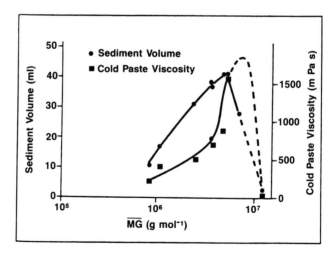

Figure 11-33. Influence of molecular degradation on hydration behavior (sediment volume and cold paste viscosity) of extruded wheat starch.

the tendency of starch molecules to associate with water molecules first increases. A more complete description of the relationship between sediment volume and the exclusive decomposition of secondary valence bonds (DSC values) is not possible for the region of larger thermal and mechanical energy input, because the enthalpy values are all zero. However, glycoside bonds can be broken under extrusion conditions only when at least some of the non-covalent bonds have been disrupted. Since this structural disruption can be monitored analytically over a wide range of energy input levels by determining average molecular weights, viscosity values for cold pastes as well as sediment volumes were plotted against average molecular weight (Fig. 11-33).

The values showed that in the first phase of molecular degradation, caused by a small energy input (right side of the diagram in Fig. 11-33), the swelling capacity and viscosity first increase, then become smaller with decreasing molecular size. In the first section of the curve, the hydration capacity of the extrudates increased with increasing energy input. The solubility values increased at the same time. Viscosity values for the hot paste were reduced. In addition, intact starch granules could be detected for SME values up to about 50 Wh kg^{-1}. Thus, with small energy input levels, the starch must be subject to an extremely nonhomogeneous strain under extrusion conditions; after the extrusion process, the material comprises partially deformed starch granules, partially gelatinized starch, and a phase that is completely disrupted structurally. A further increase in thermal and mechanical energy input levels alter this composition to favor increasing structural degradation in supermolecular and molecular structure. This result in turn leads to a decrease in swelling capacity and to a loss of viscosity.

Influence of the Forms of Energy Input on Functional Properties of Extruded Starch

The results presented thus far show that starch structure can be influenced within wide limits by the input of thermal and mechanical energy. Characterization of the relationships between the structural and functional properties indicated that the solubility, the apparent viscosity of cold and hot pastes, and swelling behavior are closely related to the degree of decomposition of the covalent and noncovalent bonds of starch. These parameters were used as suitable indicators of structural changes in the molecular and supermolecular structure of the extruded starch.

In this way, it should also be possible to derive a response relationship between the energy put into the starch and the functional properties. Since the final product properties were relatively simple to measure analytically, this response relationship was derived by means of a polynomial regression considering the analytical data for all of the extrusion experiments. The regression equation calculated for each target dimension was graphically represented in a three-dimensional diagram in which the values for the particular target dimensions were plotted against the product temperature and SME.

Solubility in Formamide–Sodium Sulfate Solution

Figure 11-34 shows the influence of energy input on the solubility of the extrudates in a formamide–sodium sulfate solution. The pertinent regression equation describes this response relationship and makes it possible to predict solubility values in the experimental region studied for any random combination of the forms of energy input.

It follows from the graph and the regression coefficients of the primary effects that increases in both product temperature and SME lead to greater product solubility. However, the higher the SME, the less is the influence of product temperature. The regression coefficient of the interactions between SME and product temperature causes the shallower slope of the surface at higher SME values.

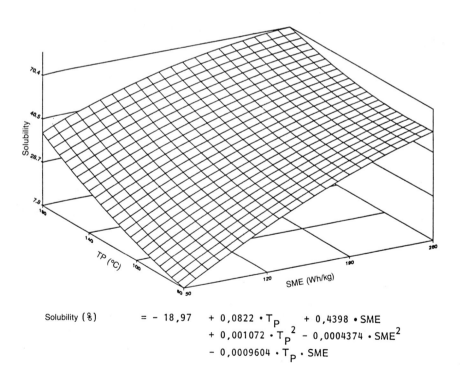

$$\text{Solubility (\%)} = -18,97 + 0,0822 \cdot T_P + 0,4398 \cdot SME$$
$$+ 0,001072 \cdot T_P^2 - 0,0004374 \cdot SME^2$$
$$- 0,0009604 \cdot T_P \cdot SME$$

Figure 11-34. Influence of energy introduction on the solubility of extruded wheat starch.

As expected, the highest solubility values were found at high simultaneous inputs of thermal and mechanical energy. In the experimental region, however, maximum values were about 75%. Solubility values were also on the same order of magnitude at a high degree of disruption of the starch molecules. Thus it can be assumed that under the extrusion conditions represented here, no further increase in solubility occurs and thus also no further cleavage of the glycosidic bonds within starch molecules.

Apparent Viscosity of the Hot Paste

Another indication of the breakage of glycosidic bonds is the apparent viscosities of hot pastes. These values are represented in Figure 11-35, as is their dependence on product temperature and SME. The shape of the surface basically indicates a loss in hot paste viscosity when both forms of energy input are high.

The regression coefficient for the SME in the regression equation indicates the direct and strong influence on hot paste viscosity. The regression coeffi-

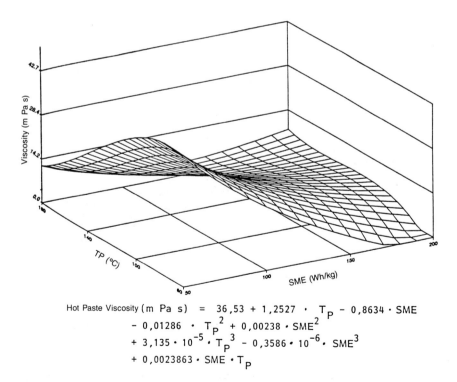

$$\text{Hot Paste Viscosity (m Pa s)} = 36,53 + 1,2527 \cdot T_P - 0,8634 \cdot SME$$
$$- 0,01286 \cdot T_P^2 + 0,00238 \cdot SME^2$$
$$+ 3,135 \cdot 10^{-5} \cdot T_P^3 - 0,3586 \cdot 10^{-6} \cdot SME^3$$
$$+ 0,0023863 \cdot SME \cdot T_P$$

Figure 11-35. Influence of energy introduction on the apparent hot paste viscosity of extruded wheat starch.

cient for the effect of product temperature is relatively small. Thus, product temperature exerts no strong effect on hot paste viscosity. Product temperature and SME show a strong interaction, however, which indicates that at high product temperatures, the influence of SME is reduced. This interaction can also be interpreted as indicating that the viscosity of the hot paste is always reduced by a temperature increase when the SME is small. At higher SME values, the viscosity of the hot paste assumes lower values and therefore can no longer be influenced by changing the product temperature. These results also support the values obtained from gelatinization curves. The high measure of precision for the fit of this curve ($r^2 = 0.91$) shows that response relationships with the system parameters can also be derived between the hot paste viscosity as a functional property and as a structure-describing parameter.

Apparent Viscosity of the Cold Paste

The apparent viscosities of the cold pastes of the extrudates are graphed in Figure 11-36. The test region used to calculate this relationship starts with

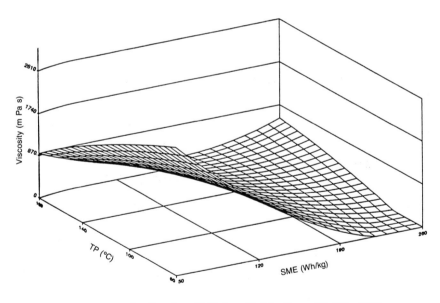

$$\text{Cold Paste Viscosity (m Pa s)} = 4781,9 - 12,92 \cdot T_P - 34,483 \cdot SME$$
$$- 0,002606 \cdot T_P^2 + 0,048376 \cdot SME^2$$
$$+ 0,10104 \cdot SME \cdot T_P$$

Figure 11-36. Influence of energy introduction on the apparent cold paste viscosity of extruded wheat starch.

the energy input at which the viscosity values assume a maximum value. The shape of the curve shows that increasing the SME or product temperature initially leads to a decrease in product viscosity. This result is an indirect indication of the increasing disruption of hydrogen bonds and glycosidic bonds.

In the region of low mass temperatures, increases in SME caused greater viscosity losses than at higher temperatures. This result can be attributed to the fact that low product temperatures lead to higher viscosities of the plasticized mass in the screw channel and cause an increase in the capability of the starch to absorb mechanical energy.

At high product temperatures, increases in SME caused only slight reductions in viscosity. This result suggests that, in the case of a disruption of covalent and noncovalent bonds that is already well under way, further disruption only occurs with less intensity when SME is increased further.

The same figure also shows the influence of mass temperature at various SME values. At low SME values, increases in product temperature led to decreases in viscosity; this effect was reduced with increasing SME, and it finally was reversed. The increase in viscosity in the region of higher energy input is caused by the statistically determined interaction between the SME and mass temperature.

Sediment Volume

The hydration behavior of an extrudate can also be determined by measuring sediment volume. It follows from the graph in Figure 11-37 that the swelling capability of the extruded starch first increased with increasing thermal and mechanical energy input. This result is also indicated by the positive sign of the two regression coefficients of the primary effects. Such an increase in swelling capability can be explained by the partial disruption of hydrogen bonds in the molecular structure of the starch. In the initial phase of energy input, this disruption leads to an increase in hydration capacity and thus to an increase in sediment volume. A mass temperature of 135°C and an SME of 125 Wh kg^{-1} lead to a maximum swelling capacity. A comparison of these values with data on residual gelatinization enthalpy makes it clear that the maximum swelling power is attained at a higher energy input than that where the measured residual gelatinization enthalpy is already totally reduced. This result is another indication that the increasing disruption of the noncovalent bonds still is not finished when the residual gelatinization enthalpy detected by DSC is zero.

Large increases in thermal and mechanical energy inputs lead, as expected, to decreases in sediment volumes and, finally, to a complete loss of swelling capability. This result indicates the complete disruption of the molecular structure of the starch. This relationship is also indicated by the negative signs of the quadratic effects in the regression equation.

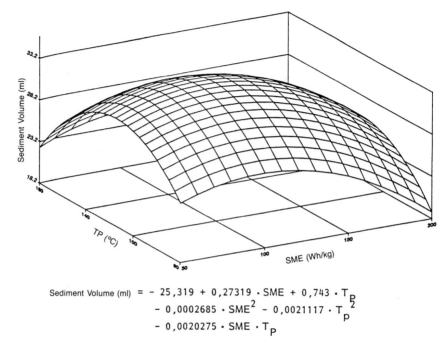

$$\text{Sediment Volume (ml)} = -25{,}319 + 0{,}27319 \cdot \text{SME} + 0{,}743 \cdot T_p$$
$$- 0{,}0002685 \cdot \text{SME}^2 - 0{,}0021117 \cdot T_p^2$$
$$- 0{,}0020275 \cdot \text{SME} \cdot T_p$$

Figure 11-37. Influence of energy introduction on the sediment volume of extruded wheat starch.

CONCLUSIONS

This work describes the experimental verification of a recently proposed system analysis model for the extrusion of starches. The model distinguishes between process parameters, system parameters, and target parameters. The process parameters correspond to the operational extrusion variables that influence the rheological properties of the starch in the extruder and therefore the introduction of specific mechanical energy. The thermal energy input is a combination of dissipated energy and heat input from the barrel wall and is expressed via the product temperature at its maximum value.

The specific mechanical and thermal energy inputs are the system parameters that govern the structural changes in the starch as a chemical and physical substance. The structural changes in turn influence target parameters such as solubility, viscosity, puffing, and hydration behavior.

Experimental examination of the model was carried out with a laboratory scale twin-screw extruder. The first part of the work describes the techniques used to measure and calculate rheological data for plasticized wheat starch during extrusion. In-line capillary viscosimetry was used to determine the shear stress on the plasticized wheat starch under various extrusion conditions. The influences of water content, barrel temperature, and degree of fill on shear stress and viscosity were investigated, and the results were presented as flow curves, viscosity curves, and iso-shear stress curves as they related to the extrusion parameters.

Shear stress in the slit die was found to depend mainly upon the water content in the product and the barrel temperature of the extruder, which in turn influences product temperature. It was also found that shear stress, at a constant degree of fill, is directly and linearly proportional to the specific mechanical energy introduction. The results indicate that at high barrel temperatures and high water contents, the amount of mechanical energy to be introduced is limited due to the loss of shear stress on the product. On the other hand, high viscous dissipation is possible only at low water contents and low barrel temperatures. The results of the rheological measurements were utilized to explain an experimentally established relationship between the extrusion parameters and the mechanical and thermal energy inputs. The tested extrusion variables were besides the above-mentioned screw speed, throughput rate, die hole diameter, and screw geometry. All quantifiable variables were investigated in the scope of statistical fractionated factorial designs, and a polynomial regression analysis was used to establish the relationship. The effect of one particular extrusion variable on the introduction of mechanical and thermal energy was found to always depend upon the set points of all remaining variables. This relationship was characterized by more or less strong interactions between the extrusion parameters and was expressed by regression coefficients. A variation of one extrusion parameter was shown to simultaneously change the mechanical and thermal energy inputs. By graphical comparison of both types of energy inputs, the degree of energy conversion during that change could be shown.

Increases in screw speed caused increases in SME values, and because of the higher frictional heat, simultaneous increases in product temperature. The same effect was found by decreasing the die hole diameter or by increasing the shear intensity of the screw configuration. Reducing the product water content increased the shear stress on the mass in the extruder; this increase resulted in a higher torque at the screw shafts and therefore increased the specific mechanical energy. As a result of this increase in energy, the higher viscous dissipation levels led to increases in product temperature. The barrel temperature had an opposite effect: increases in barrel temperature caused increases in product temperature, which led to lower viscosities of the plasticized mass in the screw chamber and a lower mechanical energy introduction.

Introduction of thermal and mechanical energy caused wide-ranging changes in the structure of the starch substance. These structural changes were characterized by X-ray diffraction, scanning electron microscopy, high-performance liquid chromatography (HPLC), DSC, and iodine binding capacity. X-ray diffraction showed a significant loss in crystallinity already at a very low energy input. At higher energy inputs a new crystalline structure was formed, which was identified as a crystalline complex between pre-existing starch lipids and the amylose portion of the starch. Scanning electron microscopy investigations showed that a partial to complete structural disturbance of the starch granules took place under extrusion conditions. Determination of molecular weight distributions by HPLC showed that part of the main valence bonds of the starch were broken down under extrusion conditions. Amylopectin molecules were the most affected. With increasing mechanical energy inputs, the molecules of the amylopectin fraction were increasingly broken down to smaller units, which permeated with the amylose molecules under the conditions of HPLC. Breakdown of the amylose molecules is either negligibly small or it does not take place during wheat starch extrusion. This lack of amylose breakdown was shown by the constant values of iodine binding capacity for all extrudates, which were extruded under the most extreme conditions. Colorimetric investigations showed that complete gelatinization takes place under most extrusion conditions, unless the thermal and mechanical energy inputs are extremely small. Starches that were extruded at an SME of 80 Wh kg^{-1} and at a product temperature of 90°C showed no residual gelatinization enthalpy under DSC conditions.

Data points in an established product temperature–SME diagram defined a window of operation in which each combination of thermal and mechanical energy could be achieved. Since each test point and all points in between could be achieved independently, a second polynomial regression was established that described the influence of thermal and mechanical energy input on the final product properties of the extrudates as the dependent target parameters.

The resulting relationship was manifested by the observed structural changes in the starch. Higher energy introduction caused increases in extrudate solubility and decreases in the viscosities of cold or hot extrudate pastes. The hydration capacity first increased with increasing energy introduction and then decreased up to a total loss of hydration capability.

REFERENCES

Eberstein, K., Hoepke, R., Konieczny-Janda, G., and Stute, R. 1980. DSC-Untersuchungen an Staerken. Teil 1: Moeglichkeiten thermoanalytischer Methoden zur Staerkecharakterisierung. *Staerke* **32**:397.

Faubion, J. M., and Hoseney, R. C. 1982. High-temperature short-time extrusion cooking of wheat starch and flour. 1. Effect of moisture and flour type on extrudate properties. *Cereal Chem.* **59**:529.

Harper, J. M. 1981. *Extrusion of Foods*, Vol. 1 and 11. Boca Raton, FL: CRC.

Henning, H.-J. 1977. Kernmagnetische Resonanzuntersuchungen zur Rolle der Wasserbindung fur die Struktur des nativen Staerkekorns. *Staerke* **29**:1.

Henning, H. -J., and Lechert, H. 1974. Messungen der magnetischen Relaxationszeiten der Protonen in nativen Staerken mit verschiedenen Wassergehalten. *Die Staerke* **26**:232.

Kim, J. C. 1982. Veraenderung von Weizen durch Extrusion. *Zeits. Lebensmitteltechnol. Verfahrenstech.* **33**:334.

Klingler, R. W. 1978. Untersuchungen uber strukturelle and funktionelle Veraenderungen der Staerke durch mechanische Beanspruchung. Diss. D 83 68/FB 13, Berlin.

Lechert, H. 1976. Moeglichkeiten und Grenzen der Kernresonanz-Impuls-Spektroskopie in der Anwendung auf Probleme der Staerkeforschung and Staerketechnologie. *Die Staerke* **28**:369.

Lechert, H., and Schwier, I. 1982. Kernresonanz-Untersuchungen zum Mechanismus der Wasserbeweglichkeit in verschiedenen Staerken. *Staerke* **34**:6.

Linko, P., Colonna, P., and Mercier, C. 1981. High-temperature–short-time extrusion cooking. In *Advances in Cereal Science and Technology*, Vol. 4, ed. Y. Pomeranz, pp. 145–235. St. Paul, MN: AACC.

Mercier, C. 1980. Veraenderungen der Struktur and Verdaulichkeit von Getreidestaerken beim Extrudieren. *Getreide, Mehl and Brot* **34**:52.

Mercier, C., Charbonniere, R., Grebaut, J., and de la Gueriviere, J. F. 1980. Formation of amylose-lipid–complexes by twin-screw extrusion cooking of maniok starch. *Cereal Chem.* **57**:4.

Mercier, C., and Felliet, P. 1975. Modification of carbohydrate components by extrusion-cooking of cereal products. *Cereal Chemistry* **52**:283.

Meuser, F., and van Lengerich, B. 1984. *System Analytical Model for the Extrusion of Starches. Thermal Processing and Quality of Food*, ed. P. Zeuthen, p. 175. London and New York: Elsevier.

Meuser, F., van Lengerich, B., and Kohler, F. 1982. Einfluss der Extrusionsparameter auf funktionelle Eigenschaft von Weizenstaerke. *Staerke* **34**:366.

Nasa, S. 1979. Study on the heat of wetting of starch. *Staerke* **31**:105.

Rabinowitsch, B. 1929. Ueber die Viskositaet and Elastizitaet von Solen. *Z. Physikal. Chem.* **145**:1.

Rumpf, H., Meuser, F., Niediek, E. A., and Klingler, R. W. 1980. *Energiedisspation in Staerken and Weizenmehl.* Tech. Univ. Berlin Schriften. Getreidetechnol. 3.

Seiler, K., and Schuy, A. 1983. Ueberpruefung der Backfaehigkeit extrudierter Weizenmehle. *Gordian* **83**:87.

van Lengerich, B. 1984. Entwicklung und Anwendung eines rechnerunterstuetzten systemanalytischen Modells zur Extrusion von Staerken and Staerkehaltigen Rohstoffen. Diss. Technische Universitaet Berlin.

Chapter 12

Rheology in
Process Engineering

Rudolph Leschke

Traditionally, rheology in the food industry has focused on the characterization of the quality of food materials. This is particularly true in dough processing. Work done by Bohn and Bailey as early as 1936 indicated that farinograph readings, essentially a measure of the shear viscosity of a material, are related to the quality of bread dough and quantify other various changes that occur during a mixing cycle (Bohn and Bailey, 1936). More recently, van Lengerich (1984) has demonstrated that shear viscosity measurements can provide information regarding the average molecular weight of extruded wheat starch. Numerous others (Bohlin and Carlson, 1980), having investigated the dynamic rheological properties of doughs, have found relationships between these properties and dough quality or chemical changes that occur during processing. Clearly, the role of rheology as a measure of physiochemical changes that occur during dough processing has been established.

While rheology can be defined as the study of the mechanical properties of materials, it is also the study of the deformation and flow of matter (Bird et al., 1960). The laws of fluid dynamics, used during the reduction and analysis of rheological data collected in the laboratory, can also be applied to the characterization of flow in large-scale processes. The level of sophistication of process engineering in the food industry is still at an early stage. By the same token, the application of much of the rheological data collected has not found widespread utility in development and design. The use of rheological data as a probe into the molecular behavior of various materials is in itself invaluable. However, an appreciation of how and where such data can be used in process design would ease the transition from laboratory to commercial applications.

This chapter discusses the role of rheology in key aspects of engineering, such as fluid dynamics, heat transfer, and reaction kinetics, all of which are pertinent to dough processing. Following this discussion, an example of the utilization of rheology in the design of extrusion dies is considered.

PROCESS ENGINEERING CONCEPTS

All process phenomena can be separated into two fundamental areas:

- The creation of a process environment
- Physical or chemical changes in a product, caused by a given process environment

While it is true that these mechanisms interact, each can be described independently of the other. How a particular process environment is created should not influence the changes that occur in a product.

The first area, the creation of a process environment, can be analyzed macroscopically. Here the mechanisms of fluid dynamics, heat transfer, and, if applicable, mass transfer, determine the specific attributes of the process environment (i.e., residence time, temperature, pressure, shear forces, mixing). Engineering principles are used to design a system that creates a desired process environment, or to analyze, quantitatively, how the process environment can be changed by varying system operation. The magnitude of these effects (momentum transfer, heat and mass transfer) depends heavily on material physical properties and, particularly, on rheological properties. Therefore, it is important that a quantitative description of the rheology of a material exists for the range of actual processing conditions.

The second area, physical and chemical change, requires analysis at the molecular level. Process engineering in this area involves determining the process environment, or process history, needed to achieve a desired product change. Knowledge of the endpoint of the reaction (based on thermodynamics) and of the reaction rate (a function of the molecular mechanism) is required, in addition to knowledge of which reaction pathways may occur. This information is difficult to obtain, particularly in the food industry, where reactions are complex. The use of rheology as a molecular probe becomes especially valuable in this analysis. Although reaction mechanisms may not be clearly understood, rheological changes can be used to monitor the progress of certain product changes. From these data, relationships describing reaction rates and endpoints in terms of rheological changes can be developed and used to control the desired product quality.

Further complications are introduced when material properties influencing transport phenomena vary with process history. This is usually the case in dough systems. Dough rheology provides a measure through which the progress of kinetic changes in a product material can be evaluated. To solve engineering problems, however, a quantitative description of how viscosity changes over time needs to be applied, not only to the development of a reaction kinetics model, but also to the evaluation of transport phenomena.

The ability to evaluate transport phenomena mechanisms separately from reaction effects makes it possible to develop and use equipment performance models that are general with respect to the type of material processed and the type of reactions occurring, if any, and are frequently general with respect to equipment size. Reaction kinetics models provide the capability to separate the evaluation of final product attributes from a particular process or equipment arrangement. For example, kinetic data can guide the scale-up from a batch process to continuous mixing even though the two processes are quite different mechanically. This approach provides the tools necessary for proper process development and design.

The application of these principles will be illustrated through analysis of the behavior of dough in an extrusion die. Discussing more complex applications is beyond the scope of this chapter. However, the principles used throughout the analysis of the die flow problem are applicable to any process operation.

EXAMPLE: DIE FLOW ANALYSIS AND SCALE-UP

The texture of many extruded products is very sensitive to slight changes in die geometry. Product variations can often be counteracted with modifications in operating conditions and screw configuration. This potential is usually unexplored at the time an extrusion process is being designed, however. It therefore is prudent to minimize die scale-up differences before a system design is complete.

Consider the following situation: A pilot plant process is developed in which a particular extrusion die processes 0.0631 kg/s (500 lb/hr) of material. Scale-up to 0.504 kg/s (4,000 lb/hr) is required. Because of the larger cross-sectional area of the production die mounting flange, the die will need to be significantly thicker than the pilot plant design to prevent mechanical damage. Specific pilot plant die data and production requirements are provided in Table 12-1. The process engineer needs to decide how to design this production die.

Table 12-1. Data on Die Scale-Up Problem

	Pilot Plant Die	Production Die
Rate (kg/s)	6.31×10^2	5.04×10^1
Number of holes	120	Unknown
Entrance contraction	60°	Unknown
Entrance diameter (m)	4.22×10^3	Unknown
Hole diameter (m)	2.36×10^3	Unknown
Hole length (m)	2.59×10^3	2.54×10^2
Die head diameter (m)	6.52×10^2	1.91×10^1
Temperature (K)	374	Unknown
Pressure drop (N/m²)	9.03×10^5	Unknown

Three basic phenomena occur as materials flow through an extruder die:

Die head pressure affects the degree of fill and the extent of back mixing in the extruder barrel. As a result, it affects the mechanical work added to the product upstream in the extruder barrel.

Work is added to the product as a result of flow through the die. This addition may alter product temperatures, which changes the process history and, consequently, the reaction effects in the die.

Product swells as it exits in response to process conditions in the die. This swelling is due to both a redistribution of radial velocities and stress relaxation upon exit from the die.

These phenomena should form the basis of the scale-up criteria.

In finding a solution to the extruder die flow problem, effects are grouped into the fundamental areas discussed earlier. This grouping is illustrated in Figure 12-1.

The creation of die head pressure and work or energy input determine, in part, the process environment. The other phenomena, reaction effects and die swell, occur only in response to the process conditions. The goal in scale-up is to manufacture product of comparable quality by preserving the process environment, despite changes required in equipment operation and design. Provided product formulation is constant, scale-up or design based on preserving the process environment ensures consistent reaction effects. It usually is not possible to match the process environment exactly. By analyzing the process mechanisms, however, it is possible to minimize the differences.

The following components of the process environment need to be evaluated. Before any design decisions can be made, these mechanisms need to be quantitatively described. In this example, performance equations for these process effects will be assembled first, and will then be used to conduct a design analysis.

- Die head pressure: This parameter is a function of the particular die entrance geometry, the shear rate distribution in the die area of constant cross section, and the length of the die.

- Material residence time in the die: This parameter is a function of the shear rate distribution and length.

- Material temperature distribution and work input in the die: These parameters are also functions of shear rate distribution and length.

Die head pressure, the most significant of these effects because it changes the process history in the extruder, will be considered first. Material residence time, temperature distribution, and work input at the die usually are less critical, because the relative magnitudes of these terms are small compared

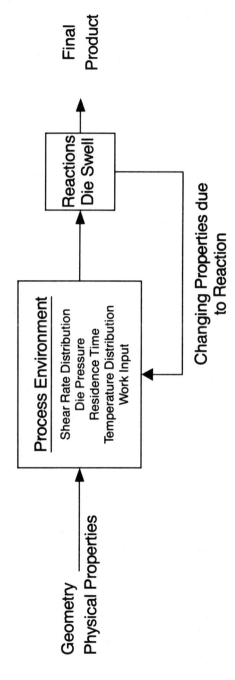

Figure 12-1. Extruder die flow problem.

477

to similar contributions upstream in the extruder. The significance of these effects will be checked, however.

Die swell is important, since it may influence product shape and size. Die swell occurs in reaction to process conditions in the die, and is not part of the process environment. Because it contributes to the overall die pressure drop, however, die swell must be analyzed with the process environment.

Evaluation of Die Head Pressure

Die pressure drop occurs in three areas: entrance or contraction losses, losses due to friction during flow through the die, and losses due primarily to velocity profile redistribution at the die exit. These effects are mathematically stated in Equation 12-1.

$$\Delta P_{total} = \Delta P_{entrance} + \Delta P_{hole} + \Delta P_{exit} \qquad (12\text{--}1)$$

Relationships derived from fluid mechanics are used to evaluate these pressure drops.

Die Hole Pressure

The value for ΔP_{hole} is determined from a simple momentum balance on the fluid inside the die (Bird et al., 1960). A momentum balance results in equations describing the material velocity distribution in the die. The shear rate, a key component in process history, is defined as the change in velocity with respect to distance across the die hole. Once a velocity distribution is known, an equation for average velocity or volumetric flow rate can be developed and related to the die pressure drop (Fig. 12-2). The momentum balance for a circular die is given by Equation 12-2.

$$\begin{aligned}
(2\pi rL\,\sigma_{rz})|_r - (2\pi rL\,\sigma_{rz})|_{r+\Delta r} + (2\pi r\Delta r\rho v_z{}^2)|_{z=0} \\
-(2\pi r\Delta r\rho v_z{}^2)|_{z=L} + 2\pi r\Delta r(P_0 - P_L) = 0
\end{aligned} \qquad (12\text{--}2)$$

where

r = die hole radial position
L = die hole length
σ_{rz} = shear stress across the r plane due to axial flow
ρ = material density
v_z = material velocity in the z direction

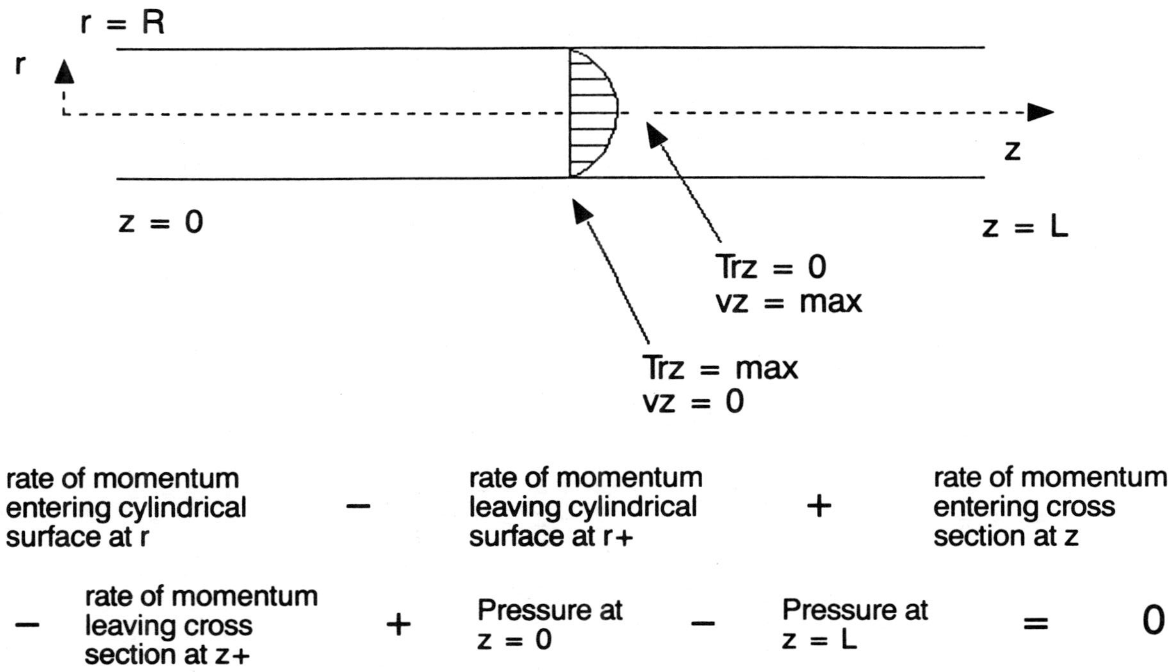
r = R
r
z
z = 0
z = L
Trz = 0
vz = max
Trz = max
vz = 0
rate of momentum entering cylindrical surface at r
−
rate of momentum leaving cylindrical surface at r+
+
rate of momentum entering cross section at z
−
rate of momentum leaving cross section at z+
+
Pressure at z = 0
−
Pressure at z = L
=
0

Figure 12-2. Momentum balance through circular hole.

P_0 = pressure just after die hole entrance
P_L = pressure after length L in the die hole

Once simplified, this equation can be written in differential form (Eq. 12-3).

$$\frac{d}{dr}(r\sigma_{rz}) = \left[\frac{P_0 - P_L}{L}\right]r \tag{12--3}$$

Integrating this equation with the boundary condition that the momentum flux must be finite at $r = 0$ (the die hole center line) results in Equation 12-4.

$$\sigma_{rz} = \left[\frac{P_0 - P_L}{2L}\right]r \tag{12--4}$$

This equation indicates that the pressure drop due to flow through the die hole can be determined from the shear stress. The solution to this problem now requires knowledge of the material rheology. Velocity profiles and, subsequently, a flow rate through the die hole can be related to a pressure drop only if shear stress as a function of shear rate is known. The simplifying assumption of isothermal flow is made so that temperature gradients which alter the rheological properties of the material do not complicate the analysis. This is a valid assumption for short dies. The assumption can be verified later for the particular scale-up example discussed in this chapter.

For a power law fluid, Remsen and Clark (1978) and several others have demonstrated that the relationship between shear stress and shear rate can be adequately described by a power law model for most dough materials. This model is given by Equation 12-5.

$$\sigma_{rz} = m\left[\frac{dv}{dz}\right]^n \tag{12--5}$$

Here m and n are material dependent rheological parameters. This expression can be substituted into the momentum balance equation to yield, upon rearrangement, Equation 12-6.

$$\frac{dv_z}{dr} = \dot{\gamma} = \left[\frac{1}{m}\frac{\Delta P_{hole}}{2L}r\right]^{1/n} \tag{12--6}$$

Solving the differential equation with the boundary condition at $r = R$, $v_z(r) = 0$ yields the velocity distribution in the tube. This boundary condition implies that there is no slip at the tube wall. A detailed discussion of slip is beyond the scope of this chapter. Slip is observed in nonhomogeneous fluids or suspensions in which the less viscous phase accumulates at the wall in a thin layer. The velocity at the wall is actually zero, but the change in velocity in the radial direction is so sudden that the fluid behavior appears to exhibit slip at the wall. This slip occurs when the two phases no longer flow in unison at some critical shear stress (Darby, 1976). The velocity distribution, based on the no-slip boundary condition, is given by Equation 12-7.

$$v_z(r) = \frac{1}{(1/n + 1)} \left[\frac{1}{m} \frac{\Delta P_{hole}}{2L} \right]^{1/n} [R^{(1/n+1)} - r^{(1/n+1)}] \tag{12-7}$$

Velocity distribution is too abstract to be useful in typical die design or performance evaluations. Die performance is usually defined in terms of a mass flow rate or volumetric throughput rate (both are easier to measure). Pressure drop through the die hole should be expressed in terms of this parameter. The volumetric flow rate is given by Equation 12-8.

$$Q = \langle v_z \rangle \pi R^2 \tag{12-8}$$

For a final solution in terms of volumetric flow rate, an average velocity, $\langle v_z \rangle$, is required. This value is obtained by integrating the velocity over the cross-sectional area of the die hole, as in Equation 12-9.

$$\langle v_z \rangle = \frac{\int_0^{2\pi} \int_0^R v_z(r) r \, dr \, d\theta}{\int_0^{2\pi} \int_0^R r \, dr \, d\theta} \tag{12-9}$$

where θ is the degree of angular rotation. Evaluation of this expression results in Equation 12-10.

$$\langle v_z \rangle = \left[\frac{1}{m} \frac{\Delta P_{hole}}{2L} \right]^{1/n} \left[\frac{R^{(1/n+1)}}{(1/n + 3)} \right] \tag{12-10}$$

Finally, pressure drop ΔP_{hole}, derived from a fundamental momentum balance, is given in terms of the volumetric flow rate by Equation 12-11.

$$\frac{\Delta P_{hole}}{2L} = m \left[\frac{Q\,(1/n + 3)}{\pi R^{(1/n+3)}} \right]^n \tag{12-11}$$

This result now needs to be added to the entrance and exit pressure drops.
    The relationship developed above is valid only for circular holes. Quite

often die cross sections are noncircular. If the velocity distribution is too complex to derive from a momentum balance, it may be possible to correct the circular solution for a noncircular hole by using a shape factor (Michaeli, 1984). Beyond this, one needs to resort to numerical analysis methods, typically finite element modeling techniques. A simpler approach is pursued here because it more clearly illustrates the utilization of rheology in process engineering analysis.

Entrance Pressure Drop

The derivation of the equation for entrance pressure as a function of flow rate from a momentum balance is too difficult to be practical. An alternate approach in developing a model for predicting these entrance pressures was adopted and applied, quite successfully, by Howkins and colleagues (1987). These authors developed the following empirical model (Eq. 12-12) based on dimensional analysis techniques, by using data for food as well as synthetic polymers.

$$\pi_1 = 2.45\,\pi_2^{-0.86}\,\pi_5^{6.14}\,\pi_6^{-0.51}\,\pi_7^{-0.24}\,\pi_8^{-0.87} \tag{12-12}$$

where Equations 12-13 to 12-18 define the pi terms.

$$\pi_1 = \frac{16 n_d^2 r_h^4 \Delta P_{entrance}}{\rho Q^2} \tag{12-13}$$

$$\pi_2 = \frac{\pi Q}{\pi n_d r_h^m [Q/(2\pi n_d r_h^3)]^{n-1}} \tag{12-14}$$

$$\pi_5 = \frac{3n + 1}{4n} \tag{12-15}$$

$$\pi_6 = \frac{16 n_d r_h^2}{D_c^2} \tag{12-16}$$

$$\pi_7 = \alpha \tag{12-17}$$

$$\pi_8 = 1 + \frac{\tau Q}{16\pi n_d r_h^3} \tag{12-18}$$

where

n_d = number of total die holes
r_h = hydraulic radius
α = entrance angle
τ = elastic relaxation time constant
D_c = die hole entrance cone radius

Note that m and n used here are the same rheological parameters used earlier for a power law fluid. This development by Howkins and co-workers represents a significant improvement over previous attempts to quantify die entrance pressures (Boger, 1982). The entrance pressure is evaluated from the calculated value of the π_1 term using Equation 12-13.

Die Exit Pressure

An exit pressure at the die discharge exists when normal stresses are developed during flow through the die. Darby (1976) has derived a relationship for the magnitude of this radial stress, in terms of a restraining pressure exerted by the wall on the fluid inside the die hole, just inside the die exit. This relationship was developed based on the constitutive equation for a Reiner-Rivlin fluid, which for viscometric flows is given in Equation 12-19.

$$\sigma_{ij} = \eta\dot{\gamma} \begin{bmatrix} 0 & 1 & 0 \\ 1 & 0 & 0 \\ 0 & 0 & 0 \end{bmatrix} + \psi_2\dot{\gamma}^2 \begin{bmatrix} 1 & 0 & 0 \\ 0 & 1 & 0 \\ 0 & 0 & 0 \end{bmatrix} \tag{12-19}$$

where

η = apparent viscosity

$\psi_2 = \sigma_{22} - \sigma_{33}/\gamma^2 = N_2/\gamma^2$ = second normal stress coefficient

$\dot{\gamma}$ = shear rate

For fully developed laminar flow in a tube, the non-zero components of the stress tensor are given by Equations 12-20 and 12-21.

$$\sigma_{zz} = \sigma_{rr} = N_2 \tag{12-20}$$

and

$$\sigma_{rz} = \sigma_{zr} = \eta\dot{\gamma} \tag{12-21}$$

Derived from a momentum balance around a discrete fluid element flowing through a cylinder (Bird et al., 1960), the equations of motion in the radial and axial directions are given by Equations 12-22 and 12-23.

$$\frac{\partial P}{\partial r} = \frac{1}{r}\frac{\partial}{\partial r}(r\sigma_{rr}) \tag{12-22}$$

and

$$\frac{\partial P}{\partial z} - \rho g_z = \frac{1}{r}\frac{\partial}{\partial r}(r\sigma_{rz}) \tag{12-23}$$

Solving these equations by using the Reiner-Rivlin definitions for the stress terms (defined by Eq. 12-19) results in Equation 12-24.

$$\underset{zz}{\Sigma} = \underset{rr}{\Sigma} = -P + \sigma_{rr} = -\left[\rho g_z + \frac{\partial P}{\partial z}\right]z - \frac{(\partial P/\partial z)^2}{4} \int_0^r \frac{\psi_2 r}{\eta^2} \, dr + C_2 \quad (12\text{--}24)$$

To evaluate this integral, however, the dependence of η and ψ_2 on shear rate or stress must be known.

Darby (1976) proceeds to simplify this expression by assuming that η and ψ_2 are constant. The expression above then is reduced to Equation 12-25.

$$\underset{zz}{\Sigma} = \underset{rr}{\Sigma} = -\left[\rho g_z + \frac{\partial P}{\partial z}\right] - \frac{\psi_2 (\partial P/\partial z)^2 r^2}{8\eta^2} - P_0 \quad (12\text{--}25)$$

At the wall condition $r = R$ such that $\Sigma_{rr} = -P + \sigma_{rr} = -P$. This allows Equation 12-25 to be written as Equation 12-26.

$$-P + P_0 = -\left[\rho g_z + \frac{\partial P}{\partial z}\right] - \frac{\psi_2 (\partial P \partial z)^2 r^2}{8\eta^2} \quad (12\text{--}26)$$

At the die exit where $z = 0$, Equation 12-26 becomes Equation 12-27.

$$P - P_0 = \frac{\psi_2 (\partial P/\partial z)^2 r^2}{8\eta^2} \quad (12\text{--}27)$$

For Equations 12-25 to 12-27, P_0 is the atmospheric pressure outside the die. Realistically, viscosity and the second normal stress coefficient vary with shear rate. However, the swell for non-Newtonian materials exhibiting elastic behavior might be approximated by evaluating these functions at shear rates existing at the wall (Howkins et al., 1986). Care must be taken when this approach is utilized for die behavior predictions, as the Reiner-Rivlin equation is applicable only to steady, homogeneous, irrotational flows (Bird et al., 1987).

If process temperatures are sufficiently high such that vaporization at the die exists, then the P_{swell} term must have added to it the vapor pressure of moisture in the system. For this particular scale-up example, the process temperature was sufficiently low so that vaporization, or flashing at the die, did not occur. Information on ψ_2 was not available. Since visual die swell was not observed, this term was assumed to be zero for the analysis.

The capability of these models to predict die pressure drop was tested for two dies of different geometries on a pilot plant scale. For a dough with a shear stress given by Equation 12-28

Table 12-2. Test of Die Pressure Change (ΔP) Prediction Techniques

Die	A	B
Rate/hole (kg/s)	5.30×10^{-4}	5.30×10^{-4}
Temperature (°K)	374	374
Die hole contractions	2	1
Contraction 1 angle	60°	60°
Hole 1 length (m)	2.69×10^{-3}	2.59×10^{-3}
Hole 1 diameter (m)	2.29×10^{-3}	2.36×10^{-3}
Contraction 2 angle	0°	—
Hole 2 length (m)	2.15×10^{-2}	—
Hole 2 diameter (m)	3.18×10^{-3}	—
Total die thickness (m)	2.54×10^{-2}	3.81×10^{-3}
Calculated pressure drop (N/m²)	1.93×10^{6}	9.03×10^{5}
Observed pressure drop (N/m²)	1.72×10^{6}	1.05×10^{6}

$$\sigma_{rz} = 2,442 \; \frac{Ns^n}{m^2} \; [\dot{\gamma}]^{0.2} \qquad (12\text{–}28)$$

the total pressure drop was predicted and compared with actual observations. Test die geometries, conditions, and results are given in Table 12-2. Although a number of simplifying assumptions were made in the development of these expressions including isothermal flow, fully developed flow, no slip at the wall, and no elastic effects, the predictions are sufficiently accurate (within 15%) to be used for design estimates.

Before the actual scale-up calculations are performed to determine production die design requirements, the expressions for residence time in the die, and for material temperature distribution and work input, need to be developed.

Material Residence Time

Pressure drop was shown earlier to be a function of volumetric flow rate as given by Equation 12-11. The average residence time in the die, excluding any transition piece, is given by the volume of the die hole divided by the volumetric flow rate (Eq. 12-29).

$$(t_r) = V/Q \qquad (12\text{–}29)$$

If the die hole entrance cones are equivalent for pilot scale and production machines, or if the die cone volume is small compared to the total die hole volume, then only the residence time per hole length needs to be evaluated during scale-up analysis. Quite often, the die residence time is small compared with the residence time within the extruder. For longer dies it may be worthwhile to check the residence time. Residence time in a cylindrical hole is found by substituting Equation 12-11 into the previous equation to obtain Equation 12-30.

$$\langle t_r \rangle = \frac{2\pi R^2}{\left[\dfrac{\pi}{1/n + 3}\right]\left[\dfrac{1}{m}\dfrac{\Delta P}{2L}\right]^{1/n} R^{(1/n + 3)}} \tag{12–30}$$

Simplification of this equation results in Equation 12-31.

$$\langle t_r \rangle = \frac{2(1/n + 3)}{\left[\dfrac{1}{m}\dfrac{\Delta P}{2L}\right]^{1/n} R^{(1/n + 3)}} \tag{12–31}$$

Material Temperature
Distribution and Work Input

Not much heat loss is expected from the die once steady operation is attained, provided that the die head used is relatively large in mass and small in exposed area. Because viscosity is strongly dependent on temperature, the shear stress–shear rate relationship is significantly influenced by nonisothermal effects. Experimental results by Gerrard and co-workers (1966) demonstrated that the wall of a capillary comes very close to this adiabatic condition under normal operation. Under adiabatic conditions, heat generation resulting from viscous dissipation can lead to large temperature variations across the flow field in the die.

The temperature distribution, a function of energy transfer, can be evaluated in a fashion similar to that shown earlier for momentum transfer. An energy balance expression must be developed. A rigorous solution would require that the momentum balance equation be solved simultaneously with the energy balance equation, because viscosity, or σ_{rz}, is a function of temperature. A worst case situation can be estimated, however, assuming that viscosity (or the m value for a power law fluid) does not decrease with temperature, but rather is constant. In this case, the results of the isothermal solution for velocity distribution can be used to evaluate the maximum viscous dissipation contribution to the total energy balance. This result will yield a ΔT for the material used. If the temperature rise determined in this manner is small ($<3°F$), then the isothermal assumption is reasonable.

In differential form, the energy balance is given by Equation 12-32.

$$\rho \langle c_p \rangle v_z \frac{\partial T}{\partial z} = k \frac{1}{r} \frac{\partial}{\partial r} \left[r \frac{\partial T}{\partial r} \right] + \eta \left[\frac{\partial v_z}{\partial r} \right]^2 \qquad (12\text{-}32)$$

where

T = temperature, a function of both z and r

v_z = velocity in the z direction, which is a function of r and was mathematically defined earlier

k = material thermal conductivity

$\langle c_p \rangle$ = material average heat capacity

η = apparent viscosity

and the other variables retain their previous definitions. Even with the assumption of the constant η or m with respect to temperature, this problem requires solution of a partial differential equation. The details of the solution can be found in Bird and co-workers (1987). Using a solution for power law fluids with a consistency factor (m) independent of temperature, developed by Bird (1955), the temperature rise at the wall, where it will be at a maximum, can be estimated. The general solution is defined in terms of the dimensionless variables given by Equations 12-33 and 12-34.

$$\theta = (T - T_0) \frac{4km^{1/n}}{D^2 \sigma_w^{(n+1)/n}} \left[\frac{3n+1}{2n} \right]^2 = \text{dimensionless temperature rise}$$

$$(12\text{-}33)$$

$$\zeta = \frac{4}{Pe} \left[\frac{n+1}{3n+1} \right] \frac{z}{D} = \text{dimensionless axial distance/Peclet number} \quad (12\text{-}34)$$

where

$\sigma_w = D\Delta P/4L$ = shear stress at the wall

$Pe = D\langle v_z \rangle \rho c_p / k$ = Peclet number

D = die hole diameter

and other variables are as previously defined. The general solution for a power law fluid with $n = 0.2$ is illustrated in graphical form in Figure 12-3. This result was obtained by interpolation from Bird's values for power law fluids with other values of n. This solution is for the temperature rise of the fluid at the wall, which represents the maximum temperature rise, hence is a worst case analysis.

The procedure to solve for temperature rise due to viscous dissipation in adiabatic circular dies where η, or m, is not temperature dependent is:

1. Assume isothermal flow and obtain expressions for shear stress at the wall, and average velocity through the die hole.

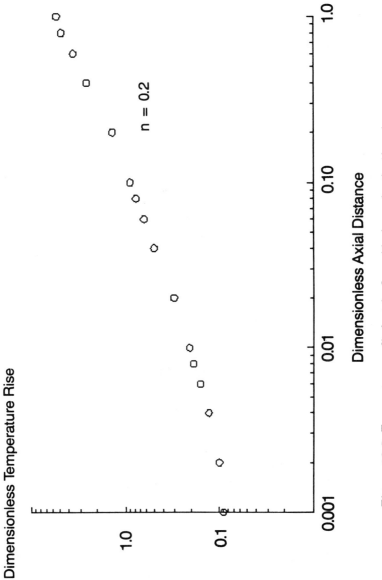

Figure 12-3. Temperature profile for tube flow with viscous heating, based on power law model and with m and n constants.

2. Using material rheological and thermal properties, and die geometry data, calculate a value for ζ.
3. Evaluate θ from Figure 12-3 (or use Bird's general solution with the appropriate n value). Then calculate ΔT from θ.

If ΔT determined in this manner is greater than 3°F, then the temperature effect on viscosity needs to be accounted for, and the actual temperature rise and pressure drop would be lower than the calculated values. A more accurate estimate of viscous dissipation heating and pressure drop would require a numerical solution of the simultaneous partial differential equations.

The temperature rise determined above is a direct result of mechanical work input. Van Lengerich (1984) has shown that the specific mechanical work input influences the quality of wheat starch extrudates. Some researchers now postulate that for other food materials, the effects of shear rate over a period of time also influence the character of an extrudate. Morgan and Steffe (1986) suggest that this influence can be quantified by a strain history as defined by Equation 12-35.

$$\phi = \langle \dot{\gamma} \rangle \langle t_r \rangle \tag{12–35}$$

where $\langle \dot{\gamma} \rangle$ = average shear rate across the tube radius and $\langle t_r \rangle$ = average residence time within the die hole. This average shear rate for a power law material is given by Equation 12-36.

$$\langle \dot{\gamma} \rangle = \left[\frac{6n + 1}{2n + 1} \right] \frac{Q}{\pi R^3} \tag{12–36}$$

where R and Q refer to the die hole radius and the volumetric flow rate, respectively. By combining these terms, the strain history can be simplified to Equation 12-37.

$$\phi = \frac{6n + 1}{2n + 1} \left[\frac{Q}{\pi R^3} \right] \frac{\pi R^2 L}{Q} \tag{12–37}$$

$$= \left[\frac{6n + 1}{2n + 1} \right] \frac{L}{R}$$

where L represents the length of the die. It is interesting to note that for a given material, this strain history is not a function of flow rate, but only of the die geometry.

All of the parameters required for die scale-up, and the equations used to determine them, have now been developed or assembled. These parameters are, in summary:

1. Pressure drop
   a. Entrance region
   b. Die hole length
   c. Swell pressure
2. Residence time
3. Viscous dissipation
4. Strain history

Estimation of each of these quantities requires rheological characterization of the material. Rheology provides the basis for all such engineering analyses.

The application of these equations to the die scale-up problem presented will now be reviewed.

Scale-Up Procedure

The many equations needed to describe the die process environment, and the complexity of these equations, make it necessary to use an iterative technique for scale-up. This procedure is initiated by identifying the limits, or range, of possible design options. The process behavior of the die is then estimated for each of the design variations by using the appropriate equations. Based on a given performance criterion, the optimum design can be selected. In this example, the design options could include variations on the following:

- Die hole length

- Die holder diameter

- Die hole diameter

- Die hole entrance angle

- Number of die holes

- Changes in diameter within the die hole

A number of these variables can be eliminated, or held constant, for the iteration sequence. The following arguments will suggest in which area variation should be explored.

Die hole length and die carrier diameter, although adjustable, have minimum limits as determined by plant equipment constraints and die strength requirements. These limits will be held constant at the minimum values as defined earlier in Table 12-1.

Final product specifications require that the die hole diameter not be changed during scale-up. To maintain product character, it is important that the final extrudate diameter not be changed. No swell behavior was noticed during pilot plant operation (i.e., the product and die radii were equal). Therefore, equivalent product diameters require equivalent die hole diameters.

Changes in the diameter within portions of the die hole provide variations in die pressure drops without requiring a change in the number of holes, or in the die hole diameter. Such changes introduce a complexity that may not be required. If simpler design options do not offer acceptable performance, then these more complex options can be considered. Simpler options, for ease of mechanical fabrication alone, should be evaluated first.

For design optimization, the options to be explored have been reduced to two; that is, the number of die holes used and the entry angle on a die hole.

Evaluation of die performance is based on the most significant contributions to the process environment. Possible contributions have been previously defined as die head pressure, material residence time in the die, and material temperature distribution and work input in the die. All of these parameters have been evaluated for this design case, with variations in die hole number and entrance angle. Table 12-3 lists the material physical properties used for the analysis. Table 12-4 lists the results of the process performance predictions given a constant die geometry. Table 12-5 tabulates the effects of variations in die hole entrance angle for a fixed number of holes.

It is clear that some aspects of process behavior are more significant than others. Based on the results in Table 12-4 and the following discussion, it can be concluded that the key criterion for this die scale-up is equivalent die pressure drop. Changes in pressure drop significantly affect upstream processing (i.e., extruder operation). Die residence time is insignificant. Calculated residence times for all of the dies considered are very small compared to the

Table 12-3. Physical Properties for Material in Die Scale-Up Example

---

Power Law Constants
$m = 2,442 \text{ N·s}^n/\text{m}^2$
$n = 0.2$

Density
$\rho = 1,092 \text{ kg/m}^3$

Heat Capacity
$c_p = 3,098 \text{ J/kg·K}$

Thermal Conductivity
$k = 0.45 \text{ J/kg·s·m·K}$

Table 12-4. Results of Die Scale-Up Calculations for Varying Die Hole Number

	Pilot Plant	Production Dies			
		1,500	1,300	1,100	900
$\Delta P(N/m^2 \times 10^6)$	0.903	0.827	0.924	1.05	1.24
Shear rate (s^{-1})	787	489	565	667	815
t_r (s)	0.00526	0.09	0.08	0.07	0.06
ϕ	9.10	44.0	45.2	46.7	48.9
ζ	0.00153	0.0260	0.0226	0.0183	0.0156
θ	0.11	0.50	0.42	0.36	0.28
ΔT (K)	0.17	0.47	0.47	0.49	0.48

30 to 45 seconds expected in the extruder. Heat generation due to viscous dissipation is also negligible. The assumption of isothermal flow, used in the development of the pressure drop equations, is valid because the change in temperature along the length of the die is so small. The strain history for the pilot plant is quite different from that for the predicted production operations. This difference can be altered only by changing the ratio of the die hole radius to the length, which is not a valid design option. Thus, in this situation the strain history differences cannot be corrected. From the magnitude of the viscous dissipation heating, however, we know that mechanical work input is low compared to work done in the extruder (1,721 J / kg in the die versus a typical 814,100 J / kg in an extruder), so any work history introduced at the die is not likely to be significant.

For this case, variations in the die hole entrance angle have little effect on the ΔP_{total} value. The only parameter useful for modifying the die process performance is the number of die holes. Based on the equivalent die pressure drop, the final die selection would be a production die with 1,300 holes.

CONCLUSION

The simple example presented, an engineering analysis of material flow through a die, should serve to illustrate the dependence of process quantifi-

Table 12-5. Effect of Die Entrance Cone Angle on Pressure Drop (1,500 hole die)

Entrance Angle	Pressure Drop (N/m^2)
45°	8.45×10^5
60°	8.25×10^5
90°	7.96×10^5

cation on accurate descriptions of various physical properties, particularly on rheological properties. Rheology provides a basis for many engineering analyses, from a study of fluid mechanics to reaction kinetics, which depend on shear rate, temperature, and residence time effects.

The example should further serve to illustrate that the processing behavior of the food material investigated could be estimated a priori, given a description of material physical properties. This makes it possible to effectively utilize existing process technology, developed for other industries, in food processing. The difference between food materials and the materials traditionally the subject of these engineering investigations is only in the necessarily complex characterization of the physical properties. It is here that the role of the food rheologist will be key in determining the success of such endeavors. Only through knowledge of material physical properties can control of our process operations be mastered.

ACKNOWLEDGMENT

The guidance provided by Ron Morgan, of Michigan State University, in transferring the results of his recent research efforts in the area of extruder die flow behavior to industrial application, is acknowledged and greatly appreciated.

NOMENCLATURE

$\langle c_p \rangle$	average material heat capacity
D	die hole diameter
D_c	die hole entrance cone radius
k	material thermal conductivity
L	die hole length
m	power law fluid material constant (sometimes referred to as material consistency factor)
n	power law flow index
n_d	number of total die holes
Pe	$D\langle v_z \rangle \rho c_p / k$ = Peclet number
$\Delta P_{entrance}$	die entrance pressure drop
ΔP_{exit}	die exit pressure drop
ΔP_{hole}	pressure drop over die land length
P_L	pressure after length L in the die hole
P_{swell}	$P_0 + \Delta P_{exit}$
ΔP_{total}	total die pressure drop

P_0	pressure just after die hole entrance when used in the momentum balance, or ambient pressure when used in the die swell pressure relationship
Q	volumetric flow rate (in this paper Q refers to volumetric flow through a particular die hole)
R	die hole radius
r	die hole radial position
r_h	hydraulic radius
Δr	change in die radial position
T	temperature, a function of both z and r
$\langle t_r \rangle$	average material residence time in the die hole
v_z	material velocity in the z direction
$\langle v_z \rangle$	average velocity in the z direction
z	axial direction vector
α	full die hole cone entrance angle
ψ_2	second normal stress coefficient of the material
$\dot{\gamma}$	shear rate
$\langle \dot{\gamma} \rangle$	average shear rate across the tube radius
ζ	dimensionless axial distance/Peclet number
η	apparent viscosity
Θ	dimensionless temperature rise
θ	angular rotation vector in cylindrical coordinates
τ	elastic relaxation time constant
μ	Newtonian viscosity
π_1	$\dfrac{16 n_d{}^2 r_h{}^4 \Delta P}{\rho Q^2}$
π_2	$\dfrac{\pi Q}{\pi n_d r_h{}^m [Q/(2 \pi n_d r_h{}^3)]^{n-1}}$
π_5	$\dfrac{3n + 1}{4n}$
π_6	$\dfrac{16 n_d r_h{}^2}{D_c^2}$
π_7	α
π_8	$1 + \dfrac{\tau Q}{16 \pi n_d r_h{}^3}$
ρ	material density
σ_{rz}	shear stress across the r plane, in the z direction
σ_w	$D\Delta P/4L = $ shear stress at the wall
ϕ	strain history
N_2	second normal stress function
$\sum$	total stress $= -P + \sigma$

REFERENCES

Bird, R. B. 1955. Viscous heat effects in extrusion of molten plastics. *Soc. Plastics Engrs. J.* **11**:35.

Bird, R. B., Armstrong, R. C. and Hassager, O. 1987. *Dynamics of Polymeric Liquids. Vol. 1: Fluid Mechanics*, 2nd ed., p. 218. New York: Wiley.

Bird, R. B., Stewart, W. E., and Lightfoot, E. N. 1960. *Transport Phenomena*. New York: Wiley.

Boger, D. V. 1982. Circular entry flows in inelastic and viscoelastic fluids. In *Advances in Transport Processes*, ed. A. S. Mujumdar and R. A. Mashelkar, p. 43. New York: Wiley.

Bohlin, L., and Carlson, T. L-G. 1980. Dynamic viscoelastic properties of wheat flour dough: Dependence on mixing time. *Cereal Chem.* **57**(3):174–177.

Bohn, L. J., and Bailey, C. H. 1936. *Effect of Mixing on the Physical Properties of Dough*, Paper 1410, Journal Series, Minnesota Agricultural Experiment Station.

Darby, R. 1976. *Viscoelastic Fluids—An Introduction to Their Properties and Behavior*. Chem. Process. and Eng. Series, Vol. 9, p. 295. New York: Marcel Dekker.

Gerrard, J. E., Steidler, F. E., and Appeldoorn, J. K. 1966. Viscous heating in capillaries. The adiabatic wall. *Ind. Eng. Chem. Fundamentals* **5**:260.

Howkins, M. D., Morgan, R. G., and Steffe, J. F. 1987. A method for predicting entrance pressure drop in food extruder dies. Paper presented by R. G. Morgan at the Baker Perkins Course on Food Extrusion Systems, Lecture 7: "Die Performance Prediction and Design," March 31.

Howkins, M. D., Morgan, R. G., and Steffe, J. F. 1986. Extrusion die swell of starch doughs. Paper presented at the ASAE Winter Meeting. ASAE Paper 86-6532.

Michaeli, W. 1984. *Extrusion Dies: Design and Engineering Computations*, p. 58. Munich, West Germany: Hanser Publishers.

Morgan, R. G., and Steffe, J. F. 1986. Rheology and extrusion of food. Paper presented at the Monterey Seminar Group. November 20, Monterey, California.

Remsen, C. H., and Clark, J. P. 1978. A Viscosity Model for a Cooking Dough. *J. Food Process Eng.* **2**:39–64.

van Lengerich, B. 1984. Entwicklung und Anwendung eines rechnerunterstuetzten systemanalytischen Modells zur Extrusion von Staerken und staerkehaltigen Rohstoffen. Diss., Technische Universitaet Berlin.

Chapter 13

Interrelationships of Rheology, Kinetics, and Transport Phenomena in Food Processing

Robert Y. Ofoli

A complete understanding of the interrelationships between rheology, kinetics, and transport phenomena is a very critical part of food process design and analysis. The need to understand the mechanisms underlying the interactions of these three areas cannot be overemphasized. To establish a common ground for discussion, the following definitions are presented.

BASICS OF RHEOLOGY, KINETICS, AND TRANSPORT PHENOMENA

Definitions

Rheology is the science of deformation and flow of matter. It is the study of the ways in which materials respond to applied stress and strain and is a necessary tool for characterizing the flow behavior of all fluids, including fluid foods.

Kinetics deals with physical or chemical changes in a system, their associated energy levels, the rates at which the changes occur, and the mode and mechanisms by which the changes are effected. In homogeneous systems, the primary variables affecting the rate of reaction are temperature, pressure, and composition; in heterogeneous systems the problem is much more complex, since more than one phase is involved (Levenspiel, 1972). This chapter will be primarily concerned with the changes affected by temperature in homogeneous systems.

The term *transport phenomena* describes the three general transfer processes and their associated mechanisms: momentum transfer (viscous flow), heat transfer (conduction, convection, and radiation), and mass transfer (convection and diffusion). These processes can be represented by the general equations of change; these equations in turn may be used to study the various phenomena that occur in fluids characterized rheologically as Newtonian or as any of the many non-Newtonian models used to describe the flow behavior of fluid foods.

Interactions of Rheology, Kinetics, and Transport Phenomena

The common bond in the interactions of rheology, kinetics, and transport phenomena is the fluid viscosity, defined as the ratio of the shear stress to the shear rate:

$$\eta = \frac{\sigma}{\dot{\gamma}} \qquad (13\text{--}1)$$

(*Note*: All variables used in this and all equations in this chapter are defined in the nomenclature.)

The apparent viscosity is usually represented as a function of shear rate. However, this parameter is affected by more than the shear rate — temperature, moisture content, time–temperature history, and shear history all have an effect. Therefore, the functionality of the apparent viscosity can be represented as

$$\eta = f(\dot{\gamma}, T, M, T - t, \Phi) \qquad (13\text{--}2)$$

where the functional terms are, respectively, shear rate, temperature, moisture content, temperature–time history, and strain history. One proposed structure of the apparent viscosity and its component parts will be given later in this chapter.

The interactions among rheology, kinetics, and transport phenomena can be demonstrated by a very simple example: the flow of a fluid food at a constant density ($1,200$ kg/m^3) in a straight tube 0.06 m in diameter and 30 m long, under an overall pressure drop of 400 kPa. Assume that the apparent viscosity of the fluid may be characterized by a Herschel-Bulkley equation of the form

$$\eta = \frac{\sigma_0}{\dot{\gamma}} + \mu_\infty \dot{\gamma}^{-0.6} \qquad (13\text{--}3)$$

The yield stress (σ_0) and the consistency coefficient (μ_∞) in the above equation have been left as variables to enable their manipulation in this illustration.

The usual quantities of interest in tube flow are the volumetric flow rate and the average velocity (or, in some situations, the velocity profile). The volumetric flow rate equation is (Skelland, 1967)

$$Q = \frac{\pi R^3}{\mu_\infty \sigma_w^3}(\Delta\sigma)^{\frac{1+n}{n}}\left[\frac{n(\Delta\sigma)^2}{1 + 3n} + \frac{2n\sigma_w\Delta\sigma}{1 + 2n} + \frac{n\sigma_w^2}{1 + n}\right] \qquad (13\text{-}4)$$

where

$$\Delta\sigma = \sigma_w - \sigma_0 = \frac{\Delta P}{L}\frac{R}{2} - \sigma_0 \qquad (13\text{-}5)$$

Now, if the yield stress in Equation 13-4 is allowed to vary from 0.0 (power law behavior) to 200 Pa, with all other variables in that equation remaining constant, the effect of rheology on the volumetric flow rate can be observed (Figure 13-1). Note the behavior of the mass flow rate curve as the yield stress goes from 0 to 200 Pa: It begins at a maximum of approximately 6,280 kg/hr at zero yield stress and ends with absolutely no flow when the yield stress reaches the value of the shear stress. In this exercise, the consistency coefficient (μ_∞) has been set at 30.6 kPa $\cdot$ s$^{0.4}$.

In rheological expressions, the effects of kinetics (primarily temperature effects) can usually be incorporated into the expression for the consistency coefficient, with an Arrhenius equation of the form

$$\mu_\infty = \mu_0 \exp\left[\frac{\Delta E}{R}(T^{-1} - T_r^{-1})\right] \qquad (13\text{-}6)$$

where T_r is a reference temperature.

In general, temperature changes will also affect the yield stress. However, erring on the side of simplicity, one can assume that the yield stress remains constant at 50 Pa over a temperature range of 25°C (298 K) to 75°C (348 K), with the reference temperature being 298 K. The effect of varying the consistency coefficient ($\Delta E = 4,000$; $R = 1.987$) is shown in Figure 13-2. Now, the mass flow rate is approximately 1,480 kg/hr at 25°C when the fluid is "cold," and 3,890 kg/hr at 75°C when the fluid is warmer. This graph demonstrates the thinning effect of temperature that is seen in many fluids. As the fluid thins out, more of it can be pumped under the same overall pressure drop.

A more realistic situation is that, in addition to the consistency coefficient, the yield stress also varies with temperature. This combination allows one to observe the full effects of thermally induced kinetics on the fluid flow rate. Again, to simplify the analysis, assume that the yield stress varies linearly with temperature, with a value of 50.0 Pa at 25°C and 0.0 at 75°C. This combined effect is shown in Figure 13-3. Observe the new minimum (1,785 kg/hr at 298 K) and maximum (16,300 kg/hr at 348 K) flow rates, again demonstrating

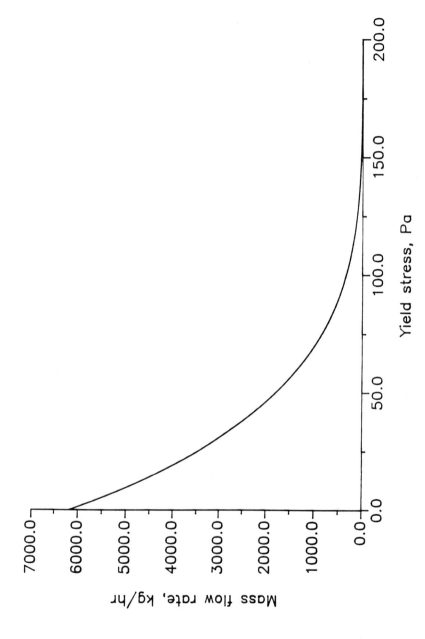

Figure 13-1. Mass flow rate versus yield stress.

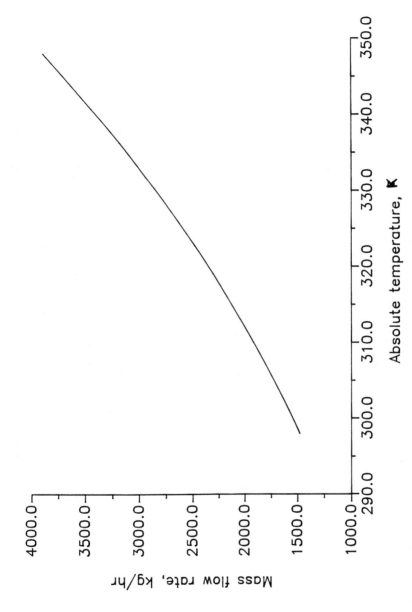

Figure 13-2. Mass flow rate versus temperature (σ_0 is a function of temperature).

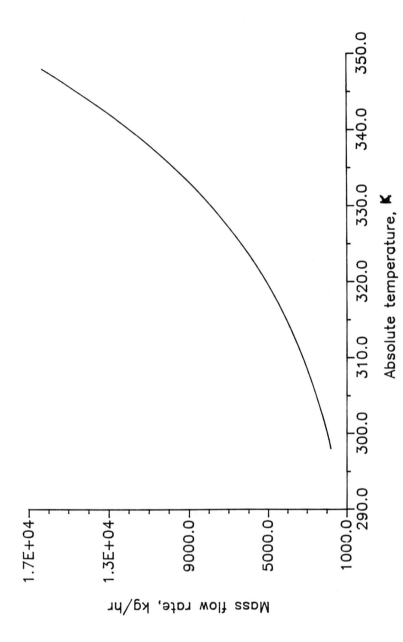

Figure 13-3. Mass flow rate versus temperature ($\mu_{\infty}\sigma_0$ are both functions of T).

that at higher temperatures the fluid is thinner (less viscous) and easier to pump.

This example is a very simplified illustration of the complex interrelationships of rheology, kinetics, and transport phenomena (in this case, fluid dynamics). The rest of this chapter presents some forms of the transport and rheological equations that make it possible to account for the interactions shown in the example.

RHEOLOGY

Fluids transported/processed/analyzed in the food industry can be classified as Newtonian or non-Newtonian, with the vast majority of fluid foods adhering to any one of several non-Newtonian models. The need for good rheological data on fluid foods in the industry is enormous. However reliable data are a requirement for process engineering analyses (extrusion, pumping, mixing, agitation, heating, coating, process control), quality control and shelf-life estimation, texture evaluation, product development, and the development of constitutive equations for rheological characterization.

Basic Rheological Characteristics of Fluid Foods

Fluid foods generally exhibit viscoelastic behavior, which may be time dependent or time independent. The two most well-documented time-dependent behaviors are thixotropy (time-dependent thinning effect) and rheopexy (time-dependent thickening effect). Another time-dependent behavior, rheomalaxis, concerns the permanent mechanical degradation of the material.

Several factors affect the flow behavior of fluid foods: temperature, the presence or absence of a yield stress, transformation kinetics (starch gelatinization and/or retrogradation, protein denaturation, caramelization, etc.), level of microbial activity, homogeneity of the fluid medium, molecular weight, particle size, and pH.

Rheological Characterization of Fluid Foods

The most common rheological models used to describe the viscoelastic flow of fluid foods are the power law or Ostwald-de Waele (Reiner, 1949), Casson (Casson, 1959), Bingham plastic (Bingham, 1922), and the Herschel-Bulkley (Herschel and Bulkley, 1926). The Heinz-Casson model (Heinz, 1959) is a generalization of the Casson model, but has been used little in analyses of foods. Another modification of the Casson model is the Mizrahi-Berk model (Mizrahi and Berk, 1972).

When a broad range of shear rates is encountered, these models (with the possible exception of the Herschel-Bulkley model) can be quite limited in their ability to adequately characterize the rheological behavior of a given food material, sometimes requiring more than one model to fit the data. For this reason, and to provide more generality for transport phenomena studies, another model was recently proposed by Ofoli et al. (1987). In this model, shear stress and apparent viscosity functions are characterized by the following equations:

$$\sigma^{n_1} = (\sigma_0)^{n_1} + \mu_\infty \dot\gamma^{n_2} \tag{13-7}$$

and

$$\eta = \left[\left(\frac{\sigma_0}{\dot\gamma} \right)^{n_1} + \mu_\infty \dot\gamma^{n_2 - n_1} \right]^{1/n_1} \tag{13-8}$$

This model has two important advantages over others:

1. It provides much more flexibility in the modeling of flow behavior over several decades of shear rates.
2. By appropriate selection of the four parameters, it represents the Newtonian model as well as a large majority of the non-Newtonian models used for fluid food analyses.

The real convenience of the generality of the model is that when it is used to develop the transport equations, they can be used for all fluids characterized as Newtonian or as any of the models listed above, simply by substituting appropriate values of the yield stress, flow behavior index, and the two power indices. The major disadvantage of the model is that it does add some mathematical complexity to process analysis.

Rheological Measurement Methods

Several methods are available for obtaining rheological data on fluid foods. Some of these methods are briefly discussed below.

Tube viscometers use pipes and capillaries to generate rheological data. Cone and plate, parallel plate, and concentric cylinder systems are used to study oscillatory and steady shear behavior. These systems are well described in most standard texts on rheological measurement, such as the text by Dealy (1982). Even though this text primarily addresses polymers, the rheological techniques are similar for food materials.

Less broadly documented techniques include the *mixer viscometer* and *back extrusion*. The mixer viscometer measures torque response during mixing. The advantages of this device are that it reduces problems related to slip, tolerates fluids with large particles, reduces sedimentation problems, and minimizes degradation of time-dependent materials during loading (Mackey et al., 1987; Castell-Perez and Steffe, 1987). The advantages of back extrusion are that it is simple and inexpensive. It allows data collection at high speeds, and it enables measurements to be made easily on thick pastes and doughs, where high forces are required (Steffe and Osorio, 1987). It is also one of the most reliable methods available for determining the yield stress of a fluid.

Role of Rheology in Process Design and Analysis

The role of rheology in process design and analysis has been well covered in another chapter of this book; therefore, no further discussion will be given here.

KINETICS

Reaction kinetics and thermodynamics together provide insight into the transformations (such as gelatinization, caramelization or browning reactions, denaturation, etc.) that occur in food products during processing. Thermodynamics gives information on the energy released/absorbed during a reaction and the maximum possible extent of the reaction, in addition to allowing for calculation of the equilibrium constant; kinetics makes it possible to identify the factors that influence the reaction rate, and it provides a means to measure the reaction rate (Levenspiel, 1972).

The study of kinetics has always been an integral part of the food industry, particularly in the area of thermal processing. In recent years, the community of food rheologists has been incorporating the knowledge obtained from reaction kinetics in the characterization of rheological properties.

Characterization of Kinetic Phenomena

A reaction may be homogeneous (taking place in one phase alone) or heterogeneous (requiring the presence of at least two phases). The reaction may be zero, first, or multiple-order. In this chapter, as is done in most food reactions, transformation kinetics will be represented as a first-order phenomenon.

A first-order reaction in a batch reactor,

$$A \rightarrow B \tag{13-9}$$

may be represented by the equation

$$-r_A = -\frac{dC_A}{dt} = kC_A \tag{13-10}$$

Equations describing other orders of reaction may be found in Levenspiel (1972) and other books on reaction kinetics.

The effects of temperature on the reaction is usually incorporated in the reaction rate constant, k, and is expressed by an Arrhenius equation of the form

$$k = k_0 \exp\left(\frac{\Delta E}{RT}\right) \tag{13-11}$$

Basic reaction kinetics in the food processing industry involve such phenomena as protein denaturation; starch gelatinization and/or degradation; and caramelization, or browning. An understanding of the modes and mechanisms associated with these phenomena is of importance to food industry personnel involved in process engineering, product development, and quality control.

Mathematical Modeling of Starch and Protein Dough Rheology

While cumbersome, perhaps, the strategy in rheological characterization should aim at developing viscosity models that incorporate the effects of temperature, moisture content, shear rate, temperature–time history, and strain history in a single relationship. Morgan et al. (1989) proposed a model that treats each of these effects as a separate constituent; the various resulting equations are "merged" to form one cohesive relationship that can be used to quantitatively describe the effects of both rheology and kinetics in transport phenomena, process engineering and/or modeling, and product development.

The shear rate dependency may be modeled by the rheological equation of Ofoli et al. (1987), given in Equation 13-8 of this chapter, or any appropriate version of it. The development of the other component parts of the generalized viscosity model is fully documented in Morgan et al. (1989), and will be presented here only in their final forms.

The effect of temperature is modeled by

$$\eta_T = \exp\left[\frac{\Delta E_v}{R}(T^{-1} - T_r^{-1})\right] \tag{13-12}$$

The moisture content or lubricating effect is modeled by

$$\eta_M = \exp\left[b(M - M_r)\right| \tag{13-13}$$

The temperature–time history effect is product specific. For starch doughs, the effect of kinetics on the temperature–time history is modeled by the expression (Mackey et al., 1989)

$$\eta_{T-t} = 1 + \beta_r(MC_s)^{\alpha(\gamma)}(1 - e^{-k_a\Psi})^{\alpha(\gamma)} \tag{13-14}$$

For protein doughs, the effect of kinetics on the temperature–time history is given by the expression (Morgan et al., 1989)

$$\eta_{(T-t)} = 1 + \beta_r(\beta_0 MC_p)^{\alpha(\gamma)}(1 - e^{-k_a\Psi})^{\alpha(\gamma)} \tag{13-15}$$

The temperature–time history function in both equations above is

$$\Psi = \int_0^t T(t)\exp\left[\frac{-\Delta E_d}{RT(t)}\right]dt \tag{13-16}$$

The relationship for the strain history effects is

$$\eta_s = A_0 + A_\infty[1 - e^{-\alpha\Theta}]^{\alpha'} \tag{13-17}$$

Collecting terms, the effects of shear rate, moisture content, strain history, and temperature–time history can be combined to yield the following expression for the apparent viscosity:

$$\eta = \left[\left(\frac{\sigma_0}{\dot{\gamma}}\right)^{n_1} + \mu_\infty \dot{\gamma}^{n_2-n_1}\right]^{1/n_1} \exp\left[\frac{\Delta E_v}{R}(T^{-1} - T_r^{-1})\right]\exp\left[b(M - M_r)\right]$$

$$\left(1 + \beta_r(MC_s)^{\alpha(\gamma)}\left|1 - e^{-k_a\Psi}\right|^{\alpha(\gamma)}\right)\left(A_0 + A_\infty[1 - e^{-\alpha\Theta}]^{\alpha'}\right) \tag{13-18}$$

for starch doughs, and

$$\eta = \left[\left(\frac{\sigma_0}{\dot{\gamma}}\right)^{n_1} + \mu_\infty \dot{\gamma}^{n_2 - n_1}\right]^{1/n_1} \exp\left[\frac{\Delta E_v}{R}(T^{-1} - T_r^{-1})\right] \exp\left[b(M - M_r)\right]$$

$$\left(1 + \beta_r(\beta_0 M C_p)^{\alpha(\gamma)}\left(1 - e^{-k_a \Psi}\right)^{\alpha(\gamma)}\right)\left(A_0 + A_\infty[1 - e^{-\alpha\Theta}]^{\alpha'}\right) \quad (13\text{-}19)$$

for protein doughs.

These equations provide useful and flexible expressions for incorporating the effects of kinetics and rheological behavior in the apparent viscosity. Because each constituent term incorporates the complete effect of only one variable, any of them can be easily replaced by another form. The model is applicable to any process or system.

The difficulty in using the above relationships involves the presence of several material constants that must be determined by experiment and/or analogy. In addition to this problem, the model requires an iterative procedure for processes in which the temperature is a function of time. Since this is the usual case, the model does present some challenges in mathematical modeling and computer simulation. Once these difficulties are overcome for a given set of material and process conditions, however, the relationships provide an excellent tool for physicochemical modeling of food behavior.

Some Practical Applications of the Kinetics Model

The generalized model described above has been used in several specific applications of interest to the food industry. Three of the applications are briefly discussed below.

In addition to providing the specific details of the development of the generalized viscosity model, Morgan et al. (1989) used the model to study the apparent viscosity of defatted soy flour dough undergoing heat-induced protein denaturation. The Heinz-Casson model was used to characterize rheological behavior. The temperature–time history effects were modeled by approximating protein denaturation effects by a pseudo first-order reaction equation. The equation was then used to model experimental data from Morgan (1979) and Luxemburg (1985), and was found to be in excellent agreement over a wide range of conditions. In addition, several of the model parameters were found to be in good agreement with those found in the technical literature.

Dolan et al. (1989) used the generalized apparent viscosity equation to model the behavior of dilute fluid foods thickened by starch gelatinization. A dilute raw corn starch solution (13.2% w.b.) undergoing various temperature–time treatments provided the fluid medium, with the temperature–time history effects being evaluated by the back extrusion method. The model was successfully used to predict flow behavior as starch gelatinized during processing. The authors assumed time-independent behavior, negligible viscoelastic effects, a fluid continuum, and first-order gelatinization kinetics. The Heinz-Casson model was used to represent the shear rate effects.

Mackey et al. (1989) used the model to predict the effects of shear rate, temperature, moisture content, temperature–time history, and strain history on the apparent viscosity of doughs of low to intermediate moisture during extrusion. The model was evaluated for pre-gelatinized potato flour doughs on an Instron capillary rheometer, and a Baker Perkins MPF-50D/25 twin-screw extruder. In a moisture content range of 22–50%, and a temperature range of 25–95°C, the model gave an excellent fit to the observed data, with a regression coefficient (R^2) of 0.95. The range of shear rates was from 1 s^{-1} to 1,000 s^{-1}. The authors found that strain history had negligible effects on the apparent viscosity.

TRANSPORT PHENOMENA

Transport phenomena are an important part of a large number of processes in the food industry. These processes range in degrees of complexity from near-isothermal pipeline transport of various food materials (purely momentum transfer) to flow processes in various types of heat exchangers (heat and momentum transfer), to processes involving highly non-Newtonian behavior in which all three transport mechanisms routinely occur simultaneously.

The apparent viscosity enables the effects of rheology and kinetics to be directly incorporated into the transport equations of heat and momentum. Mass transfer (as well as its coefficients) is also affected by rheology and kinetics, because in most processes, mass transfer depends on both momentum and heat transfer.

Characteristic Representation of the Apparent Viscosity

The characteristic representation of the shear stress for studies involving transport phenomena is

$$\overline{\sigma} = \eta \overline{\Delta} \qquad (13\text{--}20)$$

where the bar above the shear stress and the rate of deformation tensor signify their tensorial nature.

The apparent viscosity is a scalar function, and for non-Newtonian fluids, is shear dependent. Theoretically, it is a function of the three invariants; however, for most flow situations, it is only a function of the second invariant, I_2 (Bird et al., 1960). The apparent viscosity also depends on the type of rheological model used. For the model of Ofoli et al. (1987), the general form of the shear rate–dependent apparent viscosity is

$$\eta = \left[\left(\frac{\sigma_0}{\dot{\gamma}} \right)^{n_1} + \mu_\infty \dot{\gamma}^{n_2 - n_1} \right]^{1/n_1} \tag{13-21}$$

The rate of deformation tensor is a nine-component parameter that, for Cartesian coordinates, may be written in tensor notation as

$$\Delta_{ij} = \frac{\partial v_i}{\partial x_j} + \frac{\partial v_j}{\partial x_i} \tag{13-22}$$

and represents a square (3×3) matrix.

SUMMARY

Rheology, kinetics, and transport phenomena are all critical parts of food process design and analyses. The need to understand the mechanisms underlying their interactions cannot be overemphasized. The apparent viscosity provides the common link between the three areas.

A generalized model for the apparent viscosity has been presented. The model incorporates the following effects: temperature, moisture content, shear rate, temperature–time history, and strain history. Each of these effects is modeled as a separate component, with the various resulting equations being superimposed to form one equation that quantitatively describes the effects of both rheology and kinetics in transport phenomena, process engineering and/or modeling, and product development.

Several material constants must be determined by experiment and/or analogy to use the generalized model. After these constants have been determined for a given set of material and process conditions, however, the relationships provide an excellent tool for modeling food behavior.

NOMENCLATURE

A_0 material constant, dimensionless
A_∞ material constant, dimensionless
b moisture coefficient, dimensionless
C_A molar concentration of reactant A, mol m^{-3}
C_p protein concentration, dry basis, decimal
C_s starch concentration, dry basis, fraction
E activation energy, J mol^{-1}
k reaction rate constant, s^{-1}
k_a reaction transmission coefficient, s^{-1}

k_0	constant in Arrhenius equation, s^{-1}
M	moisture content, decimal
M_r	moisture content, decimal
n	flow behavior index, dimensionless
n_1	flow behavior index, dimensionless
n_2	shear/yield stress index, dimensionless
R	universal gas constant, cal gm^{-1} mol^{-1}
r_A	rate of reaction of component A, mol m^{-3} s^{-1}
T	temperature, K
T_r	reference temperature, K
T–t	temperature-time
t	time, s

Greek Symbols

α	material constant, dimensionless
β_0	strain history coefficient
β_r	material constant, dimensionless
$\dot{\gamma}$	shear rate, s^{-1}
η	apparent viscosity, Pa·s
Θ	material constant, dimensionless
μ_∞	consistency coefficient, Pa·s^{n2}
σ	shear stress, Pa
σ_0	yield stress, Pa
Φ	strain history
Ψ	temperature–time history

REFERENCES

Bingham, E. C. 1922. *Fluidity and Plasticity*. New York: McGraw-Hill.

Bird, R. B., Stewart, W. E., and Lightfoot, E. N. 1960. *Transport Phenomena*. New York: Wiley.

Casson, N. 1959. A flow equation for pigmented-oil suspension of the printing ink type. In *Rheology of Disperse Systems*, ed. C. C. Mill, pp. 82–104. New York: Pergamon.

Castell-Perez, M. E., and Steffe, J. F. 1988. Using mixing to evaluate rheological properties. In *Viscoelastic Properties of Solid, Fluid and Semisolid Foods*, M. A. Rao, ed. Barking, England: Elsevier. In press.

Dealy, J. M. 1982. *Rheometers for Molten Plastics: Practical Guide to Testing and Property Measurement.* New York: Van Nostrand Reinhold.

Dolan, K. D., Steffe, J. F., and Morgan, R. G. 1989. Back extrusion and simulation of viscosity development during starch gelatinization. *J. Food Process Eng.* In press.

Heinz, J., 1959, quoted in A. Fincke, 1961, Beitrage zur Losung rheologischer Probleme in der Schokoladentechnologie, Diss. TH Karlsruhe.

Herschel, W. H., and Bulkley, R. 1926. Konziztensmessungen von gummibensollosugen. *Kolloid-Zeitzchr* **39**:291–300; *Proc. Amer. Soc. Test. Matls.* **26**:621–633.

Levenspiel, O. 1972. *Chemical Reaction Engineering,* 2nd ed. New York: Wiley.

Luxemburg, L. A., Baird, D. G., and Joseph, E. G. 1985. Background studies in the modeling of extrusion cooking processes for soy flour doughs. *Biotechnol. Progress* **1**(1):33–38.

Mackey, K. L., Morgan, R. G., and Steffe, J. F. 1987. Effects of shear-thinning behavior on mixer viscometry techniques. *J. Tex. Stud.* **18**:231–240.

Mackey, K. L., Ofoli, R. Y., Morgan, R. G., and Steffe, J. F. 1989. Rheological modeling of potato flour during extrusion cooking. *J. Food Process Eng.* In press.

Mizrahi, S., and Berk, Z. 1972. Flow behavior of concentrated orange juice: Mathematical treatment. *J. Tex. Stud.* **3**:69–79.

Morgan, R. G. 1979. Modeling the Effects of Temperature–Time History, Temperature, Shear Rate and Moisture on Viscosity of Defatted Soy Flour Dough. Ph.D. Diss., Texas A&M University, College Station, Texas.

Morgan, R. G., Steffe, J. F., and Ofoli, R. Y. 1989. A generalized viscosity model for extrusion of protein doughs. *J. Food Process Eng.* In press.

Ofoli, R. Y., Morgan, R. G., and Steffe, J. F. 1987. A generalized rheological model for inelastic fluid foods. *J. Tex. Stud.* **18**:213–230.

Reiner, M. 1949. *Deformation and Flow.* London: Lewis.

Skelland, A. H. P. 1967. *Non-Newtonian Flow and Heat Transfer.* New York: Wiley.

Steffe, J. F., and Osorio, F. A. 1987. Back extrusion of non-Newtonian fluids: Rapid, low-cost test method has applications in quality control and product development. *Food Tech.* **41**(3):72–77.

Chapter 14

Rheological and Engineering Aspects of the Sheeting and Laminating of Doughs

Leon Levine
Bruce A. Drew

The sheeting of doughs is a commercially important unit operation in a wide variety of food processes. In the baking industry it is used for the production of cookies, pizza, bread, and pastry doughs, as well as Mexican specialties such as tacos and tortillas (Matz, 1968; Behnke et al., 1979). Sheeting used to be the method of choice for the production of pasta products, such as egg noodles (Hummel, 1950). This application is not as widespread as it once was, but it still is the only method of production for pasta shapes such as ravioli, tortellini, and bows (butterflies). The snack industry makes wide use of this unit operation for the production of fabricated snacks such as potato chips (Curry et al., 1976) and corn chips. Despite this wide variety of applications, very little has been written about the physical and engineering aspects of sheeting behavior. It is the goal of this chapter to fill this void in the available literature.

A review of the literature concerning sheeting and laminating reveals very little quantitative information. There are several notable exceptions to the lack of technical information. First, repeated sheeting of an underdeveloped dough results in its development (Kilborn and Tipples, 1974). The authors of this paper state that this development is due to the introduction of work. The amount of energy necessary to develop a dough by sheeting is about 15% of that required when the work is done by mixing. The work required to reach equivalent development appears to be the same for both processes. Further analysis of their data, some of which is shown in Table 14-1, is revealing.

Table 14-1. Effect of Repeated Sheetings
and Energy Input of a Bread Dough

Number of Sheetings*	Energy (whr/lb)	Baked Loaf Volume (cc)
12	0.07	770
22	0.17	875
32	0.25	925
42	0.39	935
62	0.47	910
82	0.61	880

*One sheeting was defined a pass between a pair of rolls with a 7/32″ gap followed by a pass through a 5/32″ gap, followed by folding of the dough in half and a turn of 90°.

The conclusion from this table is that optimum dough development occurs at a work input of 0.39 Whr/lb. Additional information can be obtained by viewing the data differently. Consider Figure 14-1, a plot of energy input/sheeting passes versus the number of sheeting passes. This curve looks very much like the development curve for a mixer, with the exception that the ordinate is the number of passes instead of time. Significantly, from the point of view of an engineer or rheologist, this curve indicates that the flow resistance (viscosity) of the dough is being reduced after the dough is developed.

Second, studies of the sheeting of durum "doughs" (Feillet et al., 1977), soft wheat doughs (Watanabe and Nagasawa, 1968), and hard wheat doughs report that sheeting seems to have an effect on the protein fraction of the dough. For the semolina dough the extractability of gluten was observed to increase, and the composition of the extracted gluten changed as the dough was processed, either by repeated passes through the rollers or by decreasing the gap between the rollers. It was suggested that the scission of disulfide bonds within and between glutenin molecules, and rearrangement may occur as a result of stretching induced by the rolls.

Photomicrographs of the hard wheat dough, processed by sheeting, revealed that repeated sheeting resulted in the organization of a protein network within the dough. Excessive sheeting tended to break down this organization. The study of the effects of sheeting on soft wheat doughs suggested that sheeting had a significant effect on the properties of noodles produced from that dough. Specifically, sheeting seemed to result in the alignment of the protein within the dough.

Third, a study (Stenvert et al., 1980) of the production of bread doughs has shown that repeated sheeting of the dough expels gas, reduces bubble size, and results in a fine-grained bread. Stronger doughs, as measured by

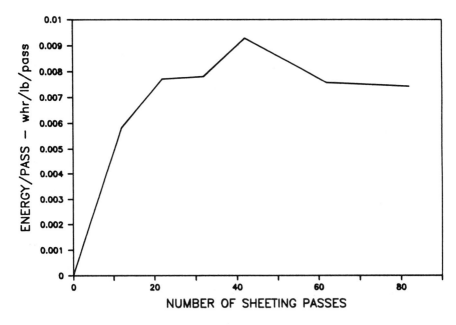

Figure 14-1. Development of dough through rolls.

extensigraph resistance, are required as the number of sheeting passes is increased.

Fourth, repeated sheeting results in the gradual reduction of the extensibility and the resistance of a wheat dough (Moss, 1980). This result suggests that doughs that are destined for repeated sheetings, as is the case in the production of multilayered pastries, be formulated to a higher strength as measured by extensigraph resistance.

Although the exact interactions between product quality and dough sheeting are not clear, the examples cited above, as well as the experience of those actually involved in production, clearly indicate that the sheeting operation plays an important role in determining the final quality of a dough. It would seem that the shear, stretching, or energy input of the rolls are key factors in this phenomenon, and any engineering analysis of sheeting must explore these factors.

There is even less information about the engineering aspects of sheeter design than there is about the qualitative effects of sheeting on dough quality. One paper (Kilborn and Tipples, 1974) reports that only about 25% of the energy used by the motors that drive a pair of sheeting rolls finds its way into the dough. The bulk of the energy is lost as friction, resistive, and windage losses in the motor, bearing, drives, and other components. The only published paper that discusses the engineering aspects of the flow of dough between rolls (Levine, 1985) covers the effects of roll speed, diameter, roll gap, and feed thickness on the power consumption of the sheeting rolls. This paper suggested that the scale-up of sheeting rolls from the laboratory or pilot plant to full scale presents a formidable problem. Increasing roll speed and increasing roll diameter to increase production rates were both found to result in increased energy input to the dough. The tendency for commercial rolls to be designed with diameters larger than those of pilot plant or laboratory rolls is a natural reaction to a design problem that will be discussed later in this chapter. The paper also indicated that the energy input of the rolls to the dough should increase with decreasing roll gap and increasing dough feed thickness. As a consequence, the scale-up problem might be very dramatic for very thin dough sheets.

The simplified model used to explain energy in the paper cited above (Levine, 1985) did not adequately predict final dough thickness or the power consumption of the rolls. The final dough thickness could be as large as four times the gap between the rolls, while theory predicted that the final sheet thickness would be only 20% greater than the gap between the rolls. It was suggested that a model incorporating the viscoelastic behavior of a dough rather than just the viscous behavior might account for the deviations in the power and final sheet thickness predictions.

Two follow-up papers (Drew et al., 1987; Drew et al., 1988), explored the introduction of dough viscoelasticity into models of sheeter behavior. These papers indicated that prediction of the final dough sheet thickness was not improved by including viscoelastic behavior in the model. These more complex rheological models do, however, improve prediction of roll power consumption and roll closing forces. The latter represent an important engineering design issue that has not been discussed previously.

COMMERCIAL SHEETING EQUIPMENT

A variety of equipment is available commercially for the sheeting and laminating of doughs. Sheeters range from a simple roll stand (one pair of rolls) to very long, sophisticated systems incorporating many pairs of rolls, such as the system illustrated in Figure 14-2. The simplest roll stands are usually found in small-scale bakeries, laboratories, and pilot plants. Two examples of such stands are the Rondo Reversible Sheeter and the Moline Table Top Sheeter. A close-up view of a four-roll system is illustrated in Figure 14-3. The use of multiple sets of rolls is a practical necessity in that it allows the baker to produce thin doughs continuously. This aspect of sheeter design will be discussed in some detail later in this chapter.

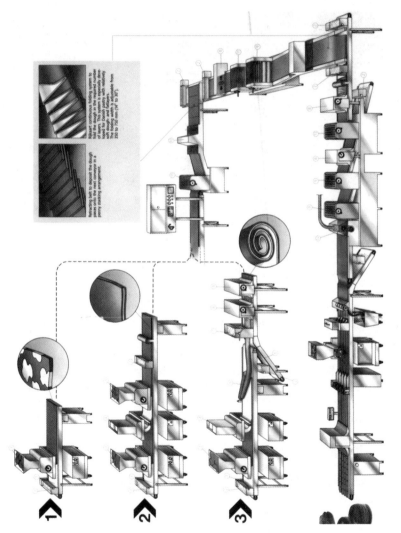

Figure 14-2. A complicated sheeting line. *(Courtesy of Rykaart, Inc., Hamilton, Ohio.)*

Figure 14-3. A four-roll system. *(Courtesy of Thomas L. Green & Company, Indianapolis, Indiana.)*

One variation on the single pair of rolls is illustrated in Figure 14-4. This configuration is normally used to form a sheet from a mass of unformed dough. Actually, the system may be considered to be two pairs of rolls in series.

Another category of sheeter replaces one of the rolls with a series of freely rotating small-diameter rolls. Examples of this type of sheeter are the Rheon SM Stretcher (Fig. 14-5), the Rykaart Multiroller (Fig. 14-6), and the Moline Cross Roller (Fig. 14-7). The manufacturers claim that these devices have the ability to make large reductions in dough thickness with minimal damage to the protein network.

Lamination of doughs with shortening is important to achieve flakiness in baked products such as pie crusts, croissants, and danish pastries. Lamination with fillings other than shortenings is used to make filled products such as cinnamon rolls, toaster pastries, dumplings, and tarts.

Several methods can be used for forming laminated doughs. The four most common methods are (1) the "Scotch," or all-in, method, which creates a pseudo-laminated product by mixing flakes of high-melting shortening directly into the dough before sheeting; (2) the sandwich method, (Fig. 14-8), which extrudes a sheet of shortening onto a sheet of dough and covers it with a second sheet of dough; (3) the roll-in method (Fig. 14-9), which extrudes

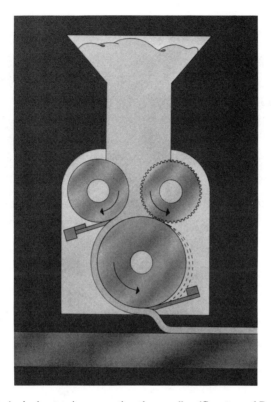

Figure 14-4. A sheeter incorporating three rolls. *(Courtesy of Rykaart, Inc., Hamilton, Ohio.)*

a layer of shortening onto a sheet of dough and rolls the result into spiral configuration; and (4) the fold-in method, which extrudes a layer of shortening or filling onto a sheet of dough, leaving sufficient uncovered dough to cover the filling when the dough is folded over. Two folding schemes are used: single fold (Fig. 14-10) and envelope folding (Fig. 14-11).

Virtually all shortening-filled laminates are subjected to subsequent rolling and folding after lamination. The folding operation is called lapping. There are two basic types of lapping methods: the cut and stack method (Fig. 14-12) and the folding method (Fig. 14-13). Lapping is also used to change the width of the dough.

One system, used to produce flaky crackers, combines the lamination and lapping operations (Fig. 14-14). In this process a mixture of shortening and flour instead of pure shortening must be used as the inside layer to improve flow characteristics.

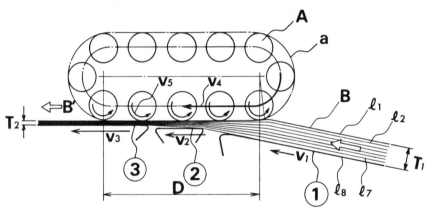

Figure 14-5. A dough stretcher. *(Courtesy of Rheon, Inc., Irvine, California.)*

ENGINEERING ISSUES IN THE DESIGN OF SHEETING EQUIPMENT

An engineer, faced with the task of designing a sheeting system, must confront a set of problems different from those that are normally presented to the cereal chemist or food scientist who develops a dough-based product.

Two primary areas must be met in a process design. First, the design of the sheeting system must produce the product that has been specified by the developer. These specifications include the physical properties of the product, such as the specific volume or texture after baking or frying. In addition, the

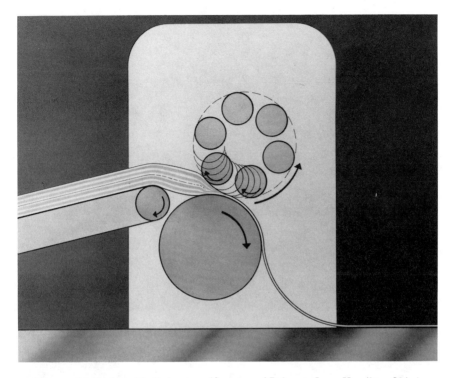

Figure 14-6. A multiroll sheeter. *(Courtesy of Rykaart, Inc., Hamilton Ohio.)*

desired physical dimensions of the product are normally specified, and the desired production rate must be met.

The specific volume or textural properties of the finished product are controlled by the shear and/or stretching that the product is exposed to during rolling. This shear may be defined by the specific energy input to the dough, as suggested by most of the literature on the mixing and sheeting of doughs (Kilborn and Tipples, 1974; Feillet et al., 1977; Watanabe and Nagasawa, 1968; Stenvert et al., 1979; Moss, 1980). This approach is useful, because a parameter such as specific energy input is measurable if the proper instrumentation is provided on the sheeting line.

Alternatively, the quality or physical properties of the dough produced by sheeting may be defined by the shear stress or total shear strain that the dough is exposed to during the sheeting operation. This has been suggested of Menjivar in Chapter 1 of this book. In these cases, the parameters are not directly measurable through instrumentation, but must be calculated through the use of measurable rheological properties and a mathematical knowledge of the fluid mechanics encountered during sheeting.

Figure 14-7. A cross roller. *(Courtesy of The Moline Company, Duluth, Minnesota.)*

Figure 14-8. The sandwich method. *(Courtesy of Rykaart, Inc., Hamilton, Ohio.)*

In either case, specifying a sheeting line adequately requires a reasonable mathematical model of the sheeting process. This model may be purely empirical, but to ensure that the predictions of the model can be extrapolated to a full-scale design, it is preferable that the model have a theoretical basis. Much of the remainder of this chapter will be devoted to the development, discussion, and application of fluid mechanical models of the sheeting operation.

Models of dough sheeting must be based on meaningful and *useable* models of the dough. Before attempting to understand the developments that follow, the reader should be familiar with the discussions of dough rheology of Menjivar and Lesche in chapters 1 and 12.

MODELS OF THE SHEETING OPERATION

Until recently (Levine, 1985; Drew et al., 1987; Drew et al., 1988), a search of the dough sheeting literature would provide little guidance in modeling the dough sheeting operation. A wealth of information, however, is available in the literature on plastics processing. This information is largely unexplored by researchers in the food processing and bakery industries. The plastics processing literature refers to the process under discussion as calendering rather than as sheeting or rolling. Any search of the literature should, therefore, begin with the subject of calendering. Many authors have analyzed the flow situation in calendering/sheeting (Middleman, 1977; Chong, 1968; Brazinsky et al., 1970; Elizarov et al., 1951; Bergen and Scott, 1951; Gaskell, 1950;

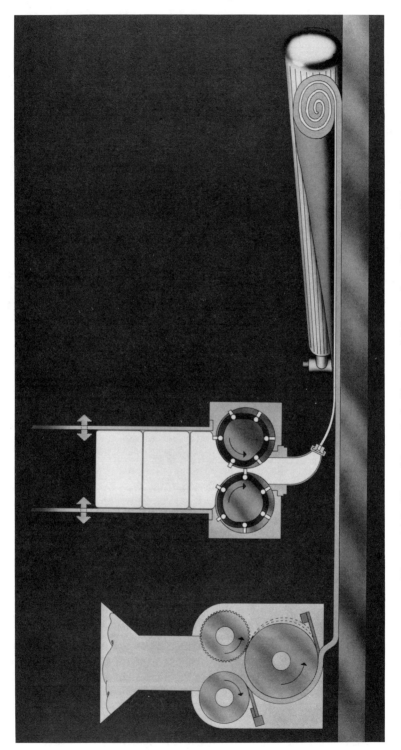

Figure 14-9. The roll-in method. (*Courtesy of Rykaart, Inc., Hamilton, Ohio.*)

Figure 14-10. A single fold. *(Courtesy of Rheon, Inc., Irvine, California.)*

McKelvey, 1962; Tadmor and Gogos, 1979; Ehrman and Vlachopoulos, 1975; Kiparissides and Vlachopoulos, 1976; Finston, 1951). This discussion reviews and extends these treatments. The material that follows is based on Middleman's text. Consider Figure 14-15, which defines the sheeting operation and sets up the coordinate system that is used in the equations that follow. These equations make some assumptions about dough—it is Newtonian (which is obviously a gross approximation) and incompressible. The following equations describe conservation of momentum and mass. (Note: All symbols in

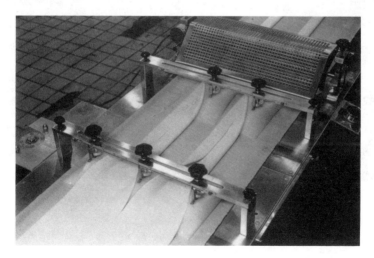

Figure 14-11. An envelope fold. *(Courtesy of Rheon, Inc., Irvine, California.)*

Figure 14-12. Cut and stack lapping. *(Courtesy of Rykaart, Inc., Hamilton, Ohio.)*

these equations, and those that follow, are defined in the nomenclature section at the end of this chapter.)

$$\rho\left(u_x\frac{\partial u_x}{\partial x} + u_y\frac{\partial u_x}{\partial y}\right) = -\frac{\partial p}{\partial x} + \mu\left(\frac{\partial^2 u_x}{\partial x^2} + \frac{\partial^2 u_x}{\partial y^2}\right) \tag{14--1}$$

$$\rho\left(u_x\frac{\partial u_y}{\partial x} + u_y\frac{\partial u_y}{\partial y}\right) = -\frac{\partial p}{\partial y} + \mu\left(\frac{\partial^2 u_y}{\partial x^2} + \frac{\partial^2 u_y}{\partial y^2}\right) \tag{14--2}$$

$$\frac{\partial u_x}{\partial x} + \frac{\partial u_y}{\partial y} = 0 \tag{14--3}$$

An analytical solution to these three simultaneous, partial differential equations is impossible, and numerical solutions are very difficult, if not impossible, to obtain. Any analysis that seeks to use them must begin with some simplifications. Doughs are very viscous, so one may safely conclude that the inertial terms (lefthand sides) in the momentum equations (Equations 14-1 and 14-2) are negligible. In addition, if the sheet thickness is assumed to be small compared to the roll diameter, then velocities and derivatives in the direction perpendicular to flow (y) are very much smaller than derivatives and velocities in the direction of flow (x). These simplifying assumptions, known as the lubrication approximation, are often applied to difficult problems in fluid mechanics. After the lubrication approximation is applied to the momentum equations, two complicated partial differential equations are reduced to Equations 14-4 and 14-5.

$$0 = -\frac{\partial p}{\partial x} + \mu\frac{\partial^2 u_x}{\partial y^2} \tag{14--4}$$

$$\frac{\partial p}{\partial y} = 0 \tag{14--5}$$

The second equation indicates that pressure is only a function of x (the horizontal coordinate). The first equation may be readily integrated subject to the following boundary conditions:

$$u_x = U \qquad \text{on } y = h(x) \tag{14--6}$$

$$\frac{\partial u_x}{\partial y} = 0 \qquad \text{on } y = 0 \tag{14--7}$$

The first boundary condition is the standard fluid mechanical assumption that fluids "stick" to solid boundaries. The second condition is obtained by examining the symmetry of the physical situation, which means that no shear

Figure 14-13. Fold lapping. *(Courtesy of Rykaart, Inc., Hamilton, Ohio.)*

stress is present on the dough at the midline between the rolls. In addition to
these boundary conditions, conservation of mass requires that the production
rate of the rolls is given by

$$Q = 2 \int_0^h u_x \, dy = 2h(x)\left[U - \frac{h^2(x)}{3\mu}\frac{dp}{dx}\right] \tag{14-8}$$

The simplified momentum equation may be integrated and combined with the
equation for the production rate of the rolls. After considerable mathematical
manipulation, the following results, Equations 14-9 and 14-10, are obtained.

$$\frac{dp'}{dx'} = \sqrt{\frac{18R}{H_0}} \frac{x'^2 - \lambda^2}{(1 + x'^2)^3} \tag{14-9}$$

$$p' = \sqrt{\frac{9R}{32H_0}}\left[\frac{x'^2(1 - 3\lambda^2) - 1 - 5\lambda^2}{(1 + x'^2)^2}x'\right.$$

$$\left. + (1 - 3\lambda^2)(\tan^{-1} x' - \tan^{-1} \lambda) + \frac{1 + 3\lambda^2}{1 + \lambda^2}\lambda\right] \tag{14-10}$$

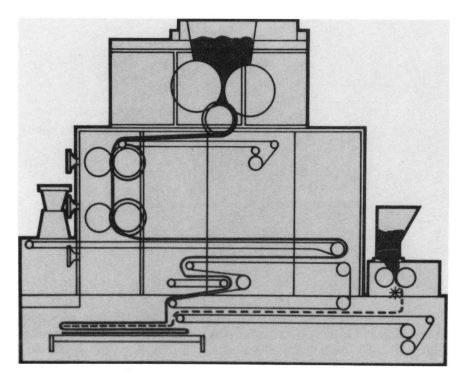

Figure 14-14. Combined lapping and laminating used for crackers. *(Courtesy of APV Baker, Inc., Grand Rapids, Michigan.)*

The production rate and λ are unknown. The literature (Middleman, 1977) makes a simple but important assumption in order to solve this problem. It is assumed, as illustrated in Figure 14-16, that at the point of separation of the dough from the roll, there are no velocity gradients in the dough. In other words, the dough leaves the rolls as if it were a rigid sheet. This assumption appears to be reasonable, since it implies that the pressure has fallen to zero. It also follows that the pressure gradient falls to zero. This assumption is critical if one desires a simple solution to the problem at hand. The possible errors associated with this assumption are discussed later in this chapter.

The assumptions about the discharge pressure and pressure gradient define the production rate of a dough. These assumptions lead to two important mathematical relationships.

$$Q = 2UH \qquad (14\text{--}11)$$

$$\lambda^2 = \frac{H}{H_0} - 1 \qquad (14\text{--}12)$$

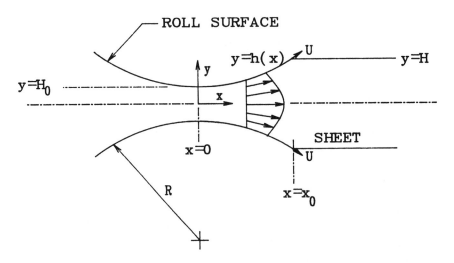

Figure 14-15. Definition of sheeting analysis.

The production rate of the sheeters is now known if the discharge thickness of the dough is known. The parameter λ is the distance between the nip and the point where the dough separates from the rolls. Since the pressure is zero at the point of separation, Equation 14-9 can be integrated to obtain a pressure distribution between the rolls.

Figure 14-17 illustrates the shape of the pressure distribution under the rolls. Note that the pressure is maximized, *not at the nip* as is frequently assumed, but at a distance λ upstream from the nip. Very far upstream from the nip, the pressure and pressure gradient are also zero. If the conditions very far upstream from the nip are placed into Equation 14-10, the following expression (Equation 14-13) is obtained:

$$p'(x' \to -\infty) = \sqrt{\frac{9R}{32H_0}} \left[\frac{1 + 3\lambda^2}{1 + \lambda^2} \lambda - (1 - 3\lambda^2)\left(\frac{\pi}{2} + \tan^{-1}\lambda \right) \right] \quad (14\text{–}13)$$

This equation may be solved numerically for the parameter λ_0. One obtains

$$\lambda_0 = 0.475 \quad (14\text{–}14)$$

or

$$\frac{H}{H_0} = 1 + \lambda_0^2 = 1.226 \quad (14\text{–}15)$$

The final discharge thickness and capacity of the system are now known. The discharge thickness is 1.226 times the gap between the rolls at the nip.

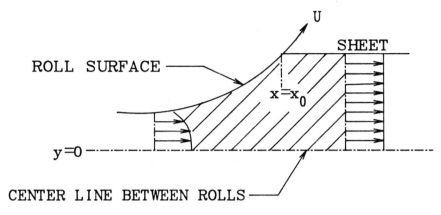

Figure 14-16. Separation region.

The production of dough is

$$Q = 2.452UH_0 \tag{14–16}$$

Now that the pressure distribution is known, the velocity distribution between the rolls may be obtained by Equation 14-17.

$$u'_x = \frac{2 + 3\lambda^2(1 - \eta^2) - x'^2(1 - 3\eta^2)}{2(1 + x'^2)} \tag{14–17}$$

Equation 14-17 is used to develop Figure 14-18, which shows the velocity distributions between the rolls at various positions. Several important points should be made about the velocity distributions illustrated: First, the velocity generally tends to increase as one moves from left to right. This increase indicates that elongation, or stretching, of the material is occurring. Second, velocity varies as one moves from the roll surface to the midline between the rolls. This variation indicates that significant shearing of the material is occurring. An order of magnitude analysis indicates that the shear strain caused by the velocity gradient is much higher than the elongational strain. Finally, at the feed side of the roll nip, some velocities are moving left to right, while others are moving right to left. This movement indicates that a circulation pattern, or mixing, is occurring in this region that may result in disruption of laminations near the surface.

If doughs were Newtonian, very important engineering conclusions could be drawn from the information about the pressure and velocity profile that was derived above. Doughs are non-Newtonian, however. One of the simplest non-Newtonian descriptions of doughs is that of a power law fluid. This description

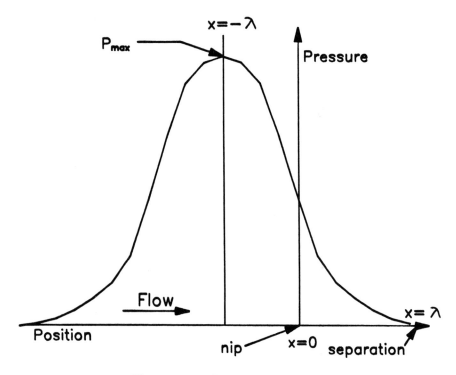

Figure 14-17. Roll pressure profiles.

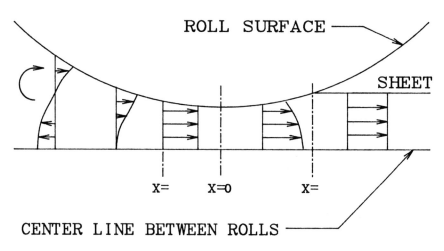

Figure 14-18. Velocity profile between rolls.

Table 14-2. Power Law Constants for Various Doughs

Material	percent H_2O	°C	K	n	Source
Wheat flour	43	43	4.45×10^3	0.35	Harper, 1981
Semolina flour	30	33	2.0×10^4	0.50	Harper, 1981
Hard wheat flour	30	38	3.33×10^4	0.41	Levine, 1983
Sweet dough	—	—	1.67×10^4	0.50	Levine, 1985

is written as

$$\tau_{xy} = K \left| \frac{\partial u_x}{\partial y} \right|^{n-1} \frac{\partial u_x}{\partial y} \tag{14-18}$$

Data exist (Harper, 1981) that suggest some food doughs can be described with such a simple non-time-dependent model. A partial list may be found in Table 14-2. The power law model allows the inclusion of shear thinning into the rheological behavior of the dough. When the derivations illustrated are repeated for power law fluids, the following equations for the pressure and velocity profiles are obtained:

$$p = K \left(\frac{U}{H_0} \right)^n \left(\frac{2n+1}{n} \right)^n \sqrt{\frac{2R}{H_0}} \int_{x'}^{\lambda} \frac{|\lambda^2 - x'^2|^{n-1}(\lambda^2 - x'^2)}{(1 + x'^2)^{2n+1}} \, dx' \tag{14-19}$$

$$u_x = U + \frac{1}{q} \left(\frac{1}{K} \frac{dp}{dx} \right)^{1/n} [y^q - h^q(x)] \tag{14-20a}$$

$$u_x = U - \frac{1}{q} \left(-\frac{1}{K} \frac{dp}{dx} \right)^{1/n} [y^q - h^q(x)] \tag{14-20b}$$

The value of λ and the discharge thickness of the dough are functions of the flow index n. These relationships are illustrated in Figure 14-19. The output of the rolls is given by Equation 14-11.

The relationships described by Equations 14-19 and 14-20 allow the estimation of several important and/or interesting parameters that describe the process. Some of the parameters are of interest to both the design engineer and the product scientist. The models allow estimation of the specific energy input of rolls to the dough from the velocity profile. This operation is useful for the specification of drive components and as an indicator of the effects of the rolls on the performance of the dough. In addition, the maximum pressure exerted on the dough can be estimated. This parameter is an indicator of the maximum compressive force that the dough experiences as it passes between the rolls. For a leavened dough one could hypothesize that it might

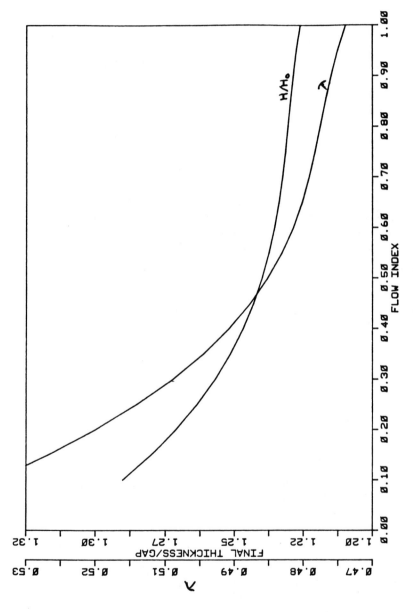

Figure 14-19. λ and final thickness/gap versus flow index.

be indicative of the extent, if any, of degassing the dough. The shear stresses and shear strain that the dough experiences as it passes between the rolls may also be estimated. If one reviews the discussions by Menjivar in Chapter 1 of this book, it is possible to reach the conclusion that one might wish to set limits on these parameters so that the strength of the gluten structure is not exceeded as the dough passes between the rolls.

Finally, the force exerted by the dough on the rolls may be estimated. Because of the pressure developed between the rolls, a force is developed that tends to bend the rolls and push them apart (Figure 14-20). In addition to needing to know these forces in order to properly specify roller bearings, the results of this bending can be quite significant for thin sheets. Significant product weight variations can result across the width of the roll. The closing forces also cause the roll shafts to deflect as illustrated in Figure 14-21. Product weight variations in the direction of flow can result from shaft deflections. Such deflections may occur when the feed viscosity changes suddenly due to small formula variations.

ESTIMATING SPECIFIC ENERGY INPUT

The specific energy input of rolls to the dough is defined by

$$\mathcal{W} = \dot{W}/QW \qquad (14\text{--}21)$$

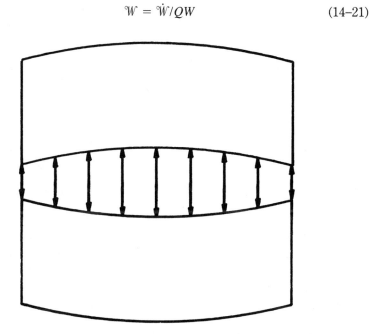

Figure 14-20. Bending of rolls under load (exaggerated).

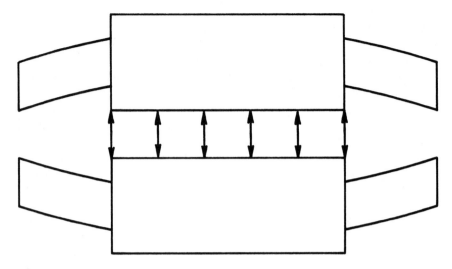

Figure 14-21. Bending of shafts under load (exaggerated).

The power consumption of the rolls is given by

$$\dot{W} = WU^2K\left(\frac{U}{H_0}\right)^{n-1}\sqrt{\frac{R}{H_0}}\,\mathscr{E}(n) \tag{14-22}$$

The function $\mathscr{E}(n)$ is illustrated in Figure 14-22. A fair approximation of the output of the rolls is given by

$$Q \approx 2.5UH_0 \tag{14-23}$$

The specific energy input is then

$$\mathscr{W} \approx \frac{UK}{2.5H_0}\left(\frac{U}{H_0}\right)^{n-1}\sqrt{\frac{R}{H_0}}\,\mathscr{E}(n) \tag{14-24}$$

Equation 14-24 is useful for exploring the qualitative aspects of sheeter design on specific energy input. The specific energy input increases with U^n, implying that increasing the production rate by increasing the roll speed has an effect on the quality of the dough. If a fixed total work input to the dough is desired, this suggests that less work may have to be done by prior dough processing, such as mixing. The specific energy input increases with the reciprocal of $H^{(n+1/2)}$. This relationship suggests that small changes in the gap between the rolls may have significant effects on specific energy input.

Finally, the equations demonstrate that specific energy input increases with the square root of the roll diameter. This relationship suggests that changing the diameter of the roll may have a significant effect on dough quality.

ESTIMATED ROLL CLOSING FORCE

The force exerted by the dough on the rolls is, as stated earlier, a consequence of the pressure developed between the rolls. The force may be calculated from

$$\frac{F}{W} = K\left(\frac{U}{H_0}\right)^n R\mathscr{F}(n) \tag{14-25}$$

The function $\mathscr{F}_0(n)$ is illustrated in Figure 14-22.

A number of conclusions can be drawn from Equation 14-25. The closing forces increase with U^n, and as a consequence, increasing line speed to increase production rate results in a greater deflection of the shafts and rolls. The closing forces increase with the reciprocal of H^n. As a result, small decreases in sheet thickness significantly increase the force and hence the roll deflection. This response becomes more and more significant as the sheet becomes thinner and thinner. The closing forces increase with the diameter of the roll. At first glance, it would seem that this increase would result in a greater deflection of the larger rolls. As will be illustrated, this is not the case, since the resistance of shafts and rolls to bending increases with a high power of diameter.

The maximum pressure exerted by the rolls on the dough is given by

$$p'_{max} = K\left(\frac{U}{H_0}\right)^n \sqrt{\frac{2R}{H_0}} \mathscr{P}(n) \tag{14-26}$$

The function, $\mathscr{P}(n)$ is presented in Figure 14-22.

Equation 14-26 allows a number of conclusions. The maximum pressure increases with U^n. The maximum pressure exerted by the rolls increases with the square root of the roll diameter. One can conclude that if pressure exerted on the dough causes the dough to degas, there will be a greater tendency for this to happen on large-diameter rolls. The maximum pressure increases with sheet thickness to the $n + 1/2$ power. If pressure plays a role in the degassing of dough, there is a greater tendency to degas thinner doughs.

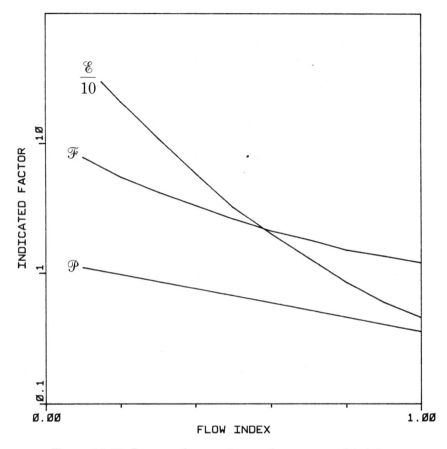

Figure 14-22. Pressure, force, and power factors versus flow index. (*After Middleman, 1977.*)

FINITE SHEET THICKNESS

All of the theoretical discussion presented above deals with feeding a dough of infinite thickness into the sheeter. This is of course only an approximation. In order to deal with a finitely thick sheet being fed to the rolls, the derivations presented above must be modified.

Instead of assuming that the pressure falls to zero at some infinite distance upstream from the nip, analysis of a finite feed sheet thickness begins with the assumption that the pressure is zero at the first point of contact between the dough and the rolls. The distance upstream of the nip where the first contact occurs is given by

$$x_f' = \left(\frac{H_f}{H_0} - 1\right)^{1/2} \tag{14-27}$$

Equation 14-26, which describes the pressure between the rolls, is now modified so that the integration occurs from the point of first contact to the discharge, rather than from $-\infty$ to the discharge.

The pressure is still assumed to be zero at the point of discharge from the rolls. This assumption implies that the following equation must be solved for the discharge point (discharge thickness):

$$0 = \int_{-x_f'}^{\lambda} \frac{|\lambda^2 - x'^2|^{n-1}(\lambda^2 - x'^2)}{(1 + x'^2)^{2n+1}}\, dx' \tag{14-28}$$

This problem has been solved numerically (Brazinsky et al., 1970).

The discharge point (thickness) is a function of the flow index. The relationships among the discharge thickness, flow index, and feed thickness are presented in Figure 14-23. The production rate of the rolls is now given by

$$Q = Ut_0 \tag{14-29}$$

Normally, the reduction ratio encountered in a pair of rolls is approximately 2. To a fair approximation (Levine, 1985) this leads to

$$Q/(4NR^2) \approx 3.3(H_0/R) \tag{14-30}$$

This relationship is illustrated in Figure 14-24.

Likewise, for a finite feed sheet thickness, the calculations for power consumption of the rolls, closing force, and maximum pressure must be modified. For a finite sheet, Equations 14-22, 14-25, and 14-26 become, respectively,

$$\dot{W} = WU^2 K \left(\frac{U}{H_0}\right)^{n-1} \sqrt{\frac{R}{H_0}} \mathscr{E}(n)C_w \tag{14-31}$$

$$\frac{F}{W} = K \left(\frac{U}{H_0}\right)^{n} R\mathscr{F}(n)C_f \tag{14-32}$$

$$\frac{p'_{max}}{\sqrt{2R/H_0}} = \left(\frac{2n+1}{n}\right)^{n} \int_{-\lambda_0}^{\lambda_0} \frac{(\lambda_0^2 - x'^2)^n}{(1 + x'^2)^{1+2n}}\, dx' = \mathscr{P}(n)C_p \tag{14-33}$$

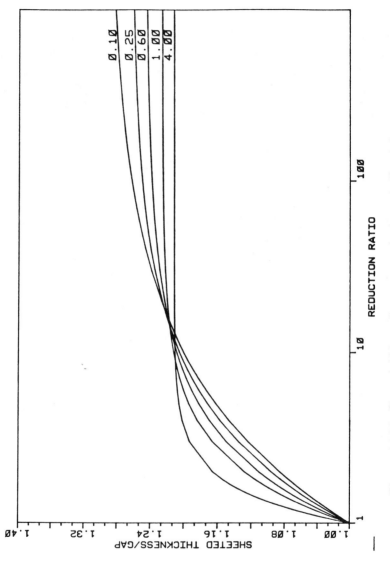

Figure 14-23. Sheeted thickness/gap versus reduction ratio (flow index as a parameter). (*After Brazinsky et al., 1970.*)

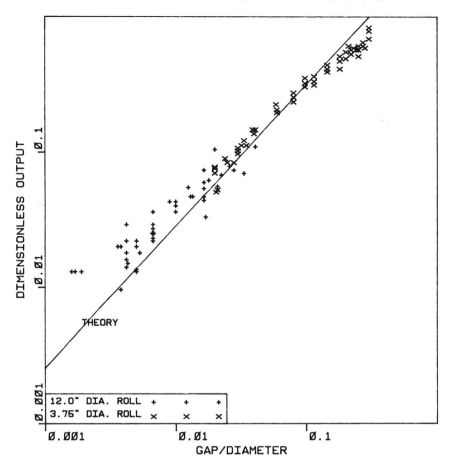

Figure 14-24. Sheeting capacity of rolls. (*After Levine, 1985.*)

The equations for an infinite sheet have been modified by the addition of correction factors (Figs. 14-25, 14-26, 14-27) which incorporate the flow index and the reduction ratio (feed thickness/roll gap). With this modification, the qualitative conclusions about roll speed, roll diameter, and roll gap still hold. The inclusion of finite feed thickness results in decreases in the closing force, power, and maximum pressure.

Both the shapes of the correction factors and the forms of the equations in which they are used in Equations 14-31 through 14-33 indicate that the pressure developed by the rolls and the energy input to the dough depend on how many steps are used to produce a given reduction. This can be illustrated by two examples.

First, consider reducing a dough sheet of 32 mm thickness through a roll gap of 2 mm. This reduction results in a final sheet somewhat greater than 2

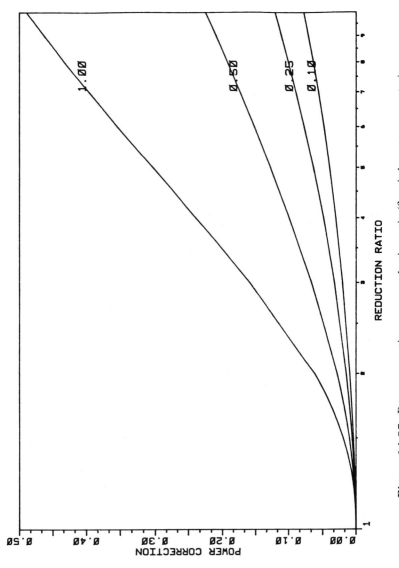

Figure 14-25. Power correction versus reduction ratio (flow index as a parameter).

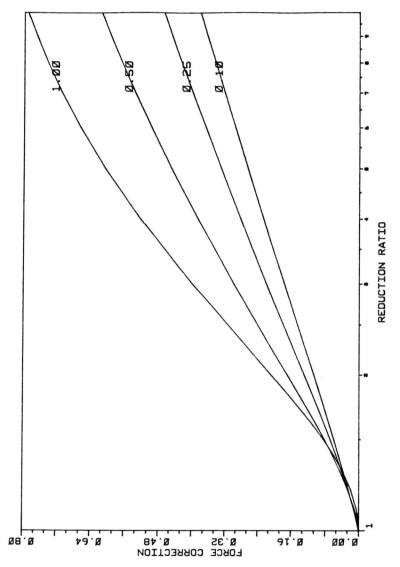

Figure 14-26. Force correction versus reduction ratio (flow index as a parameter).

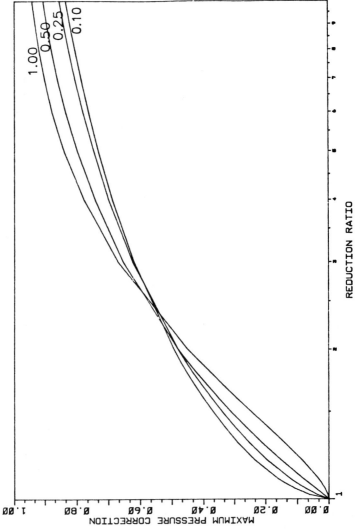

Figure 14-27. Pressure correction versus reduction ratio (flow index as a parameter).

544

mm thick and can be accomplished in a number of different ways. The simplest reduction is to feed the 32 mm sheet between a pair of rolls separated by 2 mm. This approach may or may not be possible. Many doughs tear as a result of such a severe reduction. Instead of a single pair of rolls, two or more pairs of rolls could be used, with the final pair having a gap of 2 mm. This setup would produce a sheet of approximately the same final thickness. There are, however, important differences. Consider Table 14-3, which shows a geometric progression in the reduction of the dough.

Assume that the flow index of the dough is 0.5. Equations 14-31 and 14-33 can be applied to these reduction profiles for an arbitrary viscosity and roll diameter to obtain specific energy input and maximum pressure. These results are given in Tables 14-4 and 14-5. It is clear that increasing the number of rolls reduces the maximum pressure that the dough is exposed to.

Likewise, the total energy input decreases with the number of rolls used to accomplish a reduction. Of course, economic considerations limit the number of rolls that can be realistically used. One approach is taken by sheeters, which use many rolls rotating on a conveyor. The rolls effectively make many infinitesimal reductions in the thickness of the dough. The successive reductions minimize the shear stresses, pressure, and energy input required to obtain a given reduction. In this sense these systems may truly represent a gentler sheeting system.

Through the combination of Equations 14-9, 14-17, and 14-18, and the information provided in Figure 14-23, the reader may check to see that the maximum shear stress exhibits the same behavior.

This type of analysis can be carried further by considering a set of four pairs of rolls. An unlimited number of reduction profiles can be used within such a system. Table 14-6 shows two more possibilities. Condition A is the same as that explored above (i.e., every roll uses a reduction ratio of approximately 2). Condition B uses higher reduction ratios during the initial reduction of the dough. Condition C uses higher reduction ratios during the final reduction of the dough. Application of Equations 14-31 and 14-33 demonstrates that condition B is gentler than condition A, while condition C is less gentle than condition A. The meaning of gentle becomes clear when one considers Tables

Table 14-3. Roll Gaps for Various Reduction Plans

	Roll Gap (mm)			
Number of Rolls	*#1*	*#2*	*#3*	*#4*
1	2	—	—	—
2	8	2	—	—
3	12.7	5.1	2	—
4	16	8	4	2

Table 14-4. Maximum Pressure for Various Reduction Plans

Number of Rolls	Relative Pressure			
	#1	#2	#3	#4
1	1.00	—	—	—
2	0.098	0.784	—	—
3	0.038	0.157	0.619	—
4	0.018	0.042	0.178	0.503

Table 14-5. Maximum Energy for Various Reduction Plans

Number of Rolls	Relative Energy				
	#1	#2	#3	#4	Total
1	1.000	—	—	—	1.000
2	0.042	0.334	—	—	0.376
3	0.010	0.016	0.162	—	0.188
4	0.004	0.012	0.033	0.094	0.143

14-7 and 14-8, which show the relative pressure and energy for the three plans.

Condition B minimizes the maximum pressure and the specific energy input. Condition C maximizes the maximum pressure and the specific energy. A similar statement could be made for the maximum shear stress and roll separating forces. As a general rule, in order to minimize the effects of sheeting on a dough, the reduction ratio should be maximized when the dough is thickest and minimized when the dough is thinnest.

Study of examples of the effects of the reduction path makes it clear that an optimum reduction path is likely to exist for any particular dough. That optimal path is defined by the desired final properties of the dough, so no universally optimal reduction path can be recommended. Every new product development must attempt to define some type of limits on the acceptable reduction profile that the design engineer can use in designing the sheeting line.

FINAL SHEET THICKNESS

Normally, the final desired sheet thickness for the dough has been specified. To design a sheeting system, some knowledge of the relationship between sheeter design and final sheet thickness obviously is needed. The fact that the rheological models of the sheeting system do not adequately predict the

Table 14-6. Three Different and Reduction Profiles

Condition	Roll Gap(mm)			
	#1	#2	#3	#4
A	16	8	4	2
B	12	6	3	2
C	24	12	6	2

Table 14-7. Maximum Pressure for Three Reduction Plans

Condition	Relative Pressure			
	#1	#2	#3	#4
A	0.018	0.042	0.178	0.503
B	0.035	0.065	0.116	0.326
C	0.005	0.023	0.097	0.687

final sheet thickness has been alluded to. All of these models suggest that the final sheet thickness is on the order of 1.0–1.2 times the spacing at the nip of the rolls. Actual data (Levine, 1985) indicate that "relaxed" doughs may have a much greater thickness than this. These results may be summarized by the following simple equation:

$$t_0/2H_0 \approx 0.06(H_0/R)^{-0.58} + 1 \qquad (14\text{--}34)$$

Equation 14-34 states that the "spring-back," the ratio of final thickness to nip spacing, increases significantly as the ratio of roll diameter to gap increases. The "spring-back" may have a slight tendency to increase as line speed is increased.

Anyone who has observed the behavior of a sheet of dough leaving a sheeter will recognize that the dough thickness is still changing. These changes must be caused by residual stresses in the dough. None of the current models of sheeting take this factor into account. In fact, one of the key assumptions they make is that there are no stresses at the point of separation. If this assumption is removed so that there are residual stresses at the point of separation, the final thickness will be greater. No consideration of alternative boundary conditions has yet been published, however. This is a valuable area for future investigation.

From a practical point of view, the final thickness of the dough may be established by a mass balance taking into account the input thickness of the dough and the input and output belt speeds. This relationship is stated succinctly by the following equation:

Table 14-8. Maximum Pressure for Three Reduction Plans

Condition	Relative Energy				Total
	#1	#2	#3	#4	
A	0.004	0.014	0.033	0.094	0.143
B	0.012	0.018	0.021	0.031	0.082
C	< 0.001	0.007	0.018	0.222	0.247

$$t_0 = t_i U_i / U_0 = Q/U_0 \qquad (14\text{--}35)$$

This equation states that once the volumetric flow rate is established, the output thickness depends, within limits, only on the speed of the discharge belt.

Equation 14-35 is not valid for every possible speed. The limits of this equation are defined by the tendency of the dough to buckle at low output belt speeds, and to slip on the output belt at excessive speeds.

ROLL AND SHAFT DEFLECTION/DESIGN

Estimating the degree of bending of the rolls due to the closing forces is a straightforward problem in evaluating the strength of materials. The equations governing the deflection of an object under a uniform load are (Shigley, 1977)

$$\delta = \frac{Z(F/W)(2WZ^2 - Z^3 - W^3)}{2YEI} \qquad (14\text{--}36)$$

The moment of inertia of the roll is calculated with

$$I = \frac{\pi}{4}(R^4 - R_i^4) \qquad (14\text{--}37)$$

The modulus of elasticity of the roll is a function of the construction materials of the roll. Estimates may be found in standard textbooks or in handbooks on machinery design. The rolls typically are constructed of steel. The modulus of elasticity of steel is approximately 3×10^9 kPa.

Both rolls deflect, so deviations in thickness are twice that calculated from Equation 14-36. Percent product weight deviation between the edge of the roll and center of the roll is given by

$$d = \frac{5FW^3}{38YEIH_0} \times 100\% \qquad (14\text{--}38)$$

If a cutting pattern is defined, Equation 14-36 can be used to determine the weight deviation of products cut from any point across the roll.

This problem is the cause of the designer's natural tendency to increase the diameter of the sheeting rolls as the sheeting process is scaled up. As the process is scaled up for increased production, the speed and width of the line increase. Increasing speed increases the separating forces per unit width (Equation 14-32), and the deflection increases with the first power of this force and the fourth power of the width of the roll (Equation 14-36). Increasing the diameter of the roll increases the roll stiffness (moment of inertia, Equation 14-37), which reduces the deflection of the rolls. Every designer is aware of this solution for deflection, but is not necessarily cognizant of effects of roll diameter on energy input and other parameters, and the possible effects on product quality.

The effects of roll diameter on deflection are best illustrated by example. First, consider a pilot line at some arbitrary speed. Assume that the full-scale production facility operates at the same speed, but is five times as wide in order to achieve a fivefold increase in production rate. If the deflection of the pilot plant rolls were 0.003 mm, a value that would be unnoticeable and almost impossible to measure, the deflection of the plant rolls would be 1.87 mm. The larger deflection would certainly be observable when sheeting thin doughs, such a tortillas, whose thickness is of the same magnitude as the deflection! Now, consider increasing the diameter of the rolls by a factor of 4 (75–300 mm). The separating forces on the rolls would increase by a factor of 4, but the stiffness of the rolls would increase by a factor of 256. The resultant deflection for the plant rolls would be 0.03 mm, which is still significantly greater than for the pilot plant sheeter, but is probably insignificant when compared to the target thickness of the sheet.

The shafts connected to the roll tend to be deflected because of the load on the rolls. Exact determination is impossible, without a very detailed analysis of the bearing design. As a first approximation, the bearing may be considered to be a fixed support. Each shaft supports half the total closing force. For this situation the deflection of the shaft is given by

$$\delta = \frac{(F/2)L^3}{3EI} \qquad (14\text{--}39)$$

The moment of inertia of the shaft is given by

$$I = \frac{\pi}{4}R^4 \qquad (14\text{--}40)$$

The shaft length should be estimated as the distance from the center of the bearing to the side of the sheeting rolls. Equation 14-39 suggests the obvious;

that is, the bearing should be located as close to the roll as is possible in order to minimize deflection.

A final concern for bearing/shaft design involves the performance of the sheeter. While this problem is not directly related to the forces on the rolls, it should not be overlooked.

Samples of cut products emerging from a sheeting system often exhibit a piece to piece weight variation. Careful analysis reveals that this variation is cyclical, and is not related to the loading of the roll or variations in the design of a rotary cutter. It is a consequence of the methods to assemble the rolls, shafts, and bearings. The problem is particulary noticeable in the production of very thin products.

The problem arises from the natural accumulation of tolerances in the assembly of the sheeting rolls, shafts, and bearings. These tolerances result in a very slight eccentricity in the path the roll surface follows during one rotational cycle. This is illustrated in Figure 14-28. As rotation progresses, the eccentricity of the rolls results in a change in the gap. The figure greatly exaggerates the real situation. Figure 14-29 illustrates the kind of thickness variations that can be seen if a material is sheeted with eccentric rolls.

This problem will obviously occur if the shafts are not perfectly centered on the rolls. This situation can be easily avoided by constructing the shafts and rolls as one piece. That is to say that oversized shafts and rolls should be welded together and then turned or ground to the final desired diameters. This practice will ensure that they are concentric.

The second source of the eccentricity problem is not so readily avoided. In order to readily assemble a shaft into a bearing, a tolerance is built into

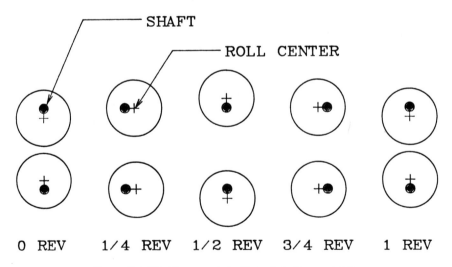

Figure 14-28. Exaggerated effect of shaft eccentricity.

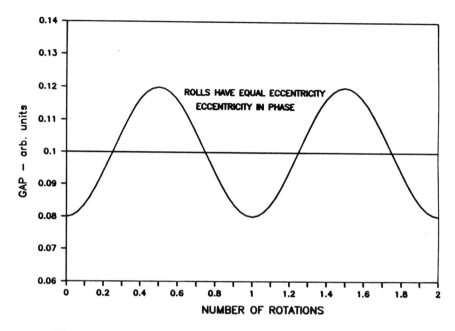

Figure 14-29. Variation of gap with rotation (eccentricity = 0.1 gap).

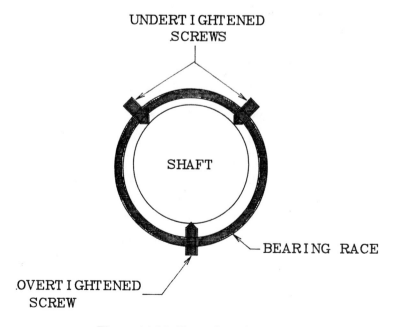

Figure 14-30. Eccentric shaft mounting

the assembly. That is to say that the diameter of the shaft may be several thousandths of an inch smaller than the diameter of the bearing race. The bearing uses set screws to "lock" the shaft into the bearing. The centering of the shaft in the bearing is controlled by how the set screws are tightened onto the shaft, as is illustrated in Figure 14-30. Some reflection will reveal that it is virtually impossible to perfectly center the shaft, so some eccentricity must be expected. Self-centering bearings should be used to avoid this problem.

SUMMARY

This chapter has presented a discussion of problems and theories that must be confronted by those responsible for the design and specification of sheeting and laminating systems. The limited literature on the subject and the mathematical complexity of published material in related fields make it difficult to cover the subject completely in a single chapter. It is hoped that this chapter has significantly increased the reader's knowledge and will stir enough interest in an important and complicated problem to encourage further research and publication in this area.

NOMENCLATURE

C_f	force correction for finite feed sheet (Figure 14-29)
C_p	pressure correction for finite feed sheet (Figure 14-30)
C_w	power correction for finite feed sheet (Figure 14-28)
d	percent product weight deviation
E	modulus elasticity of roll or shaft
$\mathscr{E}$	power function in Figure 14-25
F	closing force
$\mathscr{F}$	force function in Figure 14-25
H	half the thickness at the discharge point
H_f	half of input thickness
H_0	half gap
$h(x)$	vertical coordinate of roll surface
I	moment of inertia of roll or shaft

K	power law consistency
L	length of shaft
N	rotational rate
n	power flow index or number of layer formed by lapping
p	pressure
p'	dimensionless $= (p/K)(H_0/U)^n$
$\mathscr{P}$	pressure function in Figure 14-25
Q	volumetric flow rate per unit roll width
q	exponent used in Equations 14-20a and 14-20b $= n/(1+n)$
R	roll radius
R_i	inner radius of a hollow roll
t_i	input thickness
t_0	output thickness
U	roll surface velocity
U_i	input belt velocity
U_0	output belt velocity
u_x	velocity in horizontal direction
u_y	velocity in vertical direction
u'_x	dimensionless horizontal velocity $= u_x/U$
W	roll width
$\mathscr{W}$	specific energy
$\dot{\mathscr{W}}$	power input
x	horizontal direction
x'	dimensionless x direction $= x/\sqrt{(2RH_0)}$
x'_f	dimensionless horizontal coordinate of the point of first contact
y	vertical direction
z	distance measured from the end of the roll

Greek Symbols

δ	deflection
η	dimensionless vertical coordinate of roll surface $= y/h(x)$
λ	distance from nip to discharge
μ	viscosity
ρ	density
τ_{xy}	shear stress

REFERENCES

Behnke, J., Keller, R., Totino, R., and Westover, J. 1979. Fried dough product and method. U.S. Patent #4,170,659.

Bergen, J. T., and Scott, B. W. 1951. Pressure distribution in the calendering of plastic materials. *J. Appl. Mechanics* **18**:101–106.

Brazinsky, I., Cosway, H. F., Valle, C. F., Clark Jones, R., and Story, V. 1970. A theoretical study of liquid-film spread heights in the calendering of Newtonian and power law fluids. *J. Appl. Polym. Sci.* **2**:2771–2784.

Chong, J. S. 1968. Calendering of thermoplastic materials. *J. Appl. Polym. Sci.* **12**:191–212.

Curry, A., Levine, L., and Rose, D. 1976. Method of forming rippled chip-type products. U.S. Patent #2,959,517.

Drew, B., Levine, L., and Ramkrishna, V. 1988. Numerical solution to a problem in the flow of viscoelastic fluid between rotating cylinders. Paper read at the National Meeting of the Society for Industrial and Applied Mathematics, 13–15 July, Minneapolis, MN.

Drew, B., Levine, L., Ramkrishna, V., and Clemmings, J. 1987. Comparison of mathematical models of dough sheeted through rolls. Paper read at the National Meeting of the American Institute of Chemical Engineers. 17–19 August, Minneapolis, MN.

Ehrman, G., and Vlachopoulos, J. 1975. Determination of power consumption in calendering. *Rheol. Acta* **14**:761–764.

Elizarov, V. I., and Sirazetdinov, T. K. 1973. Non-Newtonian fluid flow in the gap between rotating cylinders. *Isvestiya, VUZ. Aviatsionnaya Teknika*, **16**(4):10–16.

Feillet, P., Fevre, E., and Kobrehel, K. 1977. Modification of durum wheat protein during pasta dough sheeting. *Cereal Chem.* **54**(3):580–587.

Finston, M. 1951. Thermal effects in the calendering of plastic materials. *J. Appl. Mech.* **18**(12):12–16.

Gaskell, R. E. 1950. The calendering of plastic materials. *J. Appl. Mech.* **17**:334–336.

Harper, J. M. 1981. *Extrusion of Foods*, vol. 1: Boca Raton, FL: CRC Press.

Hummel, C. 1950. *Macaroni Products*. London: Food Trade Products.

Kiparissides, C., and Vlachopoulos, J. 1976. Finite element analysis of calendering. *Polym. Eng. Sci.* **16**:712–719.

Kilborn, R. H., and Tipples, K. H. 1974. Implications of the mechanical development of bread dough by means of sheeting rolls. *Cereal Chem.* **51**(5):648-657.

Levine, L. 1983. Estimating output and power of food extruders. *J. Food Proc. Eng.* **6**:1–13.

Levine, L. 1985. Throughput and power of dough sheeting rolls. *J. Food Proc. Eng.* **7**:223–228.

Matz, S. A. 1968. *Cookie and Cracker—Technology.* Westport, CT: AVI.

McKelvey, J. M. 1962. *Polymer Processing.* New York: Wiley.

Middleman, S. 1977. *Fundamentals of Polymer Processing.* New York: McGraw-Hill.

Moss, H. J. 1980. Strength requirements of doughs destined for repeated sheeting compared with those of normal doughs. *Cereal Chem.* **57**(3):195–197.

Shigley, J. E. 1977. *Mechanical Engineering Design.* New York: McGraw-Hill.

Stenvert, N. L., Moss, R., Pointing, G., Worthington, G., and Bond, E. E. 1980. Bread production by dough rollers. *Bakers Digest* **53**(4):22–27.

Tadmor, Z., and Gogos, C. G. 1979. *Principles of Polymer Processing.* New York: Wiley.

Watanabe, Y., and Nagasawa, S. J. 1968. A study of the rheological properties of noodles by means of relaxation testing. *J. Food Sci. Tech. Nihon Shokukin Kogyo Dakkai-Shi* **15**(10):466–468.

Chapter 15

Practical Texture Measurements of Cereal Foods

Malcolm C. Bourne

Chapter 6 outlined the principles of food texture measurement, how to select the most promising test principle, and, finally, how to go about selecting a suitable test instrument. This chapter describes how to perform specific classes of tests based on the test principle involved. The descriptions will be representative, because comprehensive coverage of every variation of each class of test is not possible with the limited space here.

DEFORMATION

The deformation test measures the distance that a food is compressed under a standard compression force, or the force required to compress a food a standard distance. This test simulates the gentle squeezing by the hand that customers apply to bread and many other food items in the supermarket. The sensory description of this test is usually feeling "softness" or "firmness."

The first official American Association of Cereal Chemists (AACC) test of this type used the Baker Compressimeter (Platt and Powers, 1940). Recently, the AACC developed a standard method (74-09, first approved on Oct. 8, 1986) for measuring bread firmness by using the deformation principle (AACC, 1983). Briefly, the test is conducted as follows for white pan breads:

1. A Universal Testing Machine, fitted with a 36 mm diameter aluminum plunger, is set to travel with a crosshead speed of 100mm/min and a chart speed of 500 mm/min. Full-scale force is usually 1 kg, but this may be increased or decreased depending on the softness of the product. Method 74-09 presently specifies that the Instron machine be used for this test. In fact, other universal testing machines with equivalent capability (such as the Lloyd) would be equally satisfactory.

2. One slice of bread 25 mm thick or two slices, each 12.5 mm thick, are used. The slices can be cut by hand or by machine. The two or three end slices are not used, and the crust is not removed from the slices that are compressed.

3. The center of the bread slice is compressed 40%; that is, from 25 mm height to 15 mm height.

4. The force to achieve a 25% compression (6.2 mm actual compression = 31 mm along the chart) is read off the chart (see Figure 15-1). The force may be expressed in kilograms or newtons force (1 N = 101.9716 g). Since the kilogram is a measure of mass, not force, using the true S.I. unit of force, the newton (N) is recommended.

Presently, the AACC method describes two ways to measure firmness:

- CFV I is measured 31 mm from the point at which the force curve moves off the base line.

- CFV II is measured by extrapolating a straight line from the initial slope of the curve to the abscissa (x axis), and measuring 31 mm from this point.

The tail between the start of CFV I and CFV II represents incomplete contact of the metal compression plate with the bread. This incomplete contact may occur when a bread surface is not exactly horizontal or when the aluminum plunger is misaligned. In many cases the two readings will be almost equal, but on some occasions it may be large. This author prefers to use CFV II as the more reliable reading, because it eliminates errors caused by any initial incomplete contact between the plunger and the bread surface.

This test can be used for other leavened goods, such as cakes, donuts, and biscuits, and other types of loaf bread with suitable adaptations. Adaptation depends on the geometry and nature of the product and may include changing the degree of compression or thickness of the test sample, removing the crust, or changing the size of the compression plunger. Walker and co-workers (1987) reported on the use of this procedure on cake.

Table 15-1 shows the mean values from six laboratories, which measured the firmness of two different bread formulations over a period of seven days by using AACC method 74-09. Note the consistent pattern of increasing firmness with time. For the control, the firmness on day 7 was 95% greater than on day 1, while for SSL the firmness was 90% greater on day 7. The firmess on day 4 is 51–61% higher than on day 1.

Table 15-1 also shows that the sodium stearoyl lactylate (SSL) formula bread that contained 0.5% of the flour weight of SSL was always substantially lower in firmness than the control formula, which contained no added SSL.

One more notable feature in Table 15-1 is that the CFV II values are always greater than the CFV I values, the differences ranging from 8 to 27 grams.

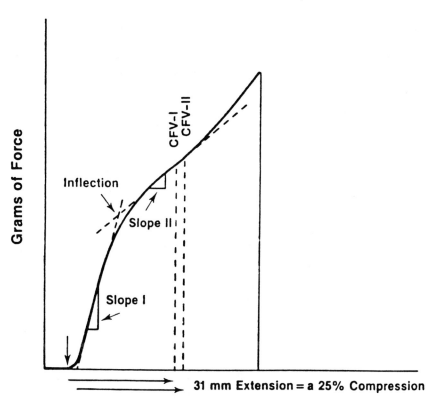

Figure 15-1. Firmness curve for white pan bread compressed in a universal testing machine. (*From AACC, 1983; copyright ©1983 by the American Association of Cereal Chemists.*)

Table 15-1. Bread Firmness by AACC Method

	CFV I		*CFV II*	
Day	*Control*	*SSL*	*Control*	*SSL*
1	291	229	301	237
4	458	348	485	358
7	567	434	587	449

Source: Data from Baker and Ponte, 1987.
Notes: Firmness is expressed in grams force using a standard sponge dough bread formula. SSL is sodium stearoyl lactylate added at 0.5% of the flour weight.

A careful study of Figure 15-1 shows that this difference is to be expected. The vertical line for CFV I is always to the left of the vertical line for CFV II, and since the force–compression curve always has a positive slope, the CFV I intercept on the force curve will always be less than the CFV II intercept.

SNAPPING–BENDING

The snapping–bending test measures the force needed to bend and snap brittle foods such as cookies and crackers. The principle was first described as a Shortometer by Davis (1921).

The sample is laid across two vertical rails that support it in a horizontal position. A third bar mounted above the sample and equidistant between the supporting rails is lowered until the sample breaks and the force is measured. The test can be satisfactorily performed in any universal testing machine fitted with a suitable triple bar assembly. The Structograph (C. W. Brabender Co.) is designed specifically to perform this test.

The force required to snap the specimen depends primarily on two factors:

1. The strength of the sample
2. The dimensions of the sample

Bruns and Bourne (1975) established the validity of the following equation for symmetrical food bars with rectangular cross section:

$$F = \frac{2}{3}\sigma_c \frac{bh^2}{L} \tag{15–1}$$

where

F = snapping force
σ_c = the breaking stress which is the material characteristic of the sample
b = width of the bar
h = thickness of the bar
L = length of the bar (distance between the supporting rails)

Since the breaking stress is the property that needs to be measured, this equation can be rearranged to give:

$$\sigma_c = \frac{3}{2}F\frac{L}{bh^2} \tag{15–2}$$

The length L of the specimen can be controlled by adjusting the distance between the supporting rails. For some products that are cut by machine, such as crackers, the width and height (b and h) may be fairly constant, in which case F is directly proportional to the breaking strength of σ_c. In some cases the height and width can be standardized by cutting the specimen to size. If b or h vary and cannot be standardized, however, it is advisable to use Equation 15-2 to correct for dimensional variations. Note that the breaking strength σ_c is proportional to the square of the thickness, and hence a small change in thickness can cause a large change in snapping force F.

For bars with cylindrical cross sections such as pretzels and bread sticks the snapping equation becomes

$$F = \sigma_c \frac{\pi R^3}{L} \qquad (15\text{--}3)$$

where R is the radius of the specimen. Note that the snapping force F is proportional to the cube of the radius, hence the force is strongly affected by small changes in specimen diameter. Loh (1985) gives an example of the use of Equation 15-3 on pretzel sticks. The snapping forces of pretzels A and B were 4.6 kg and 4.2 kg, respectively. However, the radius of pretzel A was 7.0 mm and pretzel B was 5.3 mm, and the calculated breaking stress (σ_c in 10^6Pa was 2.1 for A and 4.5 for B. Pretzel B is more than twice as strong as pretzel A, even though the snapping force for the larger diameter A is ten percent higher than for B.

TEXTURE PRESS

The texture press, formerly known as the Kramer Shear Press, is a versatile and well-known instrument consisting of a metal box with ten 1/8 inch-wide slits in the bottom. A set of ten 1/8 inch-wide blades moves down through the box, compressing, shearing, and extruding the food. Although originally designed for measuring the firmness of fruits and vegetables, it has been used for some bakery products.

Zabik's group reported texture press data for butter cakes (Gruber and Zabik, 1966), angel cake (Brown and Zabik, 1967), and sugar snap cookies (Zabik et al., 1979). Stinson and Huck (1969) found that the texture press gave a correlation coefficient of $r = 0.92$ with sensory evaluation of pastry tenderness. Matthews and Dawson (1963) used this instrument to measure the texture of pastry and chemically leavened biscuits.

CUTTING SHEAR

The cutting shear test measures the force required to cut through a product. Matsuo and Irvine (1969, 1971) used a simulated tooth to deform and cut through single strands of cooked spaghetti.

Voisey's group in Ottawa designed a system in which ten strands of cooked spaghetti are simultaneously cut through by ten 1.5 mm thick blades. This system can be mounted in a universal testing machine or in an Ottawa Texturometer. The Ottawa group reported promising results for measuring the firmness of cooked spaghetti with this device (Voisey and Larmond, 1972, 1973; Voisey et al., 1978). These investigators reported correlation coefficients between maximum force and sensory firmness ranging from $r = 0.41$ to $r = 0.69$ for 20 samples of cooked spaghetti (Voisey et al., 1978). They also reported that the time of cooking, the time that elapses between cooking and testing, and the crosshead speed each has a great effect on the results, hence these variables must be closely controlled to ensure reproducible results. The investigators did not report on the effects of temperature at the time of the test, although this must surely be another important variable that needs to be controlled.

THE PUNCTURE TEST

The puncture test measures the force required to push a probe into a food. The probe is usually cylindrical, but other shapes are occasionally used. This test is one of the easiest and fastest to perform. A number of simple commercial instruments have been developed that use the puncture principle. Universal testing machines can also be set up to perform puncture tests.

The puncture principle is widely used on fruits, and it is also used on gels, fats, vegetables, and some dairy and meat products. The force required to make the punch penetrate into the food is generally considered to be a measure of its hardness or firmness. In view of the versatility of this test, it is surprising that it has not been applied more to cereal products. The author believes that the puncture principle has the potential for much wider application to cereal foods. Table 15-2 lists the puncture force found for a number of commercial cereal products.

The puncture test operates on the assumption of a semi-infinite body geometry; that is, the size of the sample is so much larger than the punch that no edge, end, or bottom effects influence the result. Even a small item can be punctured if a small enough punch is selected.

Bourne (1966) showed that the puncture force is proportional to both the area and the perimeter of the punch and to two different textural properties of the food. The relationship is described by the following equation:

$$F = K_c A + K_s P + C \tag{15–4}$$

where

F = puncture force
K_c = compression coefficient of the food
K_s = shear coefficient of the food

Table 15-2. Puncture Force of Some
Commercial Cereal Foods

Item	Punch Diameter	
	1.02 mm	*2.36 mm*
Anisette toast	4.0	7.6
Anisette sponge	0.3	1.2
Biscotti	2.8	8.9
Breadsticks	2.5	6.3
Cheese snack sticks	6.9	—
Egg biscuits	0.7	1.9
Ginger snaps	13.5	27.8
Graham crackers	5.0	10.5
Molasses cookies	0.5	0.8
Ritz crackers	3.1	5.5
Shortbread	5.5	12.8
Sugar cookies	4.8	12.4
Vanilla wafers	7.0	13.8
White Melba toast	7.4	—

Source: Data from M.C. Bourne, unpublished.
Notes: Puncture force measured in newtons. Each
figure is the mean of 20–32 tests. The cheese
snack sticks and Melba toast fractured under
the 2.36 mm diameter punch, which voided
the test result.

A = punch area
P = punch perimeter
C = a constant

The validity of this equation has been experimentally verified for a number of foods (Bourne, 1966; Bourne, 1975a).

For most foods C is zero within the limits of experimental error. For some foods K_s is very small. In these cases the puncture equation simplifies to

$$F = K_c A \qquad (15–5)$$

That is, the puncture force is directly proportional to the punch area. It seems likely that K_s is close to zero for hard-baked goods such as pretzels and crunchy cookies and that Equation 15-5 would be applicable to these products.

Perhaps one reason why the puncture test has not been more widely used on cereal products is that hard-baked goods are prone to fracture or shatter when subjected to puncture. This problem can be overcome by using a smaller diameter punch. The author has found that even the most highly fracturable cereal product can be satisfactorily punctured if the punch diameter is sufficiently small.

Table 15-3. Shear and Compression Coefficients of 10% Starch Gels

Type of Starch	Compression Coefficient K_c (g/cm^2)	Shear Coefficient K_s (g/cm)	Ratio K_s/K_c
Unmodified corn	84.9	7.8	0.092
High-amylose corn	184.0	15.8	0.086
Modified high-amylose corn	3.5	5.2	1.47
Modified waxy maize	7.1	1.2	0.17

Source: From Bourne, 1979.

The numerical values of K_c and K_s of Equation 15-4 can be independently evaluated when a food is punctured with a series of different-sized punches (Bourne, 1966). Table 15-3 shows K_c and K_s values for some cooked starch pastes (Bourne, 1979). It seems that the values of K_c and K_s, as well as the K_c/K_s ratio, have the potential for characterizing the textural nature of different modified starch pastes in a new way.

PENETRATION

Penetrometers were originally designed to measure the distance a cone or needle sinks into a food such as margarine or mayonnaise under the force of gravity for a standard time. By selecting a suitable combination from the various cone angles and cone weights that are available, this instrument can measure a wide range of firmness, for foods ranging from a soft yogurt to a hard fat. The higher the reading, the softer the product.

This test procedure sometimes has application in the bakery industry for measuring the hardness of fats used in formulations, and for creams, frostings, and fillings that are added to baked goods. Penetrometers have the advantage of being low in cost.

The sixth edition of AACC *Cereal Laboratory Methods* describes the adaptation of a penetrometer to measure the firmness of bread by replacing the cone with a 3 cm diameter flat disc. However, the newly approved AACC Method 74-09 for measuring bread firmness renders this procedure obsolete for all except small bakeries (AACC, 1983).

TEXTURE PROFILE ANALYSIS (TPA)

Texture profile analysis compresses a bite-size piece of food two times to simulate chewing action between the teeth. This is a highly destructive test. In contrast to the deformation test, which uses a small compression, the degree of compression for TPA should be high; that is, generally 80–90%. This test can measure a number of textural properties, thus giving a more complete description of the texture than one-point measurement tests can provide (Friedman et al., 1963; Szczesniak et al., 1963).

An example of the qualitative and quantitative differences between foods is shown in Figure 15-2, which compares results for a pretzel, a corn curl, and a bread stick (Bourne, 1975). The TPA curves for the pretzel and the bread stick are qualitatively almost identical. On the first bite, both show an initial very steep rise in force, indicating a rigid nondeformable product, followed by a succession of rapid declines in force, which indicates a series of catastrophic failures. The force curves for the second bites are small, indicating very low cohesiveness and springiness. Quantitatively, the major difference between them is the peak force (hardness) on the first bite. For the pretzel this force is about 350 N, and for the bread stick it is about 180 N. The pretzel is more than three times harder than the bread stick.

The corn curl shows an initial steep rise in force but not nearly as steep as the bread stick and pretzel. As compression continues, there is a succession of small sudden drops in force indicating a series of minor fractures. The trend line of the force curve is almost horizontal, not steeply negative as for the pretzel and the bread stick. A corn curl is a highly aerated product, and as it is compressed, one layer after another collapses in succession; this response is unlike that for the pretzel and bread stick, which fracture right through into progressively smaller pieces.

The maximum force on the first bite of corn curl is 30 N, which shows that its hardness is much lower than that of the pretzel and bread stick. The force curve for the second bite of the corn curl is of medium size, indicating medium cohesiveness and springiness. To summarize, the corn curl disintegrates in a different manner, and it has a lower stiffness, much lower hardness, and greater cohesiveness and springiness than the bread stick and the pretzel.

Another example of quantitative differences between products that are qualitatively similar is shown in Figure 15-3 (page 568), which compares results for beef tenderloin, beef chuck, and spun soybean fiber beef analog. Although all three TPA curves are qualitatively similar, there are major quantitative differences, which are summarized in Table 15-4. The maximum force for the meat analog falls between those for tenderloin and chuck on the first and second bites and work on the first bite; that is, it falls within the accepted range for beef. However, the 1.8 N force required to compress the soy analog 10% on the first bite falls outside the range of 0.02–0.08 N required for the test. The soy analog is still well outside the range for beef after 20% compression

----→

Figure 15-2. Texture profile analysis curves on 10 mm high pieces of pretzel stick, bread stick, and corn curl. *(From Bourne, 1975; copyright ©1975 by AVI Publishing Co.)*

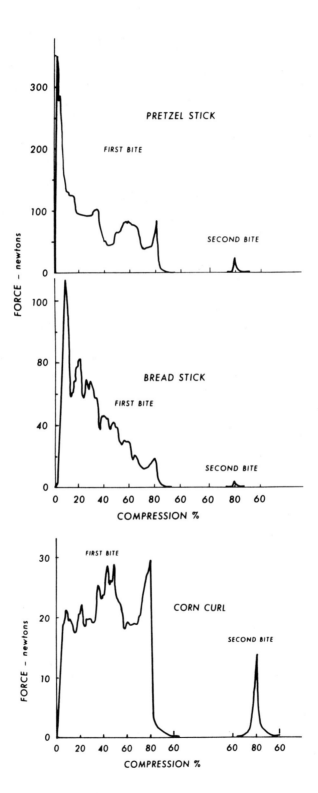

Table 15-4. Comparison of TPA Parameters of Cooked Beef and a Meat Analog

Measurement	Tenderloin	Chuck	Soy Analog
First bite			
Maximum force (newtons)	53	124	83
Work (area, arbitrary units)	0.70	1.91	1.62
Force to 10% compression (newtons)	0.02	0.08	1.8
Force to 20% compression (newtons)	0.06	0.24	3.9
Second bite			
Maximum force (newtons)	41	99	71
Work (area, arbitrary units)	0.20	0.59	0.76
Springiness (mm)	1.4	2.0	5.2
Ratio of second bite to first bite			
Peak force	0.77	0.80	0.86
Area (work done)	0.28	0.31	0.43

Source: From Bourne, 1975.
Note: From texture profile test on 1 cm diameter cylinders, 1 cm high compressed to 0.2 cm in Instron.

and for work and springiness on the second bite (Table 15-4). This test clearly shows the parameters for which soymeat analog falls within a desirable range and those parameters for which it falls outside of the desirable range. These data give the product development group the ability to measure how variations in formulation and processing affect each textural parameter and to select a combination of variables that brings the new product (soy analog) closest in textural properties to the target (cooked beef).

Texture profile analysis was originally performed with the General Foods Texturometer. Most TPA is now performed on the Instron, although it could be performed on the Lloyd machine. However, even with computer assistance it is still such a time-consuming procedure that it is generally confined to the research laboratory (Bourne, 1978). The Food Technology Corporation recently announced a package of hardware and software that can perform TPA simply on the texture press. This breakthrough makes TPA accessible as a quality control test that can be used routinely on the production line.

The author predicts that TPA will be more widely used in the future, in both research and quality control laboratories, because the multiparameter texture measurements it can make give a more comprehensive description of the complex combination of physical properties that is called "texture" than any single-point measurement.

The nature of the TPA test requires that the size and shape of the test piece be standardized and kept within narrow dimensional tolerances. Soft products can be cut with a sharp knife, but many researchers find it impossible to cut brittle cereal foods such as pretzels to a standard geometry. A small saw with a blade 6 inches long and with 32 teeth per inch that can be purchased from

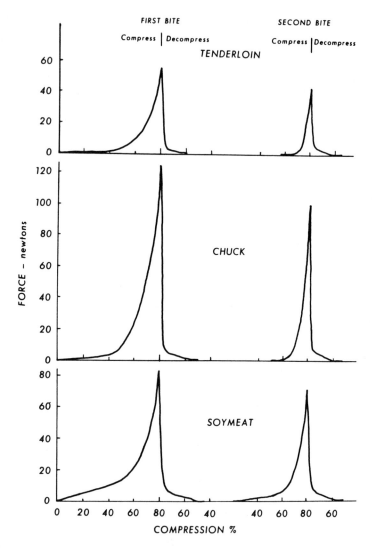

Figure 15-3. Comparison of TPA parameters of cooked beef and a meat analog. *(From Bourne, 1975; copyright ©1975 by AVI Publishing Co.)*

hobby stores is quite satisfactory for shaping hard fracturable foods (Bourne, 1982).

INHERENT VARIABILITY

Most foods, including baked goods show a high degree of variability from unit to unit, and sometimes from point to point in the same unit. For example, not only does bread exhibit a great difference in texture between the crust and the crumb, but the crumb texture within a single loaf can vary over a wide range of values, depending on its location within the loaf. This variability in texture is inherent in the product and is to be expected. It is not a defect of the instrument, provided the instrument has been correctly operated. The high inherent texture variability in food is a problem that every cereal chemist needs to be aware of and take into account when designing test protocols.

For this reason, a number of replicate tests should be made on each sample. The number of replicates required to obtain a given level of confidence is both a statistical problem and an economical problem. Statistical certainty increases as the number of replicates is increased, but the cost in terms of operator time, waiting time, and amount of product consumed also increases. A compromise has to be reached between these opposing interests. It is outside the scope of this chapter to examine this problem in detail. However, it needs to be pointed out that a high degree of variability is normal in most products and that replicate texture tests must be used in order to obtain a reasonably accurate value for the mean and at least a rough index of the degree of inherent variability in the product being measured.

APPENDIX

Some Suppliers of Texture Measuring Instruments

C. W. Brabender Instruments, Inc., 50 E. Wesley Street, South Hackensack, NJ 07606 (Structograph)

Canners Machinery Ltd., P.O. Box 190, Simcoe, Ontario N3Y 4L1, Canada (for Ottawa Texturometer and accessories)

John Chatillon and Sons Inc., Force Measurement Division, 7609 Business Park Drive, Greensboro, NC 27409 (for simple puncture testers)

Food Technology Corporation, 12300 Parklawn Drive, Rockville, MD 20852

Instron Corporation, 100 Royall Street, Canton, MA 02021

Lloyd Instruments Inc., 290 B Hansen Access Road, King of Prussia, PA 19406

Penetrometers

G.C.A. Precision Scientific Group, 3737 West Cortland Street, Chicago, IL 60647

Lab-Line Instruments, Lab-Line Plaza, Melrose Park, IL 60160-1491

REFERENCES

AACC, 1957. *Cereal Laboratory Methods*. 6th Ed. St. Paul, MN: American Association of Cereal Chemists.

AACC, 1983. *Approved Methods of the American Association of Cereal Chemists: Method 74-09*. First approved 10-8-86, revised 11-4-87. St. Paul, MN: American Association of Cereal Chemists.

Baker, A. E., and Ponte, J. G. 1987. Measurement of bread firmness with the universal testing machine. *Cereal Foods World* 32:491–493.

Bourne, M. C. 1966. Measurement of shear and compression components of puncture tests. *J. Food Sci.* 31:282–291.

Bourne, M. C. 1975a. Method for obtaining compression and shear coefficients of foods using cylindrical punches. *J. Tex. Stud.* 5:459–469.

Bourne, M. C. 1975b. Texture properties and evaluations of fabricated foods. In *Fabricated Foods*, Chap. 11, ed. G. E. Inglett, pp. 127–158. Westport, CT: AVI.

Bourne, M. C. 1978. Texture profile analyses. *Food Technol.* 32(7):62–66,72.

Bourne, M. C. 1979. Theory and application of the puncture test in food texture measurement. In *Food Texture and Rheology*, ed. P. Sherman, pp. 95–142. London: Academic Press.

Bourne, M. C. 1982. *Food Texture and Viscosity. Concept and Measurement*, 325 pp. New York: Academic Press.

Brown, S. L., and Zabik, M. E. 1967. Effects of heat treatments on the physical and functional properties of liquid and spray-dried egg albumen. *Food Technol.* 21:87–92.

Bruns, A. J., and Bourne, M. C. 1975. Effects of sample dimensions on the snapping force of crisp foods. Experimental verification of a mathematical model. *J. Tex. Stud.* 6:445-458.

Davis, C. E. 1921. Shortening: Its definition and measurement. *Ind. Eng. Chem.* 13:797–799.

Friedman, H. H., Whitney, J. E., and Szczesniak, A. S. 1963. The texturometer—A new instrument for objective texture measurment. *J. Food Sci.* 28:390–396.

Gruber, S. M., and Zabik, M. E. 1966. Comparison of sensory evaluation and shear-press measurements of butter cakes. *Food Technol.* 20:968–970.

Loh, J. 1985. Rheology of soft wheat products. In *Rheology of Wheat Products*, ed. H. Faridi, p. 210. St. Paul, MN: American Association of Cereal Chemists.

Matsuo, R. R., and Irvine, G. N. 1969. Spaghetti tenderness testing apparatus. *Cereal Chem.* 46:1–6.

Matsuo, R. R., and Irvine, G. N. 1971. Note on an improved apparatus for testing spaghetti tenderness. *Cereal Chem.* 48:554–558.

Matthews, R. H., and Dawson, E. H. 1963. Performance of fats and oils in pastry and biscuits. *Cereal Chem.* 40:291–302.

Platt, W., and Powers, R. 1940. Compressibility of bread crumb. *Cereal Chem.* 17:601–621.

Szczesniak, A. S., Brandt, M. A., and Friedman, H. H. 1963. Development of standing rating scales for mechanical parameters of texture and correlation between the objective and sensory methods of texture evaluation. *J. Food Sci.* 28:397–403.

Stinson, C. G., and Huck, M. B. 1969. A comparison of four methods for pastry tenderness evaluation. *J. Food Sci.* **34**:537–539.

Voisey, P. W., and Larmond, E. 1972. *The Comparison of Textural and Other Properties of Cooked Spaghetti by Sensory and Objective Methods.* Report No. 7008, Research Branch, Canada Agriculture.

Voisey, P. W., and Larmond, E. 1973. Exploratory evaluation of instrumental techniques for measuring some textural characteristics of cooked spaghetti. *Cereal Science Today* **18**:126–143.

Voisey, P. W., Larmond, E., and Wasik, R. J. 1978. Measuring the texture of cooked spaghetti. I. Sensory and instrumental evaluation of firmness. *Can. Inst. Food Sci. Technol. J.* **11**:142–148.

Walker, C. E., West, D. I., Pierce, M. M., and Buck, J. S. 1987. Cake firmness measurement by the universal testing machine. *Cereal Foods World* **32**:477-480.

Zabik, M. E., Fierke, S. G., and Bristol, D. K. 1979. Humidity effects on textural characteristics of sugar snap cookies. *Cereal Chem.* **56**:29-33.

Chapter 16

Texture Evaluation of Baked Products Using Descriptive Sensory Analysis

Laura M. Hansen and Carole S. Setser

Texture contributes to the overall acceptability of a food, as well as to its appearance and flavor; thus, this discussion will be focused on texture. Researchers commonly evaluate the texture of a product by measuring its physical and chemical properties. These properties, however, must eventually be related to how the product behaves in the mouth, which necessitates the use of sensory evaluation. According to Kapsalis and Moskowitz (1977),

> The relationship between mechanical measurements of texture and sensory ratings has its foundation in two inherently different measuring capabilities—(1) the machine, which is more reproducible than the human sensor but "too simple" to completely describe such a multidimensional attribute as texture; and (2) the human being, which, with its immense complexity, problems in calibration, and tendency to drift, is difficult to fit into an equation.

Texture can be defined as "the sensory manifestation of the structure or inner makeup of foods . . . as perceived by the senses of skin (tactile) and muscles (kinesthetic)" (Civille and Szczesniak, 1973). For some foods, the sounds generated during mastication are an important part of the overall perception of texture (Szczesniak, 1986). Texture is not a single attribute, but comprises many different properties. In discussions of texture, it is better to talk about "textural properties," referring to a group of related properties rather than "texture" as a single parameter (Bourne, 1982). The textural characteristics of baked products include aerated, crispy, crumbly, gummy, soft, hard, flaky, moist, tender, and tough descriptors. If the baked product contains a fluid, such as in a jelly doughnut, common terms that might apply include gluey, sandy, sticky, viscous, thick, and watery.

Evaluation of texture by sensory methods has many applications throughout the food industry. For example, in quality control/quality assurance (QC/QA), raw materials and ingredients must adhere to standard sensory reference descriptions. Products can be evaluated at initial production stages and throughout handling, shipping, and storage. Product quality can be monitored daily, weekly, or monthly by using sensory tests. In addition to its usage for QC/QA, sensory evaluation is relied upon by product development personnel to maintain desirable sensory characteristics when existing products are being modified or new food products are being formulated. Sensory evaluation is used by marketing personnel working with the product developers to determine consumer attitudes about new products and existing products. Research evaluating how accurately instrument measurements reflect sensory properties is necessary to determine the validity of the instrumental measures. Substitution of instrumental measurements, if appropriate, can be used to provide faster and more economical estimates of sensory properties (Noble, 1975). Sensory evaluation can be used further to study human perception and the nature of responses to pressure/feel sensations, as well as for studies of flavor perception. Thus, the uses of sensory evaluation are plentiful; however, the key to valid and useful results depends on choosing the appropriate test to meet the objectives of the project.

TYPES OF SENSORY TESTS

Generally, sensory test methods can be divided into two main categories: affective and analytical. Affective tests measure consumer responses such as acceptance, preference, and the hedonic, or degree of liking/disliking. Analytical tests are subdivided into the two broad areas of discrimination and descriptive tests. Table 16-1 summarizes the general categories of tests available (Anon., 1981).

Affective testing by consumers can answer questions such as "Which product is more acceptable?," "Which product is more liked?," and "Which product do you prefer"? Paired preference, ranking, and hedonic rating tests are test types that can be used to measure consumer attitudes, preferences, and/or acceptance of a product.

Difference tests are widely used and classified as discrimination tests. These tests provide the responses of "Yes, there is a significant difference" or "There is no significant difference" between the ingredients or products in question. No information as to what this difference is or how large a difference exists is given. Examples of types of difference tests are paired comparison, duo–trio, triangle, signal detection, and differences from control.

Table 16-1. Possible Sensory Test Methods

Method	No. Samples per Test	Application
Duo–trio	3 (2 alike, 1 different)	New product development, product matching, product improvement, process change, cost reduction and/or selection of new source of supply, quality control, storage stability, panelist selection, and training.
Triangle	3 (2 alike, 1 different)	Same as above
Paired comparison	2	Same as above plus consumer preference
Ranking	2–7	New product development, storage stability, consumer preference, panelist selection, and training.
Dilution	5–15	Panelist selection and training
Rating difference/ difference from control	1–18 (larger no. only with mild flavors)	Same as for duo–trio test plus correlation of sensory data with chemical and physical measurements
Attribute rating (category scaling; ratio scaling)	Same as above	Same as for rating difference test plus product grading or rating
Flavor profile analysis	1	New product development, product matching, product improvement, process change, cost reduction and/or selection of new source of supply, quality control, storage stability correlation of sensory data with chemical and physical measurements
Texture profile analysis	1–5	Same as for flavor profile analysis
Quantitative descriptive analysis	1–5	Same as for flavor profile analysis
Food action scale rating	1–18 (larger no. only with mild flavors)	New product development, product matching, product improvement, process change, consumer acceptance and/or opinions
Hedonic (verbal or facial) scale rating	Same as above	Same as for food action scale rating plus consumer preference

Source: Adapted from IFT, 1981.

Information on the advantages, disadvantages, and uses of several of the tests are summarized in ASTM 434 (1968) and in Meilgaard and co-workers (1987).

Difference testing demands precision in test design and administration. Except for the individual variables under investigation, samples must be as identical as possible. The size, shape, color, temperature, coding, and randomization, as well as the environmental conditions under which the test is conducted, all must be controlled. Difficulties can arise in the use of difference testing for texture evaluation, because frequently one cannot simply ask, "Is there a difference in texture between two products?" because texture comprises many different interrelated attributes. Specific attributes such as hardness, crumbliness, and density can be evaluated, but the researcher must be aware that if any differences exist in the attributes of the product other than the attribute in question, panelists use the other attributes as aids or clues during evaluation. In difference testing the researcher can only ask questions such as "which cracker is harder?" (paired difference test) if the researcher knows in advance which sensory characteristic, and that characteristic only, will be modified. If the sensory characteristic modified by the experimental conditions is unknown in advance, triangle or duo–trio tests can be used and the judges are asked to select or match the odd sample.

Since most foods show a composite of many different, but interrelated, textural characteristics, a method to evaluate all of the components of texture is frequently required. Descriptive analysis meets this need. In some cases, untrained panelists have been used for consumer texture profiling (Szczesniak and Skinner, 1973; Szczesniak et al., 1975). More typically, descriptive analysis relies on small, highly trained panels to qualitatively and quantitatively evaluate the characteristics of a product. The remainder of this chapter reviews techniques and considerations that are pertinent primarily to descriptive evaluations of the appearance and texture of baked goods by the trained panelists.

In texture profile analysis, the textural attributes are determined qualitatively, and the intensities of these attributes are then measured quantitatively. Profiling traditionally was based upon evaluations of a product for each of its characteristics by using open discussions after individual evaluations to establish the panel's composite findings, or consensus profiles. Open panel techniques are known to have biased results either because of a dominant personality or the presence of a senior panelist (Amerine et al., 1965). Because of the problems associated with consensus scoring, many profile panels now use individual scaled evaluations and statistical methods to analyze the data and to determine similarities and differences between two or more products.

Modifications of texture profile analysis have resulted in several hybridized attribute rating methods. Techniques such as profile attribute analysis (PAA) (Hanson et al., 1983), quantitative descriptive analysis (QDA), the Spectrum™ method, or timed intensity scaling (Meilgaard et al., 1987) make use of quantitative scales to investigate one or more attributes of interest from the complete profile. In each of the methods, judges evaluate the products independently, and the data are analyzed statistically to give mean scores and significant differences among products.

CLASSIFICATION OF TEXTURAL COMPONENTS USED IN DESCRIPTIVE ANALYSIS

Texture parameters used in sensory evaluation have been divided into three groups (Szczesniak, 1963, 1975, 1986):

1. Mechanical characteristics perceived by the kinesthetic sense as the reaction of the food to stress, including hardness, cohesiveness, brittleness, and viscosity.
2. Geometrical characteristics related to size, shape, and orientation of the particles perceived by tactile nerves in the mouth or by touch, such as gritty, grainy, flaky, stringy, and smooth.
3. Other characteristics; that is, the mouthfeel attributes related to the perception of fat and moisture during chewing and swallowing.

The mechanical characteristics are further subdivided into the following:

Primary attributes of hardness, cohesiveness, viscosity, springiness, and adhesiveness expressed popularly by terms such as soft, firm, thin, viscous, plastic, elastic, sticky, or gooey.

Secondary attributes of fracturability, chewiness, and gumminess involving interactions of more than one of the primary attributes, which include the popular terms crumbly, brittle, tender, tough, mealy, and gummy.

Particle size and shape are evaluated as geometrical characteristics based on the discrete particles that are relatively harder than the surrounding medium. Examples of such particles in order of increasing particle size are powdery, chalky, gritty, grainy, and coarse. Geometric characteristics that relate to shape and orientation are evaluated based on the different geometrical arrangements organizing the structure within each product. Examples include flaky, fibrous, puffy, cellular, aerated, pulpy, and crystalline (Brandt et al., 1963).

The fat content can be evaluated not only for the amount present, but also for the type and rate of melting and mouth-coating characteristics. Moisture content can be evaluated for the amount present and the rate and manner that it is released or adsorbed. In some baked products the amount of moisture present is more important than the rate at which it is released (Brandt et al., 1963).

COMPONENTS OF TEXTURAL
DESCRIPTIVE ANALYSIS

The classic definitions for texture profiling, which were originally developed at the General Foods Research Center, have been presented in detail by several researchers (Brandt et al., 1963; Szczesniak, 1963; Szczesniak et al., 1963). In establishing techniques for descriptive analysis of the texture of the product, the characteristics of the product, the order of appearance, and the degree to which they are present must be determined. Each of these parameters will be discussed in more detail by using baked product examples.

Characteristics Present

Panelists must determine all of the characteristics present in the product to establish a product profile. To do this, several examples in the product category of interest are examined to ensure that all possible textural properties are included. A checklist approach can provide panelists with a list of possible terms to consider and serve as a reminder of prior textural experiences. Examples of checklists are given in Szczesniak and co-workers (1975) and in Table 16-2, which has been used for some studies in the laboratories of these authors. From the list of possible textural attributes, panelists might determine that some attributes are not pertinent to the study; for example, do not differentiate among products in that category. A modified profile, or hybrid, method can be used to evaluate only the attributes that provide the information required to meet the stated objectives of a study. For example, a change in formulation might change only two or three attributes, and comparisons of how those attributes are changed with varied ingredients might be all that is necessary. In determining the attributes pertinent to a study, the panel should use just one term to define a specific characteristic, rather than words with overlapping meanings or words with the same meaning but pertaining to a different level of intensity. For instance, the panel might decide to evaluate hardness and thus would not evaluate softness as a separate characteristic, because softness is a different intensity level of hardness. In the evaluation of viscosity, the panel might score thickness and not thinness, because, again, these terms refer to different intensities of the same attribute. On the other hand, care must be taken not to confound data (Sidel et al., 1981) by using one scale for evaluating two different attributes. For example, foamy and cohesive should not be anchors for the extremes of a single scale for body, but each attribute would need to be evaluated separately. Whenever possible, decreasing the number of attributes to the fewest that will still meet the stated objective should be done to simplify data analysis and comparisons between products.

Table 16-2. Checklist of Terms Appropriate for Sensory
Evaluation of Textural Attributes in Baked Products

Deformation	Consistency	Mouthfeel	Geometrical
Hard	Thin	Dehydrating	Abrasive
Firm	Soupy	Saliva-	Rough
Cohesive	Watery	inducing	Smooth
Crunchy	Runny	Dry	Regular
Crisp	Thick	Soggy	Irregular
Adhesive	Stiff	Moist	Even
Toothpacking	Viscous	Juicy	Uneven
Tender	Peak-forming	Surface	Grainy
Tough		moistness	Coarse
Crumby		Creamy	Aerated
Gooey		Mouthfulness	Porous
Chewy			Lumpy
Elastic			Crystalline
Plastic			Granular
Springy			Fine
Gummy			

Source: Adapted from Smith and Stoneking, 1986.

Statistical techniques can be used to determine when different words are being used to evaluate the same characteristic. Multivariate techniques are designed to simplify the relationships that can exist in a data set by identifying redundancies among attribute descriptors. Factor analysis, a form of multivariate analysis, is a technique commonly used to reduce a large number of variables or descriptive terms to a smaller number. In the evaluation of texture, Moskowitz and Kapsalis (1974) used regression equations relating descriptors and found that crispness was most related to the quality of crunchiness and crunchiness was related to crispness and hardness. Vickers and Wasserman (1980) used multidimensional scaling and found that crispness and crunchiness were closely related when evaluated as a sound.

Order of Appearance

After determining the characteristics present, panelists must then determine the order of appearance of the characteristics. The order of appearance has been divided into several stages (Szczesniak, 1963). Order of appearance is product specific, and some of the stages are not appropriate for a particular product; thus the panel might choose to omit or modify that stage. Attributes are usually evaluated in the following order:

1. Surface characteristics (can be visual)
2. Initial compression (perceived on first bite)

a. Mechanical characteristics such as hardness, viscosity, and brittleness
b. Any geometrical characteristics present
c. Other characteristics present (moistness, oiliness)
(Initial compression can be subdivided into stages of partial compression, the first bite using incisors and the first bite using the molars, if the panel deems it appropriate for a specific product.)
3. Masticatory phase (perceived during chewing)
a. Mechanical characteristics of gumminess, chewiness, and adhesiveness
b. Geometrical characteristics present
c. Other characteristics present (moistness, oiliness)
4. Residual (changes made during mastication and often perceived after swallowing)
a. The rate and type of breakdown
b. Moisture absorption and mouth coating properties

In addition to the mechanical characteristics and geometrical characteristics evaluated during the initial compression and masticatory stages, auditory characteristics such as crunchiness, crackliness, or crispness might need to be evaluated. These characteristics often indicate textural properties.

The surface characteristics, or appearance factors, could be the most important factors evaluated. Appearance factors initially attract a consumer to a product, because the appearance is related to the quality of the product. For example, is the color the expected one, or is it too light or too dark? In baked products, too dark or too light a color might indicate or connote under- or overbaking. Is the length, thickness, width, or particle size what is expected? Does the surface appear wet, dry, soft, hard, or crispy? Because these questions are used by consumers for making purchasing decisions, appearance factors need to be evaluated.

Some appearance factors and the methods of evaluation used in studies on layer cakes (Deming and Bramesco, 1987) are listed below as examples.

Evenness of crust brownness—amount of blotchiness, light and dark areas (high score indicates high evenness)

Crust brownness—lightness or darkness of predominant area of crust (high score indicates very dark)

Appearance of tier elevation (**Crust ring**)—the degree of elevation of the inner tier located on the crust surface (high score indicates very predominant tier effect)

Undercrust stickiness (inconsistency)—extremely gummy appearing area immediately below the crust; the height, width, and depth of the undercrust area that is inconsistent with the rest of the crust (high score indicates extremely large sticky area)

Cell uniformity—amount of large air cells or tunnels within otherwise small air cells (high score indicates very even cell size)

Cell size—size of the majority of cells on the area of cut surface (high score indicates very large cells, open structure)

Scaling Techniques

After determining the characteristics present in a product and their order of appearance, a panel must consider the level at which each characteristic is perceived. To do so requires some type of scale with appropriate references. Originally, scales with appropriate reference products for texture profiling were established by Szczesniak and co-workers (1963); these scales have been expanded by Civille and Liska (1975) and Muñoz (1986), and they have been summarized by Meilgaard and colleagues (1987). Reference products established by those authors were intended to cover the entire intensity range found in all food categories. These scales can be expanded at any point for greater precision in a narrower range (Szczesniak et al., 1963). As with flavor profiling, the scales can be used for consensus data with texture profiling. However, Syarief and colleagues (1985) found mean scores from flavor and texture profile analyses of several food products to be superior to consensus scores. Mean scores gave smaller coefficients of variation than consensus scores for most characteristics evaluated. Mean scores further accounted for a higher cumulative proportion of variance than consensus scores based on principal component analysis.

Most scales are suitable (Lawless and Malone, 1986a, 1986b) for providing quantitative data for statistical analysis (mean scores). Such scales can include structured category to unstructured line scales, and rank order scaling to magnitude estimation or ratio scaling. Discussions of advantages and disadvantages of each type are given by Pangborn (1984). All types of rating scales can distinguish true sample differences, if they are properly used and care is taken to avoid some of the possible psychological biases that can occur (Amerine et al., 1965; Larmond, 1977). Ordinal scaling procedures require that the panelist rank the stimuli (sample) for a prescribed texture attribute from most intense to least intense; for example, hardness. The procedure is simple but inefficient because of difficulties with the assigning of numbers. Such numbers give only a relative ordering of the magnitude of the attribute, but no information about the degree of the magnitude. To be used properly, the method requires many paired comparisons $[n(n-1)/2]$, which can be time consuming.

Interval scaling, or category scaling, involves the use of equally spaced categories or intervals, labeled by numbers or word adjectives, to rate a set

BREAD FLAVOR BALLOT

	1		2		3

Thresh-old	SLIGHT			MODERATE			STRONG		
1	2	3	4	5	6	7	8	9	10

Figure 16-1. Example of a numerical category scale, which can be used to evaluate attributes in descriptive analysis. Panelists choose the number that represents their evaluation of an attribute's intensity.

of products. Examples of numerical and unstructured types of category scales are shown in Figures 16-1 and 16-2 (page 585).

Unstructured scales or linear scales involve straight horizontal or vertical lines of defined length (typically 6 inches or 15 cm), which can have anchor points at the ends and intermediate points. A position on a line is not remembered as easily as a word or a number. Concerns about the equality of intervals are lessened with this type of scale compared to the category scale. The Spectrum® technique for evaluating texture involves the use of unstructured scales (Meilgaard et al., 1987).

Ratio scaling (Kapsalis and Moskowitz, 1977) involves the use of a number, which is assigned to the first sample tested. The subsequent samples are given numerical values proportional to the first value. This technique, also referred to as magnitude estimation, tends to be utilized more for evaluation of a single attribute, for a taste or flavor, or for academic studies. The technique has been used less frequently for scoring multiple characteristics of products.

Specific work on some baked products in laboratories at Kansas State University (Bramesco and Setser, 1988) and by the Sensory Center affiliated with Kansas State University, have refined and modified the generalized scales for reliable application to baked good evaluations. For example, an extreme of hardness was originally represented by hard candy (Charms or Life Savers); however, no baked good is known at that extreme. If such an extreme is used for a scale, all product evaluations would tend to be within a small segment of the scale and product differentiation could not be achieved. Scales need to be broad enough to encompass the full range of parameter intensities; however, they also need to contain enough points to detect small intensity differences.

In addition, scales applicable to one category of baked goods, for example, cookies, might not be applicable for the evaluation of another category such as layer cakes.

In the evaluation of roughness in whole grain breads, two breads with differences in roughness (generic white and a whole multigrain variety type) might be more appropriate references than a gelatin dessert and a granola bar. Similarly, more applicable references than cream cheese and gelatin for springiness might be a soft white bread and a firmer, whole grain variety bread.

Important concepts related to the evaluation of baked goods have arisen that concern differences in evaluation based upon the moisture content of a product and the salivary flow rate of the panelist (Deming et al., 1989). For example, Bramesco and Setser (1988) have noted that products with high moisture contents are not necessarily evaluated differently by panelists with high and low salivary flow rates. On the other hand, products with low moisture contents are evaluated quite differently by persons with varied salivary flow rates. This factor must be considered in training panelists and establishing references. Further studies are planned to determine valid methods of accounting for the interactions associated with inherent panelist differences to achieve accurate evaluations.

Appropriate Scales and Evaluation Techniques for Baked Good Examples

Examples of some of the terms, order of appearance, and scales used to evaluate several types of baked goods are outlined in the following section. In work published earlier, Civille (1979) indicated appropriate terms and evaluation techniques for pound cakes, that is, cut a 1/2-inch slice into eight equal pieces, approximately 3/4 inch × 3/4 inch.

1a. Place cross section into mouth between tongue and palate and evaluate for
    Surface smoothness
    Graininess
    Moistness
1b. Compress cross section partially between tongue and palate, remove force, and evaluate for
    Springiness
2. Place cross section between molar teeth, bite down, and evaluate for
    Hardness
    Fracturability
    Moistness

       Moisture absorption
       Denseness
       Adhesiveness
       Cohesiveness
       Coarseness

3. Place cross section between molar teeth, chew, and evaluate for
       Chewiness
       Moisture absorption
       Adhesiveness of the mass
       Gumminess of the mass
       Denseness
       Graininess
       Description of breakdown

4. After chewing, swallow and evaluate for
       Ease of swallow
       Mouth-coating: type, particles
       Toothpacking

Evaluation techniques and terminology for sponge cakes used by Lee (1980) are given immediately below.

1. Place cake in mouth; feel surface with tongue and lips. Evaluate for
       Crust stickiness (adhesiveness)
       Moisture (degree of wetness or oiliness)

2. Compress partially between incisors; release, then bite through. Evaluate for
       Elasticity
       Density
       Firmness

3. Bite and pull with incisors. Evaluate for
       Extensibility

4. Chew a ⅝-inch × 1-inch × 1-inch cube with molar teeth. Evaluate for
       Chewiness
       Moisture adsorption
       Cohesiveness of mass
       Graininess of mass

5. Swallow the chewed sample. Evaluate for
       Ease of swallow
       Adhesiveness of crumb

Figure 16-2. Example of unstructured linear scales for evaluation of textural attributes. Panelists mark the ballot by placing a line perpendicular to the linear scale at the point representing their evaluation of the intensity for each attribute.

Name_____   Date_____

LAYER CAKE APPEARANCE AND TEXTURE

Crust Brownness
Pale   Dark
_____

Cell Uniformity
Uneven   Even
_____

Cell Size
Very Large   Very Small
_____

Density
Not Dense, Open   Very Dense, Close
_____

Surface Moistness of Crumb with Lips
Very Dry   Very Moist
_____

Stickiness of Crust Top-side with Lips
Very Sticky   Not Sticky
_____

Undercrust Stickiness with Lips
Very Sticky   Not Sticky
_____

First Bite Fragileness/Tenderness
Not Fragile   Very Fragile, Tender
_____

First Bite Crust Hardness
Very Hard   Very Soft
_____

Moistness of Crumb During Mastication
Dry   Very Moist
_____

Comments :

Recent work by Bramesco and Setser (1988) provides appropriate terms and evaluation techniques for layer cakes; consideration is given in this work for differences in moisture content and salivary flow discussed earlier.

Crust hardness—amount of give obtained with finger touch; flinty hard feel to touch (high score indicates very hard).

Stickiness of crust topside with lips—adhesiveness to top lip when placed on crust (high score indicates very sticky).

First bite fragility **(Crumb tenderness)**—how readily front (from center) 1-inch crumb breaks off on initial bite (high score indicates very fragile, tender, crumbly).

Crumbliness (crumb tenderness redefined)—force required to separate individual pieces with the tongue *immediately* after sample is placed in the mouth; biting, chewing, and compression are not part of this test (high score indicates very crumbly like corn bread, low anchor is Wonder™ bread).

Crumb firmness—place sample between molars; evaluate the force required to completely compress the sample (high score indicates very high firmness similar to that of pound cake; very low anchor is angel cake).

Crumb stickiness—amount of crumb adhering to teeth and palate as biting into front of cake slice (high score indicates very sticky in mouth).

Cohesiveness of mass **(Gumminess)**—how readily the cake can be prepared for swallowing during normal chewing; how much forms into ball-like mass difficult to swallow; degree to which the mass holds together after being chewed ten times (high score indicates more balling and high gumminess, cohesiveness).

Moistness—with blotted lips, the amount of moisture/cooling perceived on the surface of the sample when held between both lips (high score indicates very moist).

Overall **density**—compactness of cake overall (high score indicates extremely dense product such as a fudge brownie).

It is likely that the earlier evaluation procedures for pound and sponge cakes also could be improved, considering the same panelist–product interactions with mastication related to moisture uptake.

The approach used for cookie evaluation by Civille and Liska (1975) and Meilgaard and colleagues (1987) is given below.

1. Surface: Place between the lips and evaluate for
   Degree of smoothness of top and bottom
   Loose particles
   Dryness (absence of oil on surface)
2. First bite: Place one-third cookie between incisors, bite down, and evaluate for
   Fracturability
   Hardness
   Particle size (of crumb pieces)
3. First chew: Place one-third cookie between molars, bite through, and evaluate for
   Density
   Uniformity of chew
4. Chew down: Place one-third cookie between molars, chew ten to twelve times, and evaluate for
   Moisture absorption
   Type of breakdown (thermal, mechanical, salivary)
   Cohesiveness of mass
   Toothpack
   Grittiness
5. Residual: Swallow sample and evaluate residue in mouth for
   Oily
   Particles
   Chalky

Care should be taken to control sample size, and if the one-third cookie differs in size from one variation to another, this should be adjusted to provide similar sample sizes. Other factors that can influence the evaluations also need to be controlled. For example, it might be necessary to evaluate cookies containing fruits, nuts, or chocolate chips without the added material in some samples for attributes such as fracturability or hardness.

Scales and terms developed and used at the Kansas State University Sensory Center (Smith and Stoneking, 1986) for evaluation of pan breads are provided for guidance in evaluation of this product.

1. Optical: Observe the samples on a plate and evaluate for
   Grain
2. Surface: Place sample of product between the lips, compress partially, and repeat. Evaluate for
   Surface moisture
   Surface roughness
   Springiness

3. Tongue/palate: Place sample of product in the mouth, compress partially against the palate, release, repeat. Evaluate for

Denseness

Coarseness

Adhesiveness

4. Masticatory: Place sample in the mouth; chew with molar teeth. Evaluate for

Firmness

Graininess

Gumminess

Chew count

Type of disappearance

5. Swallowing: Swallow the chewed mass and evaluate for

Ease of swallowing

Molar packing

Mouth-coating

Mouthfeel

For other products such as muffins, attributes and techniques used in cake studies and in bread evaluations might be combined and modified. The techniques for evaluating any particular product category should be established by considering the attributes noted (such as are obtained from the checklist type procedure) and previous work done on similar products such as those noted in this section.

If only a few attributes are needed to differentiate among products being evaluated, for example, QA/QC studies, scorecards are developed for only those attributes. Training sessions for shortened versions can be modified to concentrate on the attributes of interest and their importance to overall product quality.

Panel Training

Training involves bringing panelists to a point at which they understand the terminology (characteristics) that are appropriate to describe the product being evaluated, attribute by attribute. Intensities are a part of that description. The sensitivities and memories of panelists can be increased with training. Obviously, for discrimination tests and even for some short attribute scaling studies, memory is less critical than it is for complete product descriptions and profiling. However, complete descriptions initially need to be established by panelists with extensive training; thus this training process will be described. Additional details or guidelines to panel training can be found in Civille and Szczesniak (1973).

At the early stages of training, one should orient panelists to the basic concepts of descriptive analysis and texture terminology using standard scales (Brandt et al., 1963; Meilgaard et al., 1987; Bramesco and Setser, 1988) to demonstrate various textural parameters. How basic parameters manifest themselves in some product types is examined. Fracturability, for example, is manifested and appropriately described as crumbliness in pound cakes. In baked goods, toothpacking properties would be considered instead of the term adhesiveness, which one might use to describe peanut butter. This orientation should include a demonstration of several products within a product type. All commercially available brands of the product and samples of different ingredient types used in commercial products should be examined. For example, for an attribute such as surface roughness in breads or muffins containing bran, one should provide brans of varied particle sizes that have been flaked or ground. Each type and particle size should be presented in the product (bread or muffin) being evaluated. In flavor profiling, references can be pure chemicals, but this is not possible for texture descriptions. Thus, commercial materials and products are used to establish levels, which present inherent replicability problems for the references with respect to time, geographical region, and manufacturer.

Discussions after evaluations allow panelists to interact and establish common terms or definitions for the ballot that is developed. Differences in evaluation when chewing or biting, the portion of the mouth being used, the size of the sample being evaluated, and the point in time an evaluation is made can lead to confusion and should be ascertained at this time. Techniques of evaluation and terminology used to describe the products should come from panel members themselves, because all members need to feel comfortable with any techniques and descriptive terms used. This terminology should be consistent from product to product and tied to reference materials, however. Each training session should focus on one or two specific characteristics; reference standards should be presented to alleviate questionable attributes on which panel members do not agree.

Reference standards play an important role in developing appropriate terminology and establishing intensity ranges (Rainey, 1987). Standards reduce the amount of time for training and provide documentation for terminology. Well-selected reference standards at defined points on a scale reduce panel variability across products and with time. Such standards can be used to check agreement by the panelists on the manifestation of certain terms. Some panelists might feel strongly that an attribute is relevant to the product, but if others are not finding the attribute, a problem is indicated. Panelists might not be using the same terminology, be perceiving the attribute in the same manner, or using the same technique to assess the attribute. The terminology and/or process must be clarified. The panel practices with scales developed for selected products until it is adept at using them and can rate a variety of products. Weeks, even months, of daily sessions might be necessary for this stage of the training.

Performance evaluations for panelists are an essential component of the training to determine discriminating, consistent performers. Products with known large differences and duplicate check samples should be presented occasionally and rated on scales developed for actual test sessions. Factors to evaluate include: (1) agreement between an individual panelist and a panel as a whole, (2) agreement between duplicate samples within a test period and among test periods by a panelist, and (3) ability to distinguish products with known differences by using a wide range of a scale. Graphical and statistical techniques can be utilized to evaluate the performance of panelists; discussions pertaining to the use of analysis of variance for this purpose are included in ASTM 758 (1981), Rainey (1979), and Sidel and Stone (1984).

Initial training is only a beginning; continued practice and monitoring are essential. Maintaining panelist performance requires constant surveillance and effort to sustain the interest and motivation of panelists.

Data Presentation

Once references and scales are established, the panelists evaluate the products in question. The data are collected and evaluated, and the overall profiles of the products evaluated are established. Product evaluation involves comparing similarities and differences between products. Mean scores and statistical methods provide product comparisons, attribute by attribute. If appropriate experimental designs have been used, measurement variability can be separated from product variability and should be done.

Reports detailing similarities and differences should provide starting hypotheses, experimental methods sufficiently detailed to allow the work to be replicated, data summarized in tabular or graphic format (but not both for the same data), data interpretation, and conclusions. Tabular reports of data generally are more concise, whereas graphic presentations allow trends to be noted easily and provide a simple method to convey the degrees of similarity or difference for a set of products. An example of a graphic profile obtained from the evaluation of sponge cakes in the study by Lee (1980) is given in Figure 16-3. Connecting the spokes in the "spider web" format can be misleading, however, and does not generally imply any relationship among the attributes. The two formats for presenting data for pound cake (Civille, 1979) are contrasted in Table 16-3 and Figure 16-4 (page 594). A standard white pan bread on the market is profiled in Table 16-4 (Smith and Stoneking, 1986). Interpretation of data involves relating the data to other reported research, with comparisons and contrasts in methodology and findings. Data that do not support initial hypotheses should always be included.

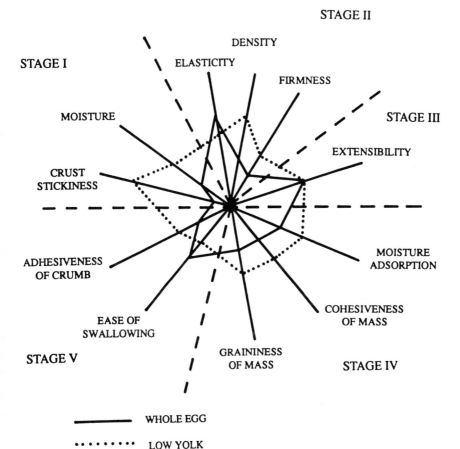

STAGE II

STAGE I

DENSITY
ELASTICITY
FIRMNESS

STAGE III

EXTENSIBILITY

MOISTURE

CRUST
STICKINESS

MOISTURE
ADSORPTION

ADHESIVENESS
OF CRUMB

COHESIVENESS
OF MASS

EASE OF
SWALLOWING

GRAININESS
OF MASS

STAGE V

STAGE IV

———————— WHOLE EGG

•••••••• LOW YOLK

Figure 16-3. Profiles in "spider web" format comparing two sponge cakes; the attributes evaluated in each of the five stages are indicated. Intensity of each attribute is indicated by the distance from zero, or the center of the spokes.

CONCLUSIONS

Sensory analysis is a useful method for evaluating the textural properties of baked products. Texture profiling and its variations can be used in research and in new product development. Changes in product formulation and processing changes can be evaluated. Descriptive textural analysis can be used to monitor product quality changes and can also be used as a tool in the evaluation of shelf life studies. Descriptors and techniques for baked goods should be product specific. To effectively use descriptive texture techniques, a well-trained panel and careful analysis and interpretation of data are necessities.

Table 16-3. Pound Cake Texture Evaluation Results

| | Sample No. | |
	Q	H
Surface smoothness	2	)(−1
Graininess	)(−1	2
Moistness	1–2	)(−1
Springiness	1–2	)(−1
Hardness	1–2	)(−1
Fracturability	)(	1–2
Moistness	1–2	)(−1
Moisture absorption	)(	)(−1
Denseness	2	)(
Adhesiveness	1	)(
Cohesiveness	1–2	)(
Coarseness	)(	2
Chewiness	13	15
Moisture absorption	1	2
Adhesiveness of mass	)(−1	)(
Gumminess of mass	2	)(−1
Denseness	2	)(−1
Graininess	)(	1–2
Description *of breakdown*	Sample compresses; mixes well with saliva; thins down to smooth slurry. Smooth and moist	Sample crumbles; takes longer to hydrate. Crumbs to end. Becomes gummy at end of chewing. Dry and crumbly
Ease of Swallow	2	1–2
Mouth-coating		
Type	Chalky; oily	Chalky; oily
Particles	Grains	
Toothpacking	)(	1

Source: Adapted from Civille, 1979.
Notes: Scale =)(, threshold to 3, intense

The reference list begins on page 595.

Table 16-4. Texture Profiles of Pan Bread

	Sample	
	A	B
Stage I		
Grain	Both breads have elliptical and round cells in approximately the same ratio	
	Grain particles present are somewhat flat	Grain particles are more apparent, are slightly larger and rounded
	Reddish brown color cast	Greenish brown color cast
	Color difference is negligible	Color difference is negligible
Stage II		
Surface moisture	8	9
Surface roughness	5	7
Springiness	9	8
Stage III		
Denseness	4	6
Coarseness	5	6
Adhesiveness	7	7
Stage IV		
Firmness	3	5
Graininess	2	5
Gumminess	6	6
Chew count (average)	32	35
Description of Breakdown		
	Forms cohesive mass which breaks into clumps of dough. Becomes thin, pasty, as mixed with saliva with some small-grain particles.	Forms cohesive mass which flattens out as mixed with saliva to form thick paste with grain particles. This paste thins when ready to swallow.
Stage V		
Ease of swallow	4	6
Molar packing	4	5
Mouth-coating	3	4
	Floury, grainy particles	Floury, grainy particles
Mouthfeel	Slight metallic	Slight metallic
	Slight numbness	Slight numbness

Source: Adapted from Smith and Stoneking, 1986.
Note: Scale = 1, threshold to 10, intense

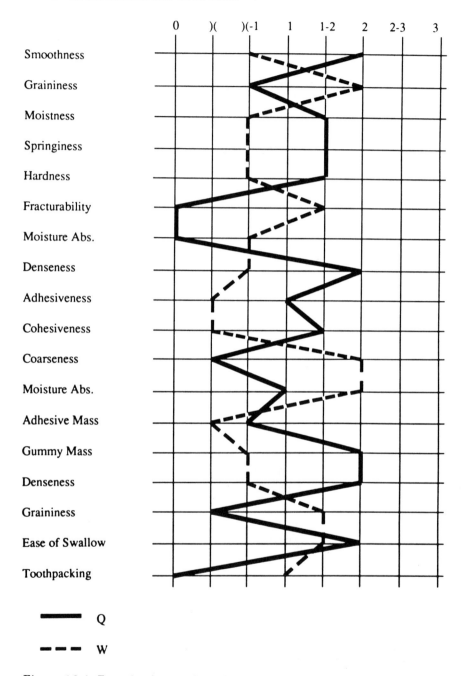

Figure 16-4. Example of a graphic profile of attributes for a pound cake presented in the order of evaluation; zero,)(, 1–3 represent no perception, threshold, slight to intense, respectively, for each attribute.

REFERENCES

Amerine, M. A., Pangborn, R. M., and Roessler, E. B. 1965. Factors influencing sensory measurements. In *Principles of Sensory Evaluation of Foods*, pp. 245–274. New York: Academic Press.

Anon. 1981. Sensory evaluation guide for testing food and beverage products. *Food Technol.* **35**:50.

ASTM. 1968. *Manual on Sensory Testing Methods*, ASTM 434. Philadelphia, PA: American Society for Testing and Materials.

ASTM. 1981. *Guidelines for the Selection and Training of Sensory Panel Members*. ASTM #758. Philadelphia, PA: American Society for Testing and Materials.

Bramesco, N. P., and Setser, C. S. 1988. Descriptive texture studies on baked products: Refinements in techniques and choosing appropriate references. *J. Text. Stud.* In preparation.

Brandt, M. A., Skinner, E. Z., and Coleman, J. A. 1963. Texture profile method. *J. Food Sci.* **28**:204–223.

Bourne, M. C. 1982. *Food Texture and Viscosity: Concepts and Measurement*. Orlando, FL: Academic Press.

Civille, G. V. 1979. *Descriptive Analysis. Sensory Evaluation Methods for the Food Technologist*. Chicago, IL: Institute of Food Technologists.

Civille, G. V., and Liska, I. H. 1975. Modifications and applications to foods of the General Foods sensory texture profile technique. *J. Tex. Stud.* **6**:19–38.

Civille G. V., and Szczesniak, A. S. 1973. Guidelines to training a texture profile panel. *J. Tex. Stud.* **4**:204–223.

Deming, D., and Bramesco, N. P. 1987. Unpublished research. Kansas State University, Manhattan, KS.

Deming, D., Bramesco, N. P., and Setser, C. S. 1989. Panelist–product interactions in descriptive sensory texture evaluations of baked products. *J. Food Sci.* In preparation.

Hanson, J. E., Kendall, D. A., and Smith, N. F. 1983. *The Missing Link: Correlation of Consumer and Professional Panel Sensory Descriptions*. Abstract 33, IFT Programs and Abstracts. Chicago, IL: Institute of Food Technologists.

IFT. 1981. Sensory evaluation guide for testing food and beverage products. *Food Technol.* **35**(10):50–59.

Kapsalis, J. G., and Moskowitz, H. R. 1977. The psychophysics and physics of food texture. *Food Technol.* **31**(4):91–94, 99.

Larmond, E. 1977. *Laboratory Methods for Sensory Evaluation of Food*. Agriculture Canada Publication 1637/E. pp. 17–19. Ottawa, Canada: Canadian Government Publishing Centre.

Lawless, H. T., and Malone, G. J. 1986a. The discriminative efficiency of common scaling methods. *J. Sensory Stud.* **1**:85–98.

Lawless, H. T., and Malone, G. J. 1986b. A comparison of rating scales: sensitivity, replicates and relative measurement. *J. Sensory Stud.* **1**:155–174.

Lee, S. H. 1980. Sensory Characteristics of Low Yolk Sponge Cakes with Stabilizers. M.S. Thesis, Kansas State University, Manhattan, KS.

Meilgaard, M., Civille, G. V., and Carr, B. T. 1987. *Sensory Evaluation Techniques*, Vol. 2. Boca Raton, FL: CRC Press.

Moskowitz, H. R., and Kapsalis, J. G. 1974. Psychophysical relations in texture. Paper presented at the Symposium on Advances in Food Texture, Guelph, Ontario, August 28–30.

Muñoz, A. C. 1986. Development and application of texture reference scales. *J. Sensory Stud.* **1**:55–83.

Noble, A. C. 1975. Instrumental analysis of the sensory properties of food. *Food Technol.* **29**:56–60.

Pangborn, R. M. 1984. Sensory techniques of food analysis. In *Food Analysis. Principles and Techniques*, Vol. 1, *Physical Characterization*, ed. D. W. Gruenwedel and J. R. Whitaker, pp. 37–39. New York: Marcel Dekker.

Rainey, B. A. 1979. *Selection and Training of Panelists for Sensory Testing. Sensory Evaluation Methods for the Practicing Food Technologist.* Chicago, IL: Institute of Food Technologists.

Rainey, B. A. 1986. Importance of reference standards in training panelists. *J. Sensory Stud.* **1**:149–154.

Sidel, J. L., Stone, H., and Blomquist, J. 1981. Use and misuse of sensory evaluation in research and quality control. *J. Dairy Sci.* **64**:2296–2302.

Smith, E. A., and Stoneking, J. 1986. Unpublished report on texture profile analysis of bread. Sensory Analysis Center, Department of Foods and Nutrition, Kansas State University.

Stone, H., and Sidel, J. L. 1985. *Sensory Evaluation Practices.* Orlando, FL: Academic Press.

Szczesniak, A. S. 1963. Classification of textural characteristics. *J. Food Sci.* **28**:385–389.

Szczesniak, A. S. 1975. General Foods texture profile revisited—Ten years perspective. *J. Tex. Stud.* **6**:5–17.

Szczesniak, A. S. 1986. Sensory texture evaluation methodology. In *Proceedings of the 39th Annual Reciprocal Meat Conference of the American Meat Science Association*, pp. 86–96. Chicago, IL: National Livestock and Meat Board.

Szczesniak, A. S., Brandt, M. A., and Friedman, H. 1963. Development of standard rating scales for mechanical parameters of texture and correlation between the objective and sensory methods of texture evaluation. *J. Food Sci.* **28**:397–403.

Szczesniak, A. S., Loew, B. J., and Skinner, E. Z. 1975. Consumer texture profile technique. *J. Food Sci.* **40**:1253–1256.

Szczesniak, A. S., and Skinner, E. Z. 1973. Meaning of texture words to the consumer. *J. Text. Stud.* **4**:378–384.

Syarief, H., Hamann, D. D., Giesbrect, F. G., Young, C. T., and Monroe, R. J. 1985. Comparison of mean and consensus scores from flavor and texture profile analyses of selected food products. *J. Food Sci.* **50**:647–650, 660.

Vickers, Z. M., and Wasserman, S. S. 1980. Sensory qualities of food sounds based on individual perceptions. *J. Tex. Stud.* **10**:319–327.

Index